AF539369

STATISTICS FOR GEOSCIENTISTS
Techniques and Applications

About the Author

Dr. Saroj K. Pal is a professor in physical geography at Delhi University and he is a leading specialist on physical and mathematical aspects of geography in India. A direct disciple of Prof S.P. Chatterjee of Calcutta University, Prof Pal started his career at Gauhati University, has lectured and researched for more than 42 years including 37 years at Delhi University as well as at Jawaharlal Nehru University. He has been a recipient of Senior Fulbright and Indo-Canadian Social Science Fellowships and he has travelled widely in Australia, Canada, U.S.A., former U.S.S.R and many European Countries.

Being a Scholar of repute as well as a keen analyst of geographical events, through his depths of research alongside their breadth, Prof Pal does not accept geography to be a generalist, holistic subject for pedagogic treatment but tries to see a reinvention of it into a front line discipline of environmental science to make contributions to the pressing human challenges for responsible, sustainable land management.

ISBN 81-7022-712-1 (HB)

First Published 1998

Published and Printed by
Ashok Kumar Mittal
Concept Publishing Company
A/15-16, Commercial Block, Mohan Garden,
New Delhi - 110059 (INDIA)

Phone : 5648039, 5649024
Fax : 091-11-5648053

STATISTICS FOR GEOSCIENTISTS

Techniques and Applications

SAROJ K. PAL
Delhi University

CONCEPT PUBLISHING COMPANY, NEW DELHI-110059

PREFACE

Any book written by a single author reflects a personal perception of the subject evolved by him over many years of teaching, research and professional experience. This book has its origin in the proposal by many friends, colleagues and student bodies to revise and enlarge the earlier book of the present author, entitled, "Statistical Techniques : A Basic Approach to Geography" , published by the Tata Mc-Graw Hill, New Delhi in 1982 but which has been out of print for past more than ten years. In the course of revision it was realized that geography has become meshed with the methodology of other disciplines and it can no longer ignore the physical knowledge of environmental sciences. These are not just a question of new information, but reflect the introduction of fresh concepts and frameworks in the subject so that not only could there have to be a change in the order and manner of the presentation of the subject matter but also a comprehensive expansion in the scope of the material included in the present book, "Statistics for Geoscientists: Techniques and Applications".

This book is written for three groups of readers. The first, and perhaps the smallest group consists of two possible audiences— the student audience, particularly those at the upper level under-graduate, post-graduate and research, as well as the professional audience in geography, especially those who are slowly but steadily embarking on statistical techniques as a tool developed for use within the value position or the research question in hand. The second group of readers includes under-graduates and post-graduates of geo-science system analysts like those in environmental geology, hydrology, ecology, pedology and agricultural science. This group always organize information (like in geography) in a spatial context and in that way geo-science systems analysis group, according to me, have become a breed of spatial information architect. Since this group always require an introductory text in geo-statistics, this book which has aimed to bring together within one volume much of the requirements of these various geo-sciences, may be able to serve this large student body of geo-science systems. Teachers in these disciplines can use this book as a guide around which they may structure a course in their speciality and I hope that the basic pedagogical tool given in the book for the beginning geo-science system students will find its better use in the classroom. The third group of readers of this book is made up of both specialists and non-specialists who have some general or specific interest in numerical approaches in geo-sciences.

This book revolves around a logical and holistic approach for application of statistical techniques in geo-science systems which are all concerned with the study of the earth's sun-lit surface. I feel the necessity of a primer-level approach for the present book that has given a fairly broad coverage of topics : the subject matter has been literally spread up for convenience sake in twenty four chapters. The book starts with the first chapter which gives a brief introduct to the multivariable system concepts in geo-science for which the study of earth's sun-lit surface forms the subject matter, the latter has close relationships to present-day ecology and the science of total environment. Geo-sciences in this first chapter is projected as apparently consisted of a number of systems but in reality it is a truth-seeking discipline, dominantly observation oriented, demanding skilled documentation and analysis for detailed description of the spatial extent of forms, their dynamicities and inter-relationships of the earth's sun-lit surface. As a practising geomorphologist over the years, I feel it very strongly that the observational skills are required to be sharpened and in seeking the validity of the interpretational arguments, the hypotheses are required to be worked out with the help of "data-handling and decision-making tool" of the statistical techniques.

The chapters two and three are concerned with the methods of central tendency and deviation measures. Frequently referred as summary statistics, in these two chapters, a detailed description of central tendency and moment measures as well as of cumulative frequency distributions are given so that at the very outset of numerical analysis of data, the properties of the data are revealed characterising a particular situation.

The classical theory of inferential statistics and hypothesis testing has the properties of a decision theory : it starts with the frequency interpretation in which probabilities are associated as relative frequencies. The chapters four to seven are concerned with the probabilistic treatment in geo-science systems. The three usual pedagogical probability distributions namely, nominal, binomial and Poisson are there in their respective chapters but since the concept of probability is central in geo-science systems so a complete discussion on axioms of probability density functions has been done at the outset in chapter four. For orientation data in geomorphology and environmental geology, a normal probability distribution is added also alongwith the log-transformation of data in this chapter four. In the subsequent two chapters one each on binomial and Poisson distributions, generalizations of their respective models with applications from geo-science systems are given in details. The chapter seven provides the probabilistic treatment to the hydrologic data through a discussion on extreme value distribution. Space and time continuity of hydrologic data has long been treated in probability distribution frames and the annual peak of a hydrologic variable has long been subjected to estimation of return periods successfully by using frequency factors designed by a number of hydrologists.

The numerical data in geo-science systems, is often termed as geo-statistics because the data is often 'geo-referenced' on a two-dimensional location space. Hence the chapter eight provides another set of summary statistics for spatial data. The spatial distributional point patterns have been analysed in this chapter as a probability distribution with respect to nearest neighbour measures. This index for spatial distribution patterning of individuals occurring across a continuous community is further worked out by arranging them as plots (quadrats). Due to the lack of sampling distribution and the other moment measures, this chapter on location measures are having a limited scope in statistical terms.

In the chapter nine on sampling designs, under area sampling is included a topic on 'quadrat count' which discusses the spatial point (or, locational) patterns under additive and multiplicative frames by variance-mean ratio test for which a number of generalizations of Poisson probability distributions are taken into account.

Statistical investigations done on the basis of sampling are centered on sampling theory. According to this theory, a prior cut-off by statistical significance level defines a critical region and then, under a specified null hypothesis, a test statistic yields a result which permits a decision on whether a sample statistic does accord or not with the properties of a parent population. For such a concept, inferential statistic is divided into parametric and non-parametric approaches. Parametric statistics in geo-science systems is discussed in three chapters, being spread up from chapter ten to chapter twelve. The list of formulae in context of significance test of sampling distribution is lengthened considerably in chapter ten in view of a number of different statistics of samples tested for standard error since in geo-science systems question of relationship between a parameter and a statistic is emphasized. Another important measure in parametric statistics is the analysis of variance test of significance of k number of independent samples ($k > 2$) which is discussed in chapter twelve.

Non-parametric statistics assume importance in geo-science systems because of their observational mode of inventories and the inability of many of the data in the subject to meet rigorousness of parametric assumptions of the sampling distribution. Accordingly, as many as seven chapters (chapters thirteen to nineteen) are included under non-parametric statistics in geo-science systems. The chapter thirteen on chi-square distribution includes the Yate's Correction and Fisher's Exact Probability test for continuity. Further, this chapter also includes the biserial correlation test.

Relations between two and more variables are dealt within statistics in the framework of correlation and regression. Last five chapters (chapters twenty to twenty four) of the book are contributed for correlation and regression analyses of geo-science systems. One important chapter is the chapter twenty second on spatial correlation. One of the prime tasks in geo-science systems is to detect whether an observed spatial distribution displays significant clustering of similar values which calls for a consideration of spatial autocorrelation for which pedagogic accounts are given in the text (in chapter twenty second) to find whether the apparent clustering is any more than we would expect as a chance process or not. In this context, this chapter is an extension of the quadrat counting under area sampling in chapter nine. This chapter on spatial correlation also discusses the scope of testing the serial autocorrelation in the time series data.

Another important addition is the application of location data of a spatial variable in a linear/polynomial regression. Known as trend surface mapping the trend surfaces show complex spatial pattern of a variable by a statistical character. Included under curve fitting and multiple regression analysis in chapter twenty four, this is an extension of curvilinear relationships. Use of the multiple regression in chapter twenty three involves matrix solution for which some familarity is given in the chapter itself.

Finally, the present manuscript is a disciplinary outcome of the author who is a geomorphologist and not a statistician. But since the statistical techniques are not the prerogative of any one method of philosophy, the author believes that adopting the statistical methodology as a tool for data handling and analysis will bring a harmony among the geoscience systems in deprecating the limitations of compartmentalized approaches and in exploring the possibility of merger of the constituents in order to create stronger and more efficient units in terms of modular course structures for enlarging the information base of the earth's sun-lit surface. In this respect, it is firmly believed that geo-science systems which have been maintaining a distinct character through its history, subject matter and analytical purposes, form a single discipline on its own to constitute the future of environmental science. Thus in spite of several books on the individual aspects of geo-sciences which provide details on the theory and uses of specific statistical methods, there remains a need for a "primer level approach" that has a fairly broad coverage of topics in geo-sciences. That is where the present book is aimed at.

And last, but not the least, the author owes a special word of thanks to his publisher, particularly, Ashok Mittal for encouraging him to write this book.

December 12, 1997

SAROJ K. PAL
Professer in Physical Geography
Delhi University
Delhi, India

ACKNOWLEDGEMENTS

As I was finalising the manuscript of this book, my memory goes back to identify all those geoscientists who positively affected me throughout my career and in many cases they were not aware of the impact they were having on me. They range from my post-graduate teachers in Calcutta University in my early days to present geoscientist colleagues and friends all over the country. To all of them I express my regard and gratitude.

In writing this book, I have got immense help from a vast range of publications. While I thank all the authors of the publications, if I recite all the publications it would make the acknowledgment very ponderous and hence I have listed all the publications I used and mentioned them separately under select bibliography of this book.

I also express my thanks to Mr. T.K. Roy who has once again drawn the diagrams like he did in my earlier book and to Mr. S.P. Nagpal for his pointing out the flaws of fact and the omissions in the manuscript. They both belong to the Department of Geography, Delhi University.

I am fortunate to have Mr. Ashok Kumar Mittal of Concept Publishing Company who has spearheaded the production efforts in shaping my hand-written manuscript of six hundred pages to this book form.

Writing a book, always takes time away from family. I wish to express my sincere thanks to my wife, Jharna for her caring me over the years lovingly in spite of the fact that she never succeeded in stopping me from writing for more than three/four hours at a time.

CONTENTS

LIST OF FIGURES

1

Geo-science Systems and Statistics

1.1 GEO-SCIENCE SYSTEMS : DEFINING SYSTEMS AND ACADEMIC LINEAGE

Geo-Science Systems collectively refer to multi-variable systems. It is concerned with the multi-faceted phenomena about the sun-lit surface of the earth. Geo-science systems are essentially a group of interrelated *natural sciences*, particularly of *geomorphology, climatology, hydrology, pedology*, and *ecology*. Its study embraces to some extent the sky above and the earth below but its primary focus is the sun-lit surface of the earth, the interface for so many complex organic and inorganic reactions as well as the stage for so much human activity. Hence geo-science systems *sensu stricto* may be defined in terms of natural processes and inorganic materials of our environment, but the term is often widened to embrace the biotic processes and organic matter that serve the close links between the natural (or physical) and biological realms such as the studies of soil, micro-climate and land productivity. Thus geo-science systems is a truth-seeking discipline whose raw materials consist of empirical observation about the sun-lit surface of the earth. And it has to be stressed that whereas the subject-matter of geo-science systems is formed by the sky above and the earth below with the sun-lit surface as its principal focus, it is neither *meteorology* nor *geology*. A climatologist may accept the values for the nature of the upper atmosphere while seeking a relation between solar radiation and aerodynamic energy processes to work out a soil-water balance for a ground station. Similarly, a geomorphologist may accept certain postulates concerning structure and tectonics while seeking to explain land forms in terms of surface and near surface materials and processes. Nevertheless, such is the nature of modern science that feedback across disciplinary bounds becomes not only desirable but also inevitable. For example, with the development of *plate-tectonics theory* over the past three decades, structural geology has become as significant to geomorphology as geomorphology has become significant to *structural geology, geophysics* and *seismology*, such as the study of baseline features like Quarternary marine terraces may indicate a rate of surface deformation which are of concern for seismic hazard. Hence, the geo-science systems in its concern about the spatial and temporal explanation of natural phenomena at (or near) the sun-lit surface of the earth, recognizes a subtle combination of subject-matter and gets meshed with methodology of neighbouring bodies of disciplines. Students of geo-science systems should be aware of the porous boundaries of its systems as well as of the feedback potential that exists with the allied disciplines.

The idea of a *system* is basic throughout science, in its simplest form, a system is a structured set of attributes and/or objects so linked together that they exhibit discernible relationships. In geo-science systems, the bounded units of systems are used for drawing up 'mass balance' as might be done for a drainage basin or 'energy balance' for a soil profile. *Eco-system* is the best example of the geo-science systems which is characterised by functional relationships of the whole by certain measurements such as energy flow and matter cycling through the *model of a trophic dynamic approach*. Hence given the energy flow and matter cycling, the chance exists that the agriculture of a place can be studied according to the balance of energy fixed by green plants (produced) and the energy spend for the food production system in the form of fertilisers, pesticides, machineries, labour engaged and transportation, all of which represent fossil fuels in some transformed state. Thus, through the system approach, not only the realities of complex natural phenomena can be captured and explained but there is also a promise of predictability, though the "truth" of nature is always elusive and the promise of predictability can never attain 100 percent probability level.

The process of translating multi-faceted natural phenomena in terms of constituent parts of a whole set can be expressed in words, as the *concept of normal cycle of erosion* by Davis (in 1899) for land form development, or as some form of diagram, commonly of process-form system for *physical geography* by Strahler (in 1980), or in a mathematical form for sediment budgeting of a drainage basin as a unitary system by Trimble (in 1981). All these three processes of conceptualisation of natural phenomena are termed 'System analysis' which provides the possibility of viewing a set of problems at different levels in a way which was never before possible and it makes available new methods for solving problems. As it is said earlier that truth is always elusive in nature, so abstractions about the earth's sun-lit surface phenomena should always be done in system concept and system analysis is the common "umbrella label" for scientific investigation of complex wholes believing that in explaining the relations between the constituent parts of the whole lies the key to the understanding of the emergent properties. Hence what is needed is that geo-scientists of all persuasions like geomorphologists, climatologists, hydrologists, ecologists, pedologists and environmentalists should work under a single umbrella, "geo-science systems" so that the development of knowledge of the earth's sun-lit surface focusing on physical environment does not lead to a fragmented scientific base and it does not have to suffer from inadequate scholarship and resources.

1.2 GEO-SCIENCE SYSTEMS AND STATISTICAL TECHNIQUES

In geo-science systems, there is the necessity of working out a carefully nurtured understanding of the multi-variate character of nature. Hence for studies in geo-science systems, two approaches are often to be seen vying for pre-eminence : (a) holism (or, holistic approach) by seeking the knowledge in the event of its own frame, and (b) reductionism or disaggregation of the event into a magnitude of component parts, the former is inductive and *a posteriori* and the later is deductive and *a priori.* The holistic approach in its concern for assembling an inventory is having an *ideographic attitude*. In it the final explanation is greatly dependent on the initial facts and on the conceptual idea about the uniqueness of the individual phenomena or events. On other hand, the reductionist approach is essentially a *nomothetic approach* which implies a desire to subsume individual cases under laws or law-like statements of very general, if not universal applicability. It follows that the nomothetic approach, as a result of its general concern to establish general laws, necessarily involves *a priori* particulars that are not perceived (or involves inductive reasoning) but are independent of perception. Hence to the extent that the scientific method embodies a commitment to the establishment of a body of knowledge which is both perfectly general and absolutely certain, it is emphasised that these two routes of holistic and reductionist approaches are interlocked. This requires the theory and the conclusion to develop from *a priori* concept confirmed by *a posteriori* analysis of empirical situation and in which the *a priori* concept is stimulated and focused into empirical questions and hypotheses. The whole provides a framework of testable theory, analytically structured, verified for internal consistency and falsifiable against observed realities in terms of their past events, notably their magnitude, frequency and spatial variation. This is a hypothetico-deductive method of Popper (1957) in which the hypotheses are proposed, analytically structured through *a priori* analysis that leads to deductive conclusions. These *a priori* constructs are then tested by empirical observations. So long the empirical conclusions are not falsified, the theory and the hypotheses are accepted. Thus, this model by Popper eschews that the subjective knowledge is required as part of the reductionist treatment of a problem in geo-science system in setting up a *null hypothesis* to test statistically whether a relationship exists or not. Figure 1.1 illustrates this interactive deduction and inductive process.

With the above statement about state-of-art regarding the geo-science systems and their methods of investigation, it is apparent that building and testing of hypotheses demands precision and for achieving precision, inclusion of statistical techniques into the set of methodological tools is imperative. It is indeed significant that the use of statistical procedures in geomorphology, climatology, hydrology, ecology and soil science have developed together. Being interwoven together as part of the geo-science systems, they all deal with the exploration of 'patterns' in their respective fields which include the spatial dispersion, relationship between and among the fields within and between their communities. Hence they all study the same processes and materials, but from different vantage points. For a long time, geo-science systems individually were considered to be merely descriptive since they deal only with the scattered surface observations of forms and processes, consequently

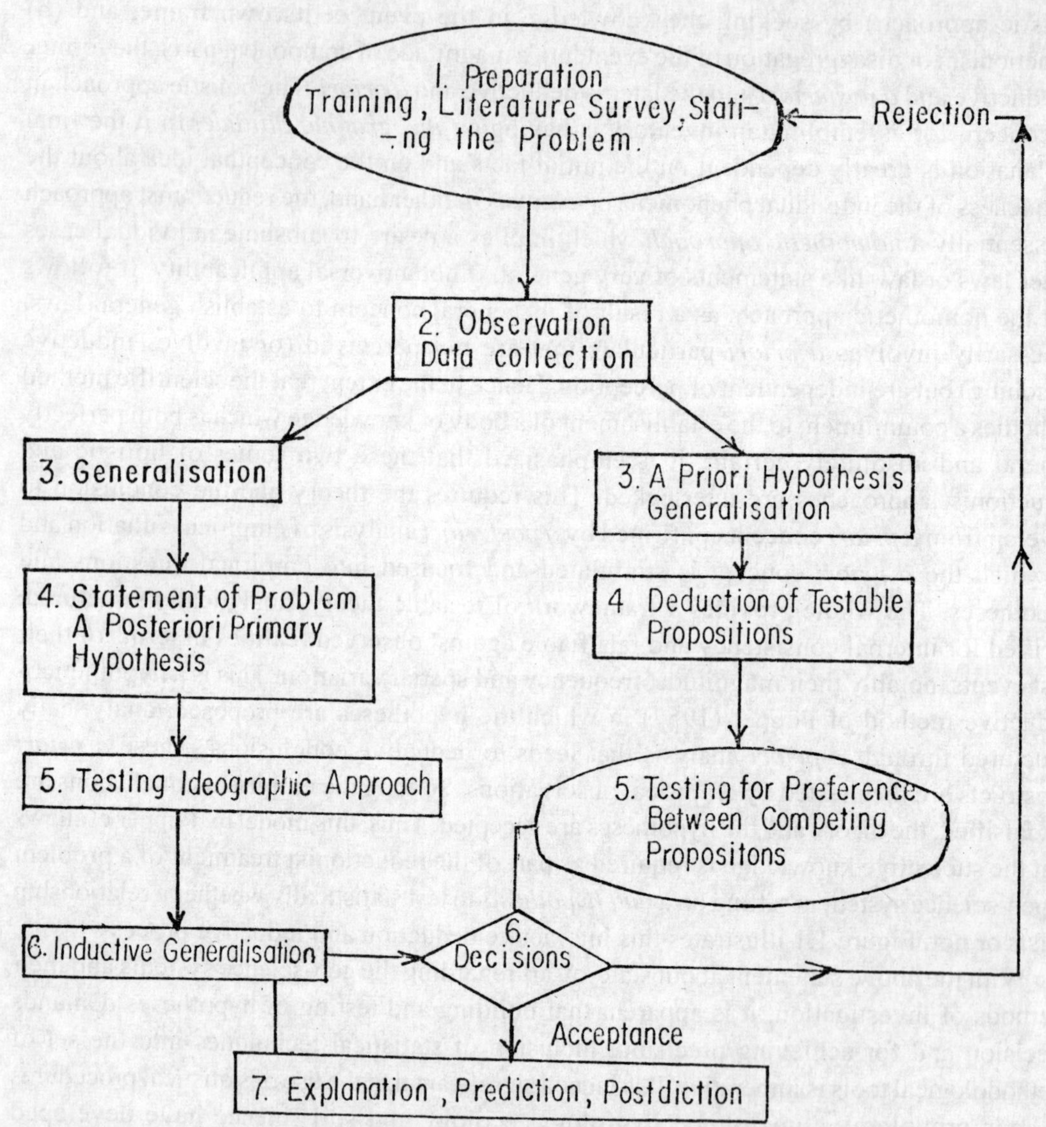

Fig. 1.1 : Statistical Reasoning in Interaction for Induction and Deduction

any quantification were felt to be unreliable or meaningless. However, with the advent of modern instruments, vast quantities of numerical information are available so that the trend in geo-science systems has been towards using a numerical approach for formulating multi-faceted phenomena of the earth's sun-lit surface and to facilitate the solutions in statistical terms. Simply put, without statistics, the logical structure of the process by which the search for trust-worthy knowledge is carried out, becomes impossible since by scientific methods we are not concerned with the content of the search, but rather with the pattern of thought and the aim to explain. It is indeed the synthesis of physical and statistical properties of past events, notably their magnitude, frequency and spatial variation, integration of these and other data into environmental equation and prediction of future events by using this information are the primary goals of geo-science systems.

The word 'statistics' was first used in Hooper's *The Elements of Universal Erudition* published in 1770. In one of the chapters of this book, Hooper defined statistics as that which 'deals with science that teaches us what is the political arrangement of all the modern states of the known world' (Till,1974). In the early 19th century, this Hooper's definition of statistics got changed by replacing the characters of states with the numerical methods. Only by the end of the 19th century, the term 'statistics' was used 'to describe the summary figures and to compare the properties of a set of observations'. During this time the theoretical basis of the subject was also laid down from three fairly distinct routes: (i) there had been the development of *probability theory* concerned with the theoretical distributions by Norman Gauss, Jacob Bernoulli and Semon Denis Poisson and much was stimulated in the early days by probabilities of various bets in gambling, (ii) as an outgrowth of probability theory there has been the development of statistical techniques, concerned with making the unbiased inference in a given multidimensional empirical situation. In the early 20th century, much of the early work in this field came from genetics concerned with breeding trials in experimental plots, deriving from the works of Sir R. A. Fisher, S. Snecdor and P. C. Mahlanobis, (iii) more recently there came the stochastic and simulation techniques by the recourse to computer and availability of ready-to-use software packages like Lotus, "SPSS" etc. Thus today we find the ideas of statistics on a firm basis and the term 'statistics' refers to the word *data* in a general sense but it also refers to the 'statistical techniques'. *Statistical techniques* are used in conducting the statistical enquiry concerning a certain phenomenon. In a statistical enquiry a set of data regarding the phenomenon is enumerated or measured or estimated according to a reasonable standard of accuracy (as required by the mode of enquiry), collected and organised in a systematic manner for a predetermined purpose, and placed in relation to each other for analysing and interpreting them by a relevant statistical technique. Thus statistical techniques are concerned with the collection, organisation, analysis and explanation of observational form arising out of the study of distribution and location of various characteristics on the surface of the earth. Hence this approach to statistical analysis of spatial distributions and interrelationships involve two aspects— (a) the collection of numerical information in terms of a set of numbers called 'data' for a particular phenomenon to be studied, and (b) the drawing together of these data into meaningful relationships/theories. These aspects of statistical techniques have become the common tool for research in geo-science systems so that statistics is not a science, it is a scientific method about techniques, the use of which constitutes a very useful and often indispensable tool for geo-science systems research. Thus statistical techniques are not merely a means to an end, but they are, however, a very useful means indeed. Hence though there is no adequate body of general theory about the nature of geo-science techniques, like which can be used in rigorous tests of statistical hypotheses, it is believed in geo-sciences that an examination of a small sample collected from a vastly large set of observations on a phenomenon can deduce information about the phenomenon faithfully.

1.3 STATISTICS AND MODEL BUILDING

Statistical techniques (or the subject of statistics) are essentially of three main types: (i) descriptive techniques, (ii) sampling and (iii) inferential techniques.

The raw material of any statistical analysis is a *set of data* (which is the plural meaning of a body of information in numerical form). The data in geo-science systems can be either primary or secondary data. For primary data, the data itself can be a product of either an 'observational' or an 'experimental' approach. In *observational approach* data are either measured by instruments or by observations over a range of conditions imposed by nature. On contrary, in *experimental approach*, the researcher divides the attribute into replicate portions on which various treatments and controls can be imposed so that any differences detected in the analysis can be attributed to the experimental treatments. If the data are of secondary nature it can be categorised into *uni-* or *bi-variate* data as well as *multi-variate* data, former are collected along a traverse or over an area for which spatial dimensions are considered so that it can be said that data have been taken along a continuum, either of time or distance. But in the multivariate data, the locations of observations are not considered though this multivariate category of data can be concerned with examination of interrelations, clustering, and classifications among data sets. *Descriptive techniques* aim to summarise raw data of any size of value ($n \cong \infty$) in terms of space (about places, areas or locations) or through time (for a time series analysis to give trends and/or fluctuations) for providing a single comprehensive index or a graph of it. The descriptive techniques facilitate accurate descriptions of events and their comparisons too. Thus desciptive techniques aim in ordering and summarising a given set of numerical data without any direct reference to any inference that may be drawn otherwise.

In contrast to the descriptive techniques, the sampling techniques are often based on data that are not a complete set of observations on a specific phenomenon or event and are known as samples. *Sampling* decides the smallest size of observations that gives the desired reliability in the results of the phenomenon/or the event studied. Thus the statistical techniques which are concerned with the methods of drawing samples from the total set of observations are known as *sampling techniques.*

If the sample is drawn from a total set of observations, some method is required to draw conclusions (or, to infer) about characteristics of the total set of observations (known as 'population'). The methods of drawing such inferences from numerical data of a phenomenon/event are known as *inferential techniques*. Thus sampling and inferential techniques are inseparable : they are the two phases of the same process.

The essential tool in inferential techniques is the *probability theory*. This theory measures the chances of selecting a particular sample from a population whose characteristics are known. The inferential techniques are used in many specific purposes, for example, we may seek to estimate the value of the parameter like estimating the population mean from sample mean or in estimating the degree to which a collection of data supports a hypothesis, in testing the correlations between the samples in bivariate or multivariate situations and in inferring the results of the analysis by certain measures of probability. Finally, it can be said that in inferential techniques we are concerned with research inferences in a probabilistic framework and significance testing for inferences to populations from samples is a rational for all of them.

Nature is complex and science has moved towards acknowledging the nature's very complexity so that as it is mentioned earlier we shall never be able to describe all the various components of our environment in detail. Consequently we must attempt to gain a more holistic view of the earth's sun-lit environment by including the main-line and bearing processes in our description. We must therefore put more emphasis on synthesis for analysis through 'modelling' as a sort of holistic approach to a scientific description of the environment. It is a synthesising tool which puts analytical results together to give an overall picture. It is to be noted that the terms 'system' and 'model' are not synonymous—a *system*, as defined earlier, is a bounded unit of a structured set of attributes and it is assumed to exist in the real world to possess the attributes whereas a *model* is an attempt to describe, analyse, simplify or display a reality, to a manageable proportion which is otherwise complex. Statistical techniques often make use of the models by structured representation of reality and geo-science systems use models more than any other disciplines among sciences for representation of a natural phenomenon/event of an area. Anyone who attempts to isolate the components of a geo-science system like runoff, sediment flow, erosion and deposition yields of a drainage basin soon realise that they are enormously complex no matter what the scale of analysis is, so that their study requires a degree of abstraction/simplification by model building.

The design and construction of a model can proceed with varying degrees of abstraction. In geo-system sciences* there are mainly three types of models: (a) scale, (b) conceptual or normative and (c) mathematical. *Scale models* of geo-system sciences are again of two types: *iconic* and *analogue* models. Aerial photographs/ satellite imageries (i.e. image like

* Because of emphasis of system. geo-science systems can also be called geo-system sciences interchangeably.

photographs) which represent the properties of the real world but at the cost of change in scale are examples of 'iconic' models. The topographic maps which show the contours can give an idea of relief of the surface and in that way they represent the 'analogue' model which shows one property by the other. While the iconic models are the miniature copies of reality, a more abstraction of the iconic model is the analogue model. A 'conceptual' or a 'normative' model is mental image of a natural phenomenon in which the supposedly essential details of the phenomenon are retained while the other details are omitted. *Conceptual models* are always, to varying degrees, abstractions or simplification of earth's sun-lit surface phenomena. They represent attempts to capture and explain aspects of reality, the 'true' nature of which are always elusive. They can be expressed in words, in graphics,

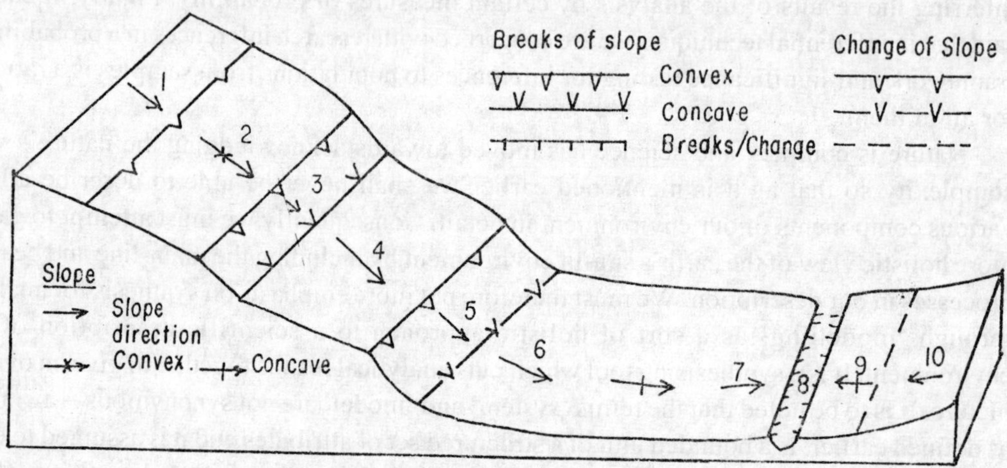

Fig. 1.2 : Model illustrating Process of a Fluvial System

as in the conceptual or graphical model of landform development or by a 'box and arrow' diagram in which the boxes represent the system components and the arrows depict the important links and relationships between the components. Figure 1.2 illustrates the ten components of process-form model of a fluvially eroded slope system in which the arrows depict the important links and relationships between the form and process variables. Since conceptual models help us how to clarify our loose thoughts about how a system is made and how it operates they are often a first step in the building up of *mathematical models.*

Thus in their purest form, the 'mathematical models' are constructed by ideas contained in a conceptual/normative model into the formal symbolic logic of mathematics. For example, in a very general mathematical model of discharge, Y in a drainage basin is stated equal to some transformed, F, value of rainfall, X that is, $Y = F(X)$. The value of F would have to be obtained from a set of equations which describes how the surface water (= rainfall-loss by infiltration and evapotranspiration) moves between each of the components of drainage basin, like the soil, groundwater and vegetation. The important point to grasp is that the mathematical model enables us to make numerical predictions about the relationship between rainfall and discharge. The prediction can be assessed by matching it against the yardstick of the observed/measured data. If a model equation is found to replicate observed data accurately, then the assumptions used to derive the equation provide a theoretical explanation of the behaviour of the system. As it is mentioned earlier, the development of theories which purport to explain observed phenomena is the essence of scientific endeavour. It is hoped that by the continual process of model building, model testing and model designing we can gradually obtain better and better explanations of the forms and processes recognised as the geo-science systems of the earth's sun-lit surface.

Three chief classes of mathematical models are of interest to geo-science systems. They are *stochastic, statistical* and *deterministic* models. Conceptual models, as mentioned earlier, are normally built as a first step in the study of the geo-system sciences regarding the laws governing the movement and storage of the components (of the systems individually). The second step is to elaborate the model by carrying out a quantitative investigation of an analytical kind in terms of measurement of the flows, storage, forces and any other components which are thought relevant to the system under scrutiny. It has been found that all processes in the soil-landscape system operating on the earth's sun-lit surface are part of a *unitary system* and they can all logically be modelled in the framework of a drainage basin. The next step is to represent components of the system by abstract

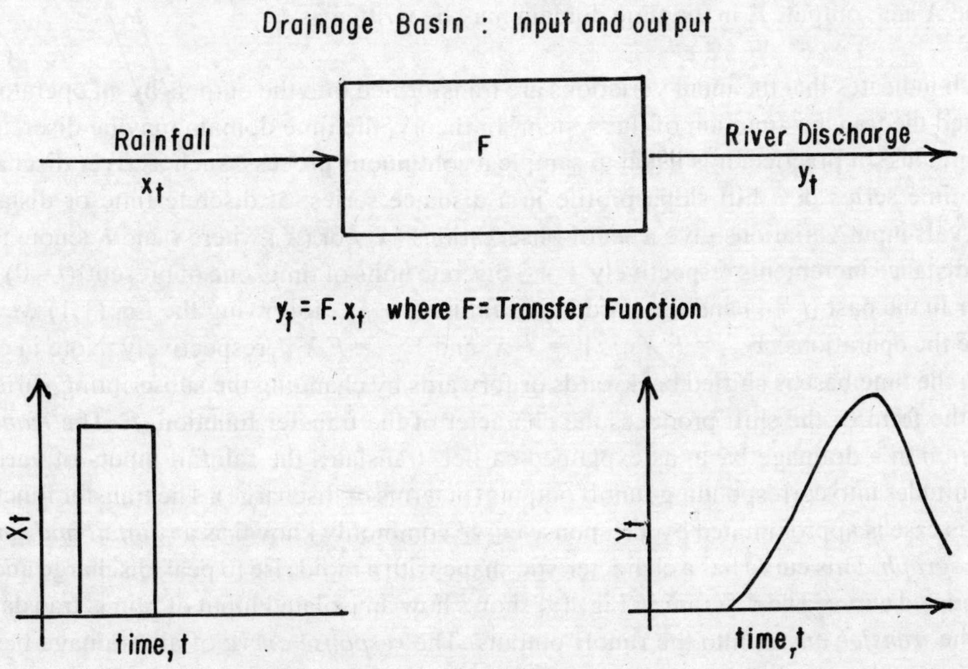

Fig. 1.3 : Box-&-Jenkins Model of a System $Y_t = F . X_t$

symbols and subject the links (of the components) to the rigour of mathematical argument by expressions containing mathematical variables, parameters and constants. Mathematics offers the most powerful and rigorous tool of investigation of the systems limited only by the creativity of the human mind. It provides a means of describing a system regarding how it functions, affords the most trustworthy base of how best to manage or control the system as well as it provides the best basis for predicting change in systems and all these are done in a symbolism which is universally acknowledged. Actually it is the predictive power of mathematical models that, in sense, sets them apart from their conceptual model counterpart since the latter remains a body of ideas and there is no formal way in which an unquantified model can be tested. A mathematical model, on the other hand, when tested by matching predictions against the yardstick of observations we come across three variations of mathematical models, which we shall describe now.

'Stochastic models' have a random process built into them which describes a system or some facet of it and these models focus attention on a set of possible outcomes, each of which has a different probability of occurring in reality. The basis of all stochastic models is the *probability theory*. Stochastic models may be subdivided into 'inductive' stochastic and 'deductive' stochastic types. *Inductive stochastic models* are applied to a variety of observed time and distance series data. The basis of the inductive stochastic model is the Box and Jenkins model of a system which defines a system by an input variable, X, an output variable, Y and a *translator operator* linking the two (Fig. 1.3). The relation between input, X and, output, Y in the time domain may be written as

$$Y_t = F \,.\, X_t \qquad \ldots (1.1)$$

which indicates that the input variations are transformed into the outputs by an operator F, termed the transfer function of the system. In theory, the time domain may be discrete or continuous. In practice, it is usual to sample a continuous process, such as river discharge in a time series or a hill slope profile in a distance series: at discrete time or distance intervals input variations give a set of observations (X_t) or (X_i) where t and i denote time and distance increments respectively. For 3 discrete units of time, one of present ($t = 0$), the other in the past ($t = 1$) and the third in the future ($t + 1$), following the Eq. (1.1) we can write the operations: $Y_{t+1} = F X_{t+1}$, $Y_0 = F X_0$ and $Y_{t+1} = F X_{t+1}$ respectively. Note in each case, the time base is shifted backwards or forwards by changing the subscripting variable and the form of the shift produces the character of the transfer function, F. The *transfer function* in a drainage basin as explained earlier, translates the rainfall inputs of various magnitudes into corresponding runoff outputs (in terms of discharge). The transfer function in this case is approximated by a response curve commonly known as a *rainfall and runoff hydrograph*. This curve has a characteristic shape with a rapid rise to peak discharge and an attenuated curve. The diagram in Fig. 1.4 shows how an isolated input of rain is translated, via the *transfer curve*, into the runoff outputs. The *response curve* of all drainage basins will be similar but differences in the drainage basin characteristics may produce transfer functions which will not be unique for each basin. Hence transfer function is very useful in characterising the parameters of a system where the input series and the output series are known. The change in output in the face of a change in the input is called the *dynamic response*. A transfer function model describes this response and characterises the inertia of the system.

The *language of probability* is the key to the *deductive stochastic models*. Since chance is involved so the predictions from the stochastic models in which events are entirely observed in independent time or spatial framework and (in which) the order of occurrence of events is important, are stated with a known degree of error or tolerance. Many of the events, phenomena like rainfall, flood, earthquakes etc. occurring at sun-lit surface of the earth are having distributions which can be generalised by their long-term behaviour as 'normal' and/or 'extreme value distribution'. *Geometric probability distributions* are defined for many of these events.

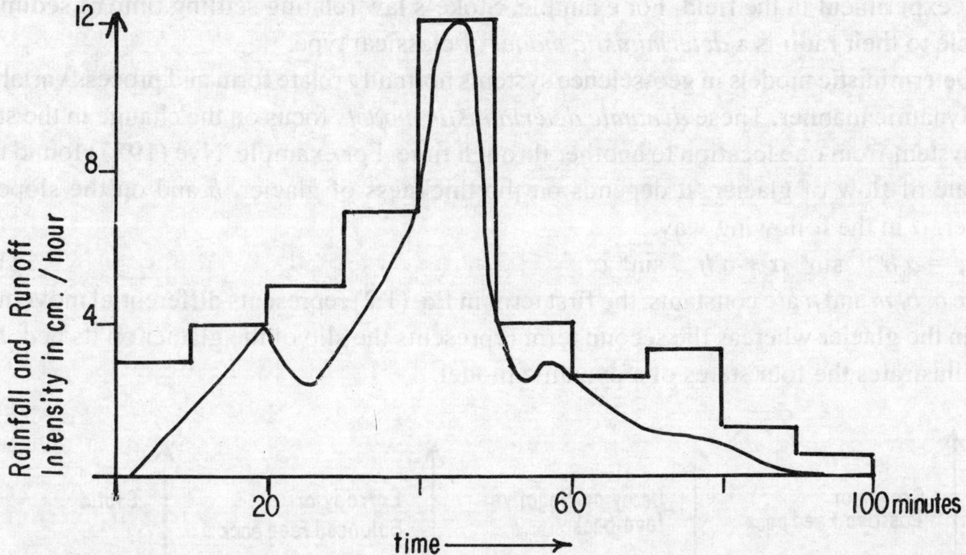

Fig. 1.4 : Rainfall and Runoff Hydrograph

"Statistical models" are like stochastic models: they too are based on probability concept and have also random components. In statistical models, the *random components* are the error components in the measurment, in the equation or in the inherent variability of the attributes of an event. In statistical models, a vast body of inferential statistical theory determines the manner in which the sample data should be collected and relationships in the data should be tested. The *statistical models* used in geo-science systems are many and varied. They may include simple tests to compare 'mean values of measured data set', 'simple correlation and regression' and sophisticated 'multivariate techniques' designed to explore complex interrelations within multi-faceted sets of variables. Statistical models can only be applied under strictly controlled conditions but they are perhaps the most useful since deterministic style of modelling is not applicable in many events whose form and processes are little understood or related in a chance-like way. Statistical models of geo-science systems of earth's sun-lit surface form the major focus of this book. For example, if we have to map the distribution of sal trees in a tropical forest, it has to be a statistical map for one to make inferences based on spatial distribution of *sal* (*shorêa robusta*) trees on a random/clustered sampling frame. It can be seen that in statistical model, as more amount of information is lost, the model becomes more abstract but more general.

Lastly, the 'deterministic' models are the conceptual models which have been translated into the language of mathematics but which contain no random or chance components; this is made possible by using the observations to evaluate or calibrate.the model equation to obtain exact predictions about reality so that in the model there is no built-in explicit assumption about the substantial differences between prediction and reality. These models are the 'classical models' derived from the principles of basic sciences without recourse to direct experiment in the field. For example, Stoke's law relating settling time of sediment particle to their radii is a *deterministic model* of classical type.

Deterministic models in geo-science systems normally relate form and process variables in a dynamic manner. These *dynamic deterministic models* focus on the change in the state of a system from one location to another through time. For example, Nye (1957) found that the rate of flow of glacier, μ depends on the thickness of glacier, h and on the slope of glacier, α in the following way:

$$\mu = a.h^{n+1} \sin^n \alpha + b.h^m . \sin^m \alpha \qquad \ldots (1.2)$$

where a, b, m and n are constants, the first term in Eq. (1.2) represents differential movement within the glacier whereas the second term represents the slip of the glacier on its bed. Fig. 1.5. illustrates the four states of a dynamic model.

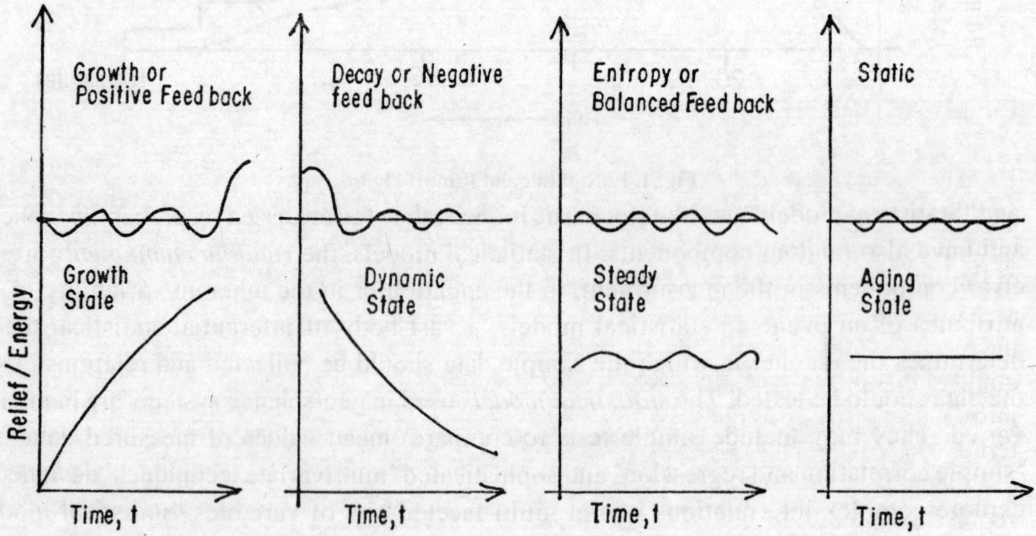

Fig. 1.5 : Four States of A Dynamic Model

Deterministic models of geo-science systems normally, relate process variables in a dynamic manner. They focus on the change in the state of a system from one location to another through time. For example, the rate of flow of glacier, μ depends as given in Eq. (1.2) on the thickness of glacier, h and on the slope of glacier, α that is, on differential movement within the glacier and the slip of the glacier on its bed. Lastly, but very importantly, these dynamic deterministic models are required to be tested by comparing the predicted results against the empirical ones for their validity. This new generation of dynamic models held out the promise of combining short-term and long-term changes of a system within a single framework. Finally, Fig. 1.6 shows the varying degrees of abstraction of the models in geo-science systems.

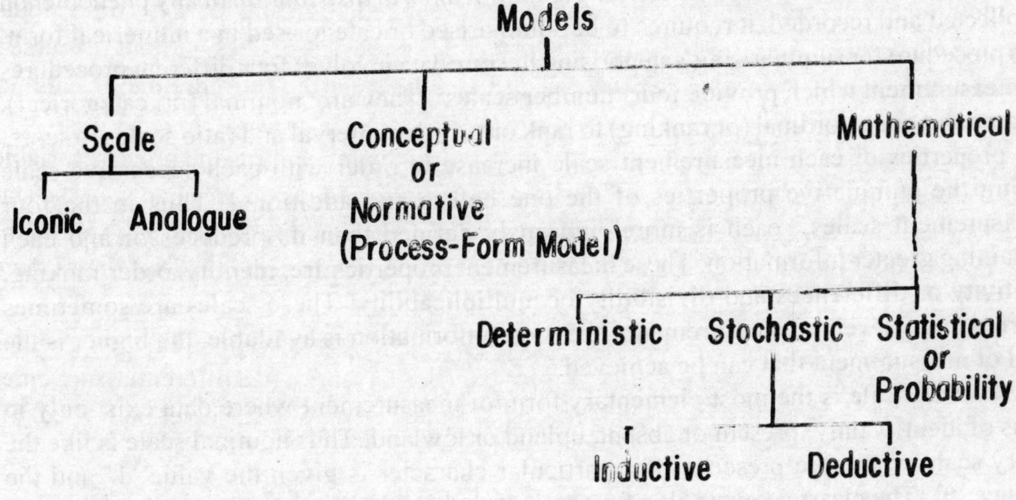

Fig. 1.6 : Models in Geo-Science Systems

1.4 MEASUREMENT SCALES IN GEO-SCIENCE SYSTEMS

The word 'measurement' includes the assignment of names to features on the ground and the computation of relationship using them. Thus the names used to describe landuse in term of cultivated land, built–up land etc. indicate the "naming function" by which elements of the set represent use of the land surface whereas density of a particular landuse in a given area uses the "counting function" of measurement. Thus measurement of an attribute provides the mechanism to describe the attribute and to communicate this description to others. The quality of information thus created and communicated about an attribute can be divided into "categorical" (or, *qualitative*) and into "continuous" (or, *quantitative*) functions in terms of mesurement.

Generally, *measurement* is a numerical value and it is taken to be the process of representing some property of an object quantitatively. Raw materials of statistics with which statistical techniques : descriptive and inferential – do work are the 'measurements' or 'data'. The term measurement or data are facts or figures recorded as a result of the counting or measuring a system, from which conclusions are drawn. The data collected for a particular system (an event, an outcome or a process) are known as 'data set'. The *data set* is always for a 'variable' which varies from one individual to another, hence individual element in a data set is known as *variate*. *Variables* can be discrete or continuous – *discrete* or discontinuous variables take on only integral i.e. the whole number values like 1, 2, 3, . . . etc. whereas *continuous* variables may be measured to an arbitrary degree of accuracy and the result of the measurement is not necessarily a whole number, e.g. 25.3 cm is a data for a continuous variable like rainfall. Thus we can make a measurement at all sorts of magnitude from counting the heads in an enumeration to the radius of the earth.

Once the data/information concerning *spatial/temporal distribution* of any phenomenon is collected and recorded, it requires to be summarised or categorised in a numerical form. This procedure for summarising/categorising the raw data involves four different procedures for measurement which provide four 'number scales'. They are: nominal (or, categorical), to name and count, ordinal (or ranking) to rank or to order, interval and ratio scales to score. The properties of each measurement scale increase in order with each successive scale having the cumulative properties of the one before and additional. Thus in the four 'measurement scales,' each is more rigorously defined than its predecessor and each containing greater information. These measurement properties are: identity, order/ranking, additivity or differences and divisibility or multiplicability. These scales are sometimes referred to as 'levels of measurement,' the more information is available, the higher is the level of measurement that can be achieved.

Nominal scale is the most elementary form of measurement where data exist only in terms of identity only: present or absent, upland or lowland. Thus nominal scale is like the binary scale where the presence of a particular character is given the value '1' and the absence '0'. The classical example of a nominal scale is the tossing of a coin which gives either head or tail. For the categories employed in nominal scale, data should be exhaustive including all phenomena under study and mutually exclusive so that no item is placed in more than one category. Nominal data cannot be manipulated by any basic mathematical operation but the nominal measurement is important in statistical analysis since the item in each category may then be counted and the total represented by a number. This is termed count data and it is in the form of frequencies. A classical example of nominal scale is counting the heads or tails in coin-flipping experiments. Many of the classifications are done in nominal scale e.g. soil, rock or mineral classification, and landuse groupings etc. all belong to nominal scale. It is a paradox that the best data available in geo-science systems is at this weak measurement scale.

Ordinal scale is the level of measurement next to nominal – at this level we have sufficient information not only to establish differences between objects by identities but also to place our data before or after another along a scale in rank order either individually or in classes. The statement $A > B > C$ implies that there are three classes of phenomena and that class A is greater than (or comes before) class B which in turn is greater than (or comes before) class C. Under this condition it is not possible for $C > A$. Some data are inherently of an ordinal nature.

The classical example of ordinal scale is Moh's scale of hardness of minerals which is given in Table 1.1.

Table 1.1: Moh's Scale of Hardness of Minerals

10	Diamond	5 Apatite
9	Corundum	4 Fluorite
8	Topaz	3 Calcite
7	Quartzite	2 Gypsum
6	Feldspar	1 Talc

Hardness is defined as resistance of mineral to scratching. A relative scale is adopted for measuring hardness. To determine this property, in the Moh's hardness scale, a set of 10 minerals are arranged in order of increasing hardness. For a quick determination of this property of minerals, a brass plate, an iron plate and a glass plate are used in addition to scratching the mineral surface with one's finger nail. Minerals that can be scratched by a finger nail have hardness less than 2.5 while those that can mark on a brass plate have values more than 3 whereas minerals that can mark an iron plate have hardness more than 4.5 and those which have hardness 6 or more can scratch a glass plate. In Table 1.1, at scale 1 we are having Talc which is characterised as being very soft and easily scratched; at the opposite end of the scale at 10, we are having Diamond which is the hardest mineral being unscratched by any mineral. But since Moh's scale is an ordinal scale the rank numbers or orders such as 1, 2, 3 etc. are not defined for any standard arithmetic operations hence we cannot say that Diamond is ten times harder than Talc. In this ranking of hardness of minerals any mineral can scratch all those beneath it in the scale and they will be scratched by those above it. Thus by observing which rock scratches other rock, one can put every rock in a sequence. So Moh's scale represents a complete ordering in terms of empirical concept of hardness. This scale also illustrates that in nominal scale of measurement, we know whether one measurement is greater than another although we do not necessarily know by how much. Thus the difference in absolute hardness between Diamond (rank 10) and Corundum (rank 9) is greater than the entire range of hardness from 1 to 9. In social sciences, occupation ranking by gross income level or caste or social rankings are examples of ordinal scale.

The "interval" and "ratio" scales are the highest scales of measurement which correspond to what we may call measurement in the real sense of it. Such measurements allow direct translation of empirical concepts of good, average or poor ranking of nominal scale into quantitative number systems, that is, these scales assign an exact arithmetical value so that the difference between any two items on the scale is known exactly.

The *interval* scale is so named because it consists of measures for which there are equal intervals (that is, the length of successive intervals is a constant) between each measurement or between each group. Thus in this scale not only the objects are given identities and ranked like in nominal and ordinal scales but also the numbers can be added or subtracted and the differences or intervals between objects in terms of that property are known. So this scale is capable of comparing differences between a number of pairs or scores to indicate the precise position of the objects along a continuous scale. Hence the interval scaled data can be added or subtracted and the measures of central tendency and measures of deviation (to be discussed in next two chapters) can be calculated.

It is sometimes important to distinguish between the interval scale for which zero is arbitrary and the ratio scale that has an absolute zero. For example, the zero of *pH* is arbitrary (since *pH* = 7 indicates neutrality on the *pH* scale whereas a *pH* below 7 indicates acidity and similarly a *pH* in excess of 7 indicates alkalinity) and *pH* scale is known an interval scale. But ratios of two different values on this scale have no meaning in interval scale.

Ratio scale is the scale in which all the requirements of interval scale are met but in addition it has a true zero scale. For example, rainfall scale (expressed in either inches or centimetres) has a true zero base. Thus if in a year a place receives 50 cm of rainfall and another place receives 100 cm, we can say that the second place receives twice as much rain as the first. But we cannot say so in case of temperature e.g. 100°C is not twice as hot as 50°C because temperature does not have a true (or natural) zero where the magnitude (temperature) is non-existent so that we can have negative temperatures which are less than zero. The starting point for the Centigrade scale as well as for the Fahrenheit scale are arbitrarily set at a point coinciding with the freezing point of water i.e. at 0°C for the former scale and 32°F for the later. Thus temperature in °C can be read in °F by a conversion factor (which is °F = 9/5. °C + 32) that is, equivalents of 100°C and 50°C in Fahrenheit scale are 212°F and 122°F respectively and this conversion in Fahrenheit scale obviously does not result in a ratio of 2:1. Thus while temperature scale in °C or °F is not a ratio scale, the Kelvin scale of temperature is an absolute scale since 0°K denotes the temperature at which all molecular motion stops and no temperature (in form of energy) can be lower than this value. Hence the Kelvin scale is a ratio scale.

The property of the ratio scale is that any two measurements bear the same proportion to each other irrespective of the unit of measurement used. Common ratio scales are measures of height, length, mass, weight, stream velocity, dip and strike, slope, income and so on. Ratio scale can be dealt with statistically in the same way as interval data. A ratio variable is also an interval variable but an interval variable is not necessarily a ratio. We need not distinguish between interval and ratio scale for most purposes. Data recorded on either of these scales are amenable to all kinds of mathematical operations and to many forms of statistical summary and analysis. Finally, it can be said that because the ratio scales have an absolute zero, so of the four scales of measurement, the ratio scale contains maximum amount of information about any entity.

It has been mentioned earlier that the type of statistical tests that can be used vary between these four different measurement scales. For example, data on nominal scale can use only the simpler tests without involving any mathematical manipulation. Thus if a town is grouped in class I town and the other in class III town, we cannot say that on average they are in class II town. But within categories in this scale, the number of occurrences can be studied and compared. Ordinal data can use a little more sophistication, while for the data on interval and ratio scales all four of the basic arithmetic operations like addition, subtraction, multiplication and division as well as a series of statistical indexes can be performed.

In Table 1.2 we summarise the above distinctions between the scales of measurement and specify measures of 'central tendency' and 'measures of variation' which are appropriate at each scale.

Table 1.2: Scales of Measurement

Scales of measurement	Characteristics	Basic Empirical operations	Application in descriptive statistics	Examples
(1)	(2)	(3)	(4)	(5)
Nominal	–Mutually exclusive categories. All observations within each category are treated the same	–Determination of equality/ difference	–Mode only –Frequency, Binomial and Multinomial expression	Presence/ Absence, landuse types, respondent's sex
Ordinal	–Direction and relative position on scale are known –Distance between any two points on the scale are not measured by a common unit of measurement	–Determination of greater or less (i.e. rankings) in empirical sense	– Mode + Median; Percentile and Inter-quartile range	A possible answer to a question: (a) Support (b) Neutral (c) Oppose
Interval	–Direction and magnitude of position on scale are known i.e. distance between any two numbers on the scale are known but there is an arbitrary point and a unit of measurement	–Determination of equality or differences of intervals in arithmetical sense	–Mode + Median, Mean and Variance, Standard deviation, Interquartile range	– Temperature in °C or °F – Time in years AD or BC – Because of an arbitrary zero point we cannot say 0°C ≠ °F
Ratio	–All characteristics of interval scale plus a true zero point so that the ratio of any two points of measurement is independent of the units	–Determination of equality or differences of ratio	–Mode + Median + Mean and Variance + Standard deviation + Coefficient of variation: Geometric mean	– Given numbers with arbitrary unit of measurement e.g. distance or weight (zero point is identical)

The above mentioned summary measures of descriptive statistics will be discussed now in the next two chapters.

2

Numerical Data in Geo-science Systems I
(Frequency Distribution and Central Tendency Measures)

2.1 FREQUENCY DISTRIBUTION

The preceding chapter has shown the collection of the numerical information for a research problem in form of data sets obtained from primary surveys or from any series of measurements. These are often so large that the data are required to be organised and summarised to manageable proportions before any study of them can be undertaken. Table 2.1 shows the annual rainfall distribution as recorded at Varanasi Observatory for 50 years, 1901-1950. This represents a data set which requires to be organised and summarised for carrying out a meaningful statistical analysis. Thus one of the first steps in studying a set of measurements/data is to arrange the raw data from the smallest to the largest observations, or *vice versa*. This arrangement is often called 'array'. An *array* helps in the simplification and arrangement of raw data. To do this values are chosen at equal intervals as limits of successive classes within the observed range. The number of individual values falling within each class is counted and termed the *frequency* of that class. The manner in which the frequencies are distributed over that class is called the *frequency distribution* of the character under that study.

Table 2.1 : Annual Rainfall Distribution
(recorded at Varanasi, 1901–50)

Year	Rainfall (in cm)	Year	Rainfall (in cm)
1901	118.0	1926	115.0
1902	37.1	1927	117.2
1903	110.8	1928	42.7
1904	136.9	1929	206.4
1905	192.4	1930	159.0
1906	92.9	1931	176.9

1907	167.2	1932	137.8
1908	112.5	1933	97.8
1909	46.0	1934	91.9
1910	114.0	1935	133.7
1911	117.7	1936	160.6
1912	154.3	1937	203.0
1913	105.5	1938	105.1
1914	93.5	1939	39.8
1915	130.9	1940	151.8
1916	168.1	1941	101.5
1917	123.3	1942	94.3
1918	158.4	1943	129.8
1919	97.1	1944	192.2
1920	106.9	1945	89.6
1921	197.7	1946	77.3
1922	91.4	1947	65.7
1923	177.7	1948	253.5
1924	82.8	1949	171.6
1925	48.8	1950	174.4

Thus the first step in a statistical analysis of a set of data often consists of arranging the data for a frequency distribution. This involves grouping the data on the basis of class intervals in a *frequency table*. So the frequency table or frequency distribution is a tabular method of counting the number of elements of the data set (termed frequency) falling within the class limits.

The frequency table has *class intervals*. The number of class intervals in a grouped data is generally a subjective judgement on the part of the research worker although certain important considerations should be borne in mind like the classes must be continuous within themselves and should be such that the characteristic features of the distribution are displayed. Hence the class interval for a frequency distribution must not be so large that the mid-point of the class interval is not the average value of the data and it should not be too small to give many classes with zero or very small frequencies. In general, any of the following formulae should be used as a rough guide for deciding the number of class intervals to be taken in any frequency table:

$$N = 2^k$$

or $$k = 1 + 3.3 \log_{10} N \qquad \text{... (2.1)}$$

where k = the number of class intervals to be used in a frequency table, N the total number of observations and $\log_{10} N$ is the common logarithm to the base 10 of the data. For example, with $N = 86$ observations this would suggest $k = 1 + 3.3\,(\log_{10} 86) = 1 + 3.3\,(1.93451) = 7.38$ which would be rounded to $k = 7$ classes. The width w of the class intervals becomes obviously:

$$w = \frac{\text{Range in the data}}{k} \qquad \text{... (2.2)}$$

In general, the width w of the class intervals in a frequency distribution is uniform. However, in certain cases of categorisation of data, the width of the class intervals cannot be uniform, particularly the distributions of those events which are characterised by uneven developments leading to few large accounting for larger share of the event and large number of smaller ones accounting for a smaller share. In this case, the class frequency is divided by the class interval to give a measure of the 'frequency density' in each class. Sometimes, if the data set contains few elements with high or low values, a separate 'open-end' is used that has only one limit — either upper or lower.

Having outlined the frequency distribution and the frequency table in abstract terms above, we now consider the frequency table in relation to the data given in Table 2.1.

From the data in Table 2.1 regarding the annual total rainfall at Varanasi, 1901-1950, we find that the maximum rainfall of 253.5 cm occurred in the year 1948 and the minimum of 37.1 in the year 1902 giving a range of 216.4 cm during the period 1901-1950. From Eq. (2.1), it appears that the number of classes can be five or six. To ease our task we take five classes with a class interval of 44 cm each as shown in column 1 of Table 2.2. Column 2 gives the mid-value of the class intervals. Column 3 shows the distribution of the 50 years of annual rainfall value among the classes by 'tally bars' from which the frequencies are counted and recorded in column 4. Finally we get the frequency table of the data grouped into five classes.

Table 2.2 : Frequency Table of Annual Rainfall at Varanasi, 1901-1950

Class intervals (col. 1)	Mid-value, m (col. 2)	Tally bars (col. 3)	Frequency, f_i (col. 4)
35 – 79 cm	57	卌 ||	= 7
79 – 123 cm	101	卌, 卌, 卌, 卌	= 20
123 – 167 cm	145	卌, 卌, |	= 11
167 – 211 cm	189	卌, 卌, |	= 11
211 – 255 cm	233	|	= 1

The information contained in Table 2.2 can also be expressed as a graph and indeed the important features of a frequency distribution gives a ready grasp in graphical representation of the frequency or frequency table. The graph is obtained by plotting the mid-values of the classes as abscissae and the corresponding class frequencies as ordinates. This curve is known as the *frequency curve*. It is actually a polygon curve.

Another graphical method of depicting frequency distribution is to measure along the horizontal axis distances proportional to the class intervals and to raise, on each of these distances, rectangles or bars proportional in height to the number of individuals falling within the class. The resulting figure is called a *histogram.*

It can be seen that in the above data of Table 2.1 when the number of class intervals is increased to nine, the frequency in the class interval of 211 to 236 cm becomes zero. Consequently, the frequency curve gets truncated as the number of class intervals increase. On the other hand, by lumping the class intervals, we get a smooth pattern of the frequency curve which helps in understanding the basic pattern of the distribution. Figures 2.1a and 2.1b illustrate the histogram and the frequency curve of the data given in Table 2.2.

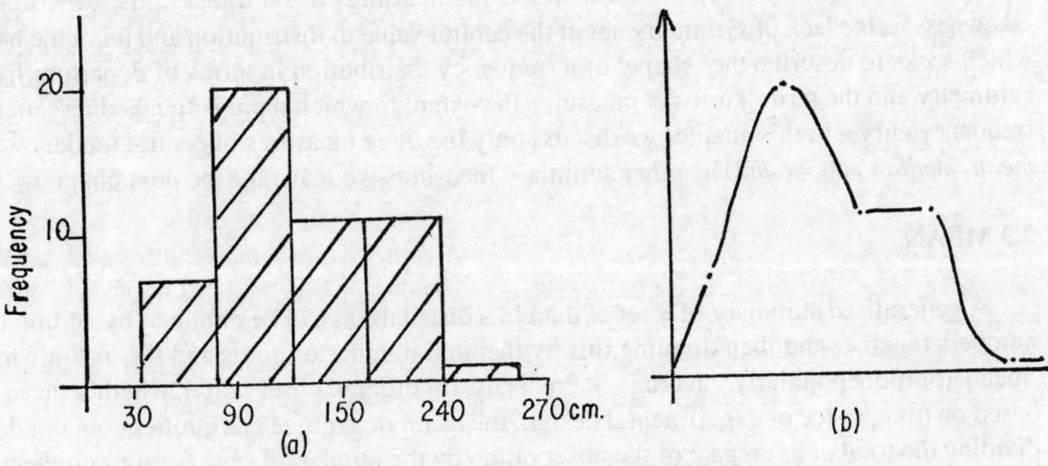

Fig. 2.1 a & b : (a) Histogram and (b) Frequency Curve

2.2 FREQUENCY CURVE AND ITS CHARACTERISTICS

The above frequency graph for any event, whether it is a frequency curve or a histogram, approaches more and more the form of a smooth curve as the number of observations increase and finer class intervals are used. For an hypothetically infinite size of data, the graph should preserve its regularity however narrow the class interval may be and, therefore, with an infinitely narrow class interval we expect to get a perfectly smooth curve rising to a single peak or *mode* (i.e. the value with the largest frequency). A smooth frequency curve is separated from one another by the indices that precisely measure the various properties of distribution of an event, known as the summary measures. There are three major summary measures. These are : (1) the 'central tendency measures', (2) the 'measures of dispersion' or 'variability' and (3) the 'measures of skewness and kurtosis.' All these three measures are collectively known as *summary measures.*

Central tendency measures are essentially the 'measures of location.' *Central tendency* is a value which in some sense is in the middle of the distribution and around which the observations tend to cluster. The 'arithmetic mean' is one of the more commonly used measures of central tendency or centrally located values. 'Mode' and 'median' are the other central tendency measures.

Another broad summary measure of frequency distribution is the measures of *dispersion* or *variability*. They measure the spread-out character of a variable in the frequency distribution or in the frequency curve around the central value. These are the 'positional measures' or the 'partition values'.

These two summary measures–the central tendency and dispersion measures take us a long way in summarising and in comparing among sets of data. For example, two areas may be having same average rainfall but the dispersion of rainfall may be much more even in one area than in the other.

The third group of summary measures is the measures of *skewness* and *kurtosis*. The 'skewness' is the lack of symmetry about the central value of distribution and it is a measure which seeks to describe the 'shape' of a frequency distribution in terms of departure from symmetry and the term 'kurtosis' measures the extent to which there is 'peakedness' in the frequency curve. In this chapter we discuss only the three measures of central tendency the *mean, median* and *mode*. The other summary measures we leave for the next chapter.

2.3 MEAN

A generalised summary of a set of data in a distribution can be obtained by adding the numbers together and then dividing this by the total number of items and this is known as 'mean' (or more popularly, "average"). Precisely, for ordered observation, whether these be based on time, space or experimental design, the *mean* or average is a quotient obtained by dividing the total or aggregate of the observation by the number of observations connected with it. Again since this mean is derived arithmetically, it is known as the *arithmetic mean*. For a set of data with observations, this can be written as

$$\bar{X}_N = \frac{X_1 + X_2 + X_3 + \ldots + X_N}{N} = \frac{\Sigma X_i}{N} \qquad \ldots (2.3)$$

where it is understood that the X_ith observation precedes the X_jth observation for $i < j$, Σ is summation of the observation. If we now add an observation and recalculate the mean based on $(N + 1)$ observations, we can write

$$\bar{X}_{N+1} = \frac{X_1 + X_2 + X_3 + \ldots + X_N + X_{N+1}}{N+1}$$

$$= \left(\frac{N}{N+1}\right)\bar{X}_N + \left(\frac{1}{N+1}\right)X_{N+1} \qquad \ldots (2.4)$$

where X_{N+1} represents the new observation. This we recognise as the recursive form of the mean. We may now look at how large or how small must a new observation be in order to affect the mean significantly. For example, for a mean to double after a new observation is added to 10 previous observations, the new observation X_{N+1} would have to be 10 times greater than the previous mean, $\overline{X}_N$; similarly for a mean to double for $N = 100$ observations, the new observation, X_{N+1} would have to be 100 times greater than the preceding mean, X_N. So we conclude that the mean becomes increasingly more difficult 'to alter with increasing sample size unless there are erratic fluctuations in the observation.'

When the raw data are arranged in a grouped form in a frequency table, a good approximation of the mean can by worked out by taking the mid-value, m of each class and their respective frequencies. Each mid-value of the classes is weighted by the respective frequency, f_i and m_i and the sum of the weighted products is then divided by the size of the data, to get the mean. This can be written as

$$\overline{X}_N = \frac{f_1m_1 + f_2m_2 + f_2m_3 + ... + f_cm_c}{N} \quad \text{where } c \text{ is the number of classes}$$

$$= \frac{\Sigma f_i m_i}{N} \qquad \text{... (2.5)}$$

This formula assumes that all the observations in any class are concentrated at the middle of the class interval. But this is not true. It can be seen that when a listed data are grouped in a frequency table and a mean is worked out this value will be a close approximate to the mean calculated from the listed data. Substituting in Eq. 2.3 the values of the annual rainfall in Table 2.1 we obtain

$$\overline{X}_N = \frac{6270.5 \text{ cm}}{50} = 125.41 \text{ cm}$$

Now calculating the mean directly from the grouped data in Table 2.2, we find that the mean annual rainfall is

$$\overline{X}_N = \frac{6326}{50} = 126.52 \text{ cm}$$

Thus Eq. (2.5) gives a mean which is different from the mean of the individuals in Eq. (2.3), but this difference can be taken as negligible.

2.3.1 Weighted Mean

Sometimes, a distinction is made between a 'weighted' and an 'unweighted' mean. An unweighted mean is a 'simple' mean in which all the weights are the same. To obtain a weighted mean, one multiplies each observation by its weights, sum these products w_iX_i and divides this quantity by the sum of weights. Symbolically,

$$\overline{X}_i = \frac{w_1X_1 + w_2X_2 + w_3X_3 + ... + w_NX_N}{\Sigma w_i}$$

$$= \frac{\Sigma w_i X_i}{\Sigma w_i} \qquad \text{... (2.6)}$$

The common application of weighted mean is an environmental index since an environmental index is not obtained by a single measurement. It is actually a combination of several sub-indices obtained by individual measurements. For example, the ambient concentrations of SO_2 of an urban area are a combined effect of emissions generated from different sources like industrial, domestic and vehicular sources etc. so that the total SO_2 concentration is, Q equal to a weighted additive form

$$\begin{aligned} Q &= w_1 q_1 + w_2 q_2 + w_3 q_3 + ... \\ &= \Sigma w_i q_i \end{aligned} \qquad ... (2.7)$$

Now if the coefficients, w_i are determined also in an additive form so that

$$\Sigma w_i = 1 \qquad ... (2.8)$$

we get the above Eq. (2.7) not only as an aggregation of the sub-indices, but also as index of the total SO_2 concentration.

2.3.2 Combined Mean

Again, if the mean of a series $X_{11}, X_{12}, X_{13}....$ is $\overline{X}_1$ and the mean of another series X_{21}, Y_{22}, Y_{23} . . . is $\overline{X}_2$ then the combined mean of two series can be calculated and it is given by

$$\overline{X}_{12} = \frac{N_1\overline{X}_1 + N_2\overline{X}_2}{N_1 + N_2} \qquad ... (2.9)$$

where $\overline{X}_{12}$ is the combined mean, N_1 and N_2 are the number of observations in each series. Note Eq. (2.9) for combined mean can be extended to any number of series. For example, for the combined mean of three series, it is

$$\overline{X}_{123} = \frac{N_1\overline{X}_1 + N_2\overline{X}_2 + N_3\overline{X}_3}{N_1 + N_2 + N_3} \qquad ... (2.10)$$

Example 2.1 The mean annual rainfall of 25 places is 61 cm and the mean annual rainfall of 35 places is 58 cm. Find the combined mean rainfall of the places.
Here $N_1 + N_2 = 60$,

$$\text{so, } \overline{X}_{12} = \frac{N_1\overline{X}_1 + N_2\overline{X}_2}{N_1 + N_2} = \frac{25 \times 61 + 35 \times 58}{25 + 35} = 59.25\text{cm}.$$

2.3.3 Computing Mean by Coding

With the above method of computing the mean Eq. (2.5) from the frequency data, the value of the product $f_i m_i$ may become very large. To avoid large numbers in the calculation, one of the mid-interval values is assumed to be the mean of the whole set of data. This assumed mean $\overline{X}_0$ is subtracted from each mid-interval value, m_i. This replacement of observations by a number d (equal to $m_i - \overline{X}_0$) is known as *coding*. This difference is multiplied by the frequency in each class, f_i and they are summed. The arithmetic mean is now calculated from the expression

$$\overline{X}_N = \overline{X}_0 + \frac{\Sigma f_i d}{N} \text{ where } d = m_i - \overline{X}_0 \qquad \text{... (2.11)}$$

or in words, True mean = Assumed mean + (sum of product of deviations from the assumed mean × frequencies) divided by the number of cases.

This method of computing the mean gives the identical result as we can have from $\overline{X}_N = \frac{\Sigma f_i m_i}{N}$ in Eq. (2.5). This is because we have, $d = m_i - \overline{X}_0$, then $f_i\, d = f_i\,(m_i - \overline{X}_0)$.

Now, summing it over i from 1 to N, we have

$$\sum_{i=1}^{N} f_i d = \sum_{i=1}^{N} f_i (m_i - \overline{X}_0)$$

Again, dividing both sides by N, we get

$$\frac{1}{N}\sum_{i=1}^{N} f_i . d = \frac{1}{N}\sum_{i=1}^{N} f_i\, m_i - \overline{X}_0$$

$$= \overline{X}_N - \overline{X}_0$$

$$\text{So, } \overline{X}_N = \overline{X}_0 + \frac{1}{N}\sum_{i=1}^{N} f_i . d \text{ proved}$$

In Example (2.2) below we compute the mean by both these methods.

Example 2.2 Compute the mean annual rainfall at Varanasi by applying both the arithmetic mean and the coding methods to the data in Table 2.2.

Class-intervals (in cm)	Mid-value m_i	Frequency f_i	$f_i\, m_i$	Mid-value m_i	Coded Mid-value d_i	$f_i\, d_i$
35 –79	57	7	399	57	–88	–616
79 –123	101	20	2020	101	–44	–880
123 –167	145	11	1595	145 $(= \overline{x}_0)$	0	0
167 –211	189	11	2079	189	44	484
211 –255	233	1	233	233	88	88

Total	$N = 50$	$\Sigma f_i m_i = 6326$	$\Sigma f_i d_i$	-924
$\bar{X}_n$ by arithmetic method		$= \frac{6326}{50}$	$\bar{X}_n$ by coding method	
		$= 126.52$ cm		$= \bar{X}_0 + \frac{\Sigma f_i d_i}{N}$
				$= 145 - \frac{924}{50}$
				$= 145 - 18.48$
				$= 126.52$ cm

2.3.4 Merits and Demerits of Arithmetic Mean

The arithmetic mean is called a measure of central tendency in the sense that it may be thought of as the 'centre of gravity' of the data. In this context, it is (i) commonly understood, (ii) rigidly defined, (iii) easy to compute and (iv) it is amenable to algebraic treatment. The two important mathematical characteristics of mean are that (a) the sum of the deviations of values from their mean is equal to zero, i.e. $\Sigma (X_i - \bar{X})$ = zero, because the values of the resulting differences are such that the negative scores exactly balance the positive ones and secondly (b) the sum of the squares of the deviations is smaller around the mean than when taken around any other value, P *i.e.* $\sum_{i=1}^{N}(X_i - P)^2$ minimum when P is the mean, $\bar{X}$, i.e. the sum of the squared deviations from the mean is smaller than the sum of squared deviations from any other single reference point, P. This is proved in the next chapter on 'Measures of Dispersion and Skewness' and this is also known as 'Least Squares' property of the mean which will be taken up in 'Regression Analysis' in Chapter 21.

Mean has also high sample stability, i.e. the variation in various values computed from different samples of the same variable is the least. For example, if we have a 100-year runoff data we can make 10 series each with a 10-year long data then the variation in the mean value computed will be the least of all other central tendencies like the median or mode. Hence mean is sometimes called 'the expected value of scores in a distribution'.

Mean is having also demerits which are that the mean (i) is effected very much by extreme values since it uses all the data. In case of extreme items, arithmetic mean gives a distorted picture of the distribution and no longer remains representative of the distribution as a central value itself. Secondly, (ii) it cannot be calculated if a single value is missing in the distribution. Finally, in certain frequency distribution, the lowest or the highest class interval or both are open-ended. For computing mean of these distributions, the total aggregate values in these open-ended classes are to be used as $f_i m_i$ in the frequency table; if the exact figures are not available, then they must be estimated in an appropriate manner. In the example below is given a distribution pattern of urban structure of India which shows unevenness in the estimation of the mean of urban places in India.

Example 2.3 Following is the distribution of urban places in India by categories, number and their total population as in the 1981 census (Table 2.3). Estimate the average size of urban places in India in 1981.

Table 2.3 : Urban Structure in India in 1981

Urban Categories	Class-intervals	Number	Total population in the category (in million)
Million Plus cities	>1,000,000	12	42.1217
Class I cities	100,000–999,999	206	52.7887
Class II cities	50,000–99,999	325	18.190
Class III cities	20,000–49,999	883	22.557
Class IV cities	10,000–19,999	1247	15.007
Class V cities	5,000–9,999	920	5.741
Class VI cities	<5,000	342	0.855
Total		N = 3935	157.2604

From the data given in Table 2.3, the arithmetic mean size of population of an urban place in India in 1981 was 39,965 persons. Now if the mean size is calculated based on this frequency table then we have to consider the aggregate for the two open ended categories i.e. the upper- and the lower-most ones and we find that due to the uneven size of the urban categories in the data it is 60,270 persons, much higher than the actual one.

2.4 OTHER MEASURES OF MEAN

The mean we have considered so far is the arithmetic mean which stands for an arithmetic series: $a + (a + d) + (a + 2d) + \dots (a + nd)$
where a is the first term and d is the common difference which can be found by subtracting any term in the series from the next term.

Among the common series in addition to arithmetic series above, we can include also the geometric series and the harmonic series. The general form of a geometric series is $a + ar + ar^2 + \dots + ar^n$ where a is the first term and r is the common ratio between any term and its preceding term in the series. Again the harmonic series is having a general progression as

$$a + \frac{a}{a + [(n_i - 1)]} + \frac{a}{a + [(n_i - 2)]} + \dots$$

where a is the first term and n_i is the position of a term in the series.

In other measures of mean, we will be considering now the 'geometric mean' and the 'harmonic mean' which are having important applications for averaging rates of change (growth/decay) and ratios.

2.4.1 Geometric Mean

Geometric mean is applicable for a sequence X, Y, Z, where Y is equal to square root of the product of X and Z and Y is called the geometric mean of X and Z. Accordingly, if there are three values, the geometric mean is the cube root of product of the terms of the series X_1, X_2, and X_3. Symbolically, for N number of terms, the geometric mean, M_G is

$$\begin{aligned} M_G &= \sqrt[N]{(X_1)(X_2)(X_3)....(X_N)} \\ &= (X_1 . X_2 . X_3 . . . . X_N)^{1/N} \end{aligned} \qquad ...(2.12)$$

For practical computation, geometric mean is best found by the logarithms. Thus taking logarithms in Eq. (2.12) we have

$$\begin{aligned} \text{Log } M_G &= \frac{1}{N} (\log X_1 + \log X_2 + \log X_3 + . . . + \log X_N) \\ &= \frac{1}{N} \Sigma \log X_i \end{aligned} \qquad ...(2.13)$$

Thus 'geometric mean' is the anti-log of the arithmetic mean of logarithms of the values in a group. Symbolically it is

$$M_G = \text{Anti-log} \frac{\Sigma \log X_i}{N} \qquad ...(2.14)$$

When frequencies are present, each logarithm must be multiplied by the corresponding frequencies. Thus for a grouped data, geometric mean is

$$M_G = \sqrt[N]{(X_1 f_1)(X_2 f_2)(X_3 f_3)....(X_N f_N)}$$

where $N = \sum_{i=1}^{N} f_i$ for discrete series

$$X_2 f_2 = (X_1 f_1 X_3 f_3) . . . . (X_N f_N)^{1/N} \qquad ...(2.15)$$

so that, taking common logarithm on both sides of Eq. (2.15) we have

$$\text{Log } M_G = \frac{1}{N} \Sigma f_i . \log m_i \qquad ...(2.16)$$

where m_i are the mid-points of the class interval for continuous series. The anti-log of Eq. (2.16) again gives geometric mean

$$M_G = \text{Anti-log } \frac{\Sigma f_i . \log m_i}{N} \quad \text{... (2.17)}$$

2.4.1.1 Combined Geometric Mean

Like combined mean, geometric mean is also suitable for combining the geometrical means $\overline{G}_1$ and $\overline{G}_2$ for two series of sizes N_1 and N_2 respectively. Thus, if combined geometric mean is $\overline{G}_{12}$ we have

$$\text{Log } \overline{G}_{12} = \frac{N_1 \log \overline{G}_1 + N_2 \log \overline{G}_2}{N_1 + N_2} \quad \text{... (2.18)}$$

$$\text{Hence } \overline{G}_{12} = \text{Anti-log } \frac{N_1 \log \overline{G}_1 + N_2 \log \overline{G}_2}{N_1 + N_2} \quad \text{... (2.19)}$$

Note, like combined mean in Eq. (2.10), the geometric mean $\overline{G}_{12}$ in Eq. (2.19) can be extended to any number of geometric series.

2.4.1.2 Merits and Demerits of Geometric Mean

Like arithmetic mean, geometric mean is (i) based upon all observations, (ii) rigidly defined, (iii) suitable for further mathematical treatment as in combining means of two series or more and (iv) not affected much by fluctuations of sampling.

It can be seen from Eqs. (2.13) or (2.16) that logarithm of geometric mean is the arithmetic mean of the logarithm of individual observations. Thus in a 'lognormal distribution' described in Chapter 4, the geometric mean has the properties analogous to the arithmetic mean of a lognormal distribution. Hence geometric mean is an average most suitable in skewed distributions where large weights have to be given of small items and small weights to the large items by logarithmic transformation of data.

Geometric mean is most frequently used in finding out the average rates of change (growth/decay) or in computing the average ratios of change. The things which grow exponentially (such as Wentworth's mm scale in grain-size analysis or in growth by compound interest principle) require the use of geometric mean so that sometimes the formuls for geometric mean follows the formulae for compound interest:

$$A = P\,(1+r)^n$$

$$\text{or } \quad r = \sqrt[n]{\frac{A}{P}} - 1 \quad \text{... (2.20)}$$

where P is the principal, A is the amount, r is the rate of interest and n is the number of years. Thus geometric mean is quite adequate when growth over a period contributes to further growth

Geometric mean is also theoretically considered to be the best average in the construction of 'index numbers'.

Among the demerits of geometric mean, it can be said that if any one of the observations in the data is zero, geometric mean becomes zero and if any one of the observations is negative, geometric mean becomes imaginary irrespective of the other magnitudes of the data since zero or negative values cannot have feasible logarithms.

Example 2.4 Find the rate of increase in population which in the first decade was 32 per cent, in the second decade 40 per cent and in the third decade 40 per cent.

The rate of growth is given by as in Eq. (2.13)

$$\text{Log } M_G = \frac{1}{3}(\log 132 + \log 140 + \log 140)$$

$$= \frac{1}{3}(2.120573 + 2.146128 + 2.146128)$$

$$= 2.13761$$

So the geometric mean is 137.28 and the average rate of increase in the three decades is 37.28 per cent.

Example 2.5 The geometric mean of 15 observations was 30.8. It was later found that one observation was recorded as 28.4 instead of 38.4. Using this correction find out the correct geometrical mean.

Geometric mean, $M_G = [(X_1)(X_2)(X_3) \ldots (X_N)]^{1/N}$

So, we have $(M_G)^N = (X_1)(X_2)(X_3)(X_N)$

So, the geometrical mean corrected is

$$M_G \text{ corrected} = \left[\frac{(30.8)^{15}.(38.4)}{28.4}\right]^{1/15}$$

$$\text{Now, by antilog, Log } M_G \text{ corrected} = \frac{1}{15}\log\left[\frac{(30.8)^{15}.(38.4)}{28.4}\right]$$

$$= 31.425606$$

$$= 31.43 \text{ approx.}$$

2.4.2 Harmonic Mean

Harmonic mean is based on the reciprocals of the numbers averaged and hence it is defined as the reciprocal of the arithmetic mean of the reciprocals of the set of observations. Thus if M_H is the harmonic mean, then for a set of values $X_1, X_2, X_3, \ldots, X_N$, this will be

$$\frac{1}{M_H} = \frac{1}{N} \Sigma \frac{1}{X_i} \quad \text{... (2.21)}$$

where X_i are individual observations. This can be written as

$$M_H = \frac{N}{\Sigma \frac{1}{X_i}} \quad \text{... (2.22)}$$

For grouped discrete series data,

$$M_H = \frac{N}{\Sigma\left(f.\frac{1}{X_i}\right)} \quad \text{... (2.23)}$$

and for grouped continuous data, the above formula will be written symbolically as

$$M_H = \frac{N}{\Sigma\left(f.\frac{1}{m_i}\right)} \quad \text{... (2.24)}$$

where m_i are the midpoints of the class intervals.

When a data set contains values which represents rate of flow or movement, the harmonic mean is an useful measure of central tendency than arithmetic mean. It is applied for averaging ratios particularly in calculating the mean speed when the rate of movement is different. For example, if a car covers 30 km in first hour and 50 km in second hour then its mean speed will be 37.50 km. determined by harmonic mean equal to

$$\left[2 \div \left(\frac{1}{30} + \frac{1}{50}\right)\right] = 2 \div 0.05333 = 37.5$$

rather than 40 km per hour.

2.4.2.1 Merits and Demerits of Harmonic Mean

Harmonic mean is the only average that can be used for averaging ratios, such as prices and in calculating time rates. The demerit of harmonic mean is that it is not easily understood and is difficult to compute.

2.4.3 Inequality of Mean

There are inequalities between the three means. Rule for inequalities is: arithmetic mean > geometric mean > harmonic mean. Let X and Y be the two positive mumbers, then

Arithmetic mean, $M_A = \dfrac{X+Y}{2}$

Geometric mean, $M_G = (X.Y)^{1/2}$

Harmonic mean, $M_H = \dfrac{2}{\dfrac{1}{X}+\dfrac{1}{Y}} = \dfrac{2XY}{X+Y}$

Now, $M_A \times M_H = \dfrac{X+Y}{2}.\dfrac{2XY}{X+Y} = XY = (M_G)^2$

On rearranging, we have

$$\frac{M_A}{M_G} = \frac{M_G}{M_H}$$

We can therefore conclude that the value of M_G is between M_A and M_H. Again

$$M_A - M_H = \frac{(X-Y)^2}{2(X+Y)}$$

which cannot be zero. Therefore $M_A > M_H$

Thus, we have $M_A > M_G > M_H$ ... (2.25)

Example 2.6 If arithmetic mean of two values is 19 and their geometric mean is $\sqrt{360}$, find the harmonic mean and the two values.

$M_A = \dfrac{a+b}{2} = 19$ i.e. $a + b = 38$

and $M_G = \sqrt{360} = \sqrt{ab}$.

so $ab = 360$.

Again $(a-b)^2 = (a+b)^2 - 4ab = (38)^2 - (4)(360) = 4$, so $a - b = 2$.

Hence the values are 20 and 18.

Example 2.7 Prove that set of readings on thermometer does not matter whether temperature is measured on Celsius or Fahrenheit for finding the arithimetic mean but it does matter when we find the geometric mean.

If, C_i is the set of readings ($i = 1, 2, \ldots, N$) in °C the arithmetic mean

$$M_A = \frac{C_1 + C_2 + C_3 + \ldots + C_N}{N} = \overline{C}_N$$

Now, from $\dfrac{°F-32}{180} = \dfrac{°C}{100}$ relationship, we have $F = 32 + \dfrac{9}{5}C$ given for geometric mean, M_G of °C equivalent of °F,

$$M_G = \frac{1}{N}\left(32+\frac{9}{5}C_1\right)+\left(32+\frac{9}{5}C_2\right)+\ldots+\left(32+\frac{9}{5}C_N\right)$$

which is $= \frac{1}{N} \{32\,N + \frac{9}{5}\,(C_1 + C_2 + \ldots + C_N)\}$

$= 32 + \frac{9}{5}\,\overline{C}_N$ proved.

2.5 MEDIAN

Median is the middle value in a set of observations when they are arranged in order of magnitude. Thus for a data set of *N* observations it divides the set into two halves and the median is

$$M_e = \frac{N+1}{2} \qquad \text{... (2.26)}$$

For example, if 81 years of annual peak discharge data are arranged in their order of magnitude, the 41st year of annual peak discharge data will be the median.

For median in a frequency table (i.e. for grouped data) the concept of the middle value in the set of observation does not held true. In this case we first determine the median class *P* equal to (N + 1) / 2 and locate it in the frequency table. Then we proceed to compute the median, M_e as follows

$$M_e = L + \frac{(P - C)}{f}.w \qquad \text{... (2.27)}$$

where *P* is already defined, the median class, *L* is the lower limit of the median class, *C* is the frequencies cumulated in the class above the median class, *f* is the frequency in the median class and *w* is the width of the interval in the median class. The median rainfall as applied to the frequency data in Table 2.2 is 119.7 cm.

If the frequencies are cumulated starting at the latest class value, the median is

$$M_e = U - \frac{(P - C)}{f}.w \qquad \text{... (2.28)}$$

where *U* refers to the 'upper' limit of the median class.

The median is easy to define and it is affected by the position in the data and not by the magnitude of the extreme values. Thus whenever a distribution is highly skewed, the median is preferable to the mean as a measure of central tendency. Hence the median is an useful measure of central tendency. An important measure of the median is that this sum is a minimum when *A* is set equal to the median, i.e.

$$\sum_{i=1}^{n} |X_i - A| \qquad \text{... (2.29)}$$

is minimum when A is the median. Finally we can say that the median is also obtained for open-end classes but this cannot be done in case of the mean.

2.6 MODE

Mode is the value which has the highest frequency, for example, in the statement, "the average height of an Indian male is 167 cm", the average referred is mode. Being the most typical the mode is easily located in the set of observations. In a frequency curve it is pointed by finding the abscissa corresponding to it's peak. When a frequency curve shows more than one peak, the curve is known 'bimodal' and 'multimodal' for a curve with more than two peaks. For bi- or multi-modal character the maximum frequency is repeated by occurring either in the very beginning and at the end or it may occur irregularly in the set of the data.

2.6.1 Mode by Grouping

For a bimodal data, the mode can be computed by the 'method of grouping' as follows:

(i) the original frequency distribution of the data is listed columnwise.
(ii) the original frequencies in the data are grouped in two's starting from the top, i.e. first and second; third and fourth and so on.
(iii) The original frequencies are grouped in two's leaving the first, i.e. second and third; fourth and fifth and so on.
(iv) Combining the frequencies in two's (i.e. 2 × 2) leaving the first two frequencies now will result in a repetition of the (ii). Hence we combine the original frequencies in three's starting from the top, i.e. first, second and third; fourth, fifth and sixth and so on.
(v) The original frequencies are grouped in threes, leaving the first i.e. second, third and fourth; fifth, sixth and seventh and so on.
(vi) The original frequencies are grouped in threes, leaving the first and second i.e. third, fourth and fifth; sixth, seventh and eight and so on.

Now to find the mode in the original frequency data, the maximum frequency for the above six columns are now listed along with their combination of values. The value (in the original data) which is repeated the maximum number of times in these noted combinations now gives the mode.

Example 2.8 Find the mode of the following frequency distribution :

Item no.	Frequency (*f*)
1	13
2	18
3	25
4	33
5	45
6	50
7	42
8	38
9	30
10	55
11	24
12	16

We start with the original data and by the method of grouping we put forward the other columns. The arrangement for grouping is as follows:

Item No.	Frequency Totals					
	(i)	(ii)	(iii)	(iv)	(v)	(vi)
1	13					
		31				
2	18			56		
			43			
3	25				76	103
		58				
4	33					
			78			
5	45			128		
		95				
6	50				137	
			92			
7	42					130
		80				
8	38			110		
			68			
9	30				123	
		85				
10	55					109
			79			
				95		
11	24					
		40				
12	16					

To find the mode we form the following table :

Column No.	Maximum frequency	Combination of values by item numbers
(i)	55	10th
(ii)	95	5th and 6th
(iii)	92	6th and 7th
(iv)	128	4th, 5th and 6th
(v)	137	5th, 6th and 7th
(vi)	130	6th, 7th and 8th

On examining the item numbers giving the maximum frequency in the six columns, we find the item 6 is repeated the maximum number of times so we say the value of the mode is 6th (for which the maximum frequency is 50) and not the 10th though the maximum frequency 55 occurs at 10.

For a grouped data in a frequency table, mode is calculated algebraically by the 'difference method'. This method is based on the assumption that since the number of class groups influence the mode, so the frequencies of the classes on each side of the modal class should be considered in the computed value according to

$$\text{Mode, } M_0 = L + \frac{d_1}{d_1 + d_2} . w \quad \text{... (2.30)}$$

where L is the lower limit of the modal class, d_1 is the difference between the frequencies in the class preceding the modal class and the frequencies in the modal class (sign neglected), d_2 is the difference between the frequencies in the class succeeding the modal class and the frequencies in the modal class (sign neglected), and w is the interval in the modal class. For example, the modal rainfall as applied to the frequency data in Table 2.2 is 105 cm.

Mode tends to be a measure of popularity because of its comprehensibility and it provides a measure of central tendency for qualitative data where mean and median are not applicable. Unlike mean, it is not affected at all by extreme values and hence mode is important as a central value in the case of distribution of extreme events like floods and droughts etc. But mode is not based upon all the observations and it is not capable of further mathematical treatment.

2.7 RELATIONSHIP BETWEEN MEAN, MEDIAN AND MODE

It is pointed earlier that mean is influenced by the size of the extremes, median by their position only and mode by their presence only. Mode is farthest from the mean and median is intermediate in its sensitivity to 'skewness'. Karl Pearson found that in a moderately skewed distribution, mode and median move towards the direction of the mean. Median tends to fall about 2/3 of its distance from the mode towards the mean. Consequently, he found that the positional measures of them can be expressed as

$$\text{Mode} = \text{Mean} \pm 3 \mid \text{Mean} - \text{Median} \mid \quad ...(2.31)$$

where '–' is for *positively skewed* and '+' is for *negatively skewed distribution*. This numerical relationship holds good for unimodal curves of moderate asymmetry. For example, the mean and median calculated to the data in Table 2.2 when substituted in Eq. (2.31) gives us a value for the mode as equal to

$$126.5 \pm 3 \mid 126.5 - 119.7 \mid = 106.1 \text{ cm}$$

which is close to the calculated value of 105 cm for mode.

This above positional measure for mode is not applicable for a highly skewed data in Table 2.3. A *positively skewed frequency curve/distribution* has a tail towards the right caused by few frequencies occurring there. For this curve, mode being the value that occurs most frequently is at the highest point of the distribution, median being the central observation point lies to the skewed side of the mode and mean being the most affected by extreme

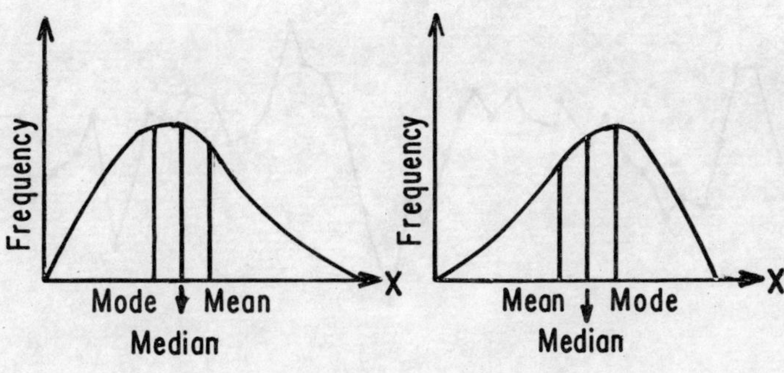

Fig. 2.2 : Positional Comparison of Central Tendency Measures

values in the distribution's tail, lies beyond the median. Hence for a *positively skewed frequency curve* the mode, median and mean are found from left to right and this relationship is *vice versa* for a *negatively skewed frequency curve* with its tail towards the left. Examples of negatively skewed curves are rare, one example may be the frequency distribution of daily rainfall in a tropical rainforest where there will be many wet days but few dry days. For this reason the mean although is the most commonly used statistical measure, it is not always the best measure of location. Figure 2.2 shows the positional comparison of central tendency measures.

2.8 AN APPLICATION IN THE CONCEPT OF MEAN

Apart, from geometric and harmonic mean (besides the arithmetic mean) there are two more means: the 'moving average' and the 'progressive average'. Moving average gives a long-term tendency in time series data whereas the progressive average gives the trend in early years of the data.

Moving average is a series of overlapping averages that serve to approximate the trend of a series by cancelling out the high and low values. Each average is a simple arithmetic mean of several values written against the middle year in the analyses of time series data. For example, if annual rainfall is plotted yearwise for a time period as a bar graph it will not show any trends or cyclic patterns in the rainfall due to wide variations in the consecutive years. In order to depict a general trend in rainfall pattern, the averages, of three or five consecutive years are found out by 'moving the group averaged' one year at a time. In Fig. 2.3 the moving average for the 50 years of annual rainfall at Varanasi in Table 2.1 is shown. The first five years of record are averaged as (118 + 37.1 + 110.8 + 136.9 + 192.4)/5 = 119.04 cm and this average is plotted at the midpoint of this group and so on

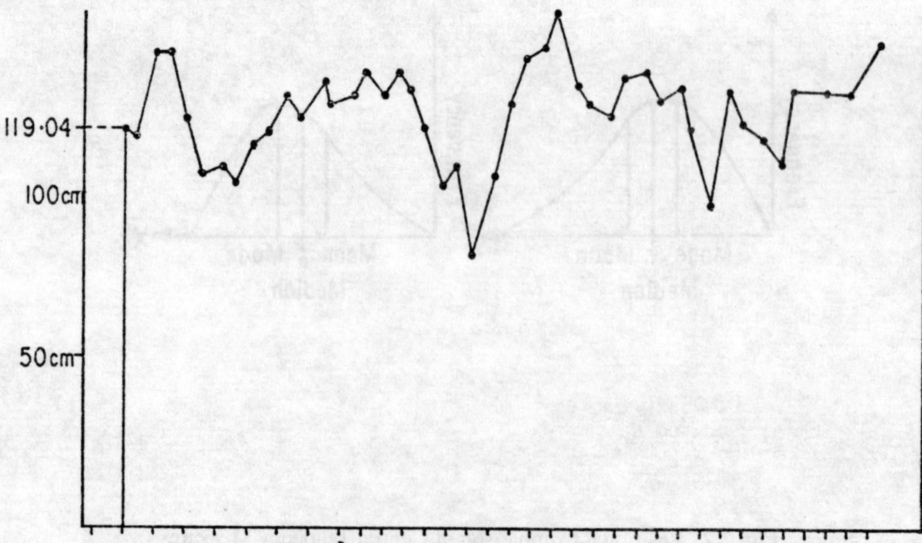

Fig. 2.3 : Moving Average

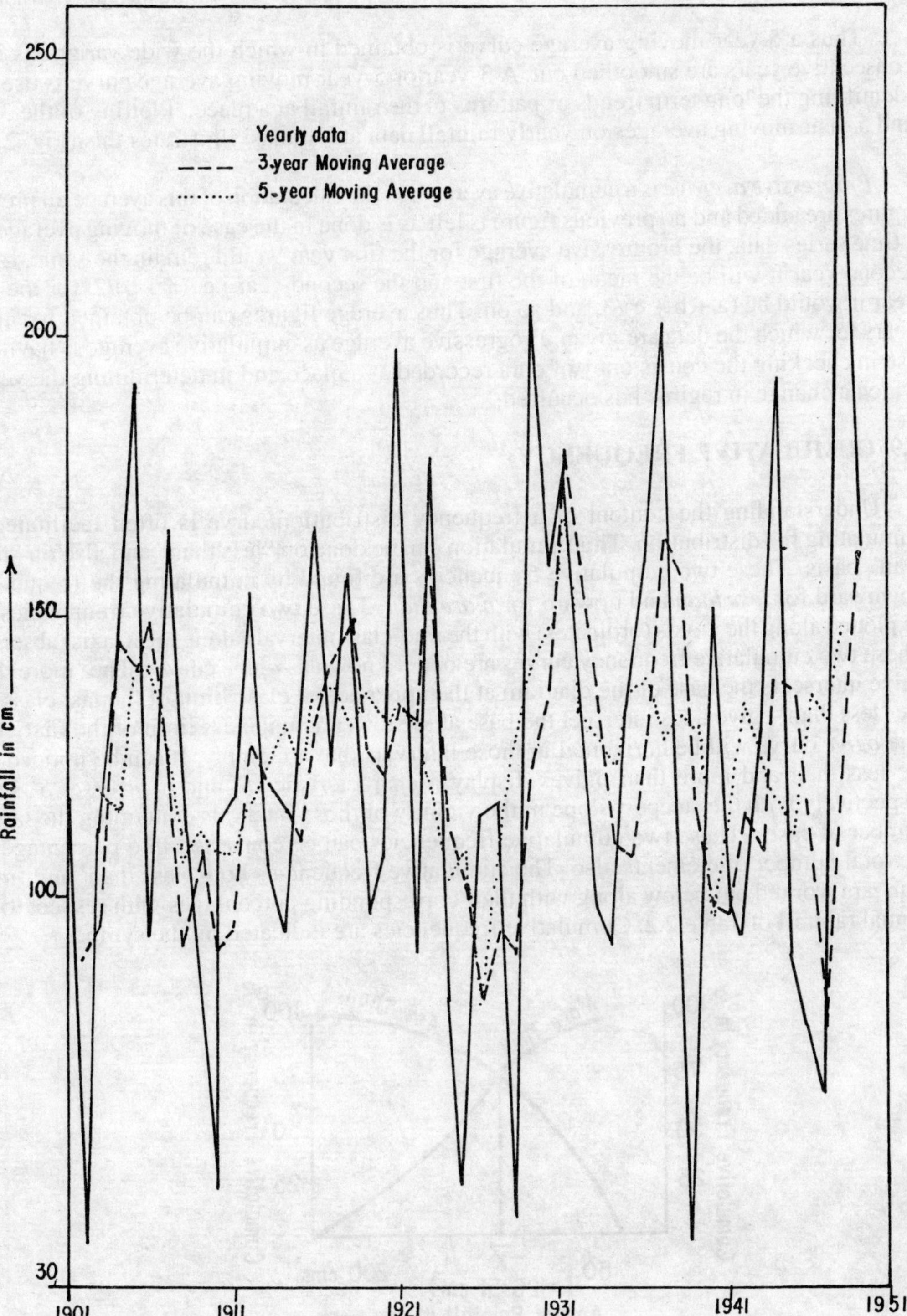

Fig. 2.3a : Plotting of the *K*-term–Moving Averages on Original Data

Thus a 5-year moving average curve is obtained in which the wide variations in the consecutive years are smoothed out. A 3-year or 5-year moving average curve is useful in identifying the long term trends or patterns in the rainfall at a place. Plotting of the 3-year and 5-year moving averages on yearly rainfall data at Varanasi illustrates this (Fig. 2.3a)

Progressive average is a cumulative average. In the calculation of this average all previous figures are added and no previous figure is left as is done in the case of moving average. For a time series data, the progressive average for the first year would remain the same, for the second year it will be the mean of the first and the second year i.e. (a + b)/2, for the third year it would be (a + b + c)/3, and so on. Thus average figures can be obtained for all the years for which the data are given. Progressive average as cumulative average is having its use in checking the consistency of data recorded at a place and in determining the year in which a change in regime has occurred.

2.9 CUMULATIVE FREQUENCY

Understanding the content of a frequency distribution/curve is often facilitated by cumulating the distribution. This cumulation can be done on "less than" and also on "more than" basis. These two cumulative frequencies are found by cumulating the frequencies downward for *less than* and upward for *more than*. These two cumulative frequencies can be plotted along the *Y*-axis (ordinates) with the mid-class intervals along the *X*-axis (abscissa). These two cumulative frequency curves are better known as *ogive* curves. The 'more than' ogive intersects the base of the diagram at the upper actual class limit of the last class and the 'less than' ogive does intersect the base at the lower actual class limit of the first class. The *ogive* curve will be horizontal in those intervals that are empty. It can be noticed that the less than and more than ogives display a characteristic '*S*' and a *reverse* '*S*' shape respectively with the steepest slope in the vicinity of those intervals containing the largest number of cases. These two cumulative frequencies can be converted into percentages of the total number of elements also. The cumulative frequencies both 'less than' and 'more than' are worked out below along with their corresponding percentages with respect to the annual rainfall of Table 2.2. Cumulative frequencies are indicated by the symbol *f*.

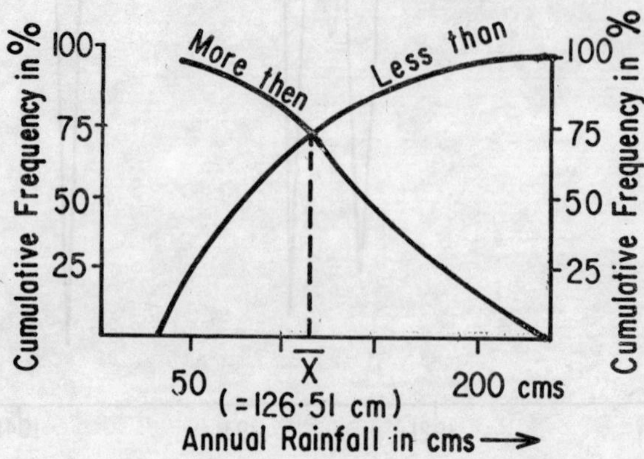

Fig. 2.4 : Cumulative Frequency Curves

Table 2.4 : Cumulative Frequency Distribution and Their Percentage Cumulatives

Class-intervals: Annual Rainfall in cm.	Cumulating up "less than", f_c	Col 2 in percentage, f_c	Cumulating down "more than", f_c	Col 4 in percentage, f_c%
(Col 1)	(Col 2)	(Col 3)	(Col 4)	(Col 5)
35–79 cm	7	14	50	100
79–123 cm	27	54	43	86
123–167 cm	38	76	23	46
167–211 cm	49	98	12	24
211–255 cm	50	100	1	2

Figure 2.4 illustrates the plotting of these two cumulative frequencies, 'less than' and 'more than'. Note the point at which these two curves intersect each other divides the frequency distribution into two equal parts, i.e. the curves intersect at the mean of the distribution (at 126.51 cm) in Fig. 2.4.

2.9.1 Characteristics of Cumulative Frequency Curve

In understanding the content of a frequency distribution, a *cumulative frequency curve* is more useful than a frequency curve. It provides a simple and quick graphical procedure for approximately determining the median and any 'percentile measures' including the 'quartiles' (these are discussed later in this chapter with 'positional measures'). In addition, it has been mentioned above that the 'more than' and 'less than' ogive curves intersect at the mean of the data.

As mentioned above, cumulative frequency curve is nearly *S*-shaped, with the relative size of two tails of *S* determined by the symmetry or skewedness of the distribution. If a distribution follows closely the pattern of a normal curve, it is possible to 'straighten out' the *S* by using a special vertical scale known as a 'probability scale'. This scale is so designed that any normal cumulative distribution will graph as a straight line. This graph paper with one arithmetic (ordinary) and one probability scale is known as 'arithmetic probability graph'. This may be compared with an exponential growth curve which becomes a straight line on semi-logarithmic graph paper. The use of cumulative frequency curve and probability paper is illustrated further in Chapter 4.

Cumulative frequency curve gets another important application in estimating the probabilities of the event of an 'annual series' like occurrence of flood in a river basin. This is discussed in detail in extreme value distribution in Chapter 7 later on.

The distributions in which the frequencies do not concentrate at central value but they are large at both end, we get a *bounded curve*. *Percentage cumulative frequency curves* are examples of bounded curves and they are an important graphical tool of dispersion to study the problem of inequality in a wide-ranging distribution of various geo-system events. We discuss below few of the applications of the cumulative frequency curves.

2.9.1.1 Applications of Cumulative Frequency Curves: Hypsometric Curve and Integral

A hypsometric (or hypsographic) curve is an ogive or a cumulative frequency curve which is obtained by plotting the contour elevation (upper limit) against the corresponding proportions of a given unit area of the earth's sun-lit surface (say a drainage basin). This curve can represent the statistical distribution of elevation dimensionally. In its simplest form heights and areas are given in absolute or relative values (Table 2.5) to compute the mean and median height of a drainage basin area.

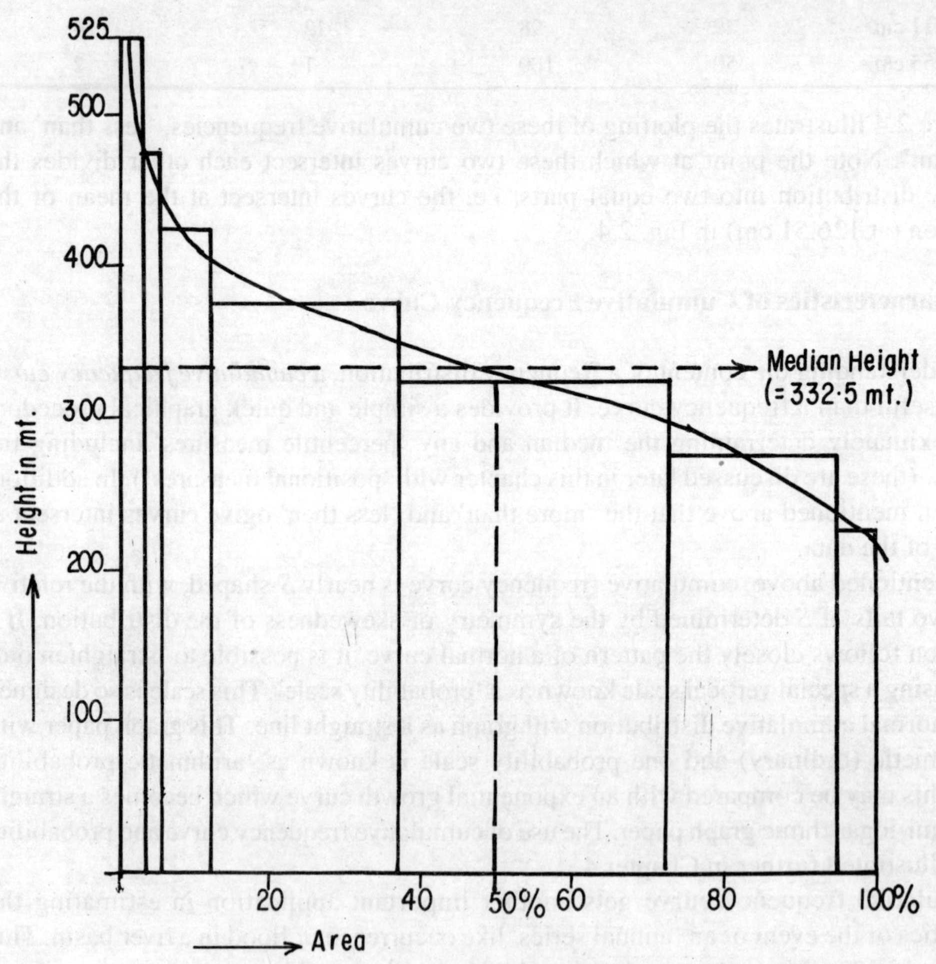

Fig. 2.5 : Area–Height Curve and Median

Table 2.5 : Area-Height Distribution of a Drainage Basin

Class-intervals for height in meteres	Mid-value h_i	Area between contours in km², a_i	Proportion of total area in %	Cumulating up: proportion of total area 'more than'
> 250	225	150	4.69	1.0000
250–300	275	720	22.50	0.9531
300–350	325	1130	35.31	0.7281
350–400	375	810	25.31	0.3750
400–450	425	210	6.56	0.1219
450–500	475	40	1.25	0.0563
< 500	say 525 (= *H*)	140	4.38	0.0438
Total		Σ 3200(= *A*)	Σ 100.00	

The mean height of the drainage basin is calculated as the weighted mean height between two adjacent contours

$$\bar{h}_c = \frac{\Sigma a_i h_i}{\Sigma a_i}$$

where $\bar{h}_c$ is the mean height of the drainage basin. The computation for $\bar{h}_c$ will be 338.91m while in Fig. 2.5 the median elevation for 50 percent of the total area is read from the curve as 332.50 m.

Table 2.5 shows the whole area (*A*) of the basin as 3200 km² and assumes the mid-value 525 m as the maximum height (= *H*) of the basin. Now we can express the area-height relationship of the drainage basin dimensionlessly if we compute the ratio between a_i and the whole area of the basin, *A* and the ratio of the mid-value of the contour-height, h_i to the maximum height, *H*. These ratios are calculated in Table 2.6 and plotted on a graph as shown in Fig. 2.6. The hypsometric curve relating these proportions of area (on *X*-axis) and height (on *Y*-axis) must pass through $X = 0$, $Y = 1.00$ and $X = 1.00$ and $Y = 0$ but its position within the graph is a measure of the state of erosion of the basin concerned, known as 'hypsometric integral'*.

---

* (Mathematically, the hypsometric integral can be found from integral calculus as integral

$$f = \frac{\text{Volume}}{\text{Total height} \times \text{Total area}} = \int_{0.0}^{1.0} a \,.\, \Delta h$$ with *a* as area and Δ *h* range in height *h*.

Table 2.6 Proportions of Hypsometric curve

h_i	h_i/H	a_i	a_i/A	Cumulating up: less than a_i/A
225	0.4285	150	0.0469	0.0469
275	0.5238	720	0.2250	0.2719
325	0.6190	1130	0.3531	0.6250
375	0.7142	810	0.2531	0.8781
425	0.8095	210	0.0656	0.9437
475	0.9047	40	0.0125	0.9562
525	1.0000	140	0.0438	1.0000

The value of the hypsometric integral can be readily found in practice by measuring the area of the surface below the hypsometric curve and it is between 0 and 1 where the whole area of the square is 1. The value of this integral is related closely to the state of erosion of the basin. A high integral value exceeding 0.6 indicates a youthful stage in development of the drainage basin, the integral between 0.60 and 0.35 indicates maturely dissected relief and if it lies below 0.35 it indicates an erosional surface. The hypsometric integrals are useful criteria for comparison of drainage basin relief.

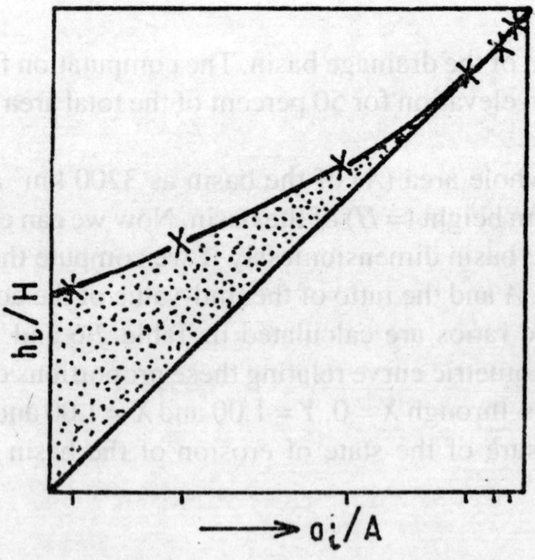

Fig. 2.6 : Hypsometric Curve

2.9.1.2 Lorenz Curve and Gini Coefficient

Another form of percentage cumulative frequency curve is the 'Lorenz Curve' (or, 'Pareto Curve'), it is a graphic method of dispersion devised by statistician Max D. Lorenz. It is an effective tool as a measure of inequality in the distribution of various items in social sciences like studies on landholdings, ethnic segregation in cities of a country, economic activities, income, wealth, expenditure etc. For example, for representation of people's income in a country, this curve is a graph of population shares against their income shares, ordered from poorest to richest. On this curve, since 0% of people have 0% of income and 100% of people have 100% of income, so this curve touches the lower left and upper right corners of the square. If everyone had the same income then *x*% of the people would have *x*% of the total income for any *x* between 0 to 100. Thus the Lorenz curve which in ideal case would be the straight line is shown in Fig. 2.7.

The techniques of drawing a Lorenz curve is as follows:

(i) The size of the item (i.e. the value of the variable) and its frequencies are both calculated.

(ii) The cumulated proportional frequencies (a_i/A cumulated) (from 0 to 1.00) are taken on the *X*-axis and the h_i/H values of the variable (from 0 to 1.00) on the *Y*-axis.

(iii) Now a diagonal line is drawn connecting the origin of the graph (0, 0) with the point (100, 100). This diagonal across the graph is the 'line of equal distribution'.

(iv) The proportions of the h_i/H values are then plotted against the corresponding proportional frequencies. The plotted points are joined with a smooth freehand curve. This curve will never cross the line of equal distribution. Any deviation of this Lorenz Curve from the diagonal will be reflected in the degree of concavity or convexity of the curve.

The extent of inequality/unevenness of the two sets of the distribution is done by seeing how far away from the equality line the Lorenz curve is. Further this Lorenz curve is from the equality line, greater are the inequalities. This inequality is measured by an index known as 'Gini coefficient'. This coefficient is indicated by expressing the area as bounded between the equality line and the Lorenz curve (marked as '×' in Fig. 2.7) in terms of a ratio to the area of the right-angled triangle ABC of the 'square' graph under and/below the equality line.

In case of uniform distribution it is apparent that the Lorenz curve will fall on the equality line and the area between the curve and the equality line would be zero. For distribution with highest concentration (i.e. when all the values in the distribution are concentrated only at one place) the curve will approach the line ABC that is, the curve will follow the *X*-axis and then the ordinate when *X* reads 1.00 (Fig. 2.7) and the Gini coefficient will be an unity. Thus Gini coefficient varies on a scale between zero and one (or of zero to 100 per cent). It involves the calculation of the shaded area shown in Fig. 2.7 in terms of

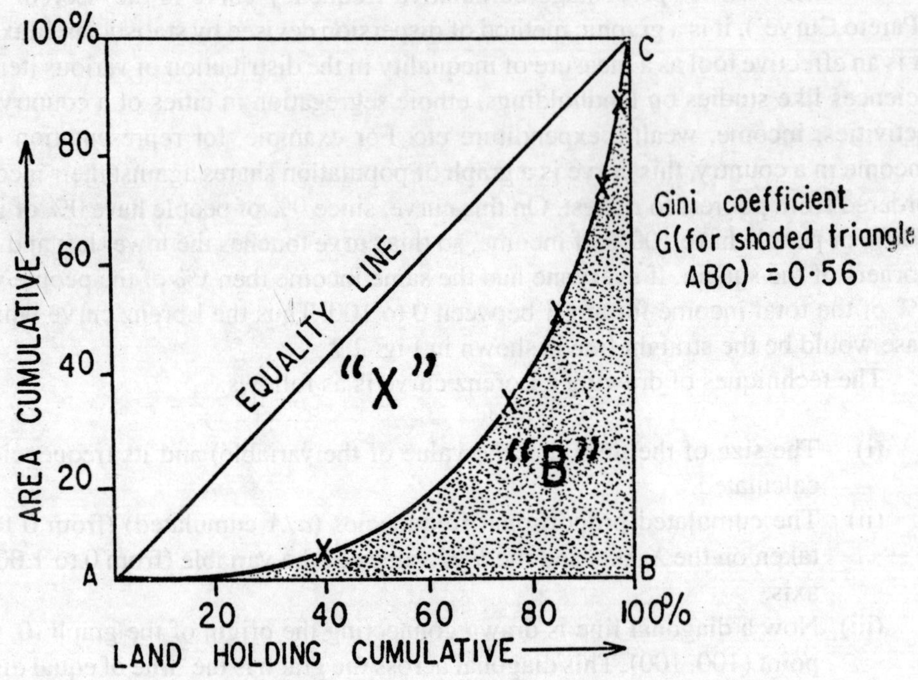

Fig. 2.7 : Hypsometric Data plotted for Lorenz Curve and Gini Coefficient

coordinates of the points on the Lorenz curve. Hence from a Lorenz curve with coordinates (X_i, Y_i) the 'Gini coefficient' will be the area under the curve expressed, as a proportion of the triangle ABC into which it fits in the graph.

It is:

$$G = \left[\sum_{i=0}^{i=n} X_i Y_{i+1} - \sum_{i=0}^{i=n} X_{i+1} Y_i \right] \quad \text{... (2.32)}$$

or,

$$G = [(X_0 Y_1 - X_1 Y_0) + (X_1 Y_2 - X_2 Y_1) + (X_2 Y_3 - X_3 Y_2) + \cdots + (X_n.\ 1 - 1.\ Y_n)] \quad \text{... (2.32a)}$$

Having outlined the form and characteristic of the Lorenz curve, we now illustrate it with reference to a specific data of geographical interest like the distribution of land holdings in India as given by size of holdings, the numbers of holdings and area of holdings below.

Size of land holdings (in hectares)	Number of land holdings (in millions)	Area of land holdings (in million hectares)
> 1	19.8	9.2
1–3	18.0	32.1
3–5	6.1	23.0
5–10	4.5	30.6
10–20	1.8	23.1
20 +	0.5	15.1

To show the inequality in size distribution of land holdings in India from the above data by Lorenz curve method, we require to have the total for both the number of land holdings and the area of land holdings, and to convert the both into their respective cumulative percentages, the latter are as follows :

Cumulative percentages of	
Number of land holdings	Area of land holdings
39.1	6.9
74.6	31.0
86.6	48.2
95.5	71.1
99.0	88.4
100.0	100.0

These cumulative percentage of both numbers of land holding and their areas can be plotted to get a Lorenz curve.

There are some problems associated with using the Lorenz curve: (i) there are data restrictions since variables must be expressed as frequencies of each unit and negative values cannot be included. Hence this technique is difficult to be applied to a study of continuous variable, (ii) it is sensitive to changes in the spatial boundaries of study units (like area of drainage basin in Table 2.5) and changes in scale. Finally, the Gini coefficient as an index of concentration should not be equated with the degree of concentration of an activity/event in an area to which the 'location quotient' is concerned with.

Location quotient (*LQ*) is another measure which determines the extent to which different areas depart from some norm. This quotient indicates the degree of concentration. Introduced by Walter Isard in 1960, it is

$$LQ = \frac{(X_i / X)}{(Y_i / Y)} \qquad \text{... (2.33)}$$

where X_i is a specific activity in an area, X is the total activity in the area, Y_i is the national total in the activity and Y the total national activitiy. Higher values of LQ in Eq. (2.33) indicate high concentration and the values of 1 indicate equal distribution.

Example 2.9 Having outlined the form and characteristic of the Lorenz curve we now compute the Gini coefficient with reference to area-height data of the drainage basin in Table 2.5 and their respective proportions given in Table 2.6.

The cumulated area ratios a_i/A and the corresponding height ratios h_i/H of the data in Table 2.6 are now used as data for calculating the Gini coefficient, G which is as follows :

$$\begin{aligned} G = [&(0.4285 \times 0.2719 + 0.5238 \times 0.6250 + 0.6190 \times 0.8781 + 0.7142 \times 0.9437 + \\ &0.8095 \times 0.9562 + 0.9047 \times 1{,}0000) - (0.5238 \times 0.0469 + 0.6190 \times 0.2719 + \\ &0.7142 \times 0.6250 + 0.8095 \times 0.8781 + 0.9047 \times 0.9437 + 1.0000 \times 0.9562)] \\ = {} &3.3401 - 3.1600 = 0.1801 \text{ or } 18.01\% \end{aligned}$$

Thus we can say that the area and height in the drainage basin are having an equal distribution in terms of their relationships. But the same does not hold for social science variables like land holdings, ethnic segregations, income and wage distributions. For the example earlier on land holdings in India, the Gini coefficient will be found to be 56.24 percent (Fig. 2.7).

2.9.1.3 Hydrologic Data : Double Mass Analysis

The variation of storm rainfall with respect to time may be shown by (i) a "hyetograph" and (ii) a "mass curve". A "*hyetograph*" is a bar graph showing the intensity of rainfall with respect to time in terms of "bars" per unit time by which the rainfall is recorded where as a *mass curve of a storm rainfall* is a plot of cumulative depth of rainfall against time. The mass curve facilitates the computation of the total depth of rainfall and intensity of rainfall at any instant of time, the former by the difference between the ordinates at the beginning and end of the time increments and the latter by slope (i.e. $\Delta R/\Delta t$, the increment of rainfall, ΔR per unit of time, Δt) of the mass curve at that time.

Being a cumulative curve, a mass curve of rainfall is always a rising curve and it may have some horizontal section which indicates periods of no rainfall.

Example 2.10 The following are the rain-gauge observations during a 210-minute storm

Time since beginning of the storm (in minutes)	Accumulated rainfall (in cm)
30	3.5
60	7.5
90	19.5
120	28.0
150	32.5
210	36.0

Construct the hyetograph and the mass curve of the storm rainfall.

For the hyetograph, intensity is calculated in cm/hour for each unit of time the storm rainfall is recorded. The computations for intensity in cm/per hour is as follows :

Table 2.7 : Intensity of Storm Rainfall

Time, t (in minute)	Accumulated rainfall (cm)	ΔR in time Δt.30 min (in cm) $\left(\frac{\Delta R}{\Delta t}.30\right)$	Rainfall intensity, i per hour (cm) $\left(\frac{\Delta R}{\Delta t}.60\right)$
30	3.5	3.5	7.0
60	7.5	4.0	8.0
90	19.5	12.0	24.0
120	28.0	8.5	17.0
150	32.5	4.5	9.0
210	36.0	1.75	3.5

The calculated intensity of storm rainfall in cm/hour is now plotted over time (t in minutes) to get the hyetograph (Fig. 2.8a), the mass curve is obtained by plotting the given accumulated rainfall against time t, in minutes (Fig. 2.8b).

Mass curve gives a trend of storm rainfall at a raingauge station. 'Double mass curve' is the curve which plots the cumulative annual (or seasonal) rainfall at station 'X' against the concurrent cumulative values of mean annual (or seasonal) for a group of neighbouring stations, for a number of years of record. In case there is a change in rainfall records of the station X, there will be change in slope of the straight line trend in the plotted data. From the plot, the year in which a change has occurred is marked, the ratio of slopes is worked out which is multiplied with the cumulative rainfall for the period before the change occurred

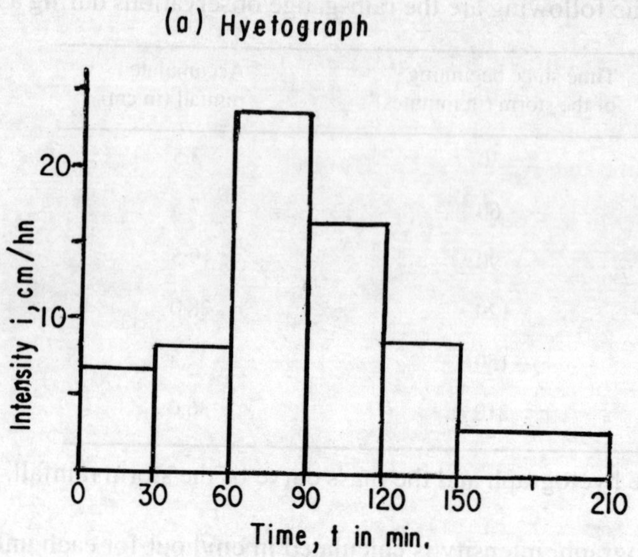

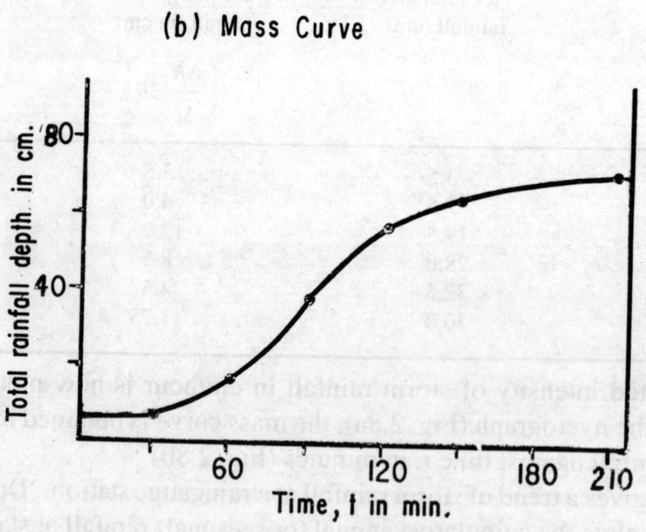

Fig. 2.8 : Storm Rainfall and Time
(a) Hyetograph (b) Mass Curve of Rainfall

to adjust the average annual rainfall to the current regime.

Example 2.11 In the following table is given the annual rainfall at raingauge station X and the mean annual rainfall for a number of neighbouring stations for a 15-year time period. Apply the double mass analysis to find out whether there has been a change in regime and if it is so, determine the average annual rainfall to be adjusted for the changed regime.

Year	Annual rainfall (in cm) at Station X	Neighbouring stations' mean
1970	26	18
1971	34	30
1972	39	25
1973	27	22
1974	22	20
1975	27	23
1976	44	31
1977	23	13
1978	20	20
1979	17	19
1980	23	28
1981	12	18
1982	17	31
1983	23	26
1984	19	18

(i) The cumulative annual rainfall are calculated for both the given yearwise rainfall data. These are :

Year	Station X	Neighbouring Stations
1970	26	18
1971	60	48
1972	99	73
1973	126	95
1974	148	115
1975	175	138
1976	219	169
1977	242	182
1978	262	202
1979	279	221
1980	302	249
1981	314	267
1982	331	298
1983	354	324
1984	373	342

(ii) The above cumulative rainfalls are now plotted (Fig. 2.9). It can be seen from Fig. 2.9 that there is a change in trend in the year 1976 which indicates that a distinct break in the rainfall ragime has occurred in this year. To make the records prior to 1976 comparable with those after a change in regime has occurred, the earlier records have to be adjusted by multiplying by the new slope in the trend now i.e. by the ratio of slopes equal to m_2 / m_1 = 1.8 / 2.5.

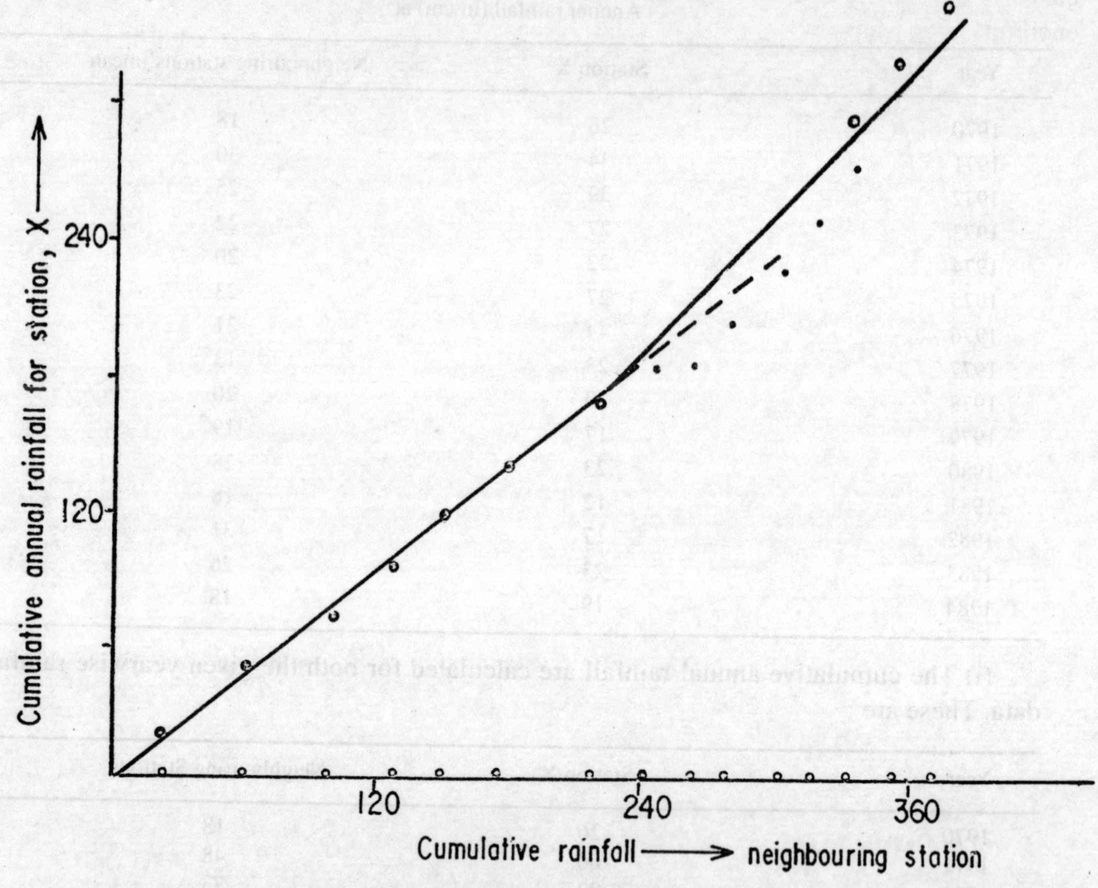

Fig. 2.9 : Double Mass Curve

Cumulative rainfall : 1984 – 1976 : 373 – 219 = 154.0
Cumulative rainfall : 1970 – 1976 :
adjusted for changing regime

$$= 219 \times \frac{1.8}{2.5} = 157.7$$

So, cumulative rainfall for the entire period as adjusted to the current regime = 311.7

Hence, average annual rainfall adjusted for the current regime is 311.7 cm /15 years = 20.78 cm.

Rainfall rarely occurs uniformly over a large area since variation in intensity and total depths of rainfall occurring from the storm centres to the peripheries of storms are common. An isohyetal pattern of consecutive 5-day storm over an area is shown in Fig. 2.10 which shows plotting of isohyets around two storm centres, A and B. The depth-area-duration curve is the plotting of the average depths of rainfall plotted over the area up to the encompassing isohyets.

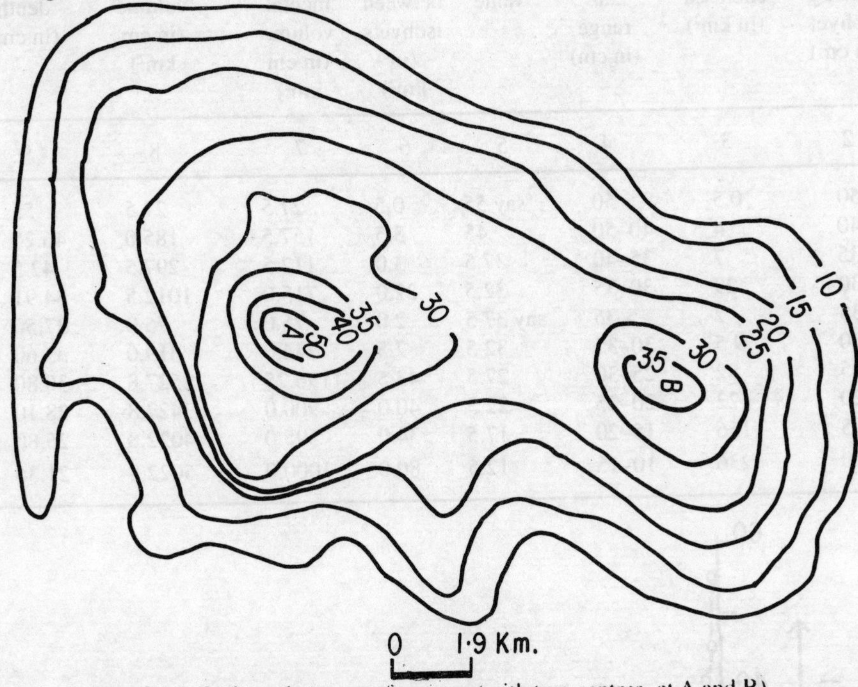

Fig. 2.10 : Isohyetal pattern of a storm (with two centres at A and B)

(i) From the above isohyetal pattern map of 5-day storm data are generated by starting with planimetering the isohyets enclosing the storm-centre areas (A and B in Fig. 2.10).

(ii) The area between the two isohyets multiplied by the average of the two isohyetal values given the incremental volume of rainfall.

(iii) The incremental volume added with the previous accumulated valume gives the total volume of rainfall.

(iv) The total volume of rainfall divided by the total area up to the encompassing isohyet gives the average depth of rainfall over that area.

(v) The computations are made for each isohyets enclosing the storm centres (A and B) and the zonal values are then combined for areas enclosed by the common (or encompassing) isohyets.

(vi) The highest average depths for various areas are obtained by dividing the total volume by area enclosed (col 8 ÷ col 3) and finally a smooth curve is drawn by plotting this average depth over area (Fig. 2.11). Table 2.8 gives the required computation.

Table : 2.8 : Computation for Depth-Area-Duration Curve (5-days storm)

Storm centre	Encompassing Isohyet (in cm)	Area enclosed (in km²)	Isohyetal range (in cm)	Mid-value	Area between isohyets (in km²)	Incremental volume (in cm km²)	Total volume (in cm km²)	Average depth (in cm)
1	2	3	4	5	6	7	8	9
A	50	0.5	> 50	say 55	0.5	27.5	27.5	55
	40	4	40–50	45	3.5	157.5	185.0	46.25
	35	7	35–40	37.5	3.0	112.5	297.5	42.5
	30	29	30–35	32.5	22.0	715.0	1012.5	34.91
B	35	2	> 35	say 37.5	2.0	75.0	75.0	37.50
	30	9.5	30–35	32.5	7.5	244.0	319.0	33.60
A	25	82	25–30	27.5	43.5	1196.25	2527.8	30.80
	20	122	20–25	22.5	40.0	900.0	3427.8	28.10
	15	156	15–20	17.5	34.0	595.0	4022.8	25.80
	10	236	10–15	12.5	80.0	1000.0	5022.8	21.30

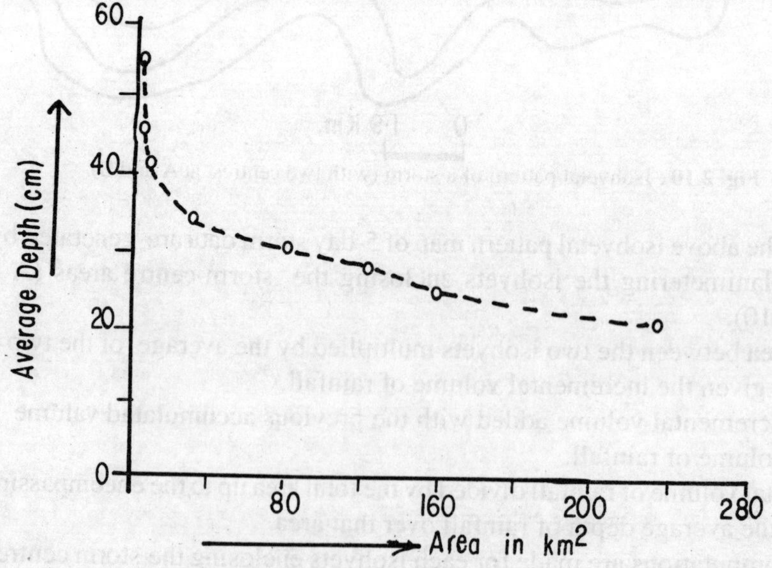

Fig. 2.11 : Depth-Area-Duration Curve

2.10 POSITIONAL MEASURES/PARTITION VALUES

The mean, median and mode are shown earlier as algebraically computed measures of central tendency particularly for grouped frequency data. Each of these measures divide the frequency curve into two halves. However, compared to the range of values in the distribution, these three values of central tendency inadequately summarises the distribution. On other hand, the 'positional measures/partition values', such as 'quartiles' (every 25%), 'quintiles' (every 20%), 'deciles' (every 10%) and 'percentiles' (every 1%) are the convenient positional measures of the scatter or dispersion of a distribution.

When the data are arranged in order of magnitude (i.e. in an array), these measures involve counting along the scales and do not require any calculation. Thus 'quartile' divides the total distribution into four equal groups by three quartiles: the first or the lower quartile (i.e. the smallest quartile), Q_1 is 1/4 $(N + 1)$th observation in a distribution in an ascending order, that is, the point below which lies the first 25% of the values. Similarly, the serial number 3/4 $(N + 1)$th observation in a distribution is the third or upper quartile (i.e. the largest quartile), Q_3, that is, the point below which lies 75% of the cases. The second quartile, Q_2 is the median itself. If in a large sample we know the three quartiles and two extreme values, we can form a good idea of the whole distribution by plotting these five points and sketching the percentage cumulative frequency curve by joining them. Any value of the sample on this curve will correspond to a particular percentile rank. In case there is a positive skewness in the distribution, the distance $Q_3 - Q_2$ will be greater than the distance $Q_2 - Q_1$. If the skewness is negative, the reverse will be true, for symmetrical distribution, they will be equal.

Likewise there are nine values called 'deciles' which divide a total distribution into 10 equal parts, and 99 values called 'percentiles' divide a distribution into one hundred equal parts. *Percentiles* are points below which a given percentage of scores fall. The serial number of any partition value like the *K*th decile (using *K* as a symbol) is $\frac{K}{10}(N + 1)$; similarly the *K*th percentile is $\frac{K}{100}(N + 1)$ and so on. Finally note that a single decile or a single percentile is not a measure of dispersion. However, when we take the average distance between two deciles or percentile, we obtain an index of how dispersed a distribution is from its mid-points.

For a grouped frequency data any partition value like deciles, quartiles or percentiles which has a proportion *P* of observations, we first find out the interval within which a desired particular value falls, then it is calculated by the interpolation formula for median given in Eq. (2.26) or Eq. (2.27) earlier. Certain percentile values and other specific positional measures are shown below:

Percentile	Decile	Quintile	Quartile
90th	= 9th		
80th	= 8th	= 4th	
75th			= 3rd
70th	= 7th		
60th	= 6th	= 3rd	
50th	= 5th		= 2nd
40th	= 4th	= 2nd	
30th	= 3rd		
25th			= 1st
20th	= 2nd	= 1st	
10th	= 1st		

Example 2.12 Calculate the lower and upper quartiles, 3rd decile and 20th percentile to the grouped frequency distribution in Table 2.2.

Lower quartile, Q_1 = size of $(N + 1) / 4$th item = 13th item, hence Q_1 lies in the class 79 – 123, similarly upper quartile, Q_3 being the 39th item, it lies in the class 167 – 211. And we have $Q_1 = 79 + (13 - 7)/20.\ 44 = 92.2$ cm, and $Q_3 = 167 + (39 - 38)/11.\ 44 = 171$ cm. Since Q_2 is the median equal to 119.7 cm, $Q_3 - Q_2$ is larger than $Q_2 - Q_1$ and the skewness is positive.

Now the 3rd decile is $(3/10)(N + 1) = (3/10)(50 + 1) = 15$th item and it is $= 79 + [(15 - 7)/20].\ 44 = 96.6$ cm and similarly the 20th percentile is $(20/100)(N + 1) = (20/100)(50 + 1) = 10$th item and it is $= 79 + [(10 - 7)/20].\ 44 = 85.6$ cm.

2.11 QUARTILE DEVIATION

The range between the first and the third quartile is the 'inter-quartile range'; thus where as the range encloses 100% of the individual measurements in a data-set, the inter-quartile range, $(Q_3 - Q_1)$ encloses 50 percent. Quartile deviation encloses the middle half of the values in the set of the data and it is

$$(Q_3 - Q_1)/2 \qquad \text{... (2.34)}$$

The coefficient of quartile deviation is a ratio between the quartile deviation and median expressed in percentage. Expressed symbolically, Coefficient of Quartile deviation is

$$C_{QD} = \frac{Q_3 - Q_1}{Q_3 + Q_1} \times 100\% \qquad \text{... (2.35)}$$

Quartile deviation is the most common dispersion measure used along with the median. It is preferable, (i) if the data forms an asymmetrical distribution, (ii) if extreme values are having a strong influence, (iii) if data are based on an ordinal scale of measurement, (iv) it is quite easy to understand and calculate, (v) it is better than range because it is not affected by extreme values, (vi) it can also be computed for a frequency distribution, having open-end classes. In this respect it is superior to all other measures of dispersion because no other measure can be calculated for such a distribution. But the most serious demerits of quartile deviation are two : (i) it is not amenable to further mathematical treatment, and, (ii) it is affected considerably by fluctuation of sampling.

The third group of summary measures is the measures of skewness and kurtosis which we will discuss now in the next chapter alongwith variance and standard deviation. But before we do so, we end up our discussion in this chapter with an example worked out with reference to quartile deviation, below.

Example 2.13 Compute the coefficient of quartile deviation for the data: 20, 28, 48, 13, 38, 15 and 50.

The given data are arranged in order as 13, 15, 20, 28, 38, 48 and 50 for which $Q_1 = (N + 1)/4$ = 2nd item (in order) = 15 and similarly $Q_3 = 3(N + 1)/4$ = 6th item (in order) = 48.

So, coefficient of quartile deviation, $C_{QD} = (48 - 15)/(48 + 15) \times 100\% = 52.38\%$

LIST OF FORMULAE

I. *Central Tendency Measures*

Individual Series	Discrete Series	Continuous Series
1. Arithmetic Mean (a) Direct Method	1. Arithmetic Mean (a) Direct Method	1. Arithmetic Mean (a) Direct Method
$\overline{X}_N = \frac{\Sigma X_i}{N}$	$\overline{X}_N = \frac{\Sigma f_i m_i}{N}$	$\overline{X}_N = \frac{\Sigma f_i m_i}{N}$
(b) Method by coding	(b) Method by coding	(b) Method by coding
$\overline{X}_N = \overline{X}_0 + \frac{\Sigma d}{N}$	$\overline{X}_N = \overline{X}_0 + \frac{\Sigma f_i d}{N}$	$\overline{X}_N = \overline{X}_0 + \Sigma f_i d$
where $d = X_i - \overline{X}_0$	where $\overline{X}_0$ the assumed mean, and $d = X_i - \overline{X}_0$	where $\overline{X}_0$ the assuned mean, and $d = m_i - \overline{X}_0$
2. Median and Positional/ Partitional Measures Size of (N + 1). Kth item where	Size of (N + 1). Kth item where $K = \frac{1}{2}$ for median, $= \frac{1}{10}$ for decile, $\frac{1}{25}$ for quartile, $\frac{1}{100}$	Size of (N). Kth item as P in the formula. [where $K = \frac{1}{2}$ for median $= \frac{1}{10}$ for decile, $\frac{1}{25}$ for quartile, $\frac{1}{100}$ for

$K = \frac{1}{2}$ for median, $= \frac{1}{10}$ for decile, $\frac{1}{25}$ for quartile, $\frac{1}{100}$ for percentile, and so on	for percentile, and so on	percentile, and so on] equal to $L + \frac{P-C}{f} . w$ (for lower limit) or $U - \frac{P-C}{f} . w$ (for upper limit)
3. Mode : Determined as the value that occurs largest number of times	Determined either by inspection or by 'method of grouping' for bi-modal or irregular data	$M_0 = L + \frac{d_1}{d_1 + d_2} w$ where $d_1 = \mid f_1 - f_0 \mid$ and $d_2 = \mid f_2 - f_0 \mid$

Empirical Mode : Mode = Mean ∓ 3 (Mean – Median)
'–' for positively and '+' for negatively skewed

4. Geometric Mean $M_G = \sqrt{(X_1)(X_2)\ldots(X_N)}$ or Anti-log $\left(\frac{\Sigma \log X_i}{N}\right)$	M_G = Antilog $\left(\frac{\Sigma f_i \log m_i}{N}\right)$	M_G = Antilog $\left(\frac{\Sigma f_i \log m_0}{N}\right)$
5. Harmonic Mean $M_H = \frac{N}{\Sigma(1/X_i)}$	$M_H = \frac{N}{\Sigma(f_i . \frac{1}{m_i})}$	$M_H = \frac{N}{\Sigma(f_i . \frac{1}{m_i})}$

Relationship : Arithmetic Mean > Geometric Mean > Harmonic Mean

6. Weighted Mean $= \frac{\Sigma w_i X_i}{\Sigma w_i}$

7. Combined Arithmetic Mean,

$$\overline{X}_{12...K} = \frac{N_1 \overline{X}_1 + N_2 \overline{X}_2 + ... + N_k \overline{X}_k}{N_1 + N_2 + ... + N_k}$$

8. Combined Geometric Mean $\overline{G}_{12...K}$ =
(for 2 to k number of sets)

$$\text{Antilog } \frac{N_1 \log \overline{G}_1 + N_2 \log \overline{G}_2 + \ldots N_k \log \overline{G}_k}{N_1 + N_2 + \ldots + N_k}$$

9. Mean for $(N + 1)$ Observations,

$$\overline{X}_{N+1} = \left(\frac{N}{N+1}\right)\overline{X}_N + \left(\frac{1}{N+1}\right)\overline{X}_{N+1}$$

II. *Positional Measures/Partition values*

1. Gini coefficient, $G = [\Sigma X_i Y_{i+1} - \Sigma X_{i+1} Y_i]$
where $i = 0, 1, 2, \ldots$

2. Quartile Deviation, $QD = \dfrac{Q_3 - Q_1}{2}$

3. Coefficient of Quartile Deviation, $C_{QD} = (Q_3 - Q_1)/(Q_3 + Q_1)$

EXERCISES

2.1. Prepare an altimetric frequency curve based on the spot height data taken from a seven and half minute quadrangle topographic map of Survey of India. The data are in metre and are given below:
948, 849, 838, 911, 997, 868, 905, 881, 999, 904, 831, 908, 834, 900, 861, 752, 875, 867, 768, 785, 888, 820, 793, 760, 857, 914, 883, 734, 886, 797, 910, 826, 837, 805, 843, 882, 876, 668, 958, 767, 972, 756, 923, 747, 771, 914, 733, 676, 704, 629, 730, 932, 643, 731, 792, 700, 701, 657, 865, 659, 730, 928, 692, 645, 737, 900, 1000 and, 851
Take 626 m as the lower boundary of the class interval. Is the data bimodal? Discuss.

2.2. Prepare a cumulative frequency curve based on the spot height data given in Exercise 2.1. Plot the percentage cumulative frequency curve also.

2.3. Find out the 1st and 3rd quartiles, 4th and 9th deciles and 3rd and 50th percentiles of the data given in Exercise 2.1. Compute the coefficient of quartile deviation.

2.4. The following is the annual rainfall data of 30 years for a place. The data are recorded in centimetres and are given here in their order of occurrence.
62.97, 63.43, 86.05, 63.70, 60.02, 70.20, 66.43, 72.25, 71.48, 63.18, 75.43, 43.13, 64.45, 65.05, 67.08, 62.18, 76.48, 79.88, 72.80, 83.35, 56.18, 64.93, 77.30, 70.00, 72.38, 87.02, 74.25, 62.88, 91.25 and 64.25.
Arrange the data in order of magnitude as well as in a frequency table. Calculate the mean, median and mode for the listed as well as the grouped data.

2.5. For the data given in the Exercise 2.4 plot the less than and more than cumulative frequency curves and interpolate the mean.

2.6. Find the median of a 80-day rainfall record at a rain gauge station by a histogram and a cumulative frequency curve. The data are in mm and are as follows :

Rainfall (mm)	Frequency (days)
100–120	6
120–140	10
140–160	18
160–180	24
180–200	10
200–220	8
220–240	4

2.7. Following are the data regarding the non-tribal and tribal population of eight regions of a country. The eight regions are marked alphabetically in the data below :

Regions	Population of the Regions	
	Non-tribal	Tribal
A	5,793	238
B	15,370	2185
C	18,070	2275
D	18,070	2300
E	8,025	395
F	14,325	3825
G	5,850	1450
H	9,300	1780

Calculate the index of ethnic segregation.

2.8. The mean of 200 values of an item is 50. Later on it was found that 2 values were misread as 50 and 10 instead of 150 and 110. Find the correct mean.

2.9. The same size of samples ($n = 432$) is found to have been distributed in three different frequency series as follows :

1st Series		2nd Series		3rd Series	
Size of Samples	Frequency	Size of Samples	Frequency	Size of Samples	Frequency
71	3	82	2	81	2
83	4	76	3	76	3
73	5	73	6	74	4
74	2	76	7	58	2
65	3	65	3	70	7
66	3	60	7	73	2
Total = 432		Total = 432		Total = 432	

Determine which of the above is the best one?

2.10. An incomplete frequency distribution is given below :

Class-interval	Frequency
10–20	10
20–30	25
30–40	–
40–50	60
50–60	–
60–70	20
70–80	12
	Total = 205

(i) Given the median, find the missing values in the above frequency table, (ii) compute the arithmetic mean of the complete frequency table by the coding method.

2.11. Calculate the arithmetic, geometric and harmonic means of the following set of data: 9.3, 9.1, 10.5, 13.2, 12.1, 10.2, 10.5, 9.8, 10.6, 14.0, 10.6, 10.1, 10.1, 9.0, 9.8, 9.7, 16.0, 10.2, 10.0 and 9.9.

2.12. In a desert region, the rainfall of a 50-minute storm was recorded at every 5th minute from beginning of the storm. The recorded data were available as accumulated rainfall (in cm) as follows:

Time since beginning of the storm	Accumulated Rainfall (in cm.)
5	0.1
10	0.2
15	0.8
20	1.5
25	2.0
30	2.3
35	2.5
40	2.7
45	2.8
50	2.8

Find the mean intensity of the storm rainfall per hour.

2.13. Following is the frequency distribution of 87 years of annual peak flood in the discharge site of a river basin. The discharge is in cumec and the data are as follows:

Class-interval	Frequency
0–2,000	0
2,000–4,000	17
4,000–6,000	27
6,000–8,000	18
8,000–10,000	18
10,000–12,000	3
12,000–14,000	0
14,000–16,000	2
16,000–18,000	1
18,000–20,000	1

Calculate the mode and the coefficient of quartile deviation.

2.14. Calculate the progressive average and moving averages in 2's and 3's for the following data: 12, 14, 15, 18, 25, 22, 25, 28, 29 and 35.

2.15 Find the mode of the following frequency distribution:

Values	Frequency
1	1
2	7
3	15
4	35
5	47
6	38
7	23
8	58
9	20
10	15
11	12
12	6

2.16. An aircraft flies round a square whose side is 10 km along at different speed: the 1st side at 100 km p.h., the 2nd side at 200 km p.h., the 3rd side at 300 km p.h. and the 4th side at 100 km p.h. Find the time taken and the average speed of the aircraft.

3

Numerical Data in Geo-science Systems II

(Measures of Dispersion and Skewness)

It has been stressed several times earlier that in any set of data the actual values differ from one another and from the mean value itself. The measurement of this spread–out character of the data is known as "dispersion" or "variability". This is just as important as finding the measures of central tendency and positional or partition values discussed in Chapter 2. *Dispersion* or *Variability* is as characteristic as similarity is in statistics. It may be measured in statistics in two basic ways: (1) in terms of distances between particular observation values, and (2) in terms of average 'deviations' of individual observations about the central value.

3.1 DISTANCE MEASURES OF DISPERSION

Dispersion is a measure or an indicator of how spread-out a distribution is to convey same idea of how far, on the average scores deviate from central tendency measures like mean, median and mode. Hence the central tendency measures should always be accompanied by an appropriate measure of dispersion. Dispersion, when measured in terms of the difference between the highest and the lowest values of the variable in a distribution is known as 'range'. It is applied when a set of values in an observation is from distances between points and individuals and when it is arranged graphically in terms of their magnitude, we get a "dispersion diagram". Figure 3.1 illustrates the 50 years of annual rainfall distribution at Varanasi in a dispersion diagram where we can see the total spread of the data within the range. However, symbolically speaking,

$$\text{Coefficient of Range} = \frac{L-S}{L+S} \qquad \text{... (3.1)}$$

where L and S are the largest and the smallest values.

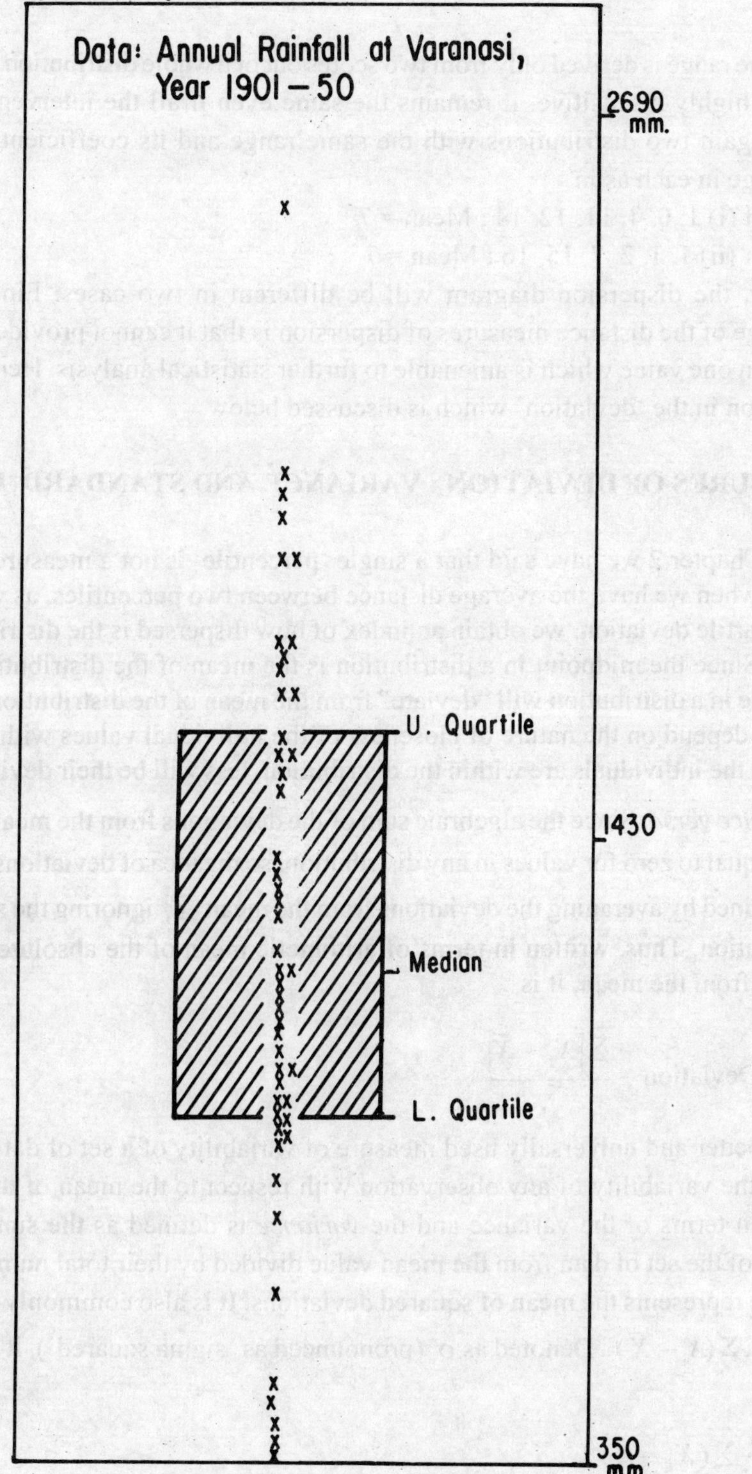

Fig. 3.1 : Dispersion Diagram

Since the range is derived only from two scores out of a whole distribution, the coefficient of range is highly insensitive. It remains the same even if all the intervening values are changed. Again two distributions with the same range and its coefficient may have the same average in each as in

Data set (i) 1, 0, 4, 11, 12, 14 ; Mean = 7

Data set (ii) 1, 1, 2, 7, 15, 16 ; Mean = 7

Clearly, the dispersion diagram will be different in two cases. Finally, the main disadvantage of the distance measures of dispersion is that it cannot provide a summary of the scatter in one value which is amenable to further statistical analysis. Hence we look for this condition in the 'deviation' which is discussed below.

3.2 MEASURES OF DEVIATION : VARIANCE AND STANDARD DEVIATION

Earlier in Chapter 2 we have said that a single 'percentile' is not a measure of dispersion. However, when we have the average distance between two percentiles, as with the coefficient of quartile deviation, we obtain an index of how dispersed is the distribution from its midpoint. Since the midpoint in a distribution is the mean of the distribution, so an individual value in a distribution will "deviate" from the mean of the distribution by an amount which will depend on the nature of closeness of the individual values within the distribution: closer the individuals are within the distribution, less will be their deviations from the mean and *vice versa*. Since the algebraic sum of the deviations from the mean, $\sum (X_i - \bar{X})$ is always equal to zero for values in any distribution, so the idea of deviations in a set of data can be obtained by averaging the deviations from the mean $\bar{X}$, ignoring the signs, known as mean deviation. Thus, written in terms of arithmetic mean of the absolute differences of each score from the mean, it is

$$\text{Mean Deviation} = \frac{\Sigma\left|X_i - \bar{X}\right|}{N} \qquad \text{... (3.2)}$$

But a better and universally used measure of variability of a set of data is "variance". Therefore the variability of any observation with respect to the mean of the set of data is measured in terms of the variance and the *variance* is defined as the sum of squares of deviations of the set of data from the mean value divided by their total number. This value of variance represents the mean of squared deviations. It is also commonly known as 'sum of squares', $\Sigma(X_i - \bar{X})^2$. Denoted as σ^2 (pronounced as 'sigma squared'), it is symbolically written as

$$\sigma^2 = \frac{1}{N}\Sigma(X_i - \bar{X})^2 \qquad \text{... (3.3)}$$

The variance in statistics is the second moment of a frequency, the first moment is the first power of deviation $\Sigma(X_i - \overline{X})/N$. The Eq. (3.3) is the second moment. The higher moments can be measured about the mean as

$$\mu_r = \frac{1}{N}\Sigma(X_i - \overline{X})^r \qquad \text{... (3.4)}$$

where r is the moment. There are four moments about the mean. The third moment is often used as to measure the "shape" of the frequency distribution while the fourth moment is used to measure the "flatness/peakedness" of the frequency distribution. They will be discussed later on in this chapter. Now coming back to the second moment, i.e. the variance, Eq. (3.3) can be written as

$$\sigma^2 = \frac{\Sigma X_i^2}{N} - \overline{X}^2 \qquad \text{... (3.5)*}$$

Both Eqs. (3.3) and (3.5) give identical values of variance. Equation. (3.5) requires less number of computations than Eq. (3.3) but both of them require that the mean be computed. Another method that can be used to calculate the variance without computing the mean is the 'sum of squares of deviation' (known hereafter as the sum of squares) divided by N, the total number of observations in the set of data. From Eq. (3.3) we have

$$\text{Sum of Squares} = \text{Variance} \times N$$

$$= \left(\frac{\Sigma X_i^2}{N} - \overline{X}^2\right) \times N$$

$$= \Sigma X_i^2 - N.\overline{X}^2$$

$$= \Sigma X_i^2 - N.\frac{\Sigma X_i}{N}.\frac{\Sigma X_i}{N}$$

---

* Expanding Eq. (3.3) we have

$$\sigma^2 = \frac{\Sigma X_i^2}{N} - \frac{\Sigma 2X_i.\overline{X}}{N} + \frac{\Sigma \overline{X}^2}{N}$$

$$= \frac{\Sigma X_i^2}{N} - 2\overline{X}^2 + \overline{X}^2 \text{ since} \left(\frac{\Sigma \overline{X}^2}{N} = \overline{X}^2\right)$$

$$= \frac{\Sigma X_i^2}{N} - \overline{X}^2 \left[\text{Note that this is also equal to } \frac{N\Sigma X_i^2 - (\Sigma X_i)^2}{N^2}\right]$$

$$= \sum X_i^2 - \frac{(\sum X_i)^2}{N} \qquad \text{... (3.6)}$$

Therefore, the equation for Variance becomes

$$\sigma^2 = \frac{1}{N}\left[\sum X_i^2 - \frac{(\sum X_i)^2}{N}\right] \qquad \text{... (3.7)}$$

The value of the variance from the grouped data differs slightly from the value obtained with the raw rata. Thus the variance for a grouped frequency distribution can be determined such as the mean in Eq. (2.5) as

$$\sigma^2 = \frac{\sum f(X_m - \overline{X})^2}{N} \qquad \text{... (3.8)}$$

where X_m is the midpoint of each class interval.

This above expression can be again simplified for the convenience of computation as

$$\sigma^2 = \frac{\sum fX_m^2 - N\overline{X}^2}{N} \qquad \text{... (3.9)}$$

When the data of a sample, X_i are to be used to estimate the variance of the population from which the sample is drawn there will be a sampling error because the small sample size usually underestimate the population parameter. Since this sampling error is a function of the ratio of sample to population size, the underestimation of sample values regarding the population parameter can be corrected by 'Bessel's Correction Factor' $(N - n)/(n - 1)$ which is included with sample standard deviation, *s* as a substitute of the "population variance estimate" s^2 : it is given by:

$$s^2 = \frac{\sum (x_i - \overline{x})^2}{n-1} \qquad \text{... (3.10)}$$

where *n* is the size of the sample and x_i is the sample data.

Similarly, for a grouped data, s^2 is given by

$$s^2 = \frac{\sum f(x_m - \overline{x})^2}{n-1} \qquad \text{... (3.11)}$$

The computational formulas for Eq. (3.10) and (3.11) are

$$s^2 = \frac{\sum x_i^2 - n\overline{x}^2}{n-1} \qquad \text{... (3.12)}$$

$$s^2 = \frac{\sum f x_m^2 - n\bar{x}^2}{n-1} \quad \text{... (3.13)}$$

One important use of variance is in finding out the "index of dispersion" in ecology. If individuals of an event are spatially differed over discrete sampling units (SUs), and if samples are taken as the number of occurrences per SU, then it is possible to summarise these data with the number of SUs with no occurrence (= 0), 1, 2, ... r occurrences i.e. in a frequency distribution. This constitutes the basic data for using in the "pattern detection methods" by calculating the mean $\bar{x}$ and variance s^2 for the sample. Now, the index of dispersion is a variance-to-mean ratio.

$$\text{ID} = \frac{s^2}{\bar{x}} \text{ where for the discrete SUs, } s^2 = \sum_{i=1}^{n} (x_i - \bar{x})^2 \quad \text{... (3.14)}$$

If the ID approaches 1.00 i.e. $s^2 \approx \bar{x}$ it indicates a random pattern, with $s^2 < \bar{x}$ and ID close to 0.00, the pattern is uniform. On other hand if $s^2 > \bar{x}$, ID values becomes more than 1.00 and it indicates a clumped pattern.

This above index of dispersion (or, clumping) is modified by R. H. Green in 1966 to a new index, Green's Index, GI that is independent on number of samples, n.

Thus

$$GI = \frac{(s^2/\bar{x}) - 1}{n-1} = \frac{ID - 1}{n-1} \quad \text{... (3.14a)}$$

in which GI varies between 0 (for random) and 1 (for maximum clustered pattern of distribution of an event or population).

Example 3.1 Following are the three sets of data which are having the same mean:

Data set (i) : 0, 1, 2, 3, 3, 3, 4, 4, 5, and 5.

Data set (ii): 0, 0, 0, 0, 0, 0, 3 , 8, 9 and 10

Data set (iii): 2, 3, 3, 3, 3, 3, 3, 3, 3 and 4

If the above data sets are the observed frequency distribution of the number of individuals per quadrats for an area regularly gridded, what inferences we can have about the spatial distribution?

The above data sets have the same mean 3.0 but have different variances as in Eq. (3.10) that is, for data set (i) $s^2 = 2.6$ for (ii) $s^2 = 18.22$ and for (iii) $s^2 = 0.22$

Now by the Index of dispersion, for data set (i) $\bar{x}$ is close to s^2 indicating a random pattern of distribution; for data set (ii) σ^2 is many times, more than $\bar{x}$ indicating a clustered distribution; for data set (iii) distribution is uniform, the resultant variance σ^2 (= 0.22) is much less than the mean, $\bar{x}$ (=3.0). Note for a perfect uniform distribution, all the quadrats will have the same number of individuals giving a variance s^2 equal to zero.

There are few statistical distributions that because of their variance-to-mean properties they have been used as models for testing the random, clustered and uniform patterns of spatial distributions. This we do later on in Chapters on Poisson and Chi-square distributions.

There are two practical difficulties associated with the use of the variance. One difficulty is that the variance being the sum of squares is not expressed in the same units as the data, e.g. the variance of heights, originally expressed in metres, will be expressed in square metres. Secondly, variance is a large number compared to the observations themselves. These difficulties are overcome simply by working with the square root of the variance, called the *standard deviation*. Thus the standard deviation is the most common measure of spread of any distribution. It is also the simplest measure of dispersion which can be manipulated statistically further. It is expressed as follows:

For individual observations:

$$\sigma = \sqrt{\frac{\Sigma X_i^2}{N} - \bar{X}^2} \quad \text{... (3.15)}$$

For a grouped frequency data:

$$\sigma = \sqrt{\frac{\Sigma f X_m^2}{N} - \bar{X}^2} \quad \text{... (3.15a)}$$

where X_m is the midpoint of the class interval. Note Eqs. (3.15) and (3.15a) give the standard deviation from the actual mean. But the Equation (3.15a) is more convenient to use when computing the 'assumed mean method' with assumed mean $\bar{X}_0$ coded as $d = X_i - X_0$ (or, $d = m_i - d_0$ for grouped data), we can have the standard deviation.

$$\sigma = \sqrt{\frac{\Sigma d^2}{N} - \left(\frac{\Sigma d}{N}\right)^2} \quad \text{... (3.16)}$$

$$\text{or } \sigma = \sqrt{\frac{\sum fd^2}{N} - \left(\frac{\sum fd}{N}\right)^2} \quad \text{... (3.16a)}$$

we use these in an example below.

Example 3.2 The method of computing the standard deviation is illustrated in Table 3.1 by using the data on India's urban structure in 1981.

Because of the unevenness of the class-intervals of the seven-tier arrangement of the urban structure of India to which the data are related, the standard deviation computed on the basis of the mid-values of the class intervals will be much higher than on size of the class intervals based on the total population divided by the total number of the urban places in the respective size group. The computation of standard deviation is done by both these methods now.

Table 3.1 Computing Standard Deviation from Grouped Data
(India's Urban Structure: See Table 2.3)

(a) By size of class intervals

Urban categories	f	Total Population in the category (in millions)	X_m (in thousands)	$f. X_m^2$ (in billions)
> 1,000,000	12	42.1217	3,510.142	147853.16
100,000 – 999,999	206	52.7887	256.256	13527.43
50,000 – 100,000	325	18.1900	55.960	1018.11
20,000 – 49,999	883	22.5570	25.546	576.24
10,000 – 19,999	1247	15.0070	12.035	180.62
5,000 – 9,999	920	5.7410	6.240	35.82
< 5,000	342	0.8550	2.500	2.14
Total :	Σ3935	Σ157.2604		Σ163193.52

$$\overline{X} = \frac{\sum fX_m}{N} = 39{,}965 \text{ persons}$$

$$\sigma = \sqrt{\frac{\sum f.X_m^2 - N\overline{X}^2}{N}} = 199{,}688 \text{ persons}$$

(b) *By mid-values of class intervals*

Urban Categories	Mid-value X'_m	$f. X'_m$ (in millions)	$f. X'^2_m$ (in billions)
>1,000,000	3,510,142	42.1217	147853.16
100,000 – 999,999	550,000	113.3	62315.00
50,000 – 100,000	75,000	24.375	1828.12
20,000 – 49,999	35,000	30.905	1081.68
10,000 – 19,999	15,000	18.705	280.58
5,000 – 9,999	7,500	6.9	51.75
<5,000	2,500	0.855	2.13
	Total	Σ 237.1617	Σ 213412.42

Now, $\bar{X} = \frac{\Sigma f.X'_m}{N} = \frac{237.1617 \text{ millions}}{3935} = 60,270$ persons

and $\sigma = 224,948.8 \approx 224,949$ persons

It can be noted that since the values in the set of data above have a wide range, its standard deviation is about five times larger than the average size of population of an urban area (the latter was 39,965 person in 1981). Again it can be noted that when the mid-values of the urban categories are taken in the computation, the above table shows that both the mean and standard deviation increase considerably. But compared to the large standard deviation, quartile deviation is small, the first quartile Q_1 is 6516 and the third quartile, Q_3 is 34983 and hence the quartile deviation is only 14234 (QD being equal to $\frac{34983-6516}{2}$ = 14234). This difference between standard deviation and quartile deviation is due to the fact that standard deviation is concerned with frequencies of each measure whereas quartile deviation takes indirect account of the "form" of the distribution of a set of data. Thus when the values in the set of data are close, as in the case of annual rainfall distribution in Table 2.1, the standard deviation is much smaller, only 45.80 cm compared to the mean annual rainfall of 125.4 cm.

3.2.1 Covariance

Computational procedure used to calculate the variance of the data of a single attribute can be extended in calculating the measure of the mutual variability of a pair of attributes that is the joint variation of two variables x and y about their common mean. This measure called *covariance* is defined as the sum of products of their deviations about their respective means. Just as the variance measures the spread of values around the central point (in a frequency curve), the covariance measures the distribution of values around a common mean. Covariance is thus analogous to the "second-order moments about the mean". Hence, if there are bivariate distributions, x and y, we can write

$$\text{Covariance, } M_{xy} = \frac{\Sigma(x-\bar{x})(y-\bar{y})}{n} \quad \text{... (3.17)}$$

which can be simplified as $= \Sigma xy - \frac{\Sigma x . \Sigma y}{n}$ ... (3.17a)

Note Σxy is defined as the 'uncorrected sum of products' and $\frac{\Sigma x . \Sigma y}{n}$ is taken as the 'correction for the means'.

Interpretation of covariance values, must proceed in the same manner as interpretation of variances. Individual values are not too meaningful because they are dependent upon the units of measurement. But covariance of two variables when related to the product of their respective standard deviations is the *correlation coefficient, r*, a degree of interrelation between the two variables x and y in terms of a ratio of the covariance of two variables to the product of their standard deviations.

$$r_{xy} = \frac{M_{xy}}{s_x \cdot s_y} \quad \text{... (3.17b)}$$

3.2.2 Combined Standard Deviation

If the standard deviation of one set containing n_1 number of observations and having a mean $\bar{x}_1$ and a standard deviation σ_1 is grouped together with another set of n_2 number of observations, a mean $\bar{x}_2$ and standard deviation σ_2, the new set of $(n_1 + n_2)$ members will have a *combined standard deviation*.

So, combined standard deviation

$$\sigma_{12} = \sqrt{\frac{n_1(\sigma_1^2 + d_1^2) + n_2(\sigma_2^2 + d_2^2)}{n_1 + n_2}} \quad \text{... (3.18)*}$$

where $d_1 = \bar{x}_1 - \bar{\bar{x}}$ and $d_2 = \bar{x}_1 - \bar{\bar{x}}$ with $\bar{\bar{x}}$ as the combined mean equal to $\frac{n_1\bar{x}_1 + n_2\bar{x}_2}{n_1 + n_2}$ (as given in Eq. 2.9)

---

* For combined variance of two sets, let us assume that the two sets are $x_{11}, x_{12}, x_{13}, \dots x_{1N}$ and $x_{21}, x_{22}, x_{23}, \dots x_{2N}$ with $\bar{x}_1 = \sum_i^{n_1} x_{1j}, \bar{x}_2 = \sum_j^{n_2} x_{2j}, \sigma_1^2 = \frac{1}{n_1}\sum_i^{n_1}(x_{ij} - \bar{x}_1)^2, \sigma_2^2 = \frac{1}{n_2}\sum_j^{n_2}(x_{2j} - \bar{x}_2)^2$ and combined mean, $\bar{\bar{x}} = \frac{n_1\bar{x}_1 + n_2\bar{x}_2}{n_1 + n_2}$

Now, combined variance $= \frac{1}{n_1 + n_2}[\sum_i^{n_1}\{(x_{1i} - \bar{x}_1)^2 + \sum_j^{n_2}(x_{2j} - \bar{\bar{x}})^2]$

$$= \frac{1}{n_1 + n_2}[\sum_i^{n_1}\{(x_{1i} - \bar{x}_1) + (\bar{x}_1 - \bar{\bar{x}})\}^2 + \sum_j^{n_2}[(x_{2j} - \bar{x}_2) + (\bar{x}_2 - \bar{\bar{x}})]^2$$

$$= \frac{1}{n_1 + n_2}[\sum_i^{n}\{(x_{1i} - \bar{x}_1)^2 + 2(x_{1i} - \bar{x}_1)(\bar{x}_1 - \bar{\bar{x}}) + (\bar{x}_1 - \bar{\bar{x}})^2\}]$$

$$+ [\sum_j^{n_2}\{(x_{2j} - \bar{x}_2)^2 + 2(x_{2j} - \bar{x}_2)(\bar{x}_2 - \bar{\bar{x}}) + (\bar{x}_2 - \bar{\bar{x}})^2\}]$$

Now since $\sum_i^{n}(x_{1i} - \bar{x}_1)$ and $\sum_j^{n_2}(x_{2j} - \bar{x}_2)$ are both zero, so the above equation reduces to:

$$(\sigma_{12})^2 = \frac{1}{n_1 + n_2}[n_1\sigma_1^2 + n_1 d_1^2 + n_2\sigma_2^2 + n_2 d_2^2] = \frac{n_1(\sigma_1^2 + d_1^2) + n_2(\sigma_2^2 + d_2^2)}{n_1 + n_2}$$

Hence, combined standard deviation, $\sigma_{12} = \sqrt{[n_1(\sigma_1^2 + d_1^2) + n_2(\sigma_2^2 + d_2^2)]/(n_1 + n_2)}$

Equation (3.18) can be extended to any number of sets of data. For example, if there are sets of data with n_1, n_2 and n_3 size of observations, we have the combined standard deviation.

$$\sigma_{123} = \sqrt{\frac{n_1(\sigma_1^2 + d_1^2) + n_2(\sigma_2^2 + d_2^2) + n_3(\sigma_3^2 + d_3^2)}{n_1 + n_2 + n_3}} \quad \text{... (3.19)}$$

where $d_1 = \bar{x}_1 - \bar{\bar{x}}, d_2 = \bar{x}_2 - \bar{\bar{x}}$ and $d_3 = \bar{x}_3 - \bar{\bar{x}}$, $\bar{\bar{x}}$ is the common mean of the total observations, $n_1 + n_2 + n_3$

Example 3.3 In two sample sets, one set has 50 items with mean 12 and standard deviation 2.5. If the whole group has 200 items with mean 12 and standard deviation 10, find the standard deviation of the second sample set.

We have $n_1 = 50$, $\bar{x}_1 = 12$ and $\sigma_1 = 2.5$ and the whole group is being a total size $n_1 + n_2 = 200$ (so $n_2 = 150$ items), $\bar{\bar{x}} = 12$ and $\sigma = 10$. So we can have

$$\bar{\bar{x}} = \frac{n_1\bar{x}_1 + n_2\bar{x}_2}{n_1 + n_2} \text{ or } 12.0 = \frac{(50)(12) + (150)(\bar{x}_2)}{50 + 150}$$

$$\text{So, } \bar{x}_2 = \frac{(200)(12) - (50)(12)}{150} = 12.0$$

Now, the standard deviation σ_2 of the second set is worked out by using the formula for the combined standard deviation

$$\sigma_{12} = \sqrt{\frac{n_1(\sigma_1^2 + d_1^2) + n_2(\sigma_2^2 + d_2^2)}{n_1 + n_2}} \text{ with } d_1 = 12 - 12 = 0 \text{ and } d_2 = 12 - 12 = 0$$

$$\text{or } 10 = \sqrt{\frac{50(6.25 + 0) + 150(\sigma_2^2 + 0)}{50 + 150}}$$

or $(100)(200) = 312.5 + 150\,\sigma_2^2$

or $150\,\sigma_2^2 = 19687.5$ or $\sigma_2^2 = 131.25$, or $\sigma_2 = 11.46$

So, the standard deviation of the second set is 11.46.

3.2.3 Merits and Demerits of Standard Deviation

The standard deviation is the most important measure of dispersion based on interval or ratio data. Like mean, it is (i) most rigidly defined, (ii) based on all the observations, and (iii) amenable to further mathematical treatment, for example, if the size, mean and standard deviation for two or more series are given, one can calculate the combined standard deviation.

When the set of data is having a symmetrical distribution (which is known as *normal distribution*, a probability distribution to be discussed in Chapter 4), we can find the percentage of observations (including their extreme values) falling within the distance scale of one, two or three standard deviations from the mean. In Chapter 4, we will see that in case of a normal distribution, the values of standard deviation of the data get a symmetrical distribution so that 68.27% values lie in the range mean ± one standard deviation, 95.45% values lie in the range mean ± two standard deviation and 99.73% value lie in the range mean ± three standard deviation. This property finds extensive use in large sample tests as shown in Chapter 10.

A number of statistical techniques, the most important being the 'Analysis of Variance' have been developed taking it as the basis. Other fields of application include 'Regression and Correlation', 'Significance tests' of various statistical techniques, skewness and kurtosis etc.

Like mean, since standard deviation is affected by every value, especially by the extreme values from the mean, due to the effects of squaring, so standard deviation is a more appropriate measure for skewed distributions also. But in case, a set of data is weighted considerably by extreme values so that the frequency of the distribution gets even truncated at zero, the standard deviation has no simple meaning and hence in this case its computation should be avoided.

Example 3.4 Compute the standard deviation σ and s for one small set of data using the definition and computation formula. The set of data is as follows:

3, 4, 4, 4, 5, 6, 7, 8, 9, 10

Computing standard deviation σ and s using data as individual observation:

x_i	$x_i - \bar{x}$	$(x_i - \bar{x})^2$	x_i^2	
3	3 − 6 = − 3	9	9	$\sigma^2 = \frac{\Sigma X_i^2}{N} - \bar{X}^2$
4	4 − 6 = − 2	4	16	
4	4 − 6 = − 2	4	16	$= \frac{412}{10} - 36 = 5.2$
4	4 − 6 = − 2	4	16	
5	5 − 6 = −1	1	25	So, $\sigma = \sqrt{5.2} = 2.28$
6	6 − 6 = 0	0	36	$s^2 = \frac{\Sigma x_i^2 - n\bar{x}^2}{n-1}$
7	7 − 6 = 1	1	49	
8	8 − 6 = 2	4	64	$= \frac{412 - 10(36)}{10-1} = 5.77$
9	9 − 6 = 3	9	81	
10	10 − 6 = 4	16	100	So, $s = 2.40$
$\Sigma x_i = 60$		$\Sigma (x_i - \bar{x})^2$	$\Sigma x_i^2 = 412$	
$\bar{x} = 6$		$= 52$		

Computing standard deviation s using data in a grouped form:

x_i	f	fx_i	$(x_i - \bar{x})$	$f\,(x_i - \bar{x})^2$	x_i^2	$f.x_i^2$
3	1	3	– 3	9	9	9
4	3	12	– 2	12	16	48
5	1	5	– 1	1	25	25
6	1	6	0	0	36	36
7	1	7	1	1	49	49
8	1	8	2	4	64	64
9	1	9	3	9	81	81
10	1	10	4	16	100	100
		$\Sigma fx_i = 60$		$\Sigma f\,(x_i - \bar{x})^2 = 52$		$\Sigma fx_i^2 = 412$

So, $\bar{x} = 6$ $\qquad s^2 = \dfrac{\Sigma fx_i^2 - n\bar{x}^2}{n-1}$

Now, $s^2 = \dfrac{\Sigma f\,(x_i - \bar{x})^2}{n-1}$ $\qquad = \dfrac{412-(10)(36)}{10-1}$

$= \dfrac{52}{9} = 5.77$ $\qquad = \dfrac{52}{9} = 5.77$

$s = 2.40$ $\qquad s = 2.40$

3.3. VARIABILITY INDICES : COEFFICIENT OF VARIATION

All the above measures of variation (including standard deviation) are in some metric units as the original data and thus the influence of the mean values precludes meaningful comparisons of variability between sets of data of the same observation. Hence when the sets of data have very different means and we want to compare the variations of one set of data with those of another, the standard deviation as an absolute measure of dispersion is of little use. To overcome this difficulty, 'variability indices' are used as relative measures of dispersion. The 'Coefficient of Variation' is an example of the variability indices. It is a descriptive statistic derived from the standard deviation and it is a measure of variability relative to the mean. Thus it is:

$$\text{Coefficient of Variation, } C_v = \frac{\text{Standard Deviation}}{\text{Mean}} \times 100$$

$$= \frac{\sigma}{\bar{x}} \times 100 \text{ (in percentage)} \qquad \text{... (3.20)}$$

Coefficient of variation is thus defined as the ratio of standard deviation to the mean expressed in percentage. The coefficient of variation remains unaltered in the change of scale of measurement when the units are on ratio scale but it is altered when the units are on interval scale. Thus the coefficient of variation remains unaltered when scale changes from feet to centimetre but a change of temperature from °C to °F would alter the coefficient of variation.

3.3.1 Merits and Demerits of Coefficient of Variation

When we have measurements which are based on a scale of measurement like rainfall, we may wish to compare the relative variability of rainfall. For example, mapping of areal distribution of annual (or of less than a year) rainfall variation uses coefficient of variation.

Secondly, coefficient of variation is an effective tool in measuring dispersion even when the unit of measurement is the same, but the mean and standard deviation differ widely. For example, in comparison of rainfall between a group of wet and dry areas, it can be shown by the coefficient of variation that in spite of higher rainfall and a higher standard deviation in rainfall for the wet areas, the rainfall variability is less among the wet areas than the dry areas.

If it can be assumed that events, for example, annual and seasonal rainfalls are 'normally' distributed, coefficient of variation is used in determining the percentage probability of annual rainfall exceeding or falling short of a specified amount.

Another use of the coefficient of variation is all used in designing the network of an area optimally for obtaining data precisely. For example, the optimum number of rain-gauge stations are required to be set up for obtaining the mean areal depth of annual rainfall of a river basin. If we assume an allowable percentage of error, ε in the estimate of the average depth of rainfall over the basin, then *N*, the optimal number of rain-gauge stations is

$$N = \left(\frac{C_v}{\varepsilon}\right) \qquad \text{... (3.21)}$$

where C_v is the coefficient of variation of the rainfall values $P_1, P_2, \ldots, P_i$.

However, when the data are markedly skewed, or where the data are grouped in a frequency distribution with an open-end class, the standard deviation has no simple meaning. In these cases, the coefficient of variation is replaced by the coefficient of quartile deviation, as discussed in Chapter 2.

Finally, the easier and the quicker calculation of this method is an advantage which makes it an obvious choice where no further calculation is needed.

Example 3.5 A catchment has six rain-gauge stations. In a year, the annual rainfall (in cm) recorded in these six rain-gauge stations was : 82, 100, 180, 110, 98 and 140. For an ε = 10% in estimation of the mean rainfall in the catchment, estimate the optimum number of rain-gauge stations in the catchment.

Coefficient of variation, $C_v = \frac{\sigma_{n-1}}{\bar{x}} \times 100$ where $\sigma_{n-1} = \sqrt{\frac{\Sigma x_i^2 - n\bar{x}^2}{n-1}}$

$$= \sqrt{\frac{90428-(6)(118.38)^2}{6-1}} = \sqrt{\frac{6416.00}{5}} = 35.82$$

Hence, if N is the optimal no. of rain-gauge stations with ε = 10% of errors, then

$$N = \left(\frac{c_v}{\varepsilon}\right)^2 = \left(\frac{35.82}{10}\right)^2 = 12.83 \approx \text{say 13 stations.}$$

3.4. MOMENT MEASURES IN STATISTICS : DISPERSION AND SKEWNESS

The term "moment" is from *mechanics* where it is referred to as moment of force as completed by multiplying the magnitude of the weight multiplied over the distance, such as a fulcrum, to the point of rotation (Fig. 3.2). In statistical moments the force of mechanics is replaced by a frequency function (e.g. the per cent of the distribution within the given class), and the point of rotation of mechanics is replaced by an arbitrary point, generally the position of the mean of the distribution (taken as the origin of the frequency curve) and the distance from the point of force to the point of rotation (of the fulcrum) are determined by the spacing from the mean to the size class (the shaded rectangle in Fig. 3.2). Thus in Fig. 3.2 the moment of force about fulcrum is x cm × 20 kg = 20x cm ^{Kg} which by moment of frequency for the size class, x is x × 20% = 20x cm.

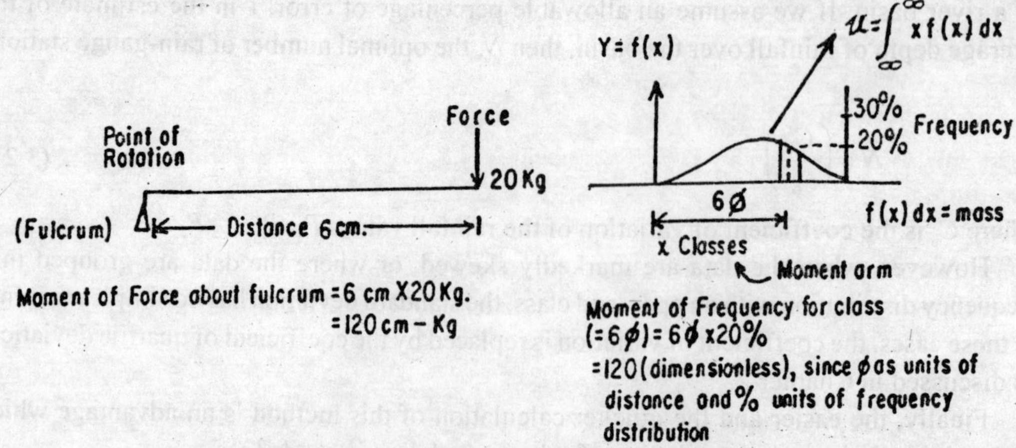

Fig. 3.2 : Moment of Measure (in mechanics and statistics)

This statistical moment is the mean of the first power of the derivation that is, the spacing of the size class or individual item in the distribution from the mean, adding them up and dividing them by the total size of the distribution. This gives the first moment about the mean which is symbolically written as

$$\mu_1 = \frac{\Sigma |x_i - \bar{x}|}{N} \quad \text{... (3.22)}$$

Successively, higher moments are defined by raising the distance term to progressively higher powers. Thus, the distance/spacing term is squared in the second moment, which is the variance (Eq. 3.3) and to the *r*th power in the *r*th moment (Eq. 3.4). Greek letter, μ is used to describe a moment, thus the first moment about the mean is μ_1, the second moment about the standard deviation is μ_2, the third μ_3 and the fourth moment μ_4. The third moment,

$$\mu_3 = \frac{\Sigma (x_i - \bar{x})^3}{N} \quad \text{... (3.23)}$$

refers to the measure of relative skewness and the fourth moment,

$$\mu_4 = \frac{\Sigma (x_i - \bar{x})^4}{N} \quad \text{... (3.24)}$$

to the measure of kurtosis. This method of four moments or the method of 'maximum likelihood' was first developed by Karl Pearson in 1902 which lead to the measures of skewness and kurtosis.

3.4.1 Measures of Skewness and Kurtosis

Measures of central tendency and position/partition are concerned with the location around which items are concentrated in a frequency distribution whereas measures of variability consider the extent to which the items vary. 'Measures of skewness and kurtosis' are another (though less important) kind of measures summarising the extent to which the items are 'symmetrically' and 'peakedly' distributed respectively. Skewness is already discussed in the earlier chapter as the lack of symmetry of the frequency and here we aim to measure the asymmetry or skewness. Similarly, kurtosis describes whether a frequency distribution is flat or peaked and here we aim to measure the flat/peakedness of the curve. In the aim to measure the skewness and kurtosis it may be analogous to refer them to 'moment' which has been briefly mentioned earlier. Below we discuss more about 'moment measures'.

3.4.1.1 Skewness

Dispersion, as we have defined in the beginning of this chapter, is concerned with the amount of variation rather than with its direction. Skewness tells us about the direction of the variation or departure from symmetry.

In the third moment about the mean, u_3 because of the cubing operation, deviation, $x_i - \bar{x}$ in a large set of data tend to dominate the sum in the numerater of μ_3 and large deviations are associated with the long tail of a frequency distribution. Hence Karl Pearson introduced a relative value in the third moment, μ_3 by squaring it first and then dividing it by the second moment from the mean cubed. Pearson termed this relative third moment as the skewness coefficient, β_1 (pronounced as 'beta-one') index which is

$$\text{Skewness Coefficient, } \beta_1 = \frac{\left[\frac{\sum (x_i - \bar{x})^3}{n}\right]^2}{\left[\frac{\sum (x_i - \bar{x})^2}{n}\right]^3} = \frac{\mu_3^2}{\mu_2^3} \quad \text{...(3.25)}$$

which can be further simplified to :

$$\text{Skewness Coefficient, } \beta_1 = \left[\frac{\sum (x_i - \bar{x})^3}{n.\sigma_x^3}\right]^2 \quad \text{...(3.26)}$$

where δ_x is the standard deviation.

In a symmetrical distribution both μ_1 and μ_3 are zero, so the skewness coefficient would be zero.

Again, since Mode = Mean ± 3 |Mean – Median| from Eq. (2.31), so a 'skewness measure', S_k can be calculated from a frequency distribution as

$$\text{Skewness Measure, } S_k = \frac{\text{Mean} - \text{Mode}}{\text{Standard Deviation}}$$

$$= \frac{\bar{x} - [\bar{x} \mp 3\,|\,\bar{x} - \text{Median}\,|]}{\text{Standard Deviation}}$$

$$= \frac{\pm 3\,|\,\text{Mean} - \text{Median}\,|}{\text{Standard Deviation}} \quad \text{...(3.27)}$$

If this above skewness measure or skewness idex exceeds ±1, the distribution is skewed, its maximum value can be ±3 which is rarely approached. For a positively skewed curve, since the mean is greater than the median, the result will be a positive number. A distribution skewed to the left will produce a negative number. And, of course, in a symmetrical distribution both the skewness coefficient and the skewness measure will be approaching zero.

Example 3.6 For the grouped data in Table 2.2 find the skewness coefficient and the skewness measure.

For a grouped data such as in Table 2.2, the rth moment from the mean is given by

$$\mu_r = \frac{\Sigma f\,(x_m - \bar{x})^r}{N}$$

From Table 2.2, we get the class interval, mid-value, x_m, and frequency, f. We calculate these values, have the mean, $\bar{x}$ = 126.52, standard deviation, σ_x = 45.8023 and do the necessary computation in the following.

Class Interval	Mid-value x_m	Frequency (f)	$f x_m$	$(x_m - \bar{x})$
35 – 79	57	7	399	– 69.52
79 – 123	101	20	2020	– 25.52
123 – 167	145	11	1595	18.48
167 – 221	189	11	2079	62.48
221 – 265	233	1	233	106.48

We calculate further by finding $r = 2$ and $r = 3$ values for $f(x_m - \bar{x})^r$. The $\Sigma f(x_m - \bar{x})^2$ and $\Sigma f(x_m - \bar{x})^3$ are 104892.47 and 1275306.7.

Therefore, the skewness coefficient, $\beta_1 = \dfrac{650.5625\,\text{millions}}{9232.5793\,\text{millions}}$

$= 0.07$

For computing skewness measure, we compute median = 119.7. which we substitute in

$$S_k = \frac{\pm 3\,|\text{Mean} - \text{Median}|}{\text{Standard Deviation}}$$

$$= \frac{\pm 3(126.52 - 119.7)}{45.8023}$$

$= 0.45$

Both the skewness coefficient and the skewness measure indicate a moderately positive skewed character and hence we can conclude that the distribution in the given data is approaching a symmetrical character.

Example 3.7 In a distribution, the skewness measure is + 0.30. If mean of the distribution is 25 and the standard deviation is 6, find the mode and median of the distribution.

From the relation for a moderately positive skewed curve: Mode = Mean – 3 (Mean – Median), we can write Mean – Mode = 3 (Mean – Median). Hence skewness measure is

$$S_k = \frac{\text{Mean} - \text{Mode}}{\text{Standard deviation}}$$ which can also be written as

$$S_k = \frac{3\,|\text{Mean} - \text{Median}|}{\text{Standard Deviation}}$$

So, for the present problem we can write

$$S_k = \frac{\bar{x} - M_0}{\sigma} \text{ and } S_k = \frac{3(25 - M_e)}{\sigma} \text{ or } 0.30 = \frac{25 - M_0}{6} \text{ and } 0.30 = \frac{3(25 - M_e)}{6}$$

and we calculate : mode = 23.20 and median = 24.40.

3.4.1.2 Kurtosis

If we know the measures of central tendency, dispersion and skewness, we still cannot form a complete idea about the distribution as it will be clear from Fig. 3.3 in which all the three frequency curves *x*, *y*, and *z* are symmetrical about the mean but they are different in height. Thus we still require to know the 'convexity of a curve' which is 'Kurtosis' (meaning, 'bulkiness' in Greek). Kurtosis describes whether the distribution in the set of a data is having a disproportionately large or small number of observations in the intermediate ranges between the mean and the extreme values and thus giving rise to a flat-toppedness or peakedness of the frequency curve.

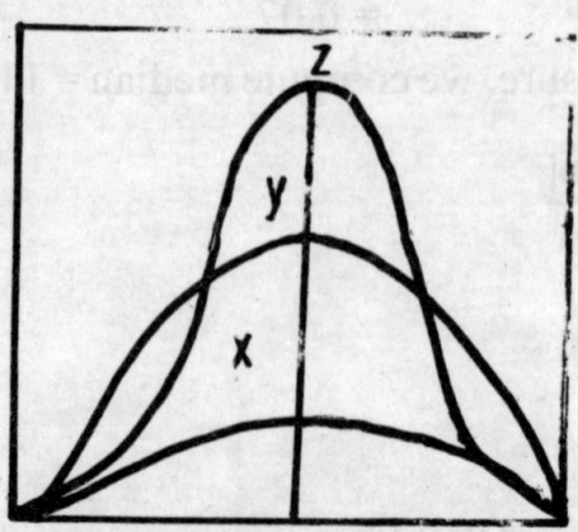

Fig. 3.3 : Three Normal Curves

Kurtosis, as mentioned earlier is the fourth moment about the mean, μ_4. In Pearsonian term, it is a coefficient, the *kurtosis coefficient,* β_2 (pronounced as beta-two). β_2 is obtained on dividing the fourth moment from the mean by the second moment from the mean, squared. Thus, writing notionally, we have

$$\text{Kurtosis Coefficient, } \beta_2 = \frac{\dfrac{\Sigma (x_i - \bar{x})^4}{n}}{\left[\dfrac{\Sigma (x_i - \bar{x})^2}{n}\right]^2} = \frac{\mu_4}{\mu_2^2} \quad \text{... (3.28)}$$

which can be further simplified to

$$\text{Kurtosis Coefficient, } \beta_2 = \frac{\mu_4}{\sigma^4} \quad \text{... (3.29)}$$

Since Kurtosis indicates the spread of the frequency curve, it is determined as

$$\text{Kurtosis Measure, } K = \frac{\text{Mean} - \text{Median}}{\text{Standard deviation}} \quad \text{... (3.30)}$$

For a symmetrically distributed curve, since median coincides with mean so the kurtosis measure, K should be zero and the kurtosis coefficient, β_2 should be 3.0, that is, $K = \beta_2 - 3$ where with $K = 0$, β_2 becomes 3.0. With $\beta_2 > 3$ and $K > 0$ the curve is more peaked than the perfect symmetrical curve and it is known as leptokurtic (i.e. the curve with narrower central position and a higher tail than a perfect symmetrical curve). When the reverse thing happens (i.e. $\beta_2 < 3$ and $K < 0$), the curve is known as 'platykurtic' (i.e. the curve with a broader central portion and a lower tail than the perfect symmetrical curve). From kurtosis point of view, the perfect summetrical curve is the curve with $\beta_2 = 0$ and $K = 0$ and it is known as 'mesokurtic'. It will be found that the kurtosis coefficient, β_2 for Example 3.7 worked out above is 2.13. Kurtosis measure, K is 0.10 indicating that the frequency curve is leptokurtic.

Thus for statistical analysis of frequency curve, it is not enough for the distribution to be merely symmetrical with zero skewness but the kurtosis coefficient for the data set must not depart widely from 3.0 (or from zero in case the kurtosis measure, K is calculated). These three different forms of the frequency curve from the kurtosis point of view are illustrated in Fig. 3.4.

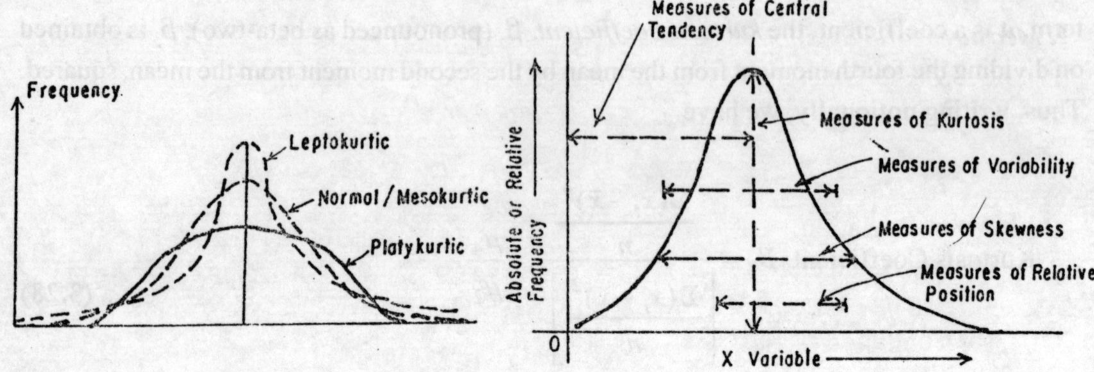

Fig. 3.4 : Different Forms of Normal Curves

Fig. 3.5 : Summary Measures

$\beta_2 < 3$: Leptokurtic

$\beta_2 > 3$: Platykurtic

$\beta_2 \rightarrow 3$: Mesokurtic

Kurtosis measures are usually applied for analysing the form of frequency curves. Also it should be noted that if the data are sample data and are of small size, the measures of skewness and the kurtosis are not reliable for statistical analysis of the frequency curve. However one important use of kurtosis is that the kurtosis of the population influences the type of average to use for a sample. If a population is platykurtic, the 'mid-range', that is, quartile deviation is appropriate for sampling; if the population is mesokurtic (or normal), the arithmetic mean should be used and if the population is leptokurtic, the median is preferable.

The measures of variation, skewness and kurtosis presented in this chapter, together with the central tendency measures and the positional/position values discussed in Chapter 2 earlier, provide the complete statistical description of a set of data. Finally, Fig. 3.5 shows the summary measures for frequency distributions.

The arithmetic mean, variance and standard deviation are the fundamental tools of descriptive statistics and they provide the basic for much of the discussion in the following chapters. So far these measures have been developed with respect to one variable measured on a single scale. We require to generalize them for handling two other situations: first, the case of distribution of independent spatial events in time; and second, the analysis of the joint variation among two or more variables. But before these situations are considered, we require to continue our discussion on distribution of the single variable further and to approximate the frequency curve to the concept of probability. This probability treatment of frequency distribution is initiated in the next chapter.

LIST OF FORMULAE

I. *Measures of Variation*

	Definitional formulae	Computational formulae
1. Mean Deviation		
(a) Individual and/or discrete series	$\frac{\Sigma \lvert X_i - \bar{X} \rvert}{N}$	
(b) Continuous series grouped into frequencies	$\frac{\Sigma f \lvert X_m - \bar{X} \rvert}{N}$	

2. Variance, σ^2 = square of standard deviation, σ

3. Standard Deviation

(a) Individual and/or discrete series

	Definitional formulae	Computational formulae
(i) for population	$\sigma = \sqrt{\frac{\Sigma (X_i - \bar{X})^2}{N}}$	$\sigma = \sqrt{\frac{\Sigma X_i^2}{N} - \bar{X}^2}$
(ii) for sample estimate	$s = \sqrt{\frac{\Sigma (x_i - \bar{x})^2}{n-1}}$	$s = \sqrt{\frac{\Sigma x^2 - n\bar{x}^2}{n-1}}$

(b) Continuous series grouped into frequencies

	Definitional formulae	Computational formulae
(i) for population	$\sigma = \sqrt{\frac{\Sigma f (X_m - \bar{X})^2}{N}}$	$\sigma = \sqrt{\frac{\Sigma f X_m^2}{N} - \bar{X}^2}$
(ii) for sample estimate	$s = \sqrt{\frac{\Sigma f (x_m - \bar{x})^2}{n-1}}$	$s = \sqrt{\frac{\Sigma f x_m^2 - n\bar{x}^2}{n-1}}$
(iii) for assumed mean method	$\sigma = \sqrt{\frac{\Sigma d^2}{N} - \left(\frac{\Sigma d}{N}\right)^2}$	$\sigma = \sqrt{\frac{\Sigma fd^2}{N} - \left(\frac{\Sigma d}{N}\right)^2}$

II. *Moment Measures in Statistics*

1. Four moments μ_r with r = 1, 2, 3 to 4

(i) *First moment*

(a) Individual and/or discrete series

$$\mu_1 = \frac{\Sigma|x_i - \bar{x}|}{N}$$

(b) Continuous series grouped into frequencies

$$\mu_1 = \frac{\Sigma f\,|x_i - \bar{x}|}{N}$$

(ii) *Second moment*

(a) Individual and/or discrete series

$$\mu_2 = \frac{\Sigma f\,(x_m - \bar{x})^2}{N} \qquad \mu_2 = \frac{\Sigma\, x_i^2}{N} - \bar{x}^2$$

(b) Continuous series grouped into frequencies

$$\mu_2 = \frac{\Sigma f\,(x_m - \bar{x})^2}{N} \qquad \mu_2 = \frac{\Sigma f\, x_m^2}{N} - \bar{x}^2$$

. . . . . .

. . . . . .

. . . . . .

(r) *rth moment*

(a) Individual and/or discrete series

$$\mu_r = \frac{\Sigma\,(x_i - \bar{x})^r}{N} \qquad = \frac{\Sigma\, x_i^2}{N} - \bar{x}^r$$

(b) Continuous series grouped into frequencies

$$\mu_r = \frac{\Sigma f\,(x_m - \bar{x})^r}{N} \qquad \mu_r = \frac{\Sigma f\, x_m^r}{N} - \bar{x}^r$$

2. Sum of Squares $SS = [\Sigma\,(x_i - \bar{x})^2]$ $\qquad SS = \Sigma\, x_i^2 - \frac{(\Sigma\, x_i)^2}{N}$

3. Covariance. $M_{xy} = \frac{\Sigma\,(x-\bar{x})(y-\bar{y})}{n} = \Sigma\, xy - \frac{\Sigma x\,.\,\Sigma y}{n}$

4. Combined standard deviation :

$$\sigma_{12} = \sqrt{\frac{n_1(\sigma_1^2 + d_1^2) + n_2(\sigma_2^2 + d_2^2)}{n_1 + n_2}}$$

$d_1 = \bar{x}_1 - \bar{\bar{x}}$ and $d_2 = \bar{x}_2 - \bar{\bar{x}}$, $\bar{\bar{x}}$ is the combined mean

5. Coefficient of Variation : C_v (in percentage) $= \frac{\sigma}{\bar{x}} \times 100\,(\%)$

6. Skewness

(a) coefficient : $\beta_1 = \mu_3^2 / \mu_2^3$ or $\left\{\sum (x_i - \bar{x})^3 / N.\sigma_x^3\right\}^2$

(b) measure : $S_k = \pm\, 3$ (Mean – Median)/Standard Deviation

7. Kurtosis

(a) coefficient : $\beta_2 = \mu_4 / \mu_2^2 = \mu_4 / \sigma^4$

(b) measure : $K = \frac{\text{Mean} - \text{Median}}{\text{Standard Deviation}}$

EXERCISES

3.1 Following are the data on acreage of food crops and oilseeds in an area.

Crops	Groundnut	Oilseeds	Jowar	Bajra	Pulses
Acreage (in 100 acres)	2054	195	928	1390	757

Identify the significant crop combinations in the region by the mean deviation.

3.2 The following tables give the random sampling of spot height on topographic map of Survey of India : one is in Udaipur district and the other in Palamau district. Spot heights are in feet.

Part of Udaipur district (latitude 24°15'N to 24°30'N and longitude 73°30'E to 73°45'E) : 2713, 2069, 2046, 2418, 1874, 2570, 2553, 1887, 3131, 3277, 2983, 2841, 3309, 2808, 2918, 2656, 2531, 2544, 2310, 2890, 2412, 2448, 2913, 2409, 2025, 3510, 3366, 2646, 3348, 2986, 2996, 2762, 2445, 2781, and 2831.

Part of Palamau district (latitude 23°45'N to 24°00'N and longitude 84° 00'E to 84°15'E) : 757, 860, 844, 1143, 834, 1371, 1128, 878, 1252, 1355, 1155, 1096, 1094, 1671, 1070, 1329 and 1634.

In which of these above regions does the surface height possess more varied relief character?

3.3 In Exercise 3.2 above, compare the heights of the two regions by quartile deviation in determining which of them possesses a more varied height? Compare the quartile deviation index with the coefficient of variation index.

3.4 In a geomorphological study on accordance of erosion surfaces from two adjoining regions of parts of Udaipur district, the following spot heights (in feet) were taken as samples.

Region 1 (latitude 24°40'N to 24°45'N and longitude 73°30'E to 73°37'30'E) : 3110, 2786, 2792, 2989, 3271, 2848, 2969, 2891, 3278, 2966, 2727, 2979, 2736, 2953, 2825, 2467, 2520, 2871, 2845, 2576, 2848, 2690, 2601, 2494, 2812, 2999, 2897, 2408, 2907, 2615, 2986, 2710, 2746, 2641, 2766, 2894, 2874, 2192, 3143, 2516, 3189, 2480, 3028, 2451, 2530, 2999, 2405, 2218, 2310, 2064, 2395, 3058, 2110, 2398, 2599, 2297, 2300, 2156, 2838, 2162, 2395, 3045, 2270, 2116, 2418, 3281 and 2753

Region 2 (latitude 24°35'N to 24°40'N and longitude 73°37' 30"E to 73°45'E) : 2825, 2005, 2044, 2015, 2005, 2074, 2021, 2625, 2064, 2713, 2057, 2832, 2467, 3025, 3078, 2031, 1975, 2001, 2044, 2011, 2398, 1919, 2405, 2635, 2543, 1886, 1863, 1805, 2415, 2461, 1985, 2543, 2500, 2749, 2320, 2536, 1962, 1946, 1939 and 1857.

Assuming that the distribution of this data is a sample set representing a normal distribution, calculate the skewness and kurtosis of both the data sets and interpret the nature of their distributions.

3.5 Of the two sets of samples, one set has 50 items with mean 15 and standard deviation 3. The total number of samples of the whole group is 150 with mean 12 and standard deviation 3.46. Find the standard deviation of the second group.

3.6 Find the missing information from the following data:

	Set 1	Set 2	Set 3	Whole group
No. of items	50	...	90	200
Mean	113	...	115	116
Standard deviation	6	7	...	7.5

3.7 A catchment area has 7 rain-gauge stations. If in a year the rainfall recorded in these rain-gauge stations are 120, 110, 140, 108, 165, 150 and 105 cm, estimate the minimum number of additional rain-gauge site the area should have for estimation of rainfall with 5 per cent error.

3.8 Calculate the moment measures μ_r with $r = 1$ to 4 for the given data : 10, 9, 8, 7, 7, 6, 5, 4, 3 and 2.

3.9 Following is the annual rainfall distribution data at a place given year-wise from 1901 to 1950 (the data is in mm) :- 1180, 371, 1108, 1369, 1924, 929, 1672, 1125, 460, 1140, 1177, 1543, 1055, 935, 1309, 1681, 1233, 1584, 971, 1069, 1977, 914, 1777, 828, 488, 1150, 1172, 427, 2064, 1590, 1769, 1378, 978, 919, 1337, 1606, 2030, 1051, 398, 1518, 1015, 943, 1298, 1922, 896, 773, 657, 2535, 1716, and 1744

(a) Arrange the above data in ascending order. Calculate the mean and median of the arranged data. Guess the mode of the data.

(b) Starting at 350 mm outline the arranged data in (a) above into a frequency table for a class interval of 200 mm. Calculate the mean, median and standard deviation of the frequency data. Find out the skewness measure of the data.

4

Probabilistic Treatment in Geo-science Systems I
(Normal Distribution)

In the earlier chapters, we were concerned with descriptive statistical techniques for collecting and summarising the raw data in the form of a frequency distribution and the various measures of central tendencies and dispersions. Now we proceed to approximate the frequency distribution by the "concept of probability theory" and to fit the frequency curve to a theoretical curve by some well-known forms of *probability distribution*. In this chapter we start with the 'concept of probability' which itself is central to many of the variables in geo-science systems for which chance seems to play a significant role.

4.1 INTRODUCTION TO PROBABILITY

Probability theory provides the fundamental concept in statistics. All statistical tests involve the calculation of probabilities, either directly or indirectly because statistics and statistical hypotheses are concerned not with 'certainties' but with 'probabilities': they are never said to be true or false but rather the probability that they are true or false is stated. Hence probability can be viewed as 'partial information either known or presumed to be known prior of an event'. In summary statistics, the mean and standard deviation are looked upon as descriptors of the past data collected but in terms of probability, our concern lies with both past and future data. Hence although there have been many definitions of statistics, it perhaps may best be considered as the determination from the possible. In any circumstance there are a variety of possible outcomes ranging from counting the "head" or "tail" i.e. 2 outcomes in a coin flipping experiment to 6 outcomes in throwing a dice. All these have an associated probability (like the probability of obtaining a head is 1/2 or 50%) which describes their frequency of occurrence. From an analysis of probabilities associated with events, future behaviour or past states of the event under study may be estimated.

Problems which involve an element of chance or uncertainty are usually termed "experiments" : games of chance like tossing a coin, casting a dice or events like occurrence of river flooding in a year are all experiments because on any single observation they may result in any one of a number of possible outcomes, for example, the coin may show either head or tail, the dice may show any number between 1 and 6, the river may either flood or stay within its banks. We refer to each of these individual outcomes as an "event" and total collection of all the possible outcomes of an experiment as the "sample space", *S* of the experiment. This *sample space, S* is a totally abstract concept not to be confused with geographical area. Thus the sample space, for casting a six-sided dice contains the numbers 1 to 6, i.e. $S = [1, 2, 3, 4, 5, 6]$. The members are "simple events" described by a single characteristic of any event. In the example of a six-sided dice each number is a simple event. Again a subset of the sample space can be a "compound event". For instance, the event that the sum of the two numbers showing after the dice have been thrown is 7 involves a compound event which is a subset of six sample points: (6,1), (5,2), (4,3), (3,4), (2,5) and (1,6). Again if an event like discharge of a river over a year is regarded as an *experiment*, then the sample space would represent all possible discharges recorded over the year and the annual peak discharge over the year can be regarded as one element or *elementry event* in the sample space. Again out of this sample space of river discharges, if we include all discharges say over mean discharge of the year, they would represent *compound events*. The probability theory has at its roots these three fundamental terms: sample space, elementary events and compound events. These three terms are conveniently included in Venn diagram (Fig 4.1).

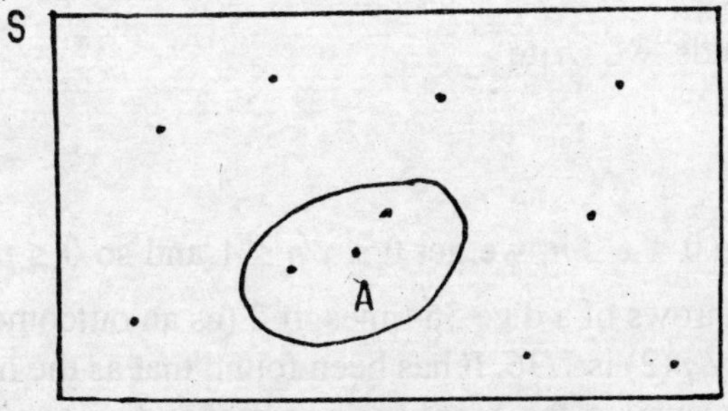

• Elementary event
S Sample space (ξ Universal set)
A Compound event

Fig. 4.1 : Venn Diagram

Tossing a coin is to count the likelihood of a specific occurrence head, "H" or tail "T". This is the simplest experiment in probability theory. It is a 'random experiment' because a random experiment is an activity that results in one and only one of possible outcomes, it can be repeated any number of times and the outcome, head, H or tail T would be unknown before each toss. The set of all possible outcomes, that will occur in tossing a coin *n* times is the sample space. Thus for tossing a coin one time, there are two probabilities: head, H or tail, T. For the second toss, for each of the H or T of the first toss, there are two probabilities which gives $2 \times 2 = 4$ possibilities for the first two tosses; for the third toss, for each of the 4 probabilities in the second toss there are $4 \times 2 = 8$ or equal to 2^3 possibilities. Thus if a coin is tossed n number of times, the total number of possibilities, p is

$$p = 2^n \qquad \text{... (4.1)}$$

that is, for a random experiment repeated n times,

Total number of possibilities/permutations = Number of possible outcomes in one trial raised to the number of times, n the experiment is repeated

The classical or, *a priori* view of probability is that all events in a sample space are equally likely to occur in any one trial or an experiment, for example, the probability of a H to occur in tossing a coin is 1/2, similarly, the probability of a dice showing 1 is 1/6. This definition of probability assumes that all random events can occur equally likely and in mutually exclusive ways. But probability is also defined as the event's long run frequency of occurrence. This is 'relative frequency view of probability' in which probability of an event (or, outcome of a random experiment) is a ratio between the actual number of times a particular outcome of an event is recorded and the total number of possible events. Thus

Probability (p) of an event (x) =

$$\frac{\text{No. of times the mutually exclusive and equally likely event occurs or can occur}}{\text{No. of times the random experiment is repeated or total No. of possible event}} \qquad \text{... (4.2)}$$

Now, if an event (x) occurs x_i times ($x_i \leq n$) we define the probability of the event as the value to which $\frac{x}{n}$ tends, we write

$$p(x) = \lim_{n \to \infty} . \frac{x}{n} \qquad \text{... (4.3)}$$

As we must have $0 \leq x \leq n$, we get $0 \leq x/n \leq 1$ and so $0 \leq p(x) \leq 1$

For example, in throws of a dice 36 times, if 2 (as an outcome or an event) is observed 7 times, the probability, $p(2)$ is 7/36. It has been found that as the number of trials increases, $n \to \infty$ the relative frequency of the event (x_i is corresponding to n) comes closer to the true probability as the event's long run frequency of occurrence. This can be verified that by tossing a coin a number of times, the probability that the head, H will show will appear to settle down towards a limiting value which is 0.5, since either of the two events, the head, H or tail, T of the coin will occur. Therefore, the 'total probability' of the head, H or tail, T to occur will be $p(\text{H}) + p(\text{T}) = 1$, where in a single toss of the coin, the range of probability of the head, H or tail, T is within 0 and 1 and we write the probability, p of an event (x_i) to occur is a number $0 \leq p(x_i) \leq 1$. If p is near 1, the event is very likely to happen, on other hand p near to 0 *indicates* that the event is unlikely to happen. We think probability $p(x_i)$ as the 'proportion of times' that the event will occur, if the experiment is repeated a large number of times; in other words, $p(x_i)$ is the event's "long-run relative frequency of occurrence". When the outcomes of a random experiment are described by a 'discrete random variable' (that is, a set of real numbers), there can be a complete listing of all possible outcomes along with the probabilities of each outcome. This listing is the *discrete probability distribution* which represents the frequency with which one would expect events (or values) to occur in the long run. Since all possible outcomes are included in the probability distribution the sum of these probabilities of all possible 'mutually exclusive outcomes' (or, 'disjoint events' that have no basic outcomes in common) should total 1, that is if x_i is an event, $\Sigma p(x_i) = 1$ and since the probability of an event is non-negative, so it is $0 \leq p(x_i) \leq 1$.

When a random variable as the outcome of random experiment of an event assumes any value within some interval of real numbers it is called a 'continuous random variable', for example, rainfall at a place can have a value 48.0054 cm. Since the amount of rainfall at a place can be infinite, so the likelihood of any one value like 48.0054 will be extremely small.

Hence with continuous probability distributions, statements are usually made regarding the probability that the random variable will assume a value within a defined interval e.g. annual rainfall between 48 and 49 cm. rather than a specific value. Note the above definition of probability given in Eq. (4.3) is known as *a priori* or "theoretical" definition of probabilities like tossing of a coin head or tail. But when the probability of an event is a ratio based on experience of over a very large number of trials rather than on a rational consideration of equally likely alternatives of occurrence of head or tail in tossing of a coin, it is known as "empirical probability". While these definitions of *a priori* and *empirical probabilities* seem fundamentally different, they are usually just different ways of looking at the same thing. If the alternatives of the event are mutually exclusive and equally likely, and if enough trials are given, we will find that both the *a priori* and *empirical definitions* yield exactly the same probability for the event.

4.2 AXIOMS OF PROBABILITY

The probability that we have considered is the simple probability of an event described by a single characteristic. Simple probability is also called the "marginal probability" in which the information about the total number of times the particular event occur comes from the appropriate margin of a contingency table e.g. in sex-wise literacy data the probability of literates in a data, comes from the margin of the table. Now for manipulating probabilities two or more, we require few basic rules or axioms. We have already defined (i) the total probability axiom, i.e. in a single trial on an experiment the total probability of events (x_i) is $\Sigma p(x_i) = 1$ and (ii) the non-negativity axiom i.e. $p(x_i) \geq 0$ and hence it is to be defined on a scale between 0 and 1 i.e. $0 \leq p(x_i) \leq 1$. Now we define the addition axiom, the multiplication axiom and the conditional probability axiom which are having applications in geo-science systems particularly in climatology and in hydrology. The axioms are stated below in the language of 'algebra of sets'.

(1) The probabilities of a number of independent and mutually exclusive events is the sum of the probabilities of the separate events like $p(x_i)$, $p(x_2)$ etc. It is the *addition axiom* so that the probability that either x_1 or x_2 happens, when written in the language of sets is the 'union' of the sets for which there is no overlap in the sample space between the subset x_1 and subset x_2.

$$p(x_1 \text{ or } x_2) = p(x_1 \cup x_2) = p(x_1) + p(x_2) \qquad \text{... (4.4)}$$

where $\cup$ is synonymous with the addition of probabilities.

For example, the probability that in a single casting of a six-sided dice, either a 4 or a 5 will occur is obtained from the addition axiom as $p(x_4 \cup x_5) = p(x_4) + p(x_5) = \frac{1}{6} + \frac{1}{6} = \frac{1}{3}$.

Two simple results follow from this axiom.

(i), this addition axiom can be used to define the idea of a complimentary event i.e. the specified event which does not occur. For example, in a particular year a river may flood, f or not flood, f' and because these two events are mutually exclusive (since their probability does not contain any outcome), we can use the addition axiom to write $p(f \cup f') = p(f) + p(f')$. Furthermore, because the union of flooding with not flooding $(f \cup f)$ is certain to occur, so it follows that the sum of the individual probabilities must be equal one. Hence we can write:

$$p(f') = 1 - p(f). \qquad \text{... (4.5)}$$

(ii), since $\Sigma p(x_i) = 1$, so the addition rules of $p(x_1 \cup x_2 \cup \ldots) = \Sigma p(x_i) = 1$ means that in any single trial one of these elementary events, x_i must occur as the outcome.

(2) The probability of two or more independent events, (but not mutually exclusive) occurring simultaneously or in succession as a result of r trials on an experiment is known as *compound events probability*. In this case the "multiplication axiom" for independent events is used which states that the independent events x_1 and x_2 or x_r happen is the product of their individual probabilities. Symbolically,

$$p(x_1\, x_2 \ldots \text{ and } x_r) = p(x_1 \cap x_2, \ldots, \cap x_r) = p(x_1) \times p(x_2) \,..\, p\,(x_r\,) \qquad \text{... (4.6)}$$

where $\cap$ is called the 'intersection' or 'joint probability' and is read 'the probability of x_1, x_2, ..., and x_r'.

For example, the probability of casting three 6's using a dice is

$$p(x_6 \cap x_6 \cap x_6) = p\,(x_6)\,.\,p\,(x_6)\,.\,p\,(x_6)$$

$$= \frac{1}{6} \times \frac{1}{6} \times \frac{1}{6} = \frac{1}{216}$$

Similarly, if a drought occurs 5 times in a 100-year record then the probability of drought (as an independent event) in two successive years is $p\,(x_1 \cap x_2) = \left(\frac{5}{100}\right)\left(\frac{5}{100}\right) = (0.05)$ $(0.05) = 0.0025$.

(3) The general axiom for the union of probabilities of any two dependent events which are not mutually exclusive or independent events is a more general form of the addition axiom. We have discussed above the compound event whose constituent events are independent. For example, if we like to find the probability of drawing two consecutive aces, it will be $\frac{4}{52} \cdot \frac{4}{52} = \frac{1}{169}$ provided we specify that the first card dealt has to be reshuffled into the deck before the second card was dealt. But if the first card is not replaced in the pack of cards before the second draw, then the probability of drawing the successive ace in the second draw is out of 3 remaining aces and 51 remaining cards, giving a chance 3/51. The probability of the compound event, drawing two aces without replacement is the product of the unconditional probability of obtaining the first ace multiplied by the conditional probability of obtaining the second ace:

$$\frac{4}{52} \times \frac{3}{51} = \frac{1}{221}$$

Formally, two events are said to be dependent if the possibility of one of the events occurring depends on the occurrence of the other event. Dependence is thus the antithesis of independence and arises when the multiplication axiom for independent events (as in Eq. (4.6) above) is seen to be incorrect, that is, when

$$p(x_1 \cap x_2) \neq p(x_1) \,.\, p(x_2)$$

To explain this addition rule for any two events that are neither *mutually exclusive* nor *independent* let us have example of flooding by a river which may or may not be associated with the avulsion of the river banks. If in a flood-prone area, the relative frequency of river flooding is 0.20 and of river avulsion is 0.25 and the chances of river avulsion when there is a flooding by river is 0.50, the last statement is a *joint probability* – it signifies the 'joint occurrence of events' and is usually written as *p* (river avulsion/flood). Thus the consideration of probabilities which are not mutually exclusive is 'conditional probability'. Hence the joint probability of the event x_1 when it is assumed that the event x_2 has happened is a 'multiplication axiom' for the dependent events. We can now give a general formula for the probability of compound events regardless of the independence of the constituent events which is the product of the unconditional probability of obtaining the first event multiplied by the conditional probability of obtaining the second. This general multiplicative rule is written as

$$p\,(x_1 \cap x_2) = p\,(x_1) \,.\, p\,(x_2 \mid x_1) \qquad \text{... (4.7)}$$

where $p(x_2 | x_1)$ is read as the probability of x_2 given x_1.

It reads if x_1 and x_2 are dependent events, the probability that they both occur is the product of probability of x_1 and the conditional probability of x_1 when x_2 has happened. Note if we have to find the joint probability of the event x_2 occurring when it is assumed that the event x_1 has happened, the expression in Eq. (4.7) will read *vice versa* that is $p(x_2 \cap x_1) = p(x_2) . p(x_1|x_2)$. ... (4.7a)

It is also possible to have a problem requiring the probability of either the event x_1 or x_2 when they are not mutually exclusive. Since the events are not mutually exclusive events we cannot obtain the probability by the addition axiom of Eq. (4.4) because if we do so the last event will get included twice. Hence we must subtract a correction factor i.e. the conditional probability of one event given the other event.

Now the addition axiom for the union of probabilities of any two dependent events becomes.

$$p(x_1 \cup x_2) = p(x_1) + p(x_2) - p(x_1 \cap x_2)$$
$$= p(x_1) + p(x_2) - p(x_1) .p(x_2|x_1) \qquad \text{... (4.8)}$$

This is the probability of disjunctive event x_1 or x_2 illustrating the additive rule. Hence going back to the previous example, the probabilities of occurrence of either the flooding or the avulsion or both flooding and avulsion is written as x_1 and x_2 in Eq. (4.8).

$$p\,(x_1 \cup x_2) = p(x_1) + p(x_2) - p(x_1) . p(x_2|x_1)$$
$$= 0.20 + 0.25 - (0.20)\,(0.50) = 0.35$$

The addition rule take the probability of flooding $p(x_1)$ and adds it to the probability of avulsion, $p(x_2)$. However, the joint probability of occurrence of flooding and avulsion together has been double counted, once in $p(x_1)$ and also in $p(x_2)$ and hence it has to be subtracted to get the correct result.

We began this section (on axioms of probability) with the addition axiom for mutually exclusive events and through the multiplication axioms for the successive occurrences of compound events (both when they are independent and dependent), to end finally with the more general case of an additive rule to apply when events are not mutually exclusive. As shown in the example above, if we have mutually exclusive events, we can still use Eq. (4.8) but the final term will be zero i.e. by definition, the occurrence of both X_1 and X_2 on the same trial has a probability of zero.

In a study of daily rainfall at a place it is found that the month of July which as a rainy month in the place is having the following transition probabilities regarding the weather of two successive days:

(a) a rainy day following a rainy day = 0.45
(b) a rainy day following a dry day = 0.28
(c) a dry day following a rainy day = 0.56
(d) a dry day following a dry day = 0.70

Calculate the probability of one day rainy and the other dry following a rainy day in the month of July at the place.

We apply here the multiplication axiom for the dependent events as given in Eq. (4.7): the probability of a rainy day following the initial rainy day is $p(x_1) = 0.45$ and the probability of a dry following a rainy day is $p(x_2 \mid x_1) = 0.56$ and we have

$$p(x_1 \cap x_2) = p(x_1).\, p(x_2 \mid x_1) = (0.45)\,(0.56) = 0.252$$

Sometimes we are having events which are "joint events" also. *Joint event* is an event that has two or more characteristics. For example, the probability of a 7 total, given that 3, 4 or 5 occurs on casting a dice is the sum of the probabilities of three events : (2, 5), (4, 5) and (3, 4) so for this event (x_1) of 7 total, we are given information about another event (x_2) in 3, 4 or 5. In this case, the probability involved is referred to as 'conditional probability' ($x_1 \mid x_2$). This conditional probability is defined as

$$p(x_1 \mid x_2) = \frac{p(x_1 \text{ and } x_2)}{p(x_2)} \qquad \text{... (4.9)}$$

where $p(x_1$ and $x_2)$ is the joint probability for x_1 and x_2 and $p(x_2)$ is an unconditional (or marginal probability) for x_2.

Thus in the above example, we have the joint probability

$$p(x_1 \mid x_2) = p\,(7 \mid 3, 4 \text{ or } 5)$$
$$= p\,[(2, 5) \mid x_2] + p[(3, 4) \mid x_2] + p\,[(4, 3) \mid x_2]$$
$$= \frac{2}{36} + \frac{2}{36} + \frac{2}{36} = \frac{1}{6}$$

From Eq. (4.9), definitions of independent and dependent events follow. The two events defined in the same sample space are said to be independent if the following relation holds

$$p\,(x_1 \mid x_2) = p\,(x_1) \qquad \text{... (4.10)}$$

which implies that

$$p(x_1 \cap x_2) = p\,(x_1)\,.\,p\,(x_2) \qquad \text{... (4.11)}$$

This rule for the multiplication of the probabilities of independent events to obtain the probabilities of their intersection is important. If this relation (Eq. 4.11) do not hold, then the events are said to be dependent.

Two further definitions are relevant to be discussed in the axioms of probability : (i) *Probability with replacement* in which if two events occur, the outcome of one can in no way influence the outcome of the other and the events are said to be mutually exclusive e.g. the probability of getting a head is the same (equal to 1/2) everytime, even if 10 heads occur in a row, the probability that the next toss will produce a head is still 1/2. (ii) *Probability without replacement* is one in which the result of one trial does influence the probability of the next trial. This happens because in this case the sample which is drawn in a trial is not put back in the lot and after each draw the next draw is made "without replacement", this latter one, we have discussed earlier with reference to the product of unconditional probability of obtaining the first with the conditional probability of obtaining the second.

4.2.1 Probability Density Functions

The discussion up to this point has been with the axioms of probability in term of any random phenomenon for which the sample space of outcomes is finite and completely specified. Probability theory, however, is developed in far more general term. Let us consider a random variable x, which takes on numerical values ($x_1, x_2, x_3, ..., x_n$) to form a frequency distribution by class intervals. To facilitate the analysis further, a 'relative frequency distribution' is formed by dividing the frequencies in each class of the frequency distribution, f_i by the total number of observations, n. In Fig. 4.2 the relative frequency proportions of the class intervals are plotted as histogram. The histograms in Fig 4.2 (a) point out a symmetrical distrbution for an unimodal data, in (b) for a bimodal data. Fig 4.2 (c) and (d) show the skewed distributons.

These frequency proportions $p(x_i)$ provide a mathematical function $f(x_i)$ defined for real numbers, which give the probability for any particular outcome. Now all the values for the function $f(x_i)$ is accomodated into "cumulative frequency function", $F(x_i)$. Thus we can have the 'probability distribution function', $F(x)$ as the theoretical limit of the cumulative frequency function. On this probability distribution function, $F(x)$ as the sample size becomes infinitely large and the class interval infinitely small so that $dF(x)$ tends to zero and we can calculate slope at any point on this probability distribution. The probability of an event $p(x_i)$ is now an area under a curve known as "probability distribution curve" for which $f(x)dx$ is a *probability density function*. This curve is the value of the slope of the distribution function for a specified value of x. The relationship between the probability distribution function, $dF(x)$ and the probability density function, $f(x)dx$ can be described by saying that the probability density function $f(x)$ represents the "density" or concentration of probability of outcomes in the interval $[x, x + dx]$ expressed as

$$\int_x^{x+dx} f(x)\, dx$$

So a probability distribution represents the complete listing of all possible outcomes of random experiment with which one would expect values (or events) to occur in the long run and this total area under the probability density function curve is

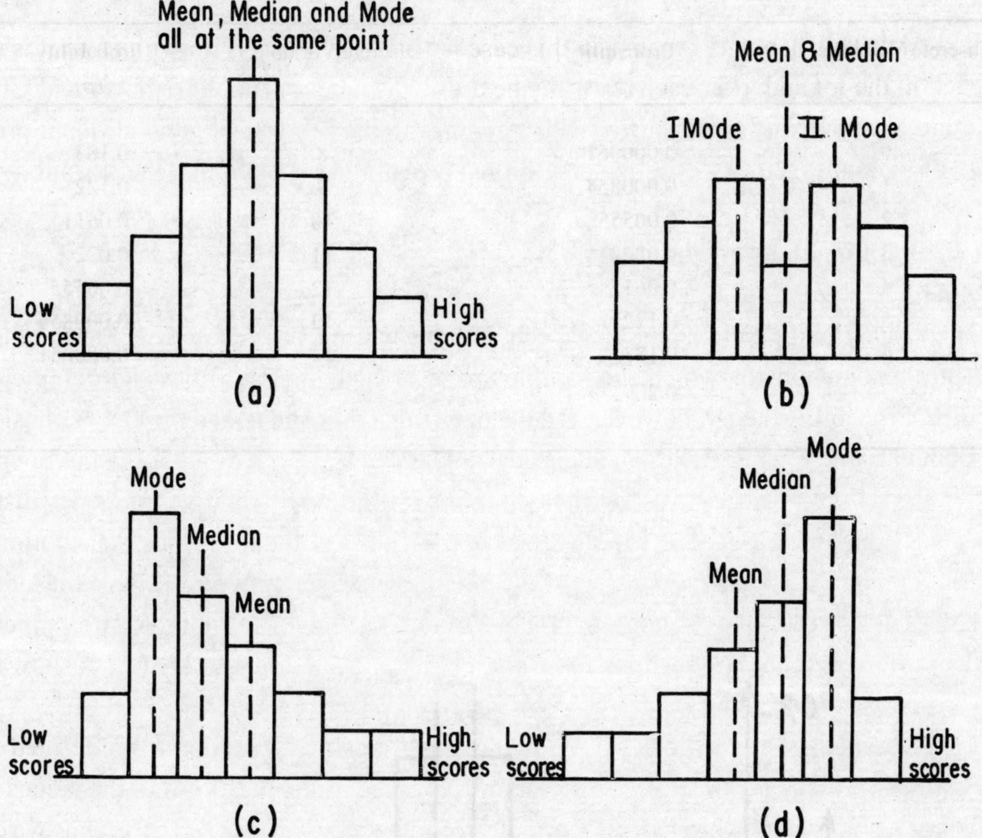

Fig. 4.2 : Histogram of Relative Frequency Proportions

$$\int_{-\infty}^{+\infty} f(x)\, dx = 1$$

that is, the total probability equal to one. It is a theoretical frequency distribution, mathematically derived and it is perfectly regular. When the outcomes are described by a discrete random variable, the probability distribution lists all possible values of the random variable and the probabilities of all mutually possible outcomes. Figure 4.3 shows the distribution and density functions for a discrete random variable which can be compared with the corresponding ones for continuous random variable. Fig. 4.3 gives the probability distribution for the discrete values from 0 to 14.

Discrete Variable	Probability (p)	Discrete Variable	Probability (p)
0	0.000061	8	0.183
1	0.000854	9	0.122
2	0.00555	10	0.0611
3	0.0222	11	0.022
4	0.0611	12	0.00555
5	0.122	13	0.000854
6	0.183	14	0.000061
7	0.209		

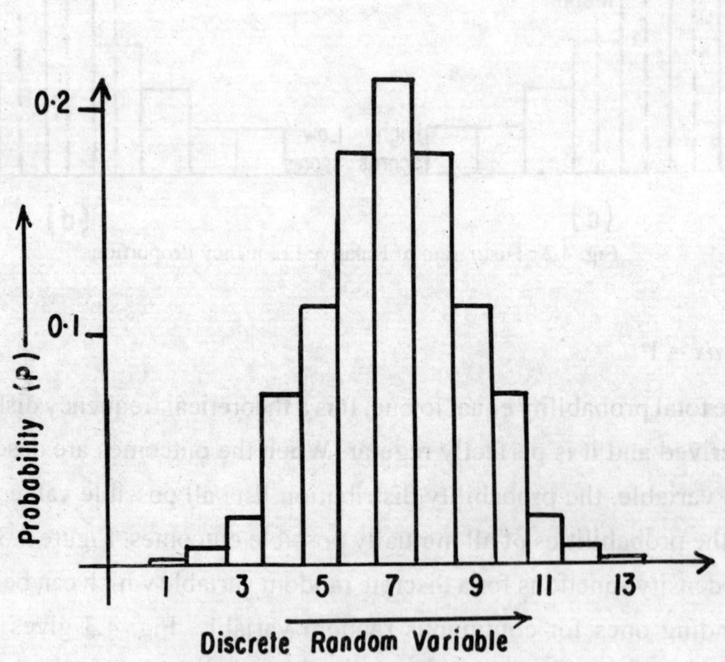

Fig. 4.3 : Distribution and Density Functon for a Discrete Random Variable

In geo-science system depending on the measurement process on various phenomena of interest, both types of distributions are having use since the task involved is to find the theoretical frequency distributions which fit the data most closely. The mathematical rationale behind each theoretical frequency distribution is however strictly governed by the nature of the probability space within which the events occur, independent or dependent state of the events and finally the size of samples.

In statistics, there exists three theoretical frequency distributions namely "normal", "binomial" and "Poisson" distributions. In this chapter we will discuss the normal distribution which will be followed by binomial and Poisson distributions in the next two chapters. Normal distribution is the only continuous variable distribution whereas the binomial and Poisson are concerned with discrete variables.

4.3 NORMAL DISTRIBUTION

The word 'normal' is a generic name applied to a well defined continuous probability distribution which is also called 'Gaussian' after the name of the mathematician Gauss. It is a mathematically derived theoretical distribution (rather than from actually conducting research and gathering data) to which a large number of naturally occurring distributions will approximate in the long run. Thus this is an ideal distribution so that many of the real world data can be approximated by it. This distribution is defined by it's mean and standard deviation – the mean is the central value having 50 per cent of the values of the observation on it's either side and standard deviation is the basic unit of measurement along the *X*-axis of the normal curve (to which generally a normal distribution is referred to). In the real world, many actual distributions like rainfall data for any raingauge station collected over a large number of years (assuming no climatic change) tends to develop normal frequency distribution with a "bell-shaped" symmetrical curve. This symmetry means that the height of the normal curve is the same if one moves equal distances to the left and right of the mean. The highest frequencies in this curve is around the mean and the frequency decreases as distance from the mean increases. This coincidence of actual rainfall distribution with the theoretical distribution happens because rainfall data over a long time period for any location appears to be affected by two groups of factors : one dominant group (e.g. latitude, longitude, nearness to sea, etc.) that acts consistently and causes, clustering about the mean and another group of factors: (e.g. depressions, storms etc.) that act randomly, to cause symmetrical variation about the mean. Following this example, it can be said that whenever there appears to be two groups of factors affecting a variable – one dominant group acting consistently, and another randomly, then there will be grounds for expecting that a frequency distribution of values will, in the long run, approximate to a *normal distribution*. Thus for many effects in nature which are a compound result of a number of separate influences, it follows that the measure of these effects are often normally distributed when the number of observations are high. Hence, the rainfall and temperature data, height of erosion surfaces, roundness of gravels on terraces/beach, crop production (under uniform conditions) are few of the many variables of geo-science systems which are likely to show a tendency towards a normal distribution provided the set of data is large. Therefore, there are many real world situations for which the actual distributions resemble the theoretical distribution of the normal curve making itself useful for application to actual research situations. Again there are phenomena which do not conform to the theoretical notion of normal distributions. For example, probability distributions encountered in river discharge studies do reveal skewed character rather than symmetrical to be normal. In these cases, the normal distribution function cannot be used as model of the data obtained and we have to look for finding another probability distribution which can fit to it. Figure 4.4 shows a relative frequency histogram and a normal curve superposed on it.

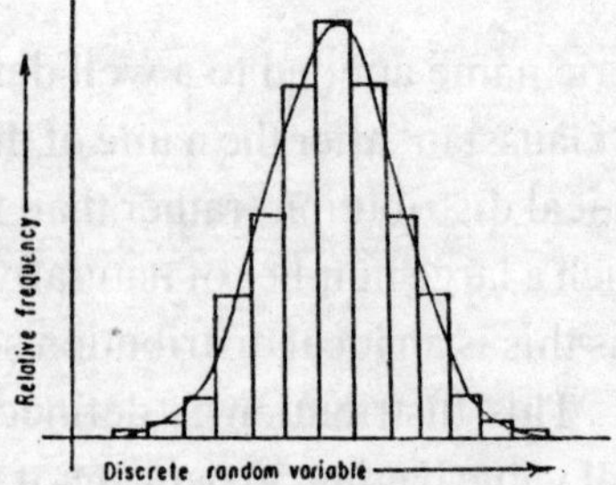

Fig. 4.4 : A Relative Frequency Histogram and Its Normal Curve

4.3.1 The Mathematical Definition of Normal Distribution

The mathematical model or equation for normal distribution represents a probability density function, *f(x)* which is the height *(Y)* of the normal curve above the base line at a given point (x_i) along the measurement scale of the random variable, *x* itself. The model used to obtain the desired probabilities is

$$Y = f(x_i) = \frac{1}{\sqrt{2\pi.\sigma_x}} . e^{-1/2 \cdot \left(\frac{x_1 - \mu}{\sigma_x}\right)^2} \qquad \text{... (4.12)}$$

where *e* is the exponent equal to a mathematical constant, 2.71828, π is the mathematical constant, 3.14159, μ and σ are the population mean and standard deviation respectively, x_i is any value of the continuous random variable, where $-\infty < x_i < +\infty$.

From Eq. (4.12) it will be apparent that since *e* and π are mathematical constants, the probabilities of the random variable *x* are only dependent upon the two parameters of the normal distribution – the population mean, μ and the population standard deviation, σ.

Everytime we specify a particular combination of μ and σ, a different normal probability distribution will be generated each having its own set of probabilities, so that they will be located at different positions along the horizontal scale (i.e. the value scale). Figure 4.5 illustrates this for three different normal distributions. Note the distributions A and B have the same mean (μ = 30) but have different standard deviations of 5 and 10 and thus they have the same location but A has a narrower spread. On other hand, distributions A and C have about the same standard deviation 2 and 5. Hence A and C have about the same form and the same location. Therefore, if the mean differs in several normal distributions but the standard deviation is the same, the slope of the normal curve will be the same in all cases but related to a different point on the magnitude scale, conversely, as it happens in Fig. 4.5 if the mean remains the same, while the standard deviation differs, different curves are indicated around the same central value, the mean.

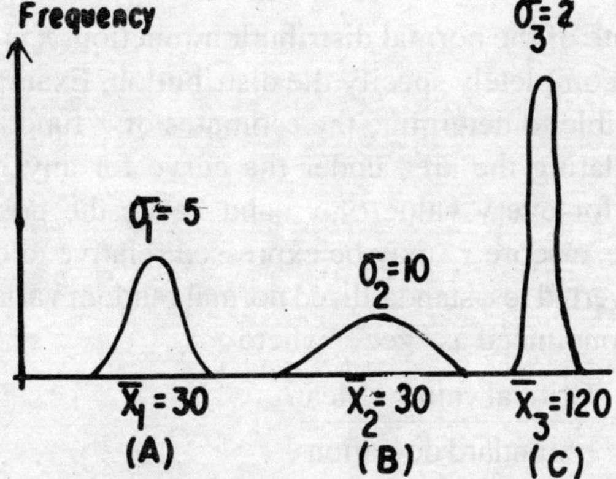

Fig. 4.5 : Three Different Normal Distributions

For given μ and σ, the substitution of several x_i values in the mathematical expression in Eq. (4.12) will yield estimates of Y which allow the normal curve to be constructed. Following is an example.

Example 4.1. If in a rainfall distribution data the mean and standard deviation are 90 cm. and 19 cm. provide the probability density function for given values of 65, 85 and 90 cm. of rainfall.

For calculating Y in $f(x)$ in the above Eq. (4.12), we have

x_i	$\frac{1}{\sqrt{2\pi}.\sigma}$	$\frac{x_i-\mu}{\sigma}$	$\frac{1}{2}\left(\frac{x_i-\mu}{\sigma}\right)^2$	$-\frac{1}{2}\left(\frac{x_i-\mu}{\sigma}\right)^2$	e^{col5}	Y
(1)	(2)	(3)	(4)	(5)	(6)	(7)
65 cm	0.02101	–1.31579	0.86565	–0.86565	0.42078	0.00884
85 cm	"	– 0.26316	0.03463	– 0.03463	0.96596	0.02029
90 cm	"	0.00000	0.00000	0.00000	1.00000	0.02101

The above values of Y are known as the "probability integrals" for the successive values of the variate, x_i.

4.3.1.1 Standard Scores and the Normal Curve

An important attribute of the normal distribution function, $f(x)$ in Eq. (4.12) is that the two parameters μ and σ completely specify the distribution. Example 4.1 shows that if we know μ and σ, it is possible to determine the estimates of Y function for constructing the normal curve and calculating the area under the curve for any interval of x_i. But the computation is tedious for every value of x_i and hence the position of any particular observation, for example, a score x_i may be expressed relative to other scores in its set of data by getting itself converted to a standardized normal random variable known as "standard score" or a "z score" (pronounced as "zee") where

$$z = \frac{x_i - \bar{x}}{\sigma} = \frac{\text{Critical value} - \text{Mean}}{\text{Standard deviation}} \qquad \text{... (4.13)}$$

Now, if we transform all of the raw scores in a distribution to standard (or z–) scores, we obtain a new standardized distribution that will always have a "mean, $\bar{x}$ equal to zero" and a "standard deviation, σ equal to one". This is an extremely useful transformation known as *standardization* which result in new values for the individuals that not only have zero mean but are measured in units of standard deviations. This is done simply by subtracting the mean of the distribution from each observation and dividing by the standard deviation of the distribution. The reason is that the mean of the z scores, $\bar{z}$ can be written as

$$\bar{z} = \frac{\Sigma z}{N} = \frac{\Sigma(x_i - \bar{x})/\sigma}{N} = \frac{1}{N} \cdot \frac{\Sigma(x_i - \bar{x})}{\sigma} \qquad \text{... (4.13a)}$$

But since $\Sigma\,(x_i - \bar{x})$ is always zero, so the mean of z should be always zero. Again for σ^2 we can write $\sigma_z^2 = \frac{\Sigma(z-\bar{z})^2}{N}$ but since $\bar{z}$ is always zero, so the equation for σ_z^2 reduces to

$$\sigma_z^2 = \frac{\Sigma z^2}{N} = \frac{\Sigma(x-\bar{x})^2/\sigma^2}{N} = \frac{\Sigma(x-\bar{x})^2}{N.\sigma^2} = \frac{\Sigma(x-\bar{x})^2}{\Sigma(x-\bar{x})^2} = 1.0$$

and hence the standard deviation of z score must likewise be 1.0.

Since the mean of z score is zero so the z score distribution has the advantage that to any point on the x-axis of the standardised normal curve, it's value can be determined by stating how many standard deviations the z score is away from the mean point of the curve. Also, since the standard deviation is one, the numerical size of a standard score indicate how many standard deviations above (+) or below (–) mean of the distribution the score is. Figure 4.6 shows a standard normal curve.

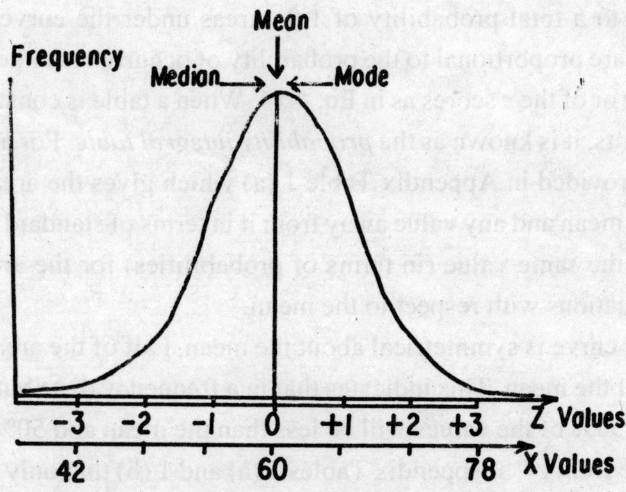

Fig. 4.6 : A Standard Normal Curve

It requires to be mentioned that the above z score transformation formula Eq. (4.13) can be substituted in Eq. (4.12) to give the probability density function of a standard normal variable z as

$$f(z) = \frac{1}{\sqrt{2\pi}\,.\sigma}.e^{-(1/2)z^2} \quad \text{... (4.14)}$$

by which we can always convert any set of data to its standardised form and then determine the probabilities. Finally it is emphasised that though the common uses of the z transformation of data is in context with the standard normal distribution, but the z scores does nothing to alter the shape of the raw score distribution. If the original data is skewed, the derived z scores will also be skewed.

4.3.2 Finding the Areas under Normal Curve

All the occurrences which contribute to the mean value are recorded within the area enclosed by any normal curve and the base line, these being accounted for in terms of both their order of magnitude and the number of occurrences at each such order of magnitude. It turns out that no matter what the values of mean and standard deviation are, giving rise to no matter what shape of the curve, the normal curve will always have a constant area and this will be equal to a total probability of 1.0. Areas under the curve between different points of a variate are proportional to the probability of occurrence either of these points as given in Eq. (4.12) or of the z scores as in Eq. 4.14. When a table is constructed lying below the successive points, it is known as the *probability integral table.* For a normal curve such a table has been provided in Appendix Table 1 (a) which gives the area under the normal curve between the mean and any value away from it in terms of standard z scores. Appendix Table 1 (b) gives the same value (in terms of probabilities) for the areas encountered in other common situations with respect to the mean.

As the normal curve is symmetrical about the mean, half of the area beneath the curve lies on each side of the mean. This indicates that in a frequency distribution approximating the normal curve, 50% of the values will be less than the mean and 50% will be more than the mean. We note from bc 'ı Appendix Tables 1 (a) and 1 (b) that only positive entries for z scores are listed. It is clear that for such a symmetrical distribution with zero mean, the area from the mean to $+z$ will be identical to the area from the mean to $-z$.

For a standard normal distribution, in the area between the curve and the standardised base line, two ordinates are drawn at a distance of one standard deviation, σ away from the mean, both below and above it (i.e. to the left and right of the mean in the normal curve). The area of the central portion so cut off is shaded in the Fig. 4.7. From the Appendix Table 1(a) as well as from column A of the Appendix Table 1(b), we find that if we go one standard deviation to the left or right of the mean, we will have 0.3413 portion of the total area is included between the $\bar{x} - 1\sigma$ (to the left) or $\bar{x} + 1\sigma$(to the right). Therefore twice the area or 68.26% of the occurrences will lie between -1σ and $+ 1\sigma$, i.e. there is roughly a 2 : 1 chance that it will. Similarly, if we were to draw ordinates at a distance equal to 2σ from the mean, the central area so cut off would amount to 95.46% of the total area, i.e. there is roughly a 21:1 chance that a value will lie between -2σ and $+ 2\sigma$. It can be seen from Fig. 4.7 that there is almost no area under the curve beyond 3σ units from the mean.

But from the Appendix Table 1 (a) or 1(b), 99.72% of the occurrences will lie between -4σ and 4σ, i.e. there is only one chance in 10,000 that a value will differ from the mean by more than this amount. Again it can be seen from Fig. 4.7 or Appendix Tables I(a) or I(b) that the form of the normal curve is such that it never really reaches the base line, but so little does it extend beyond 4σ on either side of the mean that we can often ignore these extremes of the curve.

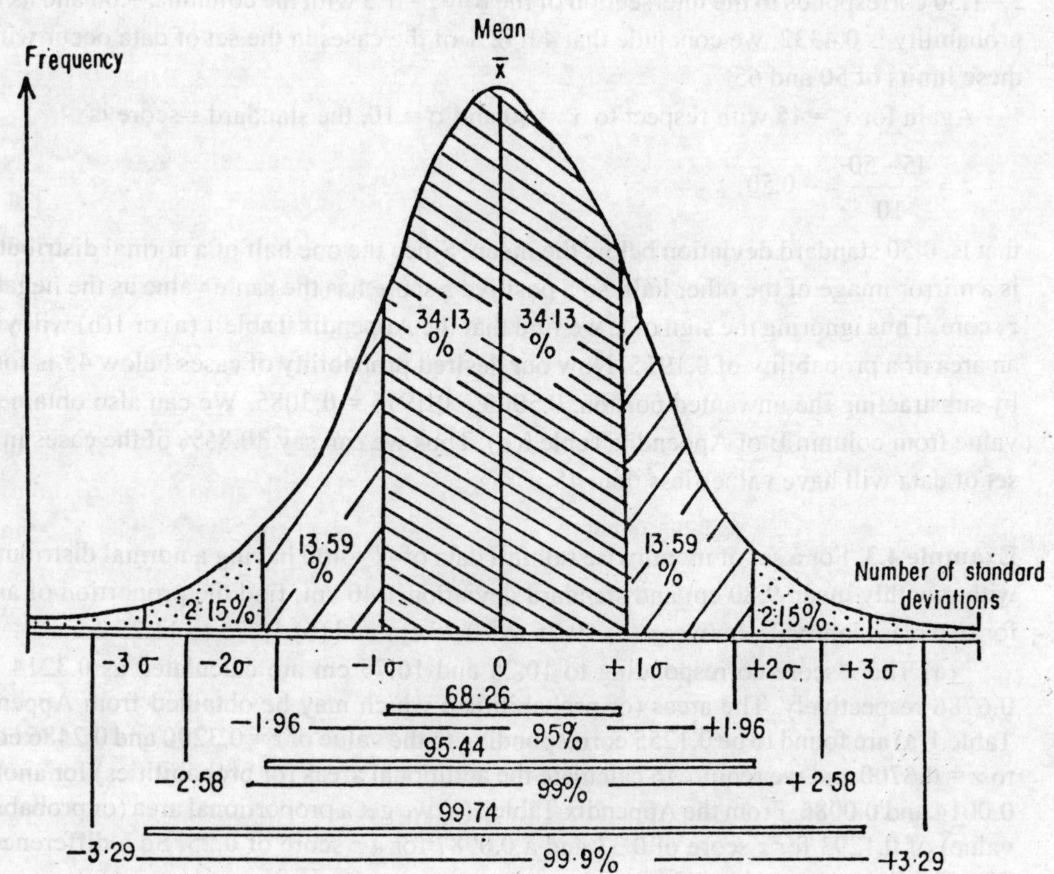

Fig. 4.7 : Areas on a Standard Curve

Example 4.2. Find the proportion of cases occurring within the interval of 50 to 65 with mean 50 and standard deviation 10. Also find out from the same set of data the proportion of cases lying below a particular value 45.

We have, $z = \dfrac{x_i - \bar{x}}{\sigma} = \dfrac{65-50}{10} = 1.50$

Therefore, the standard z score is 1.5 standard deviations from the mean. By consulting the values of z we go down the left-hand margin until we locate the z value of interest (in tenths). In the present example we stop in the row $z = 1.5$. Next we read along this row until we intersect the column from the top that contain the hundredths place of the z value, that is, 0.00. Therefore, in the body of the Appendix Table 1(a), the tabulated probability for $z = 1.50$ corresponds to the intersection of the row $z = 1.5$ with the column $z = .00$ and as this probability is 0.4332, we conclude that 43.32% of the cases in the set of data occur within these limits of 50 and 65.

Again for $x_i = 45$ with respect to $\bar{x} = 50$ and $\sigma = 10$, the standard z-score is

$$z = \frac{45-50}{10} = -0.50$$

that is, 0.50 standard deviation below the mean. Since the one half of a normal distribution is a mirror image of the other half, so a positive z score has the same value as the negative z score. Thus ignoring the sign of z, we find that the Appendix Table 1 (a) or 1(b) will yield an area or a probability of 0.1915. Now our desired probability of cases below 45 is found by substracting the unwanted portion, 0.5000 – 0.1915 = 0.3085. We can also obtain this value from column B of Appendix Table I(b). Thus we can say 30.85% of the cases in the set of data will have values less than 45.

Example 4.3. For a set of monthwise rainfall data of $N = 400$ having a normal distribution with monthly mean 9.80 cm and standard deviation 1.40 cm, find the proportion of areas for two class intervals, between (a) 10.25 to 10.75 cm and (b) 10.75 to 11.25 cm.

(a) The z score corresponding to 10.25 and 10.75 cm are calculated as 0.3214 and 0.6786 respectively. The areas (or probabilities) which may be obtained from Appendix Table 1(a) are found to be 0.1255 corresponding to the value of $z = 0.3200$ and 0.2486 equal to $z = 0.6700$ and we require to calculate the additional areas (or probabilities) for another 0.0014 and 0.0086. From the Appendix Table 1(a) we get a proportional area (or probability value) of 0.1293 for z score of 0.33 and a 0.0987 for a z score of 0.25. So a difference of 0.0014 will correspond to a difference of

$$\frac{0.0014 \times 0.0038}{0.01} = 0.0005 \text{ approx.}$$

So we may take the area corresponding to the z score of 0.3214 will be (0.1255 + 0.0005) = 0.1260

Similarly, by assuming that there is a linear relationship between the area and the normal deviate, we can calculate again the proportional area between the z score of 0.6786 and

0.6700 as

$$\frac{0.0086 \times 0.0032}{0.01} = 0.0028 \text{ approx.}$$

Hence we take the area corresponding to the normal deviate 0.6786 to be (0.2486 + 0.0028) = 0.2514. Thus the corresponding areas are found to be 0.1260 and 0.2514 and hence the fraction of the total area of the normal curve for the two class interval values of 10.25 and 10.75 is 0.2514 – 0.1260 = 0.1254. This is the relative frequency of this class in the population, the total frequency being taken to be unity. The actual total frequency being 400, the absolute frequency expected in this class is 400 × 0.1254 = 50.16 (or, 50 days to the nearest integer).

(b) For calculating the frequency in the next class we need to know the areas cut off by the ordinates at 10.75 and 11.25 cm. The former is already found and we find the *z*-score corresponding to the latter as 1.0357. The area corresponding to $z = 1.03$ is 0.3485 and for the additional 0.0057 we have proportional area of 0.0013. So the area for 11.25 cm is 0.3498.

Hence the fraction of the area in class 10.75 to 11.25 cm. is 0.3498 – 0.2514 =.0984 and the absolute expected frequency in this class is 39.36 (or 39) days to the nearest integer.

4.3.3 Characteristics of Normal Curve

Several interesting theoretical characteristics are thus apparent from normal curve. These are: (i) the normal curve is "bell-shaped" and unimodal in appearance, (ii) it is symmetrical about its centre and mean occupies the centre, (iii) because of symmetry, it's measures of central tendencies are identical so that the mean, median and mode coincide, (iv) the form of the curve in terms of it's shape (skewness) and height (kurtosis) depend upon the mean and the standard deviation, the latter two also are the basic unit of measurement along the value scale. For a normal distribution, it is already mentioned in Chapter 3 that the coefficient of skewness, β_1 is zero and the coefficient of kurtosis, β_2 is 3. (v) This normal curve is having two points of inflexion (i.e. where the curve changes curvature) at a distance of $\pm 1\sigma$ on either side of the mean, $\bar{x}$. Thus, $x_i = \bar{x} \pm 1\sigma$ gives us the points of inflexion. On account of this property, the normal curve is convex upward in the interval ($\bar{x} - 1\sigma$ and $\bar{x} + 1\sigma$) and concave upward outside this interval, (vi) Also in a normal curve, the tails extend asymptotically in both directions to infinity, never quite touching the horizontal axis. Finally, (vii) it turns out no matter what the mean and standard deviation are giving rise to whatever the shape of the normal curve is, the normal curve is having a constant area for which the total probability is one. Thus, there is not just one normal distribution but many: normal distributions may have different means and different standard deviations. But what makes them all normal is that they all have the same proportion of the area under the curve at given ordinates. This statement which means that if we draw ordinates at say $\pm 1\sigma$

from the mean for any two normal distributions, the proportions of the total distribution bounded by the vertical lines will be the same for both distributions.

4.3.4 Test of Normalcy of a Distribution

There are two ways in which the testing of the normal distribution character of a set of data can be done. The quickest way is to calculate the cumulative percentage frequency of the data and to plot it on the "arithmetic probability paper". As mentioned in Chapter 2, the cumulative percentage frequency ("less" or "more than") curve is nearly, *S*-shaped with the relative size of two tails of the *S* determined by the symmetry or skewdness of the distribution. The *arithmetic probability scale* is characterised by a special vertical scale known as a 'probability' scale and the abscissa is an arithmetic (ordinary) value scale. The ordinate and the abscissa scales are so designed that in case the data approximates a normal distribution, the plotting will give a straight line. Details of this probability paper will be discussed further in Chapter 7.

The 'goodness of fit' test is another measure for testing the normal distribution character of a set of data. Details of this test will be discussed in Chapter 13. In the following we apply the test of normalcy to rainfall distribution data.

Example 4.4 Test the normalcy of distribution to the annual rainfall data, year 1901 – 1950 at Varanasi given in Table 2.1.

The listed data of annual rainfall for Varanasi in Table 2.1 is divided into 11 classes. The lower value of the 1st class interval is 350 mm and the upper value of the 11th class interval is 2550 mm, and the intermediate intervals reach over a range of 200 mm. The class-intervals, the cumulative frequencies and the cumulative percentage frequencies are given below:

Class intervals (in mm.)	Frequencies (f)	Cumulative frequencies, f_c	Cumulative frequencies, in %
350 – 550	5	5	9.00
550 – 750	1	6	11.00
750 – 950	8	14	27.00
950 – 1150	10	24	47.00
1150 – 1350	7	31	61.00
1350 – 1550	4	35	69.00
1550 – 1750	7	42	83.00
1750 – 1950	4	46	91.00
1950 – 2150	3	49	97.00
2150 – 2350	0	49	97.00
2350 – 2550	1	50	99.00

Note for using the cumulative percentage frequencies in the probability paper (which does not attain 100 percentage exact), a small correction is done as

$$f_c\% = \frac{f_c - 0.5}{N} \times 100 \qquad \text{... (4.15)}$$

where f_c is the cumulative frequency and N = total numbers of items. These cumulative percentage frequency values are plotted on the arithmetic probability paper which do not yield a straight line exactly. (Fig. 4.8).

For calculating the cumulative probabilities, starting with 550 mm, a z score computation is done for the upper value of each class interval, the mean and standard deviation are 1265.1 mm and 458.02 mm respectively. Following is the tabulation of the z score and the cumulative probabilities.

Critical value (in mm)	z-score	cumulative probabilities (in per cent)
550	–1.4708	7.07
750	–1.0531	14.69
950	–0.6353	26.32
1150	–0.2175	41.44
1350	0.2003	57.93
1550	0.6181	73.17
1750	1.0360	84.96
1950	1.4538	92.70
2150	1.8716	96.94
2350	2.2894	98.89
2550	2.7072	99.67

These computed cumulative per cent probabilities are now plotted on the arithmetic probability paper. The plotted values give approximately a straight line (Fig.4.8).

4.4 TRANSFORMATION OF DATA

The set of data for diverse geo-science systems is quite often uni-modal but non-normal, such data are encountered in drainage basins, slope angles, annual peak discharge, wind speed in which most of the observations are small but a few are very large so that their distributions are positively skewed. Again in ecology, for a large assembly of species given, their relative abundancy are most likely a product of the interplay of many independent factors, multiplicatively compounded. In these cases, the theory of probability distributions as outlined above cannot be applied and one has to forego the possibility of the use of

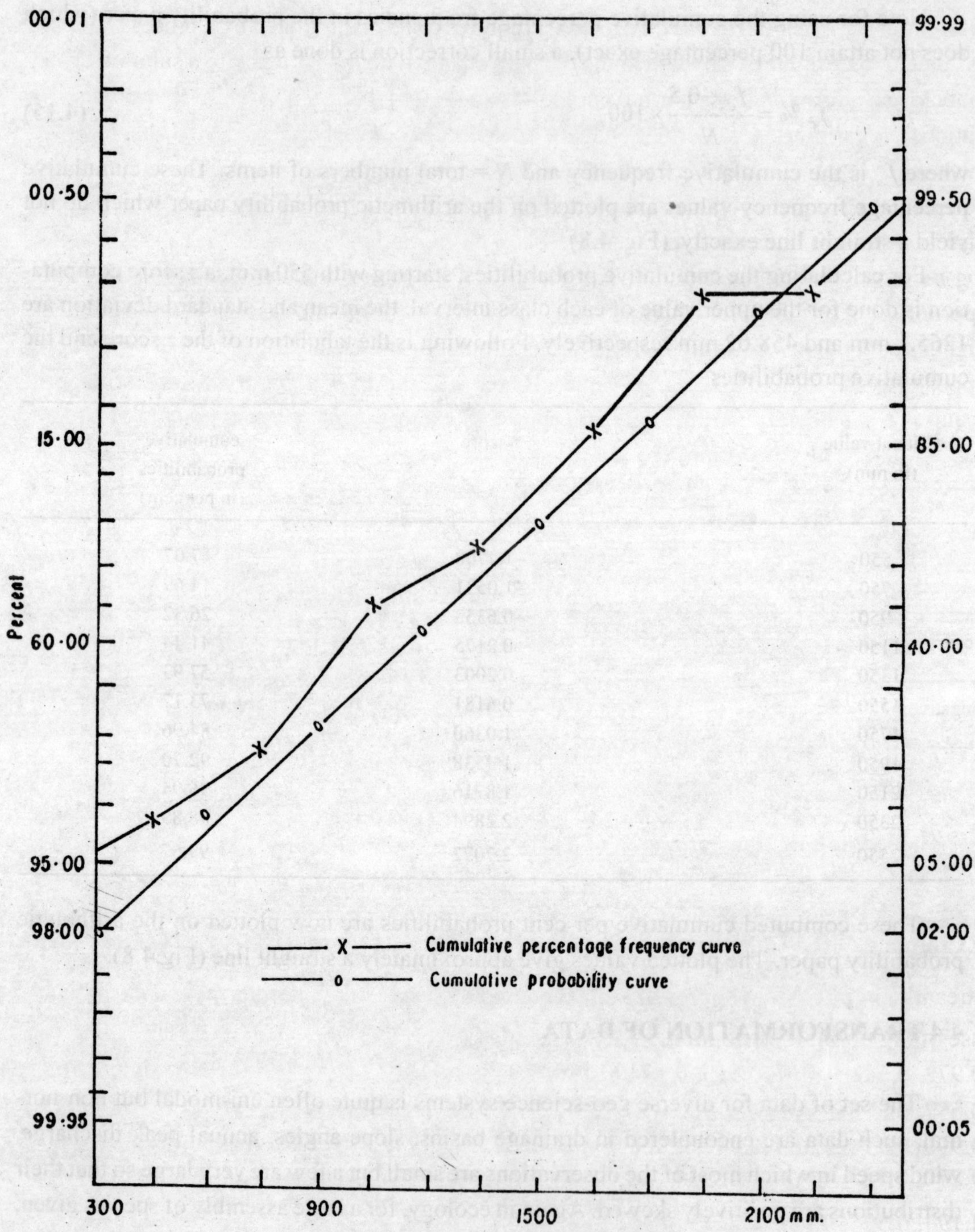

Fig. 4.8 : Cumulative Percent Probabilities

inferential statistics. To avoid this, the data requires to be manipulated by a process known as "data transformation", by which the data are changed mathematically to a suitable form that closely approximates a normal curve. It is suggested that for positively skewed distributions, 'logarithmic transformations' are effective for converting the skewed character of the data by logarithms into a normal one. For example, a random variable x may not be normally distributed but the distribution of the variable $Y = \log x$ (or $\ln x$) may be normally distributed. Thus if the random variable $Y = \log x$ is normally distributed, then x is said to be *log-normally distributed.* For mildly skewed data, a "square-root transformation" is most appropriate. For negatively skewed data, "power transformation" (either squares or even higher powers of raw data) are suitable. Transformations stretch the distribution in a direction opposite to direction of the original values. Hence selection of an appropriate transformation requires a consideration of the type and degree of skewness that is characteristic of a data set. A number of transformations are summarised below.

Types of skewness	Degree of skewness	
	Weak	Strong
Positive	$\sqrt{x}$	Log x or ln x*
Negative	x^2	Anti log x

*$\ln x = 2.302585$ times $\log x$

The log normal transformation of a set of data is now explained in Example 4.5.

Example 4.5 The following is a set of data of basin area (in sq. km.) for 174 third-order streams compiled from Survey of India topographic sheet $58\frac{A}{15}$. Interpret these data through use of log normal distribution.

0.972, 0.648, 1.017, 0.513, 0.594, 0.288, 1.323, 1.944, 0.297, 0.963, 0.765, 1.368, 0.702, 0.657, 0.693, 0.306, 0.513, 0.810, 2.007, 2.097, 0.405, 1.431, 1.215, 0.468, 2.016, 0.792, 0.999, 0.378, 0.369, 0.747, 0.306, 1.863, 1.233, 0.738, 1.179, 4.824, 0.261, 2.520, 2.835, 0.270, 0.423, 0.900, 1.323, 0.150, 0.468, 0.630, 0.252, 0.738, 0.360, 0.369, 0.486, 0.558, 4.194, 0.594, 0.702, 0.261, 0.657, 1.638, 0.648, 1.512, 0.342, 0.963, 0.378, 1.521, 0.772, 1.521, 2.034, 0.477, 0.477, 1.413, 2.286, 0.828, 0.603, 0.315, 0.675, 0.594, 0.756, 1.989, 1.035, 0.198, 0.333, 0.585, 0.288, 0.351, 0.783, 1.665, 1.593, 1.107, 1.125, 0.801, 0.297,

1.539, 0.351, 0.909, 1.305, 0.387, 0.504, 0.297, 1.035, 0.450, 0.882, 0.315, 0.225, 0.648, 0.450, 0.342, 0.324, 0.774, 1.062, 0.612, 0.612, 0.882, 0.351, 1.656, 1.143, 0.477, 0.387, 0.522, 0.819, 0.612, 0.171, 0.693, 0.522, 1.377, 1.440, 1.368, 0.396, 0.657, 0.252, 0.450, 0.828, 0.198, 0.279, 1.224, 1.944, 0.810, 1.107, 0.666, 0.648, 0.342, 0.486, 0.450, 0.180, 0.216, 1.098, 0.306, 0.162, 0.540, 0.594, 1.557, 0.702, 0.315, 1.044, 0.468, 0.828, 0.306, 0.279, 0.387, 0.414, 0.936, 0.976, 1.296, 0.747, 0.891, 0.720, 0.216, 0.912, 0.306, 0.270, 0.729, 0.648, 0.261, 0.774 and 0.333 sq. km.

For log–transformation, the given data are multiplied by 10 and logarithms are taken so that there is no negative value in the log numbers. This is because, negative logarithm values are undefined and logarithm of zero is infinity. Thus if the original data includes a decimal value without any integer (as in the data above) or a negative value or zero, an appropriate constant has to be multiplied or added to every observation. The logarithms (log.10x) are now given below for each of the corresponding basin areas:

0.9877, 0.8116, 1.0073, 0.7101, 0.7738, 0.4594, 1.1216, 1.2887, 0.4728, 0.9836, 0.8837, 1.1361, 0.8463, 0.8176, 0.8407, 0.4857, 0.7101, 0.9085, 1.3025, 1.3216, 0.6075, 1.1556, 1.0846, 0.6702, 1.3045, 0.8987, 0.9996, 0.5777, 0.5670, 0.8733, 0.4857, 1.2702, 1.0910, 0.8681, 1.0715, 1.6834, 0.4166, 1.4014, 1.4526, 0.4314, 0.6263, 0.9542, 1.1216, 0.1761, 0.6702, 0.7993, 0.4014, 0.8681, 0.5563, 0.5670, 0.6866, 0.7466, 1.6226, 0.7738, 0.8463, 0.4166, 0.8176, 1.2143, 0.8116, 1.1796, 0.5340, 0.9836, 0.5775, 1.1821, 0.8876, 1.1821, 1.3084, 0.6785, 0.6785, 1.1501, 1.3591, 0.9180, 0.7803, 0.4983, 0.8293, 0.7738, 0.8785, 1.2986, 1.0149, 0.2967, 0.5224, 0.7672, 0.4594, 0.5453, 0.8938, 1.2214, 1.2022, 1.0441, 1.0512, 0.9036, 0.4728, 1.1872, 0.5453, 0.9586, 1.1156, 0.5877, 0.7024, 0.4728, 1.0149, 0.6532, 0.9455, 0.4983, 0.3522, 0.8116, 0.6532, 0.5340, 0.5105, 0.8887, 1.0261, 0.7868, 0.7868, 0.9455, 0.5453, 1.2191, 1.0580, 0.6785, 0.5877, 0.7177, 0.9133, 0.7868, 0.2330, 0.8407, 0.7177, 1.1389, 1.1584, 1.1361, 0.5977, 0.8176, 0.4014, 0.6538, 0.9180, 0.2967, 0.4456, 1.0878, 1.2887, 0.9085, 1.0441, 0.8235, 0.8116, 0.5340, 0.6866, 0.6532, 0.2553, 0.3345, 1.0406, 0.4857, 0.2095, 0.7324, 0.7738, 1.1923, 0.8463, 0.4983, 1.0187, 0.6702, 0.9180, 0.4857, 0.4456, 0.5877, 0.6170, 0.9713, 0.9895, 1.1126, 0.8733, 0.9499, 0.8573, 0.3345, 0.9600, 0.4857, 0.4314, 0.8627, 0.8116, 0.4166, 0.8887 and 0.5224.

Now the raw score data, x and logarithm of 10x data are grouped and their central tendency measures are worked out below.

Frequency Tables for Raw Score Data and Log-Transformed Data

Original data

Classes	Mid–value	Frequency
0.150 – 0.750	0.450	102
0.750 – 1.350	1.050	44
1.350 – 1.950	1.650	18
1.950 – 2.550	2.250	7
2.550 – 3.150	2.850	1
3.150 – 3.750	3.450	–
3.750 – 4.350	4.050	1
4.350 – 4.950	4.650	1
	Total	174

Mean = 0.8569 sq km

$$\text{Median} = L + \frac{(P-c)}{f}.w = 0.150 + \frac{87.5-0}{102}\,.\,0.600 = 0.6647 \text{ sq. km}$$

$$\text{Mode} = L + \frac{d_1}{d_1+d_2}.w = 0.150 + \frac{(102-0)}{(102-0)+(102-44)}\,.\,0.600 = 0.5325 \text{ sq. km.}$$

Log - transformed data

Classes	Mid-value	Frequency	(Mid-value) x (frequency)
0.1003 – 0.3265	0.2134	6	1.2804
0.3266 – 0.5528	0.4397	34	14.9498
0.5529 – 0.7791	0.6660	37	24.6420
0.7792 – 1.0054	0.8923	52	46.3996
1.0055 – 1.2317	1.1186	32	35.7952
1.2318 – 1.4580	1.3449	11	14.7939
1.4581 – 1.6843	1.5712	2	3.1424
		Total	141.0033

Mean $= 0.8104$
Median $= 0.8249$
Mode $= 0.8762$

Taking the anti-log of the above values and dividing them by 10, we get the natural values of:

Mean $= 0.6463$ sq. km.
Median $= 0.6682$ sq. km.
Mode $= 0.7520$ sq. km.

It can be seen from above that there is an approach towards the symmetry in the frequency table and an identical nature of the values between mean, median and mode by transforming the data to logarithms. It should be emphasized that the log-transformed distribution is always skewed to the right. This negatively skewed character of the distribution is also apparent in the present example, mean being preceded by median and mode in values. Regarding skewedness character, the log normal distributions depend on parameters of mean and standard deviation just as the normal distribution depend on these parameters. If the value of the variance, σ^2 of the logarithms of the observations is small, so is the skeweness. In the present example, the variance, σ^2 is 0.08770 for the log-transformed data against 0.41374 for the original data. Hence the log-transformed data is more nearly normal than the original distribution.

4.4.1 Merits and Demerits of Transformation of Data

A normal distribution of the data is needed for inferential statistics based on probability theory. Since many of the continuous distributions in geo-science systems has a positively skewed character with an attenuated upper tail, the transformation of the data to a normal form by the above processes of data transformation e.g. *logarithmic, power,* or *square root transformation* will be required in making use of probability estimations, significance testing and regression analysis (mentioned in the subsequent chapters). However, transformations should never be made blindly because transformations may lead to biased estimates. For log normal transformation there is an advantage that this distribution is a bounded one ($x > 0$), i.e. it does not accept any negative value and the log transformations reduce the large numbers proportionally more than it does small numbers so it tends to reduce the positive skewness. The mathematical equation for normal probability density function, Y is given in Eq. (4.12) to which the equation for log normal distribution resembles. It is

$$Y = f(x_i) = \frac{1}{\sqrt{2\pi}\,\sigma_{\ln}} . e^{-1/2\left(\frac{\ln x_i - \bar{x}_{\ln}}{\sigma_{\ln}}\right)} \qquad \text{... (4.16)}$$

where $\ln x_i$ are natural logarithm or, common logarithm, $\log x_i$, $\bar{x}_{\ln}$ and $\sigma_{\ln}$ are the mean and standard deviation of $\ln x_i$ values.

Log normal distribution is important in analysing rainfall, river discharge and sediment particles data. For sediment particles data where $\varphi_i = -\log_2$. (particle diameter in mm), the above Eq. (4.16) is used with mean φ and standard deviation σ_ϕ.

4.5 CIRCULAR NORMAL DISTRIBUTION

For populations of directional measurement like slope, wind, cross-bedding directions, the resulting values are not always distributed normally. For these values, a distribution known as *circular normal* or "normal distribution on circle" can be used. This distribution is the equivalent of the normal distribution and its probability density function is

$$Y = f(x_\theta) = \frac{1}{\sqrt{2\pi}\,.\sigma_k} e^{\{\sigma_k .\cos(\theta_i - \bar{x}_\theta)\}} \qquad \text{... (4.17)}$$

with a mean $\bar{x}_\theta$ where θ_i is a circular measure between 0 and 2π radians (or 0° and 360°s), σ_k is the concentration parameter (equivalent to standard deviation) with k (Kappa) greater than zero, a statistic. Again, the mean $\bar{x}_\theta$ is calculated by treating the angular measurements θ_i in terms of $\cos\theta_i = x_i$ and $\sin\theta_i = y_i$, finding the mean of x_i and y_i then calculating the mean of the angular values, $\bar{x}$ and $\bar{y}$ in polar coordinates $(r, \bar{\theta})$, so that $r = \sqrt{\bar{x}^2 + \bar{y}^2}$ and $\cos\bar{\theta} = \bar{x}/r$ and $\sin\bar{\theta} = \bar{y}/r$. From any one of the latter two, we can calculate the mean of the measures, $\bar{x}_\theta$ in terms of radians. The mean $\bar{x}_\theta$ is the population mean, μ of the circular normal distribution. Figure 4.9 provides the circular normal distribution for various values of mean $\bar{x}_\theta$ and the angular standard deviation σ_k. Note that in this Fig. 4.9, three angular distributions are plotted with angular measurements $\mathrm{M}(\bar{\theta}, \sigma_k)$ look like normal distributions.

The value of r as equal to $\sqrt{(\bar{x}^2 + \bar{y}^2)}$ is not the concentration parameter, σ_k. It is an estimate of the spread of the angular values around a unit circle and in that way it resembles standard deviation. The closer the points are clustered, the nearer is r to unity. The angular standard deviation σ_k can be calculated as

$$\sigma_k = \sqrt{2(1-r)} \qquad \text{... (4.18)}$$

where $r = \sqrt{(\bar{x}^2 + \bar{y}^2)}$ in which $\bar{x} = r\cos\bar{\theta}$ and $\bar{y} = r\sin\bar{\theta}$

σ_k which is a statistical parameter can also be interpolated against r values from the graph shown in Fig. 4.10. An example of application of circular normal distribution is given below.

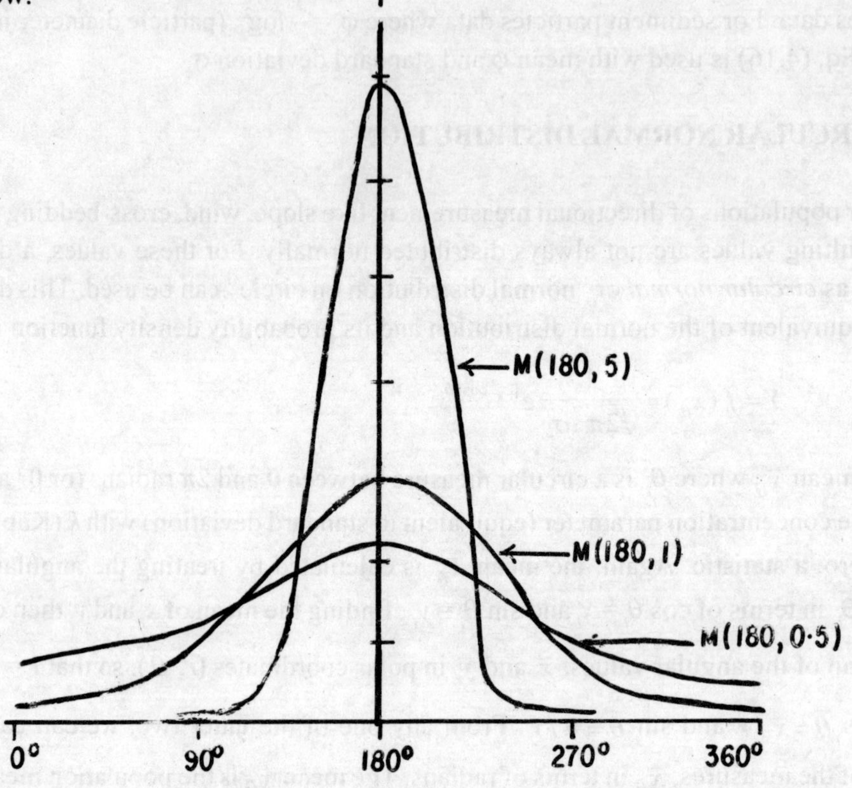

Fig. 4.9 : A Circular Normal Distribution for various values of ($\bar{x}_\theta, \sigma_k$) in degrees

Example 4.6 Following is a set of data related to slope angle measurements (Angles are measured clockwise) :

12, 353, 359, 332, 341, 299, 30, 24, 53, 284, 99, 72, 28, 193, 125, 318, 3 and 45 degrees (n = 18). Calculate the probabilities of occurrence of each slope angle value.

Below we give the slope angles, θ_i, their radian equivalents θ_i^c and their coordinates on a unit circle : $x_i = \cos\theta_i$ and $y_i = \sin\theta_i$ to calculate the mean co-ordinates ($\bar{x}$ and $\bar{y}$). We have $1° = 0.01745^c$ for conversion.

Slope angles	Radian Equivalents	Sine of angle	Cosine of angle
(θ_i) in °s	(θ_i^c)	$y_i = \sin\theta_i$	$x_i = \cos\theta_i$
12	0.20940	+0.20792	+0.97815
353	6.15985	– 0.12187	+0.99255
359	6.26455	– 0.01745	+0.99985
332	5.79340	– 0.46947	+0.88295
341	5.95045	– 0.32557	+0.94552
299	5.21755	– 0.87462	+0.48410
30	0.52350	+0.50000	+0.86603
24	0.41880	+0.40674	+0.91355
53	0.92485	+0.79864	+0.60182
284	4.95580	–0.97030	+0.24192
99	1.72755	+0.98769	– 0.15643
72	1.25640	+0.95106	+0.30902
28	0.48860	+0.46947	+0.88295
193	3.36785	–0.22495	– 0.97437
125	2.18125	+0.81915	– 0.57358
318	5.54910	–0.66913	+0.74314
3	0.05235	+0.05234	+0.99863
45	0.78525	+0.70711	+0.70711
Total		$\Sigma y_i = 2.22676$	$\Sigma x_i = 9.84291$
So, Mean:		$\bar{y}_i = 0.12371$	$\bar{x}_i = 0.54683$

The above slope angle values in degrees are grouped into class intervals of 60° and we have six classes with frequencies in brackets as: 0 to 60° (7), 60° to 120° (2), 120° to 180° (1), 180° to 240° (1), 240° to 300° (2) and 300° to 360° (5).

Now the above mean points $\bar{x}_i, \bar{y}_i$ are converted in polar coordinates $(r, \bar{\theta})$ where $r = \sqrt{\bar{x}_i^2 + \bar{y}_i^2} = \sqrt{0.12371^2 + 0.54683^2} = \sqrt{0.31433} = 0.56065$

For the other polar coordinate, $\bar{\theta}$ we have it from either $\bar{x} = r\cos\bar{\theta}$ and $\bar{y} = r\sin\bar{\theta}$. So, $\cos\bar{\theta} = \frac{\bar{x}}{r} = 0.54683/0.56065 = 0.97535$ and $\sin\bar{\theta} = \bar{y}/\text{r} = 0.22065$. We find $\bar{\theta} = 12°.7475$ = 0.22244 radians (since 1° = 0.01745 radian). This r value is not close to unity and hence the points are not clustered. Now to calculate the angular standard deviation we have Eq. 4.18 as σ_k is equal to

$$\sigma_k = \sqrt{2(1 - 0.56065)} = 0.93739 \text{ radians} = 53°.70985$$

Finally the statistic, k (Kappa) is interpolated from the graph in Fig. 4.10 in which as against r = 0.56065 along abscissa, k is taken to be 1.60. This gives us the probability function for normal distribution of the data set in the example. Following Eq. (4.17) we find:

$$Y = f(x_i) = \frac{1}{\sqrt{2\pi}\,.1.6}\,.e^{[1.6\cos(\theta_i - 12°.75)]}$$

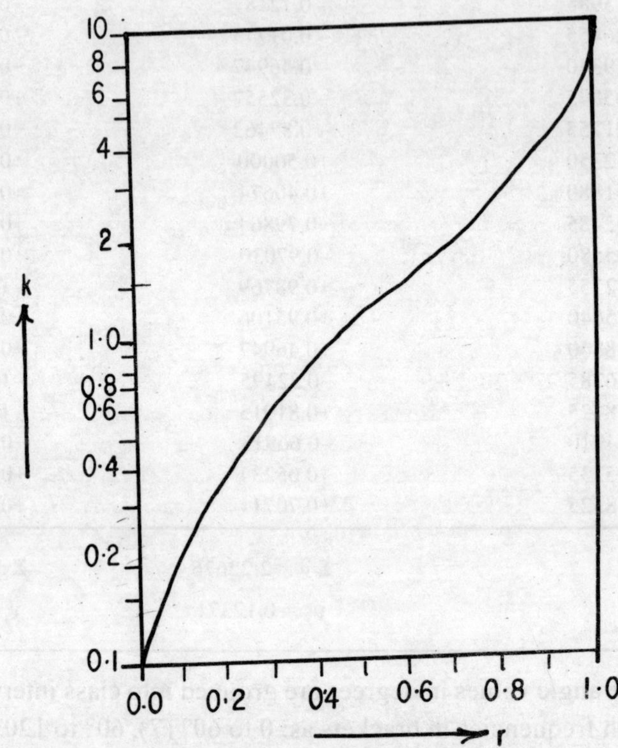

Fig. 4.10 : The values of *k* against *r*

Following the above equation, now we calculate the probabilities for individual slope angle values, θ_i . These are

θ_i	cos $(\theta_i - 12°.75)$	1.6 cos $(\theta_i - 12°.75)$	$e.^{(\text{col. 3})}$	$\frac{1}{\sqrt{2\pi}\,.(1.6)}$	Y_i
(Col. 1)	(Col. 2)	(Col. 3)	(Col. 4)	(Col. 5)	(Col 4 × 5)
1	2	3	4	5	6
12	0.99991	1.59986	4.95235	0.24933	1.23481
353	0.94118	1.50589	4.50816		1.12406
359	0.97134	1.55414	4.73102		1.17963
332	0.75756	1.21210	3.36053		0.83791
341	0.85035	1.36056	3.89838		0.97202

299	0.27983	0.44773	1.56476	0.39015
30	0.95502	1.52803	4.60909	1.18923
24	0.98079	1.56926	4.80309	1.19760
53	0.76323	1.22117	3.39115	0.84555
284	0.02181	0.03490	1.03552	0.25819
99	0.06540	0.10464	1.11031	0.27684
72	0.51129	0.81806	2.26610	0.56502
28	0.96479	1.54366	4.68169	1.16733
193	– 0.99999	– 1.59998	0.20190	0.05034
125	– 0.37864	– 0.60582	0.54563	0.13605
318	0.57715	0.92344	2.51794	0.62782
3	0.98556	1.57690	4.83991	1.20678
45	0.84573	1.35316	3.86963	0.96485

Thus with the mean vector of slope angles equal to 12°.7475 and the deviation of 53°.70985, we find from the above table the probability integrals for individual slope angles, θ_i which tells that only 5 of the 18 individual slope angle values lie within an angle of 12.75° ± 53.71°.

4.6 SUMMARY

In this chapter, we have explained the probability theory as a basic requisite for interpreting the results of statistical analysis and we have examined the normal probability distribution for continuous random variable. The normal distribution is an ideal distribution to which an indefinitely large number of observations of the "real world data" measured on ratio and interval scales only can approximate. Thus the normal distribution is useful in its own right though regarding the closeness of approximation of a real world data we could not suggest any hard and fast rule in this chapter apart from the subjective judgement such as the bell-shaped pattern of the distribution or the closeness between the central tendency measures. But we continue our investigation on normal distribution in the context of binomial distribution in the next chapter to show that the former can be an approximation of the latter. In subsequent chapters pertaining to inference, we shall see that very often the observations in geo-science systems follow or approximate the normal distribution considered here.

LIST OF FORMULAE

I. *Probability Rules or Axioms of Probability*

Rule 1 Total probability, $\Sigma p(x_i) = 1$ for all i and for any event, x_i

Rule 2 Range of probability, $0 \leq p(x_i) \leq 1$

Rule 3 Addition Rule for independent and mutually exclusive events: $p(x_1 \text{ or } x_2) = p(x_1 \cup x_2) = p(x_1) + p(x_2)$

This addition rule can be extended to cover more than two events.

Rule 4 Multiplication Rule for two or more events occurring simultaneously or in succession

$p(x_1 x_2 \ldots x_r) = p(x_1 \cap x_2 \ldots \cap x_r) = p(x_1) \cdot p(x_2) \ldots p(x_r)$

Rule 5 Multiplication Rule for dependent events or joint probability of the event x_1 when it is assumed that event x_2 has already happened

$p(x_1 \cap x_2) = p(x_1) \cdot p(x_2 \mid x_1)$.

In case the event x_1 has already happened and we consider the joint probability of the event x_2, it is

$p(x_2 \cap x_1) = p(x_2) \cdot p(x_1 \mid x_2)$, thus

$p(x_1 \text{ and } x_2) = p(x_1) \cdot p(x_2 \mid x_1) = p(x_2) \cdot p(x_1 \mid x_2)$

Rule 6 Addition Rule for dependent events

$p(x_1 \cup x_2) = p(x_1) + p(x_2) - p(x_1) \cdot p(x_2 \mid x_1)$

Rule 7 Conditional Probablitity for joint events x_1, x_2

$(p(x_2) \neq 0)$, $p(x_1 \mid x_2) = p(x_1 \text{ and } x_2) / p(x_2)$

Rule 8 Conditional Probability for independent events

$p(x_1 \mid x_2) = p(x_1)$ and $p(x_2 \mid x_1) = p(x_2)$

II. *Normal Distribution*

Probability Density Function for Normal Distribution :

$$Y = f(x) = \frac{1}{\sqrt{2\pi} \cdot \sigma_x} \cdot e^{-\frac{1}{2}\left(\frac{x_i - \mu}{\sigma_x}\right)^2} \quad \text{where } -\infty < x_i < +\infty$$

Standard Variate or z score, $z = \dfrac{x_i - \bar{x}}{\sigma}$

Probability Density Function for Standardized Normal Distribution

$$Y = f(z) = \frac{1}{\sqrt{2\pi} \cdot \sigma} \cdot e^{-\frac{1}{2} \cdot z^2}$$

Probability Density Function for Log Normal Distribution

$$Y = f(x) = \frac{1}{\sqrt{2\pi}\,.\sigma_{\ln}}.e^{-1/2\left(\frac{\ln x_i - \bar{x}_{\ln}}{\sigma_{\ln}}\right)^2}$$

for natural (and/or common) logarithm of x_i values

Probability Density Function for Circular Normal Distribution

$$Y = f(x) = \frac{1}{\sqrt{2\pi}\,.\sigma_k}.e^{\{\sigma_k \cos(\theta_i - \bar{x}_\theta)\}}$$ **with $\cos\theta_i = x_i$, $\sin\theta' = y_i$ and**

polar coordinates $(r, \bar{\theta})$ where $r = \sqrt{(\bar{x}^2 + \bar{y}^2)}$ and $\sigma_k = \sqrt{2(1-r)}$.

EXERCISES

4.1. 50 years of rainfall for a place is given below in cm :

24.2	30.9	21.6	22.2	23.3	27.5	28.8	20.2	28.4	19.7
25.8	28.8	24.7	24.1	24.9	25.7	21.5	24.0	21.2	22.4
27.9	24.2	24.1	27.1	17.8	28.6	20.7	21.7	24.5	19.8
19.6	37.0	24.8	23.5	37.0	30.3	27.6	27.5	25.8	31.8
27.6	30.1	19.2	21.4	22.4	23.7	24.1	21.3	29.3	33.6

Using the probability paper plot the *z* score value from 18 cm. of rainfall at interval of every 2 cm.

4.2. If *z* is a standard normal deviate illustrate the area for the following:-

(i) $p\,(0 \le z \le 1.37)$

(ii) $p(0 < z < 0.21)$

(iii) $p(0 < z \le 3.04)$

(iv) $p(z \le 0.43)$

and (v) $p\,(-1.07 \le z \le 0.75)$

4.3. For a normally distributed set of data $N = 10{,}000$ mean is 750 and standard deviation is 50. Show that 95% of the data has value exceeding 668 and only 5% has value exceeding 832. What is the lowest value among the highest 100 in the data?

4.4. In a set of data with normal distribution, 5% of the data are 150 and 40% are between 150 and 157.5. Find the mean and standard deviation.

4.5. In a public transport undertaking of a city, it is found that the buses have an average life of 4,000 hours to be used in the routes with a standard deviation of 1,000 hours. In the undertaking 2,000 new buses were introduced. What number of these buses might be expected to break down in the first 700 hours of the buses putting on road? After what period of working hours on road would we expect that 10% of the buses have failed ?

4.6. Following is a set of numbers having a negatively skewed distribution: 50, 49, 48, 47, 45, 43, 40, 35, 30, 20 and 10.

Work out the cumulative proportions of the data. Effect a z score conversion of the cumulative proportions of the data. Multiply the z scores with standard deviation of the raw score and add to it the mean to find a new distribution. Establish that this new distribution and the original data have the identical mean and standard deviations, but they differ in shape.

4.7. A percentage frequency distribution of scores of examinees E_1, E_2 and E_3 in a subject by 3 examiners independently is given below: -

Percentage Frequency distribution of score by 3 examiners

Marks	E_1	E_2	E_3
0 – 10	5	10	5
10 – 30	15	20	25
30 – 50	50	60	50
50 – 70	24	8	10
70 – 90	5	2	8
> 90	1	–	2

Determine the relative ranks of 3 students *A*, *B* and *C* who scored the following marks with the above 3 examiners E_1, E_2 and E_3

Students	Marks given by		
	E_1	E_2	E_3
A	25	62	73
B	48	51	35
C	78	25	50

[**Hints** : Determine the μ and σ of the 3 examiners, calculate the standard scores for students by the 3 examiners, total the standard scores and then rank each student].

4.8. A continuous random variable x operates between 0 and 10 and its probability density function is 0.02 $(10 - x)$. Prove that the area of the curve over this whole range of interval between 0 and 10 is unity.

5

Probabilistic Treatment in Geo-science Systems II

(Binomial Distribution)

5.1 BINOMIAL DISTRIBUTION

Many different types of probability models have been developed to represent specific and countable events whose distribution is discrete and not continuous so that the fractional values are impossible. In particular, one discrete probability distribution which is extremely useful for describing many phenomena in geo-science systems is the 'binomial distribution'. This distribution is also known as 'Bernouli distribution' after the name of its inventor, the learned Swiss mathematician, Jacob Bernouli in 17th century and it is concerned with the relative frequency of occurrences of only two outcomes or rather sets of conditions for each random experiment (or trial), arbitrarily called 'success' and 'failure' without inferring that a success is necessarily a desirable outcome. Thus once a particular outcome of an event is decided as a desired one, the binomial distribution deals with the probability of occurrences of the outcome in a given number of occurrences of the event. This distribution is associated with the repetition of the event where in each occurrence there are only two possible outcomes which are mutually exclusive (that is, either head or tail but not both of them simultaneously in a single tossing of coin) and which together represent the sum total of probability (always equal to one). When an event can occur in one of the two possible ways it is said to be dichotomous. A binomial distribution finds considerable use in problems relating to a "set of dichothomous distribution" i.e. a two-class segregation of success or failure, good or bad, head or tail etc., each class is characterised by certain constant probabilities p and q referring to the probabilities of the two classes in sets of n independent events.

A binomial distribution is illustrated most simply by using the example of tossing a coin. For example, when a coin is tossed, it is either of only two events i.e. head H or tail T and in a single toss of an unbiased coin the probability of H or T is equal to 0.5. We know from probability theory that this distribution applies when (i) a number of independent trials are made with replacement (ii) an experiment is performed under the same conditions, for a fixed number of trials, say, *n* and (iii) the probability of the outcome remains the same. In the toss of a coin experiment, the event space will be [head 'H' tail T]. For tossing a coin one time there are two possibilities: head H or tail T. For the second toss for each of the head H or tail T of the first toss, there are two probabilities which give $2 \times 2 = 4$ possibilities for the first two tosses. Thus in tossing a coin three times, for each of the four probabilities in the second toss there are $4 \times 2 = 8$ possibilities. When these eight possibilities are arranged in order we have "permutations" or possible outcomes (i.e. head, head, head; head, head, tail; etc.). The permutation of tossing the coin three times is as follows: (with head as H and tail as T):

HHH, HHT, HTT, HTH, THH, TTH, THT, TTT

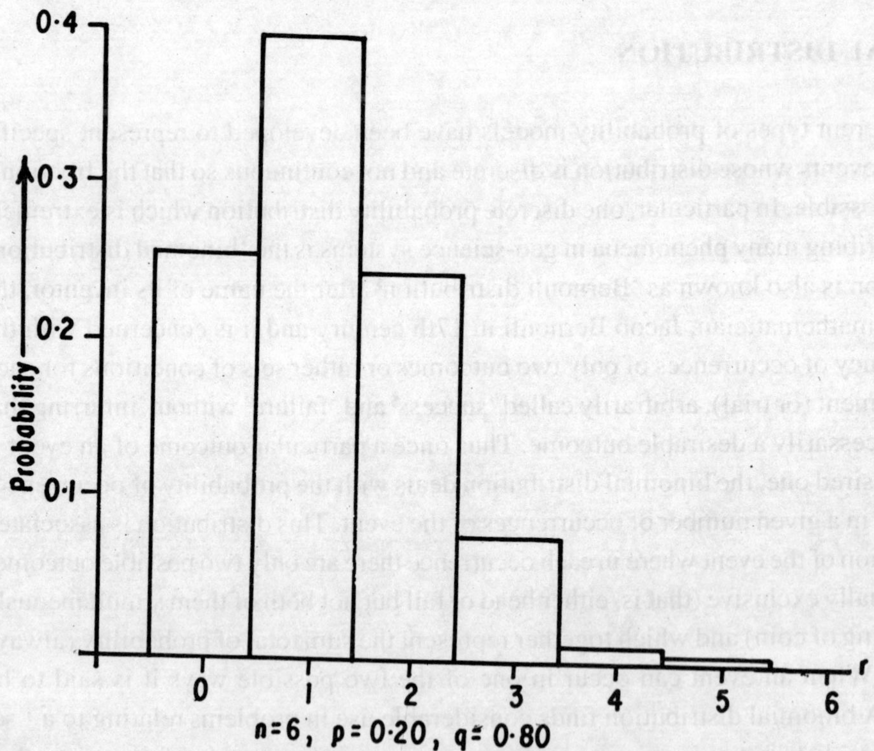

Fig. 5.1: A Skewed Binomial Distribution

where there are $2^3 = 8$ permutations (or occurrences) and the probability of one permutation $(1/2)^3$ i.e. $1/2 \times 1/2 \times 1/2 = 0.125$. Therefore, the probability of one permutation for an event where there can be only two outcomes is given by $(1/2)^n$ where there are *n* number of trials. In the above sequence of permutation, the combination (i.e. the arrangement without and order of sequence) of 3H, 2H, 1H and no head occur one, three, three and one time respectively out of eight permutations in trial of tossing the coin three times. Note in the above what we have found is the number of possible 'combinations' of 3H, 2H, 1H and no head. These can be generalised to the number of possible combinations of n items taken r at a time. Symbolically this is represented as

$$\binom{n}{r}$$

It can be demonstrated that the number of possible combination, taken r items at a time, is

$$\frac{n!}{r!\,(n-r)!}$$

where the "!", exclamation points stand for 'factorial' which means that the number preceding the"!" is multiplied by one less the number, then by two less the number and so on down to 1. so that

n factorial or $n!$ $= n\ (n-1)\ (n-2), ..., (n-r), ..., 1$

again 3! $= 3.2.1 = 6$

note, 0 factorial, 0! $= 1$

Now if we ask how many tosses will show exactly three heads, H H H if a coin is tossed 3 times we will have

$$\binom{3}{3} = \frac{3!}{3!(3-3)!} = \frac{3!}{3!0!} = \frac{3.2.1}{(3.2.1)\ (1)} = 1$$

Similarly, in tossing the coin 3 times, the possible number of combinations that contain no head (i.e. three tails) is

$$\binom{3}{0} = \frac{3!}{0!(3-0)!} = \frac{3.2.1}{1.\ (3.2.1)} = 1.$$

Thus, with three flips of a coin there is one way (combination) we can get all 3 Hs, three ways we can get 2Hs and 1H each and one way we get no head.

Now since the number of total possible permutations (or, outcomes) are eight, we can convert the above frequencies of occurrences into probabilities. These probabilities are $\frac{1}{8}, \frac{3}{8}, \frac{3}{8}$ and $\frac{1}{8}$ (note sum of these probabilities is one). These are known as the *binomial coefficients* and the probability distribution that governs experiments such as this coin tossing experiment is known as Binomial Probability Distribution which are characterised by (a) only two possible outcomes for each trial: success or failure, (b) the probability of outcome

outcome (success or failure) does not change from trial to trial, (3) each trial is independent of all others and (4) the trials are performed a fixed number of times. If the probability of occurrence of the head in tossing of the coin, p (H) is denoted by p and the non-occurrence is denoted by q (which is equal to $1 - p$, as $p + q = 1$) then the general formula for obtaining the individual terms of the distribution in n observations is according to the *binomial theorem*, $(p+q)^n$. The algebraic expression of the expansion $(p+q)^n$ becomes, the "binomial probability distribution". We now discuss below the "binomial distribution function".

5.2 BINOMIAL DISTRIBUTION FUNCTION

In the binomial theorem, $(p + q)$, when $n = 2$, we have the expansion of the elementary binomial which is as follows:

$$(p + q)^2 = p^2 + 2pq + q^2$$

Similarly, the expansion of $(p + q)^3$ and $(p + q)^4$ are as follows:

$$(p + q)^3 = p^3 + 3p^2q + 3pq^2 + q^3$$

$$(p + q)^4 = p^4 + 4p^3q + 6p^2q^2 + 4pq^3 + q^4$$

In each expansion, it can be noted that as the value of n increases, so too does the number of terms in the expansion and the total number of terms in each expansion is always equal to $n + 1$. The Binomial coefficients are also symmetrical about the central point in each term: 1, 2, 1 for $n = 2$; 1, 3, 3; 1 for $n = 3$; and 1, 4, 6, 4, 1 for $n = 4$.

These binomial coefficients can also be computed from an "arithmetic triangle" known as Pascal's triangle. This is set out in Table 5.1 for a portion of it for $n = 1$ to $n = 5$

Table 5.1 : Pascal's Triangle

No. of events n	Binomial Equation $(p + q)$	Coefficients in the binomial expansion	Probability of permutation p or q
Apex	$(p + q)^0$	1	...
$n = 1$	$(p + q)^1$	1 1	$\frac{1}{2}(=1/2^1)$
$n = 2$	$(p + q)^2$	1 2 1	$\frac{1}{4}(=1/2^2)$
$n = 3$	$(p + q)^3$	1 3 3 1	$\frac{1}{8}(=1/2^3)$
$n = 4$	$(p + q)^4$	1 4 6 4 1	$\frac{1}{16}(=1/2^4)$
$n = 5$	$(p + q)^5$	1 5 10 10 5 1	$\frac{1}{32}(=1/2^5)$

This above triangle of values are built by adding each successive pairs of values on the previous line to generate new numbers in the row below them. To begin, we start with 1 for the apex of the triangle and allocate 1 to the first and 1 to the last number in every row. The 1s in the second row add to 2, which is allocated to the middle of the second row. This 2 in turn adds to 3 with each of the adjoining1s to produce two 3s in the third row. Hence when $n = 3$, the binomial expansion can be done by adding each successive pair of values together above them (i.e. from row $n = 3$ in Table 5.1) giving coefficients 1, 1 + 2 = 3, 2 + 1 = 3 and 1. This method of calculating the coefficients of the binomial expansion can go on indefinitely but they are somewhat laborious, particularly when n is large. Consequently, a set of values of the binomial coefficients for n from 2 through 20 and number of successes desired, r from 2 through 10 is given in the Appendix Table II.

For establishing the powers to which p and q must be raised for any expansion of the binomial formula $(p + q)^n$, the rule is that in the first case the powers of p is equal to n and that of q is nil, the powers of p then steadily decrease by one each time for terms from left to right in the expansion while that of q increase equally from nil to n in the same direction. Thus, the powers of p and q in the expansion $(p + q)^n$ is:

$$p^n, p^{n-1}q, p^{n-2}q^2, p^{n-3}q^3, p^{n-4}q^4, \ldots\ldots, p^{n-r}q^r, \ldots\ldots q^n \qquad \ldots (5.1)$$

Now we can express the above terms of the binomial expansion $(p + q)^n$ in a form as follows: from tossing of coin n number of times, with the trials being independent of one another, we immediately evaluate the probability, p of getting r heads and the probability of getting $n - r$ tails q. Now since the trials are independent, we can simply multiply the unconditional probabilities. Hence the probability of getting r heads H in a row and then tails, T for $(n - r)$ terms is

$$\underbrace{(p, p, p) \ldots}_{r \text{ terms}} \quad \underbrace{(q, q, \ldots, q)}_{(n-r) \text{ terms}} = p^r.q^{n-r} \ldots \qquad \ldots (5.2)$$

this expression is also the probability of obtaining r heads H and $n - r$ tails T, in any order. So, in order to obtain the probability of getting exactly r heads in any order it is only necessary to count the number of ways we get r heads and $n - r$ tails. But this counting of the number of ways/combinations of choosing the same probability of getting r head H and $n - r$ tails T i.e. $p^r\, q^{n-r}$ is tedious when n is large. Luckily a mathematical formula is available which makes this counting unnecessary.

This is

$$\binom{n}{r} = \frac{n!}{r!(n-r)!} \qquad \ldots (5.3)$$

so that the random variable having the binomial probability distribution function, $f(r)$ for obtaining r number of successes (or event) desired, r occurring in n number of trials (or, occurrences) can be written in a generalised way as:

$f(r)$ = (number of ways/combinations of getting r successes out of n trials/observations possible) × (probability of a particular sequence)

or, writing symbolically

$$f(r) = \frac{n!}{r!(n-r)!} . p^r . q^{n-r} \quad \text{... (5.4)}$$

where p is probability of success or occurrences and q (= $1 - p$) is probability of failure or non-occurrence.

Thus the complete binomial expansion can be written with the help of Eq. (5.4) as

$$(p+q)^n = p^n + \frac{n!}{(n-1)!} p^{n-1} q + \frac{n!}{2!(n-2)!} p^{n-2} q^2 + ... + \frac{n!}{r!(n-r)!} p^{n-r} q^r + ... + q^n \quad ...(5.5)$$

For example, if there are eight occurrences, i.e. $n = 8$, it can be seen that the terms of expansion of $(p + q)^8$ will be

$$p^8 + 8p^7q + 28p^6q^2 + 56p^5q^3 + 70p^4q^4 + 56p^3q^5 + 28p^2q^6 + 8pq^7 + q^8$$

Example 5.1. A sample of eight stones was observed in a river for their movability. What is the probability that (i) all the stones are moving and (ii) only four stones are moving?

From the above expansion of $(p + q)^8$ where p is the probability of stones that are moving and q the probability of stones that are not moving, it can be said that there can be $2^8 = 256$ permutations and the probability of one permutation is 1/256 . So there is only a 1/256 = 0.0039 probability that they are all moving or they are all stationary but there is a 70/256 = 0.2734 probability that four stones are moving and the remaining four are stationary.

5.2.1 Characteristics of Binomial Distribution

The Binomial distribution possesses four essential properties:

(i) The possible observations in binomial distribution may be obtained either by considering each observation being selected (a) from a finite population with replacement, or, (b) from an infinite population without replacement.

(ii) Each observation may be classified into one of two mutually exclusive and collectively categories, usually called 'success' and 'failure' (or in case of tossing a coin, head H or tail T, voting a candidate yes or no etc.)

(iii) The probability of an observation being classified as success, p, is constant from observation to observation. Thus the probability of an observation's being classified as failure, $1 - p$ is constant over all observations.

(iv) The outcome (i.e. success/failure) of any observation is independent of history of the outcome itself and of any other observation. For example, the flood events are independent from year to year so that one can find the probability of occurrence of flood.

The shape of a binomial distribution may be symmetric or skewed and it depends on the probability p and the number of occurrences n. Whenever $p = q = 0.5$ regardless of how large or small the value of n is, the binomial distribution will be symmetrical in shape like normal distribution. But rarely do we come across an event in geo-system sciences to have mutually exclusive equal probabilities* and this involves marked skewness to the binomial distribution as has been shown in Fig 5.1. On the other hand, Fig 5.2 illustrates a symmetrical binomial distribution. From both these figures, irrespective of the size of p, when n increases it will be apparent that the binomial distribution is represented by histogram (i.e. by blocks) since it deals with discrete values in contrast to normal distribution which uses values on continuous scale and is represented by a curve.

5.2.2 Approximating the Binomial Distribution

It is stated earlier that whenever $p = q = 0.5$ the binomial distribution is symmetric, otherwise the distribution is skewed to the right or left depending on whether $p < 0.5$ or $p > 0.5$ respectively. However the closer the p is to 0.5 and the larger values of n, the more symmetrical the distribution becomes.

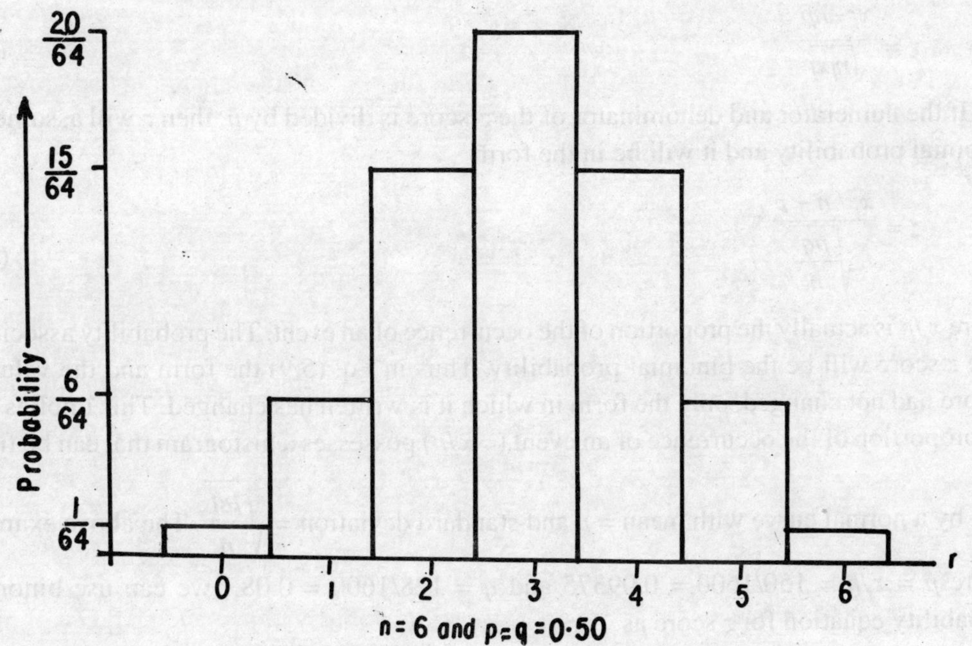

Fig. 5.2 : A Symmetrical Binomial Distribution.

On the other hand, the larger the values of n is, the more tedious it is to compute the

* More commonly we have a bias one way or the other, for example, bankful discharge.

exact probabilities of success by the use of binomial probability function in Eq. 5.4. Because of the similarity between normal and binomial distributions (particularly when $p = 0.5 = q$) and the product $n(1-p)$ equals or exceeds 10, the binomial distribution can be approximated to the normal distribution by defining two measures of the binomial distribution, the mean and standard deviation.

These measures are

$$\text{Mean } \bar{x} = np \qquad \text{... (5.6)}$$

$$\text{Standard deviation} = \sqrt{npq} \qquad \text{... (5.7)}$$

For example, for $n = 1600$ and $p = 0.08$, the mean $\bar{x}$ would be $np = 128$ and standard deviation $\sqrt{npq} = 117.76$. Now if we like to find the probability in the set of condition the number of values for p will not be more than 150 we can get the solution by normal approximation method as follows.

If we substitute the mean and standard deviation of binomial distribution (that is, Eqs. 5.6 and 5.7) in the z-score formula of the normal distribution i.e. Eq. (4.13), we have the standard variable z which assumes the form

$$z = \frac{x_i - np}{\sqrt{npq}} \qquad \text{... (5.8)}$$

If the numerator and denominator of the z-score is divided by n, then z will assume the binomial probability and it will be in the form

$$z = \frac{x_i / n - p}{\sqrt{\frac{pq}{n}}} \qquad \text{... (5.9)}$$

where x_i/n is actually the proportion of the occurrence of an event. The probability associated with z score will be the binomial probability. Thus in Eq. (5.9) the form and the value of z score had not changed, only the form in which it is written has changed. This implies that the proportion of the occurrence of an event ($= x_i/n$) possesses a histogram that can be fitted well by a normal curve with mean $= p$ and standard deviation $= \sqrt{\frac{pq}{n}}$. The above example where $p = x_i/n = 150/1600 = 0.09375$ and $q = 128/1600 = 0.08$, we can use binomial probability equation for z score as

$$z = \frac{(x_i / n) - p}{\sqrt{\frac{pq}{n}}} = \frac{0.09375 - 0.08}{\sqrt{(0.08)\,(0.92)/1600}} = \frac{0.013750}{\sqrt{0.00046}} = \frac{0.01375}{0.00678} = 2.0273$$

Using Appendix Table 1(a), we find that the area under the curve between the mean and z-score 2.0273 is 0.4787, so that the approximate probability that $p = 0.08$ will not be more than 150 is 0.5000 + 0.4787 = 0.9787.

Now we can appreciate the amount of labour saved by using the normal approximation to the binomial model in lieu of the exact probability by the binomial function. Imagine by using the Eq. 5.4 we require to make 151 computations and then summing up the results:

$$\frac{1600!}{0!1600!}(0.08)^0(0.92)^{1600} + \frac{1600!}{1!\,1599!}(0.08)^1(0.92)^{1599} + \ldots + \frac{1600!}{150!1450!}(0.08)^{150}(0.92)^{1450}$$

Another example of this normal approximation to the binomial model is given below.

Example 5.2. A politician claims that a survey in his constituency showed that 60% of the electorate agreed with his vote on an important piece of legislation in the assembly. If it is assumed that this is correct for the time being and an impartial survey of 400 voters is taken in this constituency, what is the probability that the sample will yield less than 50% in agreement?

Assuming that the probability of agreement for 400 voters is 0.6, this problem can be treated as a binomial distribution problem with $p = 0.6$ and $n = 400$. For such a large value of n, since $n(1 - p)$ [which is 160] > 10, we apply the normal approximation to binomial probability of Eq. 5.9

$$z = \frac{0.5 - 0.6}{\sqrt{\dfrac{(0.6)(0.4)}{400}}} = -4.08$$

Since the probability of $z = -4.08$ is too small as in Appendix Table 1(a), the claim of the politician of 60% backing is to be discredited.

5.2.3 Binomial Model

The Binomial Model can be developed in the following way if we assume that probability of occurrence or success is p and non-occurence or failure is q (= $1 - p$) so that we can have:

1. The probability that n successive events will all be failure is $q^n = (1 - p)^n$

2. The probability that the nth event will be a success but all the preceding ones will be failure is

$$f(n\text{th}) = (1 - p)^{n-1}.\,p$$

3. The probability of one success in a series of events is

$$f(1) = n\,(1 - p)^{n-1}.\;p$$

since the success can occur on any of the n events.

4. The probability that $(n - r)$ failures occur, followed by r successes, is

$$p = (1 - p)^{n-r} . p^r$$

5. However the $(n - r)$ failures and r successes may be arranged in $\binom{n}{r}$ combinations, or equivalently in $\frac{n!}{(n-r)!r!}$ different ways.

So, the probability that r successes will be placed in n events is

$$f(n - r) = (1 - p)^{n-r}.p^r$$

$$f(r) = \frac{n!}{(n-r)!r!}(1-p)^{n-r}.p^r \quad \text{[as in Eq. (5.4)]}$$

Thus, we have noted above that when the binomial distributions are near the centre (i.e. those with p around 0.5), a few values of n will produce a symmetrical distribution and as the p becomes smaller or larger than 0.5, we can still bring about an approximate symmetry to the distribution, but we require more and more large values of n. Thus the binomial distribution of (0.2 + 0.8) produces a noticeably skew form (as in Fig. 5.1), but that of $(0.2 + 0.8)^{100}$ produces one nearly symmetrical about the mean as the binomial distribution limits to a normal distribution. The problem is that such large expansions become unwieldy and the advantage of similarity between the distributions should be taken. One rule of thumb for using the normal approximation to binomial distribution is that both the mean np (or nq) and the variance, npq of a binomial distribution are greater than 10.

5.2.4 Uses of Binomial Distribution

One common use of the binomial distribution is in the situation where there is no *a priori* knowledge as to the values for the probability of an occurrence or non-occurrence. In this situation, the required probabilities are estimated from the data. Let us suppose that we have superimposed a grid of unit size over a rectangular area of interest. We count the number of occurrences of a particular item (or event) in each grid square. For example, let the number of grids are 12 in 3×4 form and total number of items occurred in the area are 16. Now if in the grid network, there are 3 grid squares with no (zero) occurrence, 5 grid with one occurrence, 1 grid with two and 3 grid with three occurrences, we can find the mean number of items per grid square as equal to

$$= \frac{\text{Number of items observed}}{\text{Number of grid squares counted}} = \frac{16}{12} = 1.33$$

Now, if we are concerned with the probability of occurrence of three items per grid we have

$$f(3) = \frac{\text{Mean number of items per grid square}}{\text{Maximum number of items in any grid square}} = \frac{1.33}{3} = 0.44$$

Hence, the probability of non-occurrence $q(3) = 1 - p(3) = 0.56$.

Using the expansion $(p + q)^3$ and substituting $p = 0.44$ and $q = 0.56$ we can find the probability of occurrence of three items per grid square as

$$f(3) = 1 - p^3 = (0.44)^3 = 0.0852$$

Now, the multiplication of the above value of the p^3 with the total number of grid squares (which is 12 in the present example) gives us the expected number of grid squares with the maximum number of items (equal to 3) as 1.0224 ≈ one grid square.

Finally, it should be borne in mind that in normal distributions, the probability values obtained apply strictly to the total population but the binomial distributions predict the occurrence (or, non-occurrence) of an event within a limited range of data. For example, a 20% probability of a particular value being exceeded in a normal distribution is applicable to the total population. This suggests that in 10,000 occurrences, the condition will be met in 2,000 cases but it does not imply that out of 10 cases in occurrence this value will be exceeded. Since the binomial distribution enables us to calculate the probability of the occurrences of an event within a limited range of data, this method of analysis is having wide scope of applications for geo-science systems, particularly for summarizing the long-term behaviour of sequences of events which are independent of the history of occurrences and non-occurrences in the sequence of events. This also leads to various generalisations of the binomial model which we will be discussing now.

5.3 GENERALISATION OF BINOMIAL MODEL

Many different types of probability functions have been developed to represent various discrete phenomena which are various generalizations of binomial probability functions. The generalisation of binomial model include:

(1) Multinomial distribution
(2) Hypergeometric distribution
(3) Negative binomial distribution
(4) Geometric distribution

5.3.1 Multinomial Distribution

A binomial distribution is concerned with the two possible outcomes on each experimental trial of an event like tossing a coin, occurrence of an event or non-occurrence etc. In 'multinomial distribution' there is the involvement of more than two outcomes on each trial. For example, casting a dice is having six possible outcomes i.e. 1, 2, 3, 4, 5 or 6. The assumptions under the multinomial are the same as in binomial. Thus, with a r-sided dice, the probability of obtaining r_1, occurrences of state 1, r_2 occurrences of state 2.........and r occurrences of state x when the dice is cast n times is the multinomial distribution. So when an event can occur in more than two alternative ways, the probabilities of different results that might be obtained in sets of n such independent events, are given multinomial distribution, e.g. if weather at a place in year is 50% scorching, 30% muggy and 20% fair, then finding the probability of sample of 6 days selected at random having 3 muggy, 2 scorching and 1 fair weather is a multinomial distribution issue. Hence treating this probability distribution function as a particular way of binomial distribution, we have:

$$f(r_1, r_2, r_3) = \frac{n!}{r_1!r_2!r_3!} p_1^{r_1}.p_2^{r_2}.p_3^{r_3} \quad \text{... (5.10)}$$

In the present case, $f(r_1, r_2, r_3) = \frac{6!}{2!3!1!}(0.5)^2.(0.3)^3.(0.2)^1$

where $r_1 = 2$, $r_2 = 3$ and $r_3 = 1$ and $n = 6$

$= (60)\,(0.25)\,(0.09)\,(0.2) = 0.27$

Again, if we want to obtain the probable frequencies of the various outcomes in N sets of n trials, the following expression will be used

$$N(p+q)^n = N[p^n + \frac{n!}{(n-1)!}p^{n-1}.q + \frac{n!}{2!(n-2)!}p^{n-2}.q^2 + \ldots$$

$$+ \frac{n!}{r!(n-r)!}p^{n-r}.q^r + \ldots + q^n] \quad \text{... (5.11)}$$

Example 5.3 For a place during the rainy season half of the total number of days is assumed to be rainy days. Now if a 100 random sets of samples of 10 days' records of occurrences of rains are taken, how many sets will report that rain did not occur on 3 days or less?

We have the binomial equation as $100\,(p+q)^{10}$ in which $p = q = 0.5$ given.

Now for finding the sets not reporting rainy days as 3 or less we have $p_x = 100p^{10} + 100(10)p^9q + 100(45)p^8q^2 + 100(120)p^7q^3$

$$f_x = \frac{100}{1024} + \frac{1000}{1024} + \frac{4500}{1024} + \frac{12000}{1024}$$

$$f_x = \frac{17600}{1024} = 17.2$$

Hence 17 sets of the total of 100 sets will be expected to report that 3 days or less were without rain out of 10 days in the rainy season.

5.3.2 Hypergeometric Distribution

This distribution is a random variable distribution which arises in the context of sampling without replacement from a relatively small population. Thus whereas in binomial distribution the samples are replaced everytime keeping the probability of the event constant from trial to trial, in the hypergeometric distribution, there is no replacement making the probability changing each time a draw is made. Hence unlike binomial distribution in hypergeometric distribution it cannot be said that the outcomes of one event is independent of the other outcomes.

This distribution is formed by the ratio of the number of ways an event of interest can occur over the total ways any event can occur. In this condition, the binomial distribution cannot be satisfied. Thus in a total sequence of N event let there be two dichotomous occurrences X and Y so that $N = X + Y$ and $Y = N - X$. From the N events, n are randomly selected, one after the other, without replacement. Now if we assume that r is the random variable denoting the number of x (or, y) obtained then from "combinational counting technique", we have

$$\frac{X!}{r!(X-r)!} \cdot \frac{Y!}{r!(Y-r)!} \qquad \text{... (5.12)}$$

Again by combination, the n objects can be taken from the total sequence of N events in

$$N_{c_n} \text{ or } \frac{N!}{n!(N-n)!} \text{ ways} \qquad \text{... (5.13)}$$

Finally, we get the hypergeometric distribution by finding the probability of obtaining x occurrences (or, events) precisely in a sample of n events from the total sequence of N events which contains X occurrences and $N - X (= Y)$ non-occurrences.

The formula is

$$f(X = x) = \frac{\dfrac{X!}{r!(X-r)!} \cdot \dfrac{Y!}{r!(Y-r)!}}{\dfrac{N!}{n!(N-n)!}} \qquad \text{... (5.14)}$$

for X (i.e. r) $= 0, 1, 2$ ...... $[n, X]$. The symbol in $[n, X]$ means the smaller of n or X. In case n increases without limit, n/N becomes small and the hypergeometric distribution approaches binomial distribution. Hence in Eq. (5.14), the numerator is the product of the number of ways one can select X occurrences (or, successes) out of the $(X - r)$ number of occurrences in the population times the number of ways one can select Y successes out of the $(Y - r)$ number of ways in the population. In the chapter on χ^2 test, we shall take up a small sample test for 2×2 tables, Fisher's Exact test, which is based on hypergeometric distribution involving only 2 distinct type of events.

Example 5.4 Housing colonies are to be set up in 4 areas of a city. If the areas are marked P, Q, R and S, in area marked P, 5 blocks can be constructed, similarly in Q, R and S areas 4, 6 and 4 blocks can be constructed. Now if 6 housing blocks are being built by 4 promoters and it is observed that no block was constructed in the area P, 3 blocks were selected by the promoters in area Q, 2 blocks in R and 1 in S. If the selection was entirely random without replacement (from a final population), what is the probability of this selection mix?

Now, we have

$$f(0,3,1,2=6)=\frac{{}^5C_0.{}^4C_3.{}^6C_2.{}^4C_2}{{}^{19}C_6}$$

$$=\frac{\left(\frac{5!}{0!5!}\right).\left(\frac{4!}{3!(4-3)!}\right).\left(\frac{6!}{2!(6-2)!}\right).\left(\frac{4!}{2!(4-2)!}\right)}{\frac{19!}{6!(19-6)!}}$$

$$=\frac{1.4.15.6}{27132}=0.0133.$$

Example 5.5 A sample of 10 is taken from a population of 20 rainy days. If this population contains 2 days without rain, determine the probability distribution which states that the 2 days without rain will not be found in the sample.

If we consider that the probability that the samples of 10 rainy days should not have the dry days then with $n = 10$ and $p = 2/20 = 0.10$, we can have the probability:

$$f(r=0)=\frac{10}{0!10!}.(0.10)^0(0.90)^{10}=0.34868.$$

But in the present example, the sampling has to be done without replacement (so that the sample of rainy days do not contain the 2 non-rainy days), probability of dry days does not remain equal to 0.10 on each trial. The value of probability, p on any particular trial depends on what has been already selected on previous trials. Now the probabilites that all the 10 samples are with dry days are calculated by combinational analysis as equal to

$$\frac{18}{20}\times\frac{17}{19}\times\frac{16}{18}\times\frac{15}{17}\times\frac{14}{16}\times\frac{13}{15}\times\frac{12}{14}\times\frac{11}{13}\times\frac{10}{12}\times\frac{9}{11}=0.2368$$

Note the above example is for hypergeometric distribution with more than 2 possible outcomes per trial for which hypergeometric distribution with only two outcomes

$$f(x = X) = \frac{{}^{(N-X)}C_{(n-x)} \cdot {}^{X}C_{x}}{{}^{N}C_{n}}$$

is extended to any number of possible outcomes on a given trial by employing

$$f(x_1, x_2, ..., x_k) = \frac{X_{1_{C_{x1}}} . X_{2_{C_{x2}}} ... X_{k_{C_{xk}}}}{N_{C_n}} \tag{5.15}$$

where $\sum_{i=1}^{k} X_i = N$ and $\sum_{i=1}^{k} x_i = n$, N and n are the population and total sample size respectively; X_i and x_i are the number in population or sample with possible outcomes.

In the present case the Eq. (5.15) can be written as:

$$\frac{{}^{18}C_{10} \cdot {}^{10}C_{0}}{{}^{20}C_{10}} = \frac{\frac{18!}{10!(18-10)!} \cdot \frac{10!}{0!(10-0)!}}{\frac{20!}{10!(20-10)!}}$$

$$= \frac{43758}{184756} = 0.2368$$

and we get the same probability as it was from the earlier method of combinational analysis.

5.3.3. Negative Binomial Distribution

If the regional success ratio in an oil-drilling operation is assumed to be 10%, the probability of oil being struck in two wells bored is

$$f_2 = \frac{(r+x-1)!}{(r-1)!x!}(q)^x . p^r$$

$$= \frac{(2+0-1)!}{(2-1)!0!}(0.90)^0 (0.10)^2$$

$$= 0.01$$

The possibility that the five oil wells are required to be dug to achieve 2 successes is

$$f_2 = \frac{(2+3-1)!}{(2-1)!3!}(0.90)^3 . (0.10)^2 = 0.029$$

The probabilities calculated are low because they relate to the likelihood of obtaining 2 successes and exactly x dry wells. It may be more useful to consider the distribution of the probability that more than x wells must be drilled before the goal of r discoveries is achieved. In this case, the experiment is performed under the same conditions, till a fixed number of successes is achieved. This is found by first calculating the negative binomial distribution in "cumulative" form which gives the probability that the goal of 2 successes will be achieved in $(x + r)$ or fewer wells.

Negative binomial distribution is used where kth occurrences ($k \geq 2$ events) are expected. To investigate its probability we require to have 2, 3, 4 ...., up to n events.This negative binomial has 2 parameters: (1) u, the parameter estimated from sample mean and (2) k, a parameter as a measure for degree of clumping. This negative binomial distribution is the distribution in which one could obtain the same binomial distribution by expanding

$$\left[\frac{p}{1+p}+\frac{1}{1+p}\right]^{-k} \qquad \text{... (5.16)}$$

which is a binomial term with negative exponent.

The negative binomial distribution defines the probability that the kth occurrence is at the xth time. It is written

$$f_x = \left({}^{x-1}C_{k-1}\right).p^k q^{x-k} \qquad \text{... (5.17)}$$

with $x = k,\ k + 1$, ...

It has a mean, $\mu_x = \dfrac{k}{p}$ ... (5.18)

and a standard deviation, $\sigma_x = \dfrac{\sqrt{kq}}{p}$ ... (5.19)

To find the probability of the third occurrence ($k = 3$) of a 5-year flood ($p = 0.2$) in the tenth year ($x = 10$), the answer is

$$f_{10} = \left(10 - {}^1C_{3-1}\right)(0.2)^3(0.8)^{10-3} = \frac{9!}{2!(9-2)!}(0.2)^3(0.8)^7$$

$$= (36)\ (0.008)\ (0.20972)$$

$$= 0.06040$$

In this distribution both the parameters, mean and standard deviation are estimated from the data. The mean, μ in this example is $\mu = \frac{k}{p} = \frac{3}{0.2} = 15$ and hence k is small relative to the mean which will extend the right side of the negative binomial distribution. This negative binomial distribution can be used to model the occurrence of clustered points in space in a manner equivalent to use the Poission distribtuion (discussed in next chapter) to model randomly arranged points.

5.3.4 Geometric Distribution

The geometric distribution enables the probability of the first occurrence happening at the *x*th time (trial) to be found. Thus whereas the negatve binomial distribution is concerned with the *k*th occurrence at the *x*th time the geometric distribution is concerned with the first occurrence at the *x*th time. Thus geometric distribution is a special case of the negative binomial, where interest is focused on the number of trials prior to the initial success. Both the distributions are having their applications in hydrological sequences and the parameters, mean and standard deviation are computed from the data. Hence the geometric distribution defines its probability function, *f(x)* as

$$f(x) = pq^{x-1} \quad \text{...(5.20)}$$

where $x = 1, 2, 3, ...$

The geometric probability distribution has parameters:

$$\text{means, } \mu_x = \frac{1}{p} \quad \text{... (5.21)}$$

$$\text{and standard deviation } \sigma_x = \frac{q}{p} \quad \text{... (5.22)}$$

Since the mean is given by $1/p$ it can be interpreted by saying that, on average, a *T*-year event occurs on the *T*th year which is also the return period. Now by using the geometric distribution function p_x the probability of, say, a 5-year flood ($T = p = 0.2$) occuring in the third year ($x = 3$) of a sequence of years is calculated as

$$f(3) = (0.2)\,(0.8)^{3-1} = (0.2)\,(0.8)^2 = 0.128$$

LIST OF FORMULAE

I. *Binomial Distribution*

Probability Function for Binomial Distribution

$$f(r) = \frac{n!}{r!(n-r)!} . p^r . q^{n-r}$$

or,

$$f(r) = \binom{n}{r}^r .p^r .q^{n-r}$$

Binomial Expansion

$$(p+q)^n = p^n + \frac{n!}{(n-1)!} p^{n-1} q + \frac{n!}{2!(n-2)!} p^{n-1} q^2 +$$

$$\ldots + \frac{n!}{r!(n-r)!} . p^{n-r} . q^r + \ldots + q^n$$

Parameters of Binomial Distribution

Mean: $\bar{x} = np$ and standard deviation $\sigma = \sqrt{npq}$

Approximation of Binomial Distribution to a *z*-score

$$z = \frac{\frac{x_i}{n} - p}{\sqrt{\frac{pq}{n}}} = \frac{x_i - np}{\sqrt{npq}}$$

II. *Extensions of Binomial Model*

(1) Multinomial Distribution

Probability Function for Multinomial Distribution

$$f(r_1, r_2, .., r_k) = \frac{n!}{r_1! r_2! ... r_k!} . p_1^{r_1}, p_2^{r_2}, ... p_k^{r_k}$$

Multinomial Expansion for *N* sets of *n* trials of two dichotomous events

$$N(p+q)^n = N[p^n + \frac{n!}{(n-1)!} p^{n-1} . q + \frac{n!}{2!(n-2)!}$$

$$p^{n-2} q^2 + \ldots + \frac{n!}{r!(n-r)!} p^{n-r} . q^r + \ldots + q^n]$$

(2) Hypergeometric Distribution

Probability Function for Hypergeometric Distribution

$$f(X = x) = \frac{\frac{X!}{r!(X-r)!} . \frac{Y!}{r!(Y-r)!}}{\frac{N!}{n!(N-n)!}}$$

For k number of possible outcomes

$$f(x_1, x_2, ..., x_k) = \frac{X_{1C_{x1}} . X_{2C_{x2}} ... X_{kC_{xk}}}{N_{C_n}}$$

III. *Negative Binomial Distribution*

Probability Function for Negative Binomial Distribution

$f(x) = \left({}^{x-1}c_{k-1}\right).p^k.q^{x-k}$ with $x = k, k+1, ...$

Parameters of Negative Binomial Distribution

Mean $\mu_x = \frac{k}{p}$ and standard deviation $\sigma_x = \frac{\sqrt{kq}}{p}$

IV. *Geometric Distribution*

Probability Function for Geometric Distribution

$f(x) = pq^{x-1}$ with $x = 1, 2, 3, ...$

Parameters of Geometric Distribution

Mean $\mu_x = \frac{1}{p}$ and standard deviation $\sigma_x = \frac{q}{p}$

EXERCISES

5.1 The annual rainfall for five years are drawn at random from rainfall data of a place. In the actual rainfall of the place annual rainfall is found below normal in 10% of years. Find the probabilities of 0, 1, 2, 3, 4 and 5 years with rainfall below normal in the sample.

5.2 Assuming the probability that rain occurs in football sesson at a place is 0.01, find the probability that out a group of 10 games played in sesson, (a) none, (b) exactly one, (c) not more than one, (d) more than one and (e) at least one will be played on a ground soaked with rain water.

5.3 In the rainfall distribution of an area, it is found that years with crop failure and years with no crop failure are equally distributed. If the rainfall of two years are taken, in what proportions would you expect successive years (a) with crop failure, (b) with no crop failure and (c) one of each.

5.4 A binomial distribution has $n = 10, p = 0.3$. Find the mean and variance of the distribution.

5.5 A 50-year record of a place regarding annual rainfall consists of the years with more than the normal rainfall and the years with less than normal. The probability of the years having a rainfall more than normal is believed to be 0.60. What is the probability that the number of years having more than normal rainfall will be greater than 25?

5.6 For a place, the probability of any day in a week being wet is 0.57. Calculate the probability of (a) there being exactly 2 wet days out of any 7, (b) either day of a weekend is wet (c) at least 2 days wet out of 7 days in a week being wet.

5.7 In a housing area 250 houses were constructed 15 years ago. A survey regarding the present ownership of the houses reveal that because of bad construction 130 houses have undergone at least one change of ownership. If a sample of 10 houses are taken in this area, what would be the probability that at least one house has undergone change of ownership. Calculate this probability by the binomial as well as by normal distribution methods.

5.8 A sample of 365 days of weather data of a place is found to have 40 days foggy. What are the probable limits of foggy days of the place?

5.9 Ten days' weather record at a place shows that 3 days are rainy days and 7 days are without any rain. If a random sample of 6 days' weather record is taken without replacement what will be the probability that the sample will contain 2 rainy days. Deduce the probability when the same samples are taken with replacement.

5.10 Analysis of data on maximum 1 day rainfall depth at a place indicates that a depth of 30 cm. rainfall has a return period of 50 years. Determine the probability of 1-day rainfall depth equal to or greater than 30 cm. at the place occuring (a) once in 20 successive years, (b) 2 times in 15 successive years and (c) at least once in 20 successive years.

5.11. A community consists of 50% Hindus, 30% Muslims and 20% Christians. If a sample of 6 individuals is selected randomly, what is the probability that there are 2 Hindus, 3 Muslims and 1 Christian.

6

Probabilistic Treatment in Geo-science Systems III (Poisson Distribution)

6.1 POISSON DISTRIBUTION

'Poisson distribution' deals with the calculation of probability of very rare or isolated events which occur instantaneously and independently on a time horizon and are detected only when the samples are very large. This distribution takes only the events which are isolated events in a continuum of time or distance, area or volume, so that it is not possible to say how often the event did not occur. Seen in this way, the *Poisson distribution* is the limit of the binomial when the probability of an event (p) is very small, whilst the total number of trials (the sample size, n) is very large. Thus this distribution is most commonly applied to problems concerned with discrete events in time or space particularly when occurrence of the events is low relative to the maximum possible that could occur. A French mathematician Simeon Denis Poisson invented this distribution in 1837.

Classical examples of Poisson distribution is occurrences of floods in an alluvial plain or road accidents in a city. In either of these two cases p (= the actual number of occurrences of flood or road accidents in very small) and n (= the number of years of 'record' flood or the numbers of people travelling by the road) is very large, so that np = the average numbers of floods over a period or road accidents on an average day is a finite constant. This mathematical limit is found useful in explaining the behaviour of rare events and it is named Poisson distribution. This distribution thus becomes limiting form of binomial distribution with 'n' becoming large and 'p' approaching zero. Again like binomial distribution, this distribution is a discrete variable distribution and so histograms are used. But since n is large and p is very small approaching zero so unlike binomial, the normal curve approximation to this distribution is unsatisfactory. In this respect, the distribution of the event is markedly skewed. Again the Poisson distribution is the limit of the binomial as n

$\to \infty$ with np = constant. As the number of units increase, np remaining constant, p must get smaller and smaller and the binomial distribution gradually approaches the Poisson.

6.2 POISSON DISTRIBUTION FUNCTION

The Poisson distribution model is based upon two assumptions: (i) that the probability of n occurrences in time or space is exactly the same no matter where consideration of an interval begins or ends, (ii) events which occur in any interval of time or space are independent of events that occur in other intervals of time or space. These two basic assumptions are the foundation for the Poisson series.

Poisson distribution is based on the natural law of growth known as the 'exponential law', $(1 + 1/n)^n$ which can be simply obtained from the expansion of binomial series, $(p + q)^n$ as

$$(p+q)^n = p^n\left(1+\frac{q}{p}\right)^n = p^n\,[1+n\left(\frac{q}{p}\right)+\frac{n(n-1)}{2!}\left(\frac{q}{p}\right)^2+..$$

$$+\frac{n(n-1)\ldots(n-r+1)}{r!}\left(\frac{q}{p}\right)^r+\ldots+\left(\frac{q}{p}\right)^n] \qquad \ldots (6.1)$$

Now in Eq. 6.1 as n tends to infinity (i.e. n is very large), fractions $\frac{n-1}{n}$ etc. tends to one and thus by substituting $p = 1$ and $q = 1/n$ in Eq. (6.1) above, we get

$$\underset{n\to\infty}{Lt}\left(1+\frac{1}{n}\right)^n = 1+n.\frac{1}{n}+\frac{n(n-1)}{2!}.\left(\frac{1}{n}\right)^2+\ldots+\frac{n(n-1)\ldots(n-r+1)}{r!}.\left(\frac{1}{n}\right)^r+\ldots+\left(\frac{1}{n}\right)^n$$

$$=1+1+\frac{1}{2!}+\frac{1}{3!}+\ldots\frac{1}{r!}+\ldots+\frac{1}{n!} \qquad \ldots (6.2)$$

$$=\frac{1^0}{0!}+\frac{1^1}{1!}+\frac{1^2}{2!}+\frac{1^3}{3!}+\ldots+\frac{1^r}{r!}+\ldots+\frac{1^n}{n!}$$

$= 2.71828^* = e$ (a constant) ... (6.3)

By a similar process, this value e can be raised to any power z to give

$$e^z = 1 + \frac{z^1}{1!} + \frac{z^2}{2!} + \frac{z^3}{3!} + \ldots + \frac{z^r}{r!} + \ldots + \frac{z^n}{n!} \quad \ldots (6.4)^{**}$$

Now to make use of a theoretical probability distribution, the total of the terms in this exponential equation must sum to unity. This can be done by multiplying Eq. (6.4) by e^{-z} as

$$e^{-z} . e^{z} = e^0 = 1 \quad \ldots (6.5)$$

Thus the relative frequencies or probability of the Poisson distribution are given by the successive $n + 1$ terms of Eq. (6.4) as

$$1 = e^{-z}\left(1 + \frac{z^1}{1!} + \frac{z^2}{2!} + \frac{z^3}{3!} + \ldots + \frac{z^r}{r!} + \ldots + \frac{z^n}{n!}\right) \quad \ldots (6.6)$$

where z can be taken as the expected number of occurrences of the event (= the mean value for the set of data). In this series of Eq. (6.6), the 1st term records the probability of the occurrence of the 0th event e^{-z} [0] and the following $(n + 1)$ terms of the expansion give the probabilities of occurrence of the 1st to nth event.

The above arrangement fulfils our requirement that the successive terms will add to unity. Unfortunately, this series is infinitely long, but since the series is convergent so the values of successive terms quickly become very small. Earlier in the expansion of $\underset{n\to\infty}{Lt}\left(1+\frac{1}{n}\right)^n$ we find by totalling the first nine terms we almost reach the value of the exponent, $e = 2.71828$.

Thus from the above exponential expansion in Eq. (6.6) we get the probability of any

---

* This expression for $\underset{n\to\alpha}{Lt}\left(1+\frac{1}{n}\right)^n$ is a continuous one but since the series $1 + 1 + \frac{1}{2!} + \frac{1}{3!} + \frac{1}{4!} + \ldots + \frac{1}{r!} + \ldots + \frac{1}{n!}$ is 'convergent' and so when n is large, it is possible to aggregate the partial sum of manageable terms as

$$\underset{n\to\alpha}{Lt}\left(1+\frac{1}{n}\right)^n = 1 + 1 + 0.5 + 0.1\dot{6} + 0.041\dot{6} + 0.0083 + 0.00138 + 0.000198412 + 0.000024801$$

$= 2.7182781$ (≈ 2.71828)

$= e$ (a constant)

** similarly e^{-z} produces an oscillating series : $e^{-z} = 1 - \frac{z}{1!} + \frac{z^2}{2!} - \frac{z^3}{3!}$

number of occurrences r in a Poisson distribution. Hence the Poisson distribution function is

$$f(r) = e^{-z}.\frac{z^r}{r!} = 2.71828^{-z}.\frac{z^r}{r!} \qquad \text{... (6.7)}$$

for r is any number of occurrences equal to 0, 1, 2, ... and z is the empirical density figure, an estimate of u, the mean. In this discrete probability distribution function, both the mean, np and the variance $\sqrt{npq}$ are equal to z and less than 5. So the successive values of $p(r)$ depends on the number of occurrences r and the mean value of the set of the data. Thus using z as an estimate of u, the mean, the probabilities of occurrences of $r = 0, 1, 2, ..., r$

$$f(0) \quad = e^{-z}$$

$$f(1) \quad = z^1.\frac{e^{-z}}{1!} \quad \text{or} \quad \frac{z^1}{1}.p(0)$$

$$f(2) \quad = z^2.\frac{e^{-z}}{2!} \quad \text{or} \quad \frac{z}{2}.(p)(1)$$

. .

. .

. .

$$f(r) \quad = z^r.\frac{e^{-z}}{r!} \quad \text{or} \quad \frac{z}{r}.p(r-1)$$

In the above, note that once $p(0)$ is computed, the simplified form of computing these probabilities (shown on the far r.h.s. of the above equations) eliminates the need to compute factorials. The two sets of computed probabilities of the Poisson distribution is given in Table 6.1 with $z = 1$ and $z = 2$ for a number of occurrences. The computation of the exponential function e^{-z} (or 2.71828^{-z}) involves the use of logarithms which is worked out in the List of Mathematical and Statistical Symbols. However, for various values of z up to 7 the value of exponential function e^{-z} is given in Appendix Table III, otherwise it can be calculated also with the help of a calculator. For example, if $z = 0.437$, $e^{-z} = \text{In}^{-1(-0.437)} = 0.64597$* where 'In' denotes a logarithm to the base e (natural logarithm).

---

* If $e^{-z} = y$ we write $-z = \log_e y = \ln y$

so, $y = \frac{1}{\text{In}}(-z) = \text{In}^{-1(-z)}$

Table 6.1: Computed Poisson probabilities for 0, 1, 2....*r*, occurrences per quadrat (or, sampling unit $z = 1$ and $z = 2$)

(*r*)	for $z = 1$, $p(r)$	(*r*)	for $z = 2$, $p(r)$
0	0.36788	0	0.13531
1	0.36788	1	0.27067
2	0.18394	2	0.27067
3	0.06131	3	0.18045
4	0.01533	4	0.09022
5	0.00307	5	0.03609
6	0.00051	6	0.01203
7	0.00007	7	0.00344
		8	0.00002
		9	0.00000037
	Σ 0.99999		Σ 0.99890037

If the Poisson distribution values for $z = 1$ and $z = 2$ given in Table 6.1 are plotted, the histograms will be skewed. It is also apparent from Table 6.1 that as z increases, the point at which the maximum probability occurs also moves towards successively higher values of occurrence.

The *Expected Poisson Frequencies*, E_x is calculated from the above Poisson Probability model by multiplying each probability by the total number (n) of sampling units in the sample. Thus, letting E_x as the expected Poisson frequencies of $x = 0, 1, 2 \ldots r$ individual occurrences, we have

$$E_0 = n.p\,(0)$$
$$E_1 = n.p\,(1)$$
$$E_2 = n.p\,(2)$$
$$\vdots$$
$$E_r = n.p\,(r)$$

Each of these equations represent a frequency class, resulting in a total of $q = r + 1$ frequency classes of expected occurrences. The expected number for occurrences should not be less than 5 individuals ideally and no expected value less than 1. In case E_r be less than these desired minimums, the frequency classes should be pooled.

Example 6.1 Records over a long period has shown that in a city the maximum number of fatal road accidents that have taken place in a day is 8 and the mean number of fatal accidents per day is 1. What is the probability that the number of total accidents taking place in a day will be more than 8?

The probability of occurrences of 8 or less fatal accidents is the cumulative probability function for the Poisson model. This is

$$f(z \leq 8) = \sum_{r=0}^{8} e^{-z} \cdot \frac{z^r}{r!}$$

equal to

$$f(z \leq 8) = p(0) + p(1) + p(2) + p(3) + \ldots + p(8)$$

$$= e^{-1}\left(1 + \frac{1}{1!} + \frac{1}{2!} + \frac{1}{3!} + \ldots + \frac{1}{8!}\right)$$

$$= e^{-1}(1 + 1 + 0.5 + 0.1\dot{6} + 0.041\dot{6} + 0.008\dot{3} + 0.0013\dot{8}$$

$$+ 0.000198 + 0.0000248)$$

$$= e^{-1}(2.71827) = (0.36788)(2.71827)$$

$$= 0.9999956$$

Hence $f(z > 8) = 1 - 0.9999956 = 0.0000044$ is the probability that the number of accidents will exceed 8 in a day.

6.2.1 Characteristics of Poisson Distribution

The Poisson distribution series is an infinite series, it can be extended infinitely though only the first eight or nine terms practically sum to unity (Table 6.1)

The Poisson distribution is usually strongly skewed so that the small number of occasions on which a rare event (such as a flood) occurs, is against the large number of occasions on which it only occurs rarely. Therefore, any estimation of probabilities by the normal distribution function overestimates the frequency of the event with few occurrences. Any estimate by the binomial distribution will be equally too divorced from reality to be of any real value. Figure 6.1 illustrates how with increasing estimate of the mean of an event the Poisson curve changes from an extremely skewed position at $z = 0.6$ to a balanced one at $z = 6$.

Like the binomial distribution, the mean and variance of the data can be a factor for identifying the Poisson distribution as a more fitting one in preference to normal or binomial distribution. One of the most remarkable qualities of the Poisson distribution, almost its hallmark is the equality of the mean and the variance i.e. the mean and variance each equal to z, the average density of events, so standard deviation is equal to $\sqrt{z}$. The variance to mean ratio in a sample, $s^2/\bar{x}$ is usually referred to as "index of dispersion" in ecology. If it is the question of spatial pattern based on the number of individuals per sampling units, SU (= quadrats), where the SUs are discrete, we have index of dispersion,

$\text{ID} = s^2 / \bar{x} = \left(\sum_{i=1}^{n} \frac{(x_i - \bar{x})^2}{n} \right) / \bar{x}$ where x_i = number of occurrence in the *i*th SU and *n* is the total number of SUs. Finally, a notional guide can be given regarding the use of a particular probability distribution among normal, binomial and Poisson which are closely related to each other to a data size for determining the probability of an event occurring in the data. The guide is as follows.

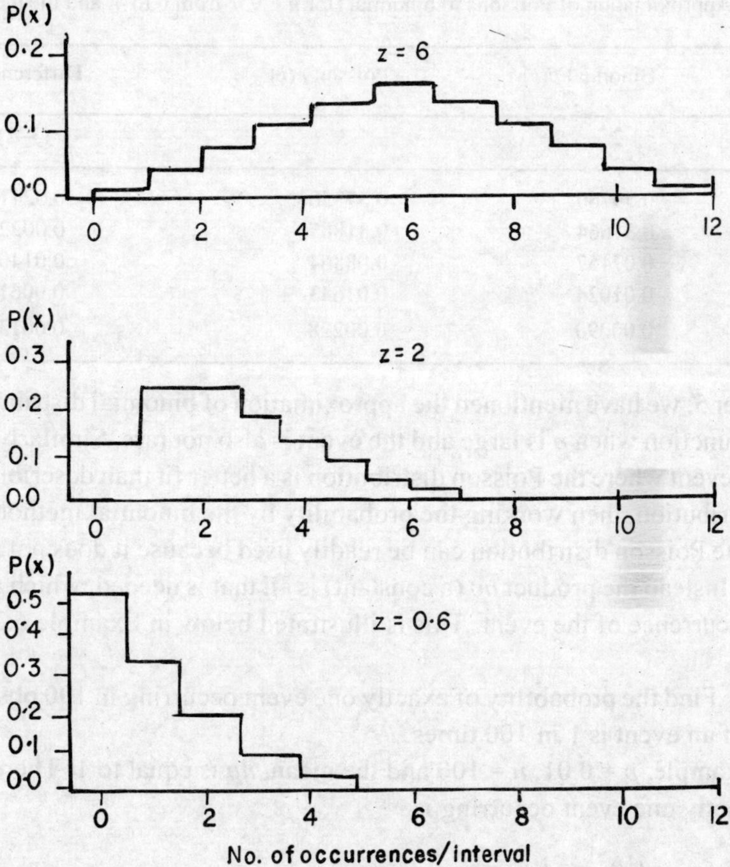

Fig. 6.1 : Poisson Probability Distributions with Different *z* values

1. In the set of data, when *n* is sufficiently large so that both $np \geq 5$ and $nq \geq 5$, the use of the normal distribution will provide a good approximation.
2. In the set of data when mean *np* or $nq \geq 5$ and variance $npq \geq 9$ but *n* is small and when neither *p* nor *q* are not rare events, then the binomial distribution provides a good approximation. In Table 6.2 here, the values of Poisson probabilities are given with the equivalent binomial probabilities for $n = 9$, $r = 0$ to 4 and the mean $z =$

0.556. The closeness of the Poisson's approximation to the binomial distribution can be noted in the difference column of Table 6.2.

3. The Poisson distribution is used in preference to the binomial distribution when n is large, $n > 20$ and $p \leq 0.1$, and the probability of the event is so rare that the mean of the distribution, $np \leq 5$. Also most statistical tables (Appendix Table II) do not give the binomial distribution for n larger than 20, so that approximation by Poisson distribution then becomes necessary.

Table 6.2 : Approximation of Poissons to binomial (for $n = 9$, r from 0 to 4. and the mean $z = 0.556$)

r	Binomial $p(r)$	Poisson $p(r)$	Difference
1	2	3	(3–2)
0	0.59760	0.57350	−0.02410
1	0.31664	0.31887	0.00223
2	0.07457	0.08864	0.01407
3	0.01024	0.01643	0.00619
4	0.00090	0.00228	0.00138

In Chapter 5, we have mentioned the approximation of binomial distribution by normal distribution function when n is large and the event is also not rare. Similarly due to the rare nature of the event where the Poisson distribution is a better fit than describing the event by binomial distribution, then working the probability by the binomial method becomes very tedious but the Poisson distribution can be readily used because it does not require either n or p directly. Instead the product np (a constant) is all that is needed, which also is given by the rate of occurrence of the event. This is illustrated below in Example 6.2.

Example 6.2 Find the probability of exactly one event occurring in 100 observations if the probability of an event is 1 in 100 times.

In this example, $p = 0.01$, $n = 100$ and the mean, np is equal to 1. Therefore, the probability of exactly one event occurring is

$$f(1) = \frac{1!e^{-1}}{1!} = \frac{1}{e} = 0.36788$$

Now, if instead of the Poisson distribution, we apply the binomial probability function, we face the following cumbersome situation:

$$f(1) = \frac{100!}{1!(100-1)!}.(0.01)^1 (0.99)^{99} = 0.36973$$

It may also be noted that when the mean of the distribution is known, the Poisson distribution is also known. In this respect, a binomial distribution requires two things to be known: the probability of an event's occurrence p and the total number of occurrence, n. On the other hand, a normal distribution requires two attributes to be known: the mean and the standard deviation. Hence when more than one probability model fit a set of data well, it is necessary that model conceptualisation should precede the data fitting, the latter should be pursued always as an end in itself.

6.2.2 Merits And Demerits of Poisson Distribution

As it is mentioned earlier, the Poisson distribution is useful to events (a) in time or (b) occurrences in space which are rare and independent of each other. For events in time the Poisson distribution provides a crude model in the form of frequency distributions by extracting very occasional flood from a large record of data. But since a flood discharge can recur on two or three consecutive days as the result of a major storm, so independence of flood events in time cannot be assured. Hence, for estimating the flood events in a given time interval by certain other probability density functions which we have treated as 'extreme value distribution' in the next chapter separately. But other rare events like the probability distribution of occurrence of major earthquakes or incidence of a rare disease, or of road accidents in a city in a time period can be realistically approximated with the Poisson distribution.

Secondly, Poisson distribution is suited in describing the actual frequencies of independent event accorded by gridded units over an area. Thus the Poisson distribution can be used to model random placement of points in space. "Quadrat" as mentioned earlier is the name given to gridding an area by a consistent size and shape. More often square quadrats are chosen. After the quadrat size and shape is chosen, the quadrats are placed on the map, the points contained in each are counted and results are placed in a tabulated form for representing the observed frequencies. This is known as "quadrat counting". The quadrats form the *Samping Units* (SU). Note that it is incorrect to consider the sampling units (or, quadrats) as samples since a sample consist of a collection of sampling units. For example, using a 1m^2 SU, a sample might consist of 20 such SUs randomly located throughout the study area. The method is best understood by an example. Suppose an area is divided into a lattice of $5 \times 5 = 25$ square quadrats and 5 points are located independently over this area. Now if we ask what is the probability that a quadrat will contain one point, we have to solve it by the binomial formula as

$$f'(r=1) = \frac{5!}{1!(5-1)!} \cdot p^1 q^4$$

where p is the probability of occurrence one point in a quadrat to 1/25 or 0.04 and q is the non-occurrence equal to $1 - p = 0.96$. Hence we can infer $p_{(r=1)} = 0.16987$. Calculating p_r for all possible values of r (from 0 to 5) gives, the binomial probability distribution of this experiment on 'point patterns'.

When the area is mosaiced into a large number of quadrats, so that each quadrat forms a minute area and the probability of a single point being independently located in any specified quadrat becomes very small, so that as the number of quadrats tend to be infinity ($n \to \infty$), the probability of a specified quadrat receiving an independently located point will tend towards zero ($p \to 0$). As p tends to zero for large number of quadrats, the probability generated by the binomial distribution function becomes more and more closely approximated by the Poisson distribution function,

$$f(r) = e^{-z} . \frac{z^r}{r!}$$

where $r = 0, 1, 2$, and z is the average number of points per quadrat in a spatial context.

For example, for the Poisson probability of a quadrat containing one spatial point, it will be $p(r = 1) = 0.16375$. Note for probability distribution of point patterns in this Poisson approximation model, we require to find a quadrat size which minimizes the variance of the observed set of frequencies so that we can apply the thumb rule that mean equals square root of variance (i.e. the standard deviation) in a Poisson distribution. Once the quadrat size is fixed, we use the number of points per quadrat to predict how many quadrats should contain specified number of points 0, 1, 2, ..., by Poisson method and then test them in a Chi-square test, χ^2 described in Chapter 13 to determine whether the points are distributed at random over the area.

6.3. GENERALISATION OF POISSON MODEL

The Poisson distribution model has been discussed so far for rare events as point patterns in quadrat counting. But this distribution is clearly inadequate for many situations. For example, the point patterns like agglomeration, diffusions etc. which do not conform to the two limiting assumptions of Poisson model like (i) for n points placed in a region where each possible location for a point is likely to be chosen equally and (ii) the location of each point is independent of the location of any of one point so that the selection of location of one point does not bear any influence for location of any other point. There may be another situation like the point patterns may be based on one, two or more independent processes each of which act in separate but spatially contiguous areas when they are mapped. We discuss two generalised models of Poisson distribution:

(a) Double Poisson distribution
(b) Gamma distribution

6.3.1 Double Poisson Distribution

In "Double Poisson model", random variables more than one, each associated with a probability function (representing an independent location process) are required to be combined into one compound distribution.

Thus *double Poisson distribution* is essentially a 'compound model' applied to heterogeneous nature of the location process in a study area. This distribution model depends on two or more independent processes each of which acts in separate but in spatially contiguous areas. Map pattern produced by this distribution represents a model of 'additive process' that is, two (or more) independent location processes added together. If the process in the two (or more) spatially separated locations is assumed to be Poisson, the resulting distribution is called a *double Poisson process model*. The double Poisson model is given by

$$f(r,\bar{z}_1,\bar{z}_2)=\frac{1}{2}\cdot\frac{e^{-\bar{z}_1}.\bar{z}^r}{r!}+\frac{1}{2}\cdot\frac{e^{-\bar{z}_2}.\bar{z}^r}{r!} \qquad \text{... (6.8)}$$

for $r = 0, 1, 2...$ and the two distributions are equal. In Eq. (6.8) since the two distributions are equal, so the *double Poisson coefficient* of 1/2 is assigned to each term/location process. The two parameters $\bar{z}_1$ and $\bar{z}_2$ are estimated by

$$\bar{z}_1 = \bar{x} + \sqrt{\sigma - \bar{x}} \qquad \text{... (6.9)}$$

$$\bar{z}_2 = \bar{x} - \sqrt{\sigma - \bar{x}} \qquad \text{... (6.10)}$$

where $\bar{x}$ and σ are the mean and standard deviation of the observed frequencies for the entire pattern. This model as a frequency distribution was proposed by *N.* Schilling in 1947. H. Mc Connell and I.M. Horn in 1972 justified the use of this model in working out the probability of occurrences of two types of karst features: (i) dolines which are smaller features formed above a water table along zones of weakness in limestone and successive ponding of surface run off and (ii) the collapse sink which are formed by the collapse of cavern roofs caused by surface drainage. They argued that both types of depressions should be randomly distributed over the karst area but at different densities. This double Poisson model predicts closely the observed frequency distribution of the karst depressions. We give below a specific example of application of double Poisson model.

Example 6.3 The map pattern of a heterogeneous point feature over an area is spaced into 50 quadrats, 5 quadrats in each row. The quadrats are divided into the equal groups. The upper half in 25 quadrats contain 25 points of group A of the heterogeneous event and the lower half of the remainning 25 quadrats contain another 25 points of group B of the event.

A frequency distribution of number of points per quadrat is worked out groupwise from the map and it is given in Table 6.3.

Apply the Double Poisson distribution model and find out the expected frequencies quadratwise.

Table 6.3 : Groupwise Frequency Distrbution of Number of Points per Quadrat

No. of Points per Quadrat f_i	Observed frequencies for the Point Feature		
	Group A in Top 25 Quadrats	Group B in Bottom 25 Quadrats	Total x_i
0	0	4	4
1	1	6	7
2	5	6	11
3	7	5	12
4	4	3	7
5	2	1	3
6	1	0	1
7	4	0	4
8 or more	1	0	1
Total (=N)	Total = 25	Total = 25	Grand Total 50

Mean of the data for the entire heterogeneous point pattern, is

$$\bar{x} = \frac{\sum fx_i}{N} = \frac{150}{50} = 3.00$$

and the standard deviation is

$$\sigma = \frac{\sum f(x_n - \bar{x})^2}{N} = 3.21$$

Hence $\bar{z}_1 = \bar{x} + \sqrt{\sigma - x} = 3.00 + \sqrt{3.21 - 3} = 3.45$

and, $\bar{z}_2 = \bar{x} - \sqrt{\sigma - x} = 3.00 - \sqrt{3.21 - 3} = 2.54$

Now we substitute these estimates, $\bar{z}_1$ and $\bar{z}_2$ of the mean of the two groups in the double Poisson equation model (Eq. 6.8) for various occurrences of points per quadrat to get the expected frequencies and Poisson double probabilities are shown in Table 6.4.

Table 6.4 : Expected Frequency and Poisson Double Probabilities

No. of Points per quadrat	Double Poisson probabilities	Expected frequencies
0	0.05531	2.77
1	0.15492	7.75
2.	0.22167	11.08
3	0.21633	10.82
4	0.16238	8.10
5	0.09939	4.97
6	0.05188	2.59
7	0.02366	1.18
8	0.00960	0.48

Note that in the above double Poisson model, the variance is greater than the mean and there is a tendency for the points to cluster. The clustering in a point pattern is the result of the greater density of point in few quadrats and the majority in very few or none. The goodness of fit of the above expected distribution with the observed numbers of distribution of points per quadrat can be tested by Chi-square which is explained in Chapter 13.

6.3.2 Gamma Distribution

An additional generalisation of Poisson distribution model is the 'Gamma Distribution' which is applied to a set of data where values are always positive and the mean length exceeds the modal length. Hence, gamma distribution is always a positively skewed distribution. The density function of this distribution is given by

$$Y = f(r) = \frac{e^{-x/b} . r^{a}}{\gamma\,(a+1) b^{a+1}} \qquad \text{... (6.11)}$$

where $\gamma\,(a + 1)$ is the *gamma function, a* and *b* are two parameters, with $b > 0$ and $a \geq -1$ that is an integer value. The probability $p(0)$ is itself zero in gamma distribution and thus gamma and the Poisson are very much distinct from each other. Again where as Poisson distribution is changed by one parameter, the mean density function *z*, the gamma distribution is characterised by two parameters, the

mean, $\bar{x} = b\,(a + 1)$ ... (6.12)
and variance, $\sigma^2 = b^2\,(a + 1)$

As stated above, the gamma distribution approximates observed distributions where the vast majority of occurrence are small but not zero. For examples, the frequency distribution of short-period rainfall intensities within a longer rain period or the sediment particle analysis all belong to the gamma distribution because in both the cases the occurrences may be small but not zero. Experience tells us in both these cases for large sample sizes, there will be none of zero diameter in range of sizes for particles and the short-period rainfall intensities in a long rain period may not be zero.

Example 6.4 In a sediment analysis, mean particle size is 1.2 mm and the standard deviation is 2.8 mm. Find the gamma probabilities of sediment particles of 1, 2, 3, ... , 10 mm diameter.

We have the two parameters; mean = 1.2 = $b(a+1)$ and variance σ^2 = 7.84mm = $b^2(a+1)$. From these we have $b = \frac{1.2}{(a+1)}$. Now substituting this value of b in $b^2(a+1) = 7.84$, it becomes $\frac{1.44}{(a+1)} = (7.84)$ from which we have $a = -0.82$ by computation. But since the gamma function $\gamma(a+1) = a!$ where a is an integer value, ≥ -1, we must approximate a to the nearest integer value, hence we have $a = -1$ and then $b = 2.8$.

Now substituting the values of $a = -1$ and $b = 2.8$ in gamma probability function we have

$$f(r) = \frac{e^{-r/2.8}.r^{-1}}{2.8^{(-1+1)}} = \frac{e^{-0.36r}.r^{-1}}{1} = e^{-0.36r}.r^{-1}$$

so that gamma probabilities of r = 1, 2, 3, ... , 10 mm diameter are

$f(1) = e^{-0.36}.1^{-1} = 0.69967$

$f(2) = e^{-0.72}.2^{-1} = 0.24338$

$f(3) = e^{-1.08}.3^{-1} = 0.11320$

.

.

.

$f(10) = e^{-3.6}.10^{-1} = 0.00273$

similarly, $f(0) = 0.00000$

6.4. POISSON DISTRIBUTION IN A GEOGRAPHIC AREA CONTEXT

In a geographic area context, maps represent the spatial relationship of the area. The map relationships are expressed almost always in terms of points located upon the map. Most maps are estimates of continuous functions, based on discrete observations (of the event/attribute) as points, like a contour map, However, in a real context, mapping an attribute may lead to few points of observation and hence for a map with point observations, its quality determine on density (as well as on uniformity) of the point distribution. Now for a map area to be tested for uniformity of distribution, we can divide the map area into a number of equal sized subareas called earlier as quadrats, as such that each subarea contains a numbers of points. Now if we assume that the point distribution on the map is uniform, we expect at the specific quadrat size taken, the each quadrat (as a subarea) contains the same numbers of points and this can be tested by Chi-square test.

But since both regular and random patterns are expected to be homogeneous, so establishing that pattern is uniform does not specify the nature of uniformity. Hence when points are distributed at random across a map area, even though the coverage is uniform, we do not expect exactly the same number of points to lie within each subarea or quadrat. For example, in a sedimentary basin with area A, m number of oil wells ($m = 168$) are discovered and spaced randomly over the basin area A. Now the basin is gridded into 160 quadrats, each of area, a which again is divided into n extremely small, equal-sized sub-areas, which we might regard as potential drilling sites. The probability that any one of these extremely small subareas contain a potential oil well tends towards zero as n becomes infinitely large.

Now, the area of each sub-area (*i.e.* drilling site) $\frac{A}{n}$ and the probability that a site contains a potential oil well is $p = \lambda \,.\, A/n$ where λ is the density of oil wells discovered in the basin equal to $\lambda = m/A$. So, the probability that the site does not contain a potential oil well is:

$$1 - p = \left(1 - \lambda.\frac{A}{n}\right) \qquad \text{...(6.13)}$$

Now, the probability of the specific combination of r number of potential oil wells and $(n - r)$ number of non-potential oil well sites within a tract that is a quadrat A is

$$\left(\lambda.\frac{A}{n}\right)^{r}\left(1-\lambda.\frac{A}{n}\right)^{n-r} \qquad \text{... (6.14)}$$

Therefore that the quadrat area will contain exactly r number of potential oil wells is:

$$f(r) = \binom{n}{r}\left(\lambda.\frac{A}{n}\right)^{r}\left(1-\lambda.\frac{A}{n}\right)^{n-r}$$

which on expanding into factorials, is

$$f(r) = \frac{n(n-1)(n-2)...(n-r+1)}{r!}\left(\frac{\lambda A}{n}\right)^{r}\left(1-\frac{\lambda A}{n}\right)^{n-r}$$

On rearranging and cancelling

$$f(r) = \left(1-\frac{1}{n}\right)\left(1-\frac{2}{n}\right).....\left(1-\frac{r-1}{n}\right)\left(1-\frac{\lambda A}{n}\right)^{n-r}\left[\left(1-\frac{\lambda A}{n}\right)^{n}.\frac{(\lambda A)^{r}}{r!}\right] \quad ... (6.15)$$

Now as n becomes infinitely large → ∞ so all of the fractions that contain n in their denominator become infinitesimally small and vanish, so all terms inside parentheses simply become equal to 1, and the terms inside the brackets simplify to:

$$f(r) = e^{(-\lambda A)}.\frac{(\lambda A)^{r}}{r!} \quad ... (6.16)$$

Equation (6.16) is an expression of the Poisson distribution, as applied to the probability of rare, random events (like discovery of oil wells) occurring within geographic areas. Note in this equation. λA is simply the mean number of oil wells per quadrat so in practice, $\lambda A = m/\tau$. Note also n the number of drilling sites has vanished from this equation. Now for example if the basin is divided into 10 × 16 grids of 160 quadrats, each containing 10 sq. miles, the mean number of wells per quadrat is m/τ = 168/160 = 1.505. Now we can estimate the number of quadrat that contain no oil well, exactly 1 oil well, 2 oil wells, and so forth.

Now, once we compute the mean number of the event per quadrat we can compute the variance, s^2 also

$$s^2 = \frac{\sum_{i=1}^{\tau}(o_i - m/\tau)^2}{\tau - 1}, \quad ... (6.17)$$

where o_i is the No. of points per quadrat

Now we can compare the estimated mean $\bar{x}$ and variance s^2 which leads to following alternative conclusions

$\bar{x} = m/\tau > s^2$ Point pattern more uniform than random

$\bar{x} = m/\tau = s^2$ Point pattern random

$\bar{x} = m/\tau < s^2$ Point pattern more clustered than random

Now, because of the particular set of quadrats chosen, there can be random variations in the quadrats for which there may arise some difference between mean, m/τ and variance, s^2. So the statistical significance of the observed difference (between m/τ and s^2) may be tested by a *t test* based on *the standard error of the mean*, which is the variance that would be expected in values of m/τ if a basin were repeatedly sampled by different sets of quadrat. The standard error in the mean number of oil well discoveries per quadrat is

$$s_e = \sqrt{2/(\tau - 1)} \quad ... (6.18)$$

Now, the t test compares the ratio between m/τ and s^2, which should be equal to 1.0, if the two statistics, the mean and the variance are the same:

$$t = \frac{\left(\frac{m/\tau}{s^2}\right) - 1.0}{s_e} \qquad \text{... (6.19)}$$

where $d.f = \tau - 1$

Now if for a $d.f.$ at 0.05 level of significance the critical value is ± 1.96 and if the computed t by Eq. (6.19) is – 8.9, we can conclude that the spatial distribution is not random. Note the standard error and the t test have been discussed in Chapters 10 and 11 respectively.

The initial step in a spatial pattern detection for any event or for any community in ecology should involve testing the hypotheses whether the distribution of the number of individuals per quadrat unit is uniform by following the Poisson distribution. If this hypothesis is rejected, then the distribution may be in the direction of clustered/clumped (usually) or uniform (rarely). If the direction is toward a clustered dispersion then it can be tested by indices of dispersion which are based on the ratio of the variance to mean as mentioned above. If the direction is not towards clustered pattern by this test, an uniform pattern can be suggested. Since the uniform pattern is not common, no test for this pattern is considered here.

LIST OF FORMULAE

I. *Poisson Distribution*

Poisson Distribution Series:

$$e^{-z}\left(1+\frac{z^1}{1!}+\frac{z^2}{2!}+\ldots+\frac{z^r}{r!}+\ldots+\frac{z^n}{n!}\right) = \text{Total probability}$$

Poisson Probability Function $f(r) = e^{-z} . \frac{z^r}{r!}$ where e = a constant = 2.71828

Parameters of Poisson Distribution: Mean $np \leq 5$, variance npq is equal to mean np

II. *Generalisations of Poisson Distribution*

Double Poisson Distribution, $f(r, \bar{z}_1, \bar{z}_2) = \frac{1}{2} . \frac{e^{-\bar{z}_1} . \bar{z}_1^r}{r!} + \frac{1}{2} . \frac{e^{-\bar{z}_2} . \bar{z}_2^r}{r!}$ **for** $r = 0, 1, 2, \ldots$

Parameters of Double Poisson

Mean, $\bar{z}_1 = \bar{x} + \sqrt{\sigma - \bar{x}}$ and $\bar{z}_2 = \bar{x} - \sqrt{\sigma - \bar{x}}$

Gamma Distribution

$$Y = f(r) = \gamma(p) = \frac{e^{-x/b}.r^a}{a!b^{a+1}}$$ where $a! = \gamma(a+1)$, is the gamma function

Parameters of Gamma Distribution

Mean, $\bar{x} = b(a+1)$

Variance, $\sigma^2 = b^2(a+1)$

with $b > 0$ and $a \geq -1$, an integer

III. *Test for randomness of point patterns*

$$\bar{x} = m/\tau \text{ and } s^2 = \frac{\Sigma(O_i - m/\tau)^2}{\tau - 1}$$

Now, if $\bar{x} > s^2$: Uniform

$\bar{x} = s^2$: Random

$\bar{x} < s^2$: Clustered

EXERCISES

6.1 Assuming that the annual floods are independent events, compute the probability of flood occurring 0, 1, 2, 3, 4, and 5 times a year from recorded data on annual floods for 140 years as tabulated below.

No. of floods	No. of years
0	33
1	43
2	32
3	18
4	6
5	8

6.2 Fit a Poisson distribution to the data given below :

No. of occurrences, r:	0	1	2	3	4
Frequency, f:	30	62	46	10	2

6.3 Find the probability of exactly 2 events occurring in 10 observations if the probability of an event is 5 in 100 times by (a) the binomial and (b) the Poisson distribution.

6.4 In an area of 10 river valleys the number of floods recorded over a period of 20 years was 50. Supposing that each river is as likely to have a flood as any other, calculate the frequency of 0, 1, 2 and 3 floods per year per river.

6.5 If the probability of a flood Q being exceeded at a place is 0.02325, calculate the probability for 4 floods greater than Q to occur in the next 10 years as well as the probability that none would occur. Assume the annual flood as an independent event.

7

Probabilistic Treatment in Geo-science Systems IV (Extreme Value Distribution)

7.1 EXTREME VALUE DISTRIBUTION

'Extreme value distribution' (or extremal distribution) is the distribution of the extreme number of values (largest or the smallest like flood or drought), each value being selected out of the p values contained in each of n samples which approaches an asymptotic limit as p is increased indefinitely. The type of limiting form (asymptotic) depends on the type of the initial distribution of the np values. For example, daily rainfall at a station is recorded for every year. If the maximum rainfall for the year is taken as an observation and the process is repeated for 50 years, 50 maximum rainfall observations will follow an extrene value distribution. This term of "extreme value distribution" was first introduced by E.J. Gumbel in 1941 and hence it is known as Gumbel's distribution. He was the first to realize that the annual maximum flood data (or the annual maximum storm rainfall or maximum river flows etc.) are nothing but the extreme observed values in different years and hence they should follow the law of extreme value distribution. The law deduced by him is one of the most widely used probability distribution functions for extreme values in hydrology and climatology for prediction of flood peaks, maximum rainfall, the longest period without any rain, maximum temperature, maximum wind speed etc. Since the extreme value distribution was initially applied to estimation of the magnitude of flood for the desired 'return period of T years' and latter extended to these other attributes, so the extreme value distribution is explained below with reference to flood frequency analysis. Note that these extreme values, are selected from observations made at equal interval period which is generally taken as one *water year* and the series so selected is the annual series. If the time interval selected is less than one year, the seasonal variation will introduce nonhomogeneity to the data. However, homogeneity of the data may be maintained if data are selected from a particular season or month.

7.1.1 Flood Frequency Analysis

The method of dealing with the runoff data measured at a site across a river is known as the *flood frequency* analysis. The analysis does not provide a hydrograph of the river flow data but gives only a peak discharge for a particular frequency or a return period. Hence in the case of a river basin for which a past record of runoff data are available of sufficient number of years, they can yield satisfactory estimates in designing hydraulic structures for a designed flood provided that one value of the runoff data is independent of any other data so that the values are individual values and the variable in question is a random variable. Other conditions are that the factors influencing the character of each runoff value remains unaltered and the measurement technique as well as the discharge site are identical. Thus, as a preliminary step the basic data should be screened and adjusted so that any nonconformities that may exist is removed from the data.

For the purpose of statistical analysis, a convenient sequence of data called a 'series' is used. If flood is defined an unusually high stage of a river due to runoff from storm rainfall in too great quantities to be confined in the normal water surface elevations of the river (so that the river overflows its banks and inundates the adjoining area), the magnitudes of this flood at a given location in the river vary from year to year. These varying magnitudes of the flood constitute a statistical array, a "hydrological series" which enables one to assign a frequency to a given flood-peak values. When flood-peaks are arranged in the descending order of magnitudes, there are two methods of compiling this flood-peak data. One, the "annual (flood) series" and two, the "partial duration series". The values of annual peak flood from a given river basin for a large number of successive years constitute the *annual series*. In the annual series only the highest flood in each year is used thus ignoring the next highest in any year which sometimes may exceed many of the annual maximum, the other way to assemble the data is to prepare a list of all floods above a selected basic stage (usually the lowest value of the annual series), regardless of the time interval, so that in some (wet) years there may be a number of floods above the basic stage whereas in the other (than wet) years there may not be any such floods at all. Such a series is the *partial duration series*. In the partial duration series though all the larger floods are used (which are omitted in the annual series because they were not the highest annual flood in any year considered) but the disadvantage is that the flood events may be sequential leading to "autocorrelation effect" and hence the data obtained do not furnish a proper frequency or true distribution series and so a reasonable statistical analysis cannot be worked out in this series. On other hand, the annual flood series data enable one to assign a frequency to a given flood-peak value. Thus while the use of partial duration series is restricted to estimate within the range of observations (it is more so for flood studies than for storm rainfall as the different flood events are less independent than the corresponding storm rainfall), the annual are used in the estimation of a 'design flood' with an assigned frequency say 1 in 100 years that is, the maximum peak flood flow that any hydraulic structure like dams or bridges etc. are supposed to experience and to accommodate the effects.

7.2 PROBABILITY DISTRIBUTION OF FLOOD FREQUENCY

In the study of flood frequencies over a period of time some basic concept of probability are used so that the concept of theoretical distributions can be applied. Once a probability distribution is established it is a simple and straight-forward process to calculate the required probabilities. There are two methods to fit the theoretical distributions: (i) one is a straight-forward 'plotting technique' for the ranked sequence of annual flood-series data and the (ii) second considers the use of hydrologic 'frequency factor'. Both the methods try to fit the data to the distributions.

7.2.1 Weibull's Plotting Positions

The technique in plotting the distribution is to rank the data. "Ranking" is a procedure that order a data set in a decreasing or increasing order of magnitude and to assign order number. When the ranking is in a descending order, the largest flood is assigned a rank of one ($m = 1$), the next highest flood is given rank two ($m = 2$) and so forth. Now it can be asked what is the probability that any given rank will be equalled or exceeded in a given year? This question motivates study of the distribution of the number of exceedences.

The ranking of the data is a very clever idea for this "stochastic or probabilistic study" of annual maximum values like flood. By ranking, we determine the number of the times any particular flood is equalled or exceeded during the period of record, N. Thus the largest flood is equaled or exceeded once. If we have 30 years of record and we want to estimate the number of times, the largest flood would be equalled or exceeded in the next 30 years, we could guess that if the data are representative (that is, the individual values are independent), the mth largest flood in the next 30 years of record. To formalize this let us refer to the period of record of the annual maxima series as N, the rank of the flood discharges of descending values as m, with largest equal to 1. Now if we define the 'recurrence interval' as the mean time in years for the mth largest value among annual maxima series of length N to be equalled or exceeded once on the average in n future trials we can have the mean number of exceedances in future n years as

$$\overline{X} = n.\frac{m}{N+1} \qquad \text{... (7.1)}$$

If the mean number of exceedences $\overline{X} = 1$, we can have the 'probability', p of a flood of given mth rank, occurring in the next n years simplified to 'relative frequency' as

$$p = \frac{m}{N+1} \quad \text{... (7.2)}$$

the percent probability of which is $p \times 100$ per cent. The probability that it will not occur in a given year is the probability of non-exceedance, p' which is

$$p' = 1 - p \quad \text{... (7.3)}$$

Equation (7.2) is commonly referred to as Weibull's (1939) *plotting position formula* for unspecified distributions. Now the *recurrence interval* (also called the "return period") as defined earlier is the average number of years during which a flood of a given magnitude, m will be equalled or exceeded. This denoted as T-year or T is

$$T = \frac{1}{p}\ \text{year} = \frac{N+1}{m} \quad \text{... (7.4)}$$

T is $N+1$ year for the highest flood and T is 1-year for the lowest flood.

Now for the annual peak discharge data of any river, flood frequencies in terms of return years T for every mth rank in the data is scaled and plotted along the abscissa of a semi-logarithmic graph paper (starting with common logarithm value 1) with the discharge along the ordinate. On the horizontal axis on the top of this graph the corresponding percent probability values, p in per cent are plotted. We now illustrate this plotting of the extreme value distribution by Weibull's formula by an example of the maximum daily discharge yearly for Kosi river below.

Example 7.1 Calculate the plotting positions on extreme value graph paper of maximum daily discharge yearly for Kosi river at Barahkshetra, Nepal from the year 1947 year to 1984. The data are in cumecs.

Year	Date of Occurrence	Peak discharge (in Cumecs)
1947	July 31	8842.906
1948	July 13	13410.282
1949	July 10	11201.637
1950	August 20	9899.847
1951	August 24	7256.149
1952	August 24	8676.229
1953	July 30	5433.717
1954	August 24	24214.069

Year	Date of Occurrence	Peak discharge (in Cumecs)
1955	August 7	7078.175
1956	August 29	5436.039
1957	August 12	7531.179
1958	August 25	10560.637
1959	August 10	5973.980
1960	September 28	6625.172
1961	August 20	8295.622
1962	August 2	10504.012
1963	August 16	7644.429
1964	August 22	7927.556
1965	August 14	5804.104
1966	August 24	10815.451
1967	August 9	8833.563
1968	October 5	25849.496
1969	July 28	8134.805
1970	July 15	13861.898
1971	July 12	12174.461
1972	July 28	10708.005
1973	October 13	9200.212
1974	August 5	12740.715
1975	July 28	9197.381
1976	August 23	9479.942
1977	August 27	7775.800
1978	July 28	9827.509
1979	July 24	13340.831
1980	July 30	7703.773
1981	August 22	7989.278
1982	July 18	6967.926
1983	July 4	8816.660
1984	September 17	13149.466

The above data are ranked in descending order and for every *m*th rank, the return period and the corresponding percent probabilities are calculated. Since the logarithmic value to start with along the abscissa at the origin of the graph is 1 and if the successive values 2, 4, 8, ... are taken at interval of 2 cm apart, the plotting positions of the *T*-year values are required to be calculated which we show in Table 7.1.

Table 7.1 : Computation of Weibull's Plotting Positions

Ranked Discharge (in cumecs)	Rank/Order No. (m)	Percent Probability (p in %)	Return-Period (T)	Plotting Positions (in cms.)
25849.496	1	2.56	39.00	10.44
24214.069	2	5.13	19.50	8.44
13861.898	3	7.69	13.00	7.25
13410.282	4	10.26	9.75	6.44
13340.831	5	12.82	7.80	5.90
13149.466	6	15.39	6.50	5.25
12740.715	7	17.95	5.57	4.79
12174.461	8	20.51	4.88	4.40
11201.637	9	23.08	4.33	4.16
10815.451	10	25.64	3.90	3.90
10708.005	11	28.21	3.55	3.55
10560.637	12	30.77	3.25	3.25
10504.012	13	33.33	3.00	3.00
9899.847	14	35.89	2.79	2.79
9827.509	15	38.46	2.60	2.60
9479.942	16	41.02	2.44	2.44
9200.212	17	43.59	2.29	2.29
9197.381	18	46.15	2.17	2.17
8842.906	19	48.71	2.05	2.05
8833.563	20	51.28	1.95	1.90
8816.660	21	53.85	1.86	1.72
8676.229	22	56.40	1.77	1.54
8295.622	23	58.96	1.70	1.40
8134.805	24	61.54	1.63	1.26
7989.278	25	64.10	1.56	1.12
7927.556	26	66.66	1.50	1.00
7775.800	27	69.25	1.44	0.88
7703.773	28	71.79	1.39	0.78
7644.429	29	74.35	1.35	0.70
7531.179	30	76.92	1.30	0.60
7256.149	31	79.49	1.26	0.52
7078.175	32	82.03	1.22	0.44
6967.926	33	84.60	1.18	0.36
6625.172	34	87.18	1.15	0.30
5973.980	35	89.77	1.11	0.22
5804.104	36	92.34	1.08	0.16
5436.039	37	94.88	1.05	0.10
5433.717	38	97.47	1.03	0.06

Now the above observation of the ranked discharge data reveal that there are a greater number of discharge data below the mean discharge of 9970.600 cumecs than those above it and the variations above the mean of the annual peak discharge data are greater than those below the mean. Hence, the plotting of return-years or the probability values against the discharge data on the semi-log graph will not be a straight line curve but the line bends off towards the tails of the data giving a skew curve. It happens so because the distribution of extreme value observations depends upon the tails of the frequency distribution from which the observations are drawn. The data in Table 7.1 calculates the plotting position for each return period of ranked discharge and these plotting positions are located against the discharge data in Fig. 7.1. Inspection shows that except the first three ranked discharge data, the plotted points lie near to a straight line, and a line drawn in by eye (Fig. 7.1) has to be acceptable, otherwise we require the use of the second approach of carrying out the extreme value distribution analysis through the use of *frequency factors*.

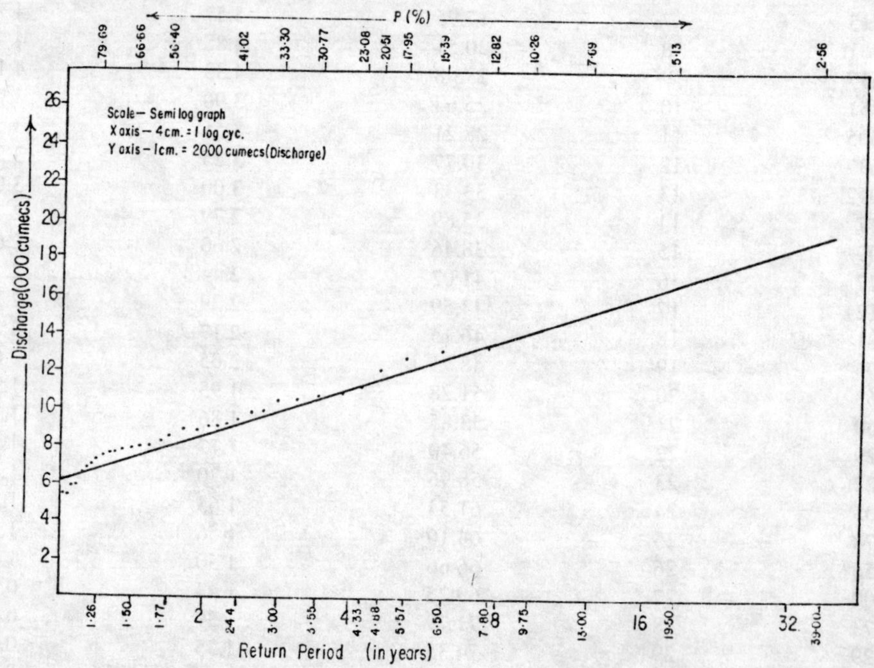

Fig. 7.1 : Weibull's Plotting Positions

7.3 FREQUENCY FACTOR OF EXTREME VALUE DISTRIBUTION

One of the important aims of hydrological studies is to present estimates for computing the magnitude-frequency relationship of an event. T. Ven Chow in 1964 proposed the use of a general equation for hydrologic frequency analysis as

$$x = \bar{x} + k\sigma \qquad \text{... (7.5)}$$

where k is the "*frequency factor*", x is the flood magnitude of given return period T, $\bar{x}$ is the mean of the recorded floods and σ is the standard deviation. k as frequency factor is literally the number of standard deviations above and below the mean to attain the desired probability. The value of k for different methods of distributions are different. But in the case of skewed distributions, its value varies with the coefficient of skewness and can be affected greatly by the number of years of record. If $k.\sigma$ is taken as the departure of the variate x, a random flood magnitude may be represented by the mean of the recorded floods $\bar{x}$ plus $k.\sigma$ the departure of the variate. There are several methods for estimation of discharge of particular frequency like 1-in-100 years etc based on this technique of frequency factor. We will discuss two methods (i) the "Gumbel's Extreme Value Distribution method" and (ii) "Pearson type III" (arithmetic or logarithmic) Distribution method. In Gumbel method as the probability, p is increased indefinitely, the extreme value distribution reaches an asymptotic limit but in the Pearson type III (arithmetic or logarithmic), the frequency factor, k is a function of both the recurrence interval and skewness.

7.3.1 Gumbel's Extreme Value Distribution

It is mentioned earlier that E.J. Gumbel was the first to realize that the annual peak flood data (or annual maximum one day storm rainfall and similar type of data) are nothing but the extreme values in different years' observations and he introduced the extreme value distribution first which is commonly known as Gumbel's extreme value distribution. It is one of the most widely used probability distribution with a bearing on the natural phenomena like peak discharge, storm rainfall, low flows and other similar events.

Gumbel's extreme value distribution is based on the argument that distribution of an extreme event is unlimited and hence the most suitable distribution for fitting to the extreme value data is of the 'double exponential type', Hence the probability of occurrence of a peak discharge, x equal to or larger than a value, x_0 is given by

$$f(x \geq x_0) = 1 - p' = 1 - e^{-e^{-y}} \qquad \text{... (7.6)}$$

in which e is the base of natural logarithm and y is a dimensionless reduced variate given by

$$y = \frac{1}{0.78\sigma}(x_T - \bar{x}) + 0.577 \qquad \text{... (7.7)}$$

where x_T = the event (or, magnitude of the discharge with probability of occurrence) in T-year return period, $\bar{x}$ the arithmetic mean and σ the standard deviation of all the data in the series.

Now if Eq. (7.6) is transposed we have the reduced variate, y as

$$y = -\ln[-\ln(1-p)] \quad \text{... (7.8)}$$

and since $T = \frac{1}{p}$ (from Eq. 7.4) we have the Eq. 7.6 is rewritten for computing the return period, T as

$$T = \frac{1}{1 - e^{-e^{-y}}} \quad \text{... (7.9)}$$

Also we can re-write the Eq. 7.8 for the reduced variate, y as

$$y = -\ln\left[-\ln\left(1 - \frac{1}{T}\right)\right] \quad \text{... (7.10)}$$

Since the event x of T-year return period is x_T then from Eq. (7.7) we can compute x_T as

$$x_T = \bar{x} + \sigma(0.78y - 0.45) \quad \text{... (7.11)}$$

in which the frequency factor k is $0.78\,y - 0.45$ with

$$y = -\ln\left[-\ln\left(1 - \frac{1}{T}\right)\right] \quad \text{... (7.11a)}$$

Thus the magnitude of the peak discharge, x (= an event) for the desired return period of T-years can be estimated from the Eq. (7.11) with y known from the Eq. (7.11a).

Now the plotting positions for the variate, x (the annual peak discharge here) are again like those in Weibull's Plotting position that is, the discharge, x is plotted on the ordinate and the return period T on the abscissa but while the ordinate of a Gumbel probability graph may have an arithmetic scale (like in Weibull's), the reduced variate y in Eq. 7.9 is linear with the variate, x itself in Gumbel, so that both the abscissa and the ordinate varies linearly on this plotting scale. Hence Gumbel distributions will plot as a straight line so that a value of T-year equal to 2.25 years would plot 0.5341 linear units away from zero (the mean position) whereas a value of T-year equal to 1.354 would plot – 0.2930 on this linear scale.

Hence for the given data, after the rank numbering, m and return period, T of the annual data series are completed, the plotting positions for corresponding reduced variates, y are calculated. Now the plotting of the discharge against the corresponding reduced variates, if joined together will yield the straight line for the Gumbel distribution. This straight line can then be extrapolated to read the peak discharge magnitude against any desired return period, $T < n$.

The Gumbel probability paper, mentioned above, thus consists of an abscissa specially marked for various convenient values of the return period, T. To construct the T-scale on the abscissa, first construct an arithmetic scale of y values, say from −2 through 0 to +5 (as illustrated in Fig. 7.2). Return period, T-years calculated from the reduced variate is now ready for use against the ordinate scale on which the value of the variate (the annual peak discharge) can be plotted by linear units.

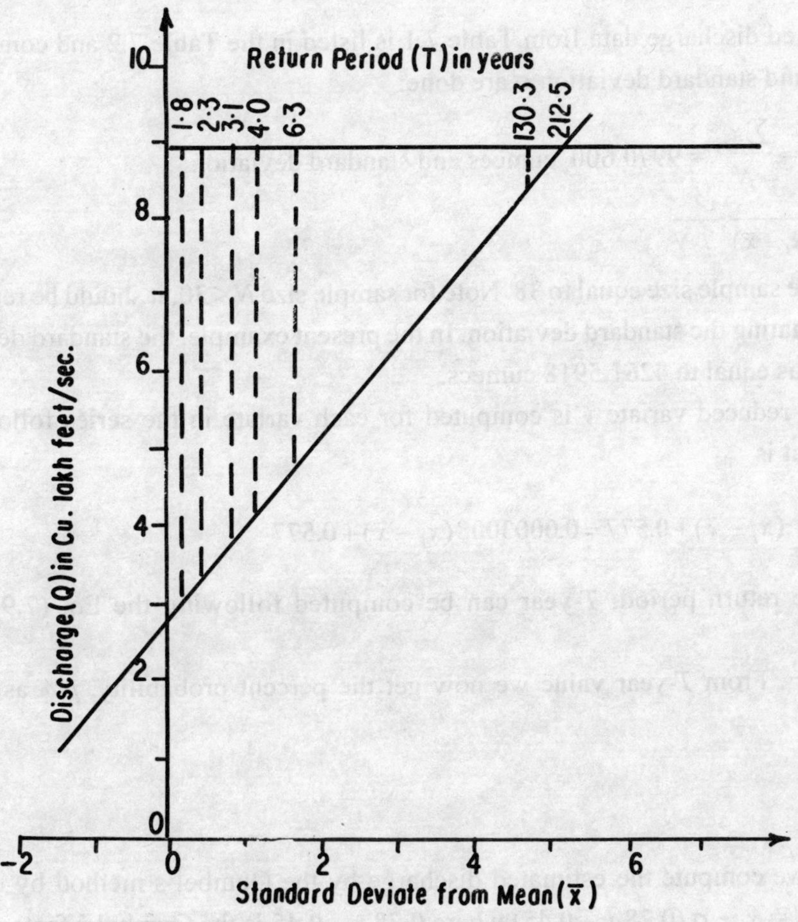

Fig. 7.2 : A Gumbel Flood Frequency Curve

There are two approaches to the solution of extreme value distribution by Gumbel's method. The first approach is to find the recurrence interval for each of the variate in the given annual series so that the fitting of the observed data to the estimated frequency line can be tested. The second approach is to find the annual peak discharge for a given recurrence interval, T. We use the ranked annual peak discharge data to Table 7.1 for the Kosi river at Barahkshetra site for illustration of both these approaches in Example 7.2.

Example 7.2. Compute the return period T-year and prepare the Gumbel probability graph for Kosi river at Barahkshetra from the available ranked annual peak discharge data in Table 7.1.

The ranked discharge data from Table 7.1 is listed in the Table 7.2 and computations for mean $\bar{x}$ and standard deviation σ are done:

Mean $\bar{x} = \dfrac{\sum x_i}{N} = 9970.600$ cumecs and standard deviation

$$\sigma = \sqrt{\Sigma(x_i - \bar{x})^2 / N}$$

where N is the sample size equal to 38. Note for sample size $N < 30$, it should be replaced by $N - 1$ in estimating the standard deviation. In the present example, the standard deviation σ is computed as equal to 4261.5918 cumecs.

Next the reduced variate y is computed for each variate in the series following the Eq. (7.7), that is

$$y = \frac{1}{0.78\sigma}(x_i - \bar{x}) + 0.577 = 0.0003008(x_i - \bar{x}) + 0.577$$

Now, the return period, T-year can be computed following the Eq. (7.9), that is

$T = \dfrac{1}{1 - e^{-e^{-y}}}$. From T-year value we now get the percent probability, $p\%$ as equal to

$\dfrac{1}{T} \times 100\%$.

Finally, we compute the estimated discharge by the Gumbel's method by using the Eq. (7.11), $x_T = \bar{x} + \sigma(0.78\,y - 0.45)$ where $0.78\,y - 0.45$ is the Gumbel k factor.

All the above sequence of tabulation follows now in Table 7.2.

Table 7.2 : Flood Frequency Analysis of Kosi river (Gumbel method)

Year	Ranked Discharge x_i (in cumecs)	$(x_i - \bar{x})$	Reduced variate, y	Return Period T (in years)	Percent Probability p%	Computed x_T (in cumecs)
(1)	(2)	(3)	(4)	(5)	(6)	(7)
1968	25849.496	15878.896	5.353984	211.95	0.004718	25847.715
1954	24214.069	14243.469	4.861985	129.78	0.007705	24214.325
1970	13861.898	3891.298	1.747652	6.26	0.159856	13862.154
1948	13410.282	3439.682	1.611789	5.53	0.180885	13410.538
1979	13340.831	3370.231	1.590895	5.52	0.180846	13341.007
1984	13149.466	3178.866	1.533325	5.15	0.194117	13149.772
1974	12740.715	2770.115	1.410357	4.62	0.216556	12740.971
1971	12174.461	2203.861	1.240006	3.98	0.251274	12174.717
1949	11201.637	1231.037	0.947343	3.11	0.321431	11201.637
1966	10815.451	844.851	0.831163	2.83	0.353086	10815.706
1972	10708.005	737.405	0.798839	2.76	0.362276	10708.260
1958	10560.637	590.037	0.754505	2.67	0.375149	10560.893
1962	10504.012	533.412	0.737470	2.63	0.380177	10504.268
1950	9899.847	– 70.753	0.555714	2.29	0.436542	9900.102
1978	9827.509	–143.097	0.533950	2.25	0.443609	9827.758
1976	9479.942	– 490.658	0.429391	2.09	0.478426	9580.198
1973	9200.212	– 770.388	0.345237	1.97	0.507397	9200.467
1975	9197.381	– 773.219	0.344385	1.96	0.507694	9197.636
1947	8842.906	– 1127.694	0.237746	1.83	0.545429	8843.161
1967	8833.563	– 1137.037	0.234935	1.83	0.546437	8833.818
1983	8816.660	– 1153.940	0.229850	1.82	0.548261	8816.915
1952	8676.229	– 1294.371	0.187603	1.77	0.563489	8676.484
1961	8295.622	– 1674.978	0.073101	1.65	0.605251	8295.877
1969	8134.805	– 1835.795	0.024722	1.60	0.623026	8135.060
1981	7989.278	– 1981.322	– 0.019058	1.56	0.639130	7989.533
1964	7927.556	– 2043.044	– 0.037626	1.55	0.645959	7927.812
1977	7775.800	– 2194.800	– 0.083280	1.51	0.662721	7775.921
1908	7703.773	– 2266.827	– 0.104949	1.49	0.670656	7704.028
1963	7644.429	– 2326.171	– 0.122802	1.48	0.677180	7644.878
1957	7531.179	– 2439.421	– 0.156871	1.45	0.689585	7531.469
1951	7256.149	– 2714.451	– 0.239611	1.39	0.719380	7256.387
1955	7078.175	– 2892.425	– 0.293153	1.35	0.738324	7078.431
1982	6967.926	– 3002.674	– 0.326320	1.33	0.749890	6968.123
1960	6625.172	– 3345.428	– 0.429433	1.27	0.784842	6625.473
1959	5973.980	– 3996.620	– 0.625337	1.18	0.845703	5974.014
1965	5804.104	– 4166.496	– 0.676442	1.16	0.860105	5804.414
1956	5436.839	– 4534.561	– 0.787170	1.13	0.888883	5436.295
1953	5433.717	– 4536.883	– 0.787869	1.12	0.889053	5433.973

Now using the extreme value probability paper we plot the x_T value from the computation form and join them with a straight line to obtain the required frequency curve (Fig 7.2). Note that since this *Gumbel frequency curve* is a straight line, so it is not necessary to plot the entire frequency curve if the *T*-year value for a given x_i value, or the x_i value for a given *T* value is required. Since the estimated x_T value for a given return period, *T* is $x_T = \bar{x} + \sigma(0.78y - 0.45)$ so when $\bar{x} = x_T$ the equation for x_T becomes $0.78y = 0.45$ or $y = 0.577$ which corresponds to $T = 2.33$ year.

Now if we compare the Table 7.1 with the Table 7.2 we find that for the larger annual events, the Weibull's method gives lower recurrence interval i.e. largest floods of increased frequency. For example, for two first ranking floods of not less than 24,000 cumecs, Weibull's method indicate return period $T = 39$ and 19.5-years while by the Gumbel's method these values are 211- and 129-years,

Now for the second approach, we can compute the various return periods like 5, 10, 20, 50, 100, 200, 500, and 1000 years and construct the frequency curve for Kosi river at Barahkshetra by the Gumbel method. In this context we first compute the reduced variate by Eq. 7.10, that is

$$y = -\ln\left[-\ln\left(1 - \frac{1}{T}\right)\right],$$

the value of which is substituted in Eq. (7.11) that is $x_T = \bar{x} + \sigma(0.78y - 0.45)$ and we get the estimated value for the various return periods. It may be noted that since the mean and standard deviation are known so theoretically there is no requirement of the continuous record of annual peak value. Hence, if annual peak values of some years are missing, it is not a problem and it is not necessary to fill the missing data or discard the broken data from analysis. Now in the Table 7.3 below we give the computed values for the above return periods.

Table 7.3 : Construction of Gumbel Frequency Curve for Kosi river at Barahkshetra at various return periods, *T*-years

Return Period (*T*-years)	Reduced Variate (y)	Computed x_T Values (in cumecs)
5	1.4994	14,232.19
10	2.25036	15,533.19
25	3.19863	18,684.95
50	3.90194	21,023.09
100	4.60015	23,343.97
200	5.29581	25,656.38
500	6.21361	28.707.17
1,000	6.90726	31,012.88

7.3.2 Pearson Type III Distribution

The Gumbel distribution as a distribution of n extreme values (largest or smallest) approaches an asymptotic limit as the probability is increased indefinitely. This distribution can be called the "Type-I distribution" with origin at the mean. The "Type II distribution" is concerned with the cumulative probability function. "Pearson's Type III distribution" is a skew distribution so that the distribution has a limited range in the left in which direction, the probability curve gets truncated. Hence below a certain value of the variate the probability is zero, but it is infinite converging asymptotically to the axis of the variate. This means even infinitely large (or small) values have a certain probability of occurrence. The distribution is bell-shaped but the origin of the variate is at the mode.

To fit the *Pearson type III distribution* as the standard method for hydrological frequency analysis, a procedure is adopted to convert the annual data series to logarithms for reducing skewness and then to compute. In this form the distribution is known as "log Pearson type III", to distinguish it from the Pearson type III distribution which is arithmetic. In actual computation all the three moments (the mean, standard deviation and skewness of the logarithm data) are required to fit the distribution. Yet it is extremely flexible in the sense that a zero skew will reduce the *log Pearson type III distribution* to a log-normal and a *Pearson (arithmetic) type III distribution* to a normal. In the following we give the formulae for the three moments for the log Pearson type III distribution

Let $Q = \log x_i$,

$$\text{so Mean} = \frac{\Sigma f \log x_i}{N} = \frac{\Sigma Q}{N} = \overline{\log x} \qquad \text{... (7.12)}$$

$$\text{and, } s = \sqrt{\frac{\Sigma f(Q)^2 - (\Sigma f Q)^2 / N}{N-1}}$$

$$\text{so, Standard deviation: } \sigma_{\log x} = \sqrt{\frac{\Sigma f\,(\log x_i - \overline{\log x})^2}{N-1}} \qquad \text{... (7.13)}$$

$$\Rightarrow \sqrt{\frac{\sum f(\log x_i^2) - (\sum \log x_i)^2 / N^2}{N-1}}$$

$$\text{Skew coefficient } \beta_1 = \frac{\Sigma f(\log x_i - \overline{\log x})^3}{(N-1)(N-2)(\sigma_{\log x})^3} \qquad \text{... (7.14)}$$

Note for the Pearson type III distribution, the above formulas will do except the fact that the variate is x and not $\log x$.

Now the values of x_T for various return period, T are computed from

$$\text{Log}\, x_T = \overline{\log x} + k\,\sigma_{\log x} \qquad \text{... (7.15)}$$

and the frequency factor k is obtained from Appendix Table IV for the computed value of β_1 and the desired recurrence interval, T.

It may be noted from the above Eq. (7.15), the frequency factor in Pearson type III distribution, is a function of both the recurrence interval and skewness, i.e. $k = f(T, \beta_1)$. Since skewness has greater variability than the mean, transforming the data to logarithms is an effective means of reducing variability as done in log Pearson type III distribution. For small size of samples, $N < 30$, in the Pearson type III analysis following modification for skewness coefficient is suggested

$$\overline{\beta}_1 = \left(1 + \frac{8.5}{N}\right)\beta_1 \qquad \text{... (7.16)}$$

For example, for an annual discharge data, if the skew is found to be –1.91, logarithm of std. deviations = 0.309 and of mean = 3.541, find out the probability of peak discharge being exceeded is 10%. Now for the exceedance probability of 0.1, k is interpolated to be 1.088. Hence $x_{10} = \text{Log}\, Q_{10} = \overline{\log x} + k\,.\,\sigma_{\log x} = 3.541 + 1.088\,(0.309) = 3.877$

So, $Q_{10} = 10^{3.877} = 7533$ cumecs

Example 7.3. Following is a frequency table of annual peak flow of a river for a period of 87 years.

Annual peak flow (in 1000 cumecs)	Frequencies (f)
0–2	0
2–4	17
4–6	27
6–8	18
8–10	18
10–12	3
12–14	0
14–16	2
16–18	1
18–20	1

Fit a log Pearson Type III distribution to the data and determine the peak discharges for return periods of 2, 5, 10, 25, 50, 100 and 200 years.

We have the data given log-transformed and we determine the mean, standard deviation and skewness coefficient, β_1 as per the Equations 7.12, 7.13, and 7.14. Accordingly for the grouped frequency data we have

$$\text{Mean}: \overline{\log x} = \frac{\Sigma f \log x_i}{N} = \frac{67.3856}{87} = 0.7750$$

$$\text{Standard deviation, } \sigma_{\log x} = \sqrt{\frac{\Sigma f (\log x_i - \overline{\log x})^2}{N-1}} = \sqrt{\frac{3.3315}{87-1}} = 0.1962$$

$$\text{Skewness coefficient, } \beta_1 = \frac{\Sigma f (\log x_i - \overline{\log x})^3}{(N-1)(N-2)(\sigma_{\log x})^3}$$

$$= \frac{87(0.5165)}{(87-1)(87-2)(0.1962)^3} = 0.81$$

Note N being fairly large = 87, we do not apply any modification for skewness coefficient, β_1, here.

Now corresponding to the skewness coefficient, β_1 of (0.81), for different recurrent intervals (T = 2, 5, 10, 25, 50, 100 and 200 years), the values of frequency coefficients, $k = f(T, \beta_1)$ can be read from Appendix Table IV for Log Pearson Type III distribution. The computed discharge values, x_i for different return periods are now shown in the following Table 7.4.

Table 7.4 : Discharge Frequency Analysis (Log Pearson Type III Method)

Return Period, (T)	Skew Co-efficient (β_1)	Frequency Factor, k (from Appendix Table IV)	$\sigma_{\log x}$	$k \cdot \sigma_{\log x}$	$\overline{\log x}$	$\log x_T$	x_T (in '000 cumecs)
(1)	(2)	(3)	(4)	(5)	(6)	(7)	(8)
2	0.81	– 0.132	0.1962	– 0.0259	0.7750	0.7491	5.61
5		0.780		0.1530		0.9281	8.47
10		1.336		0.2621		1.0371	10.89
25		1.993		0.3910		1.1660	14.66
50		2.453		0.4813		1.2563	18.04
100		2.891		0.5672		1.3422	21.99
200		3.312		0.6498		1.4248	26.59

This information is plotted now in Fig 7.3 as Log Pearson Type III distribution. Note the plotting is done on a probability graph paper with abscissa showing the relative frequencies $\left(=\dfrac{m}{N+1}\right)$ and discharges being plotted along the ordinate on logarithmic scale. Thus the computed discharges, x_i versus relative frequencies plotted on this probability paper will yield a straight line so that the plotted points will lie evenly on each side of the line. On this probability paper, from the straight line the values of the discharges corresponding to any frequency can be read. Thus, the empirical frequency distribution follows a *log normal distribution* with mean = 5.96 (thousand cumecs) and standard deviation = 1.57 (thousand cumecs). Note both of these are antilog of their respective values.

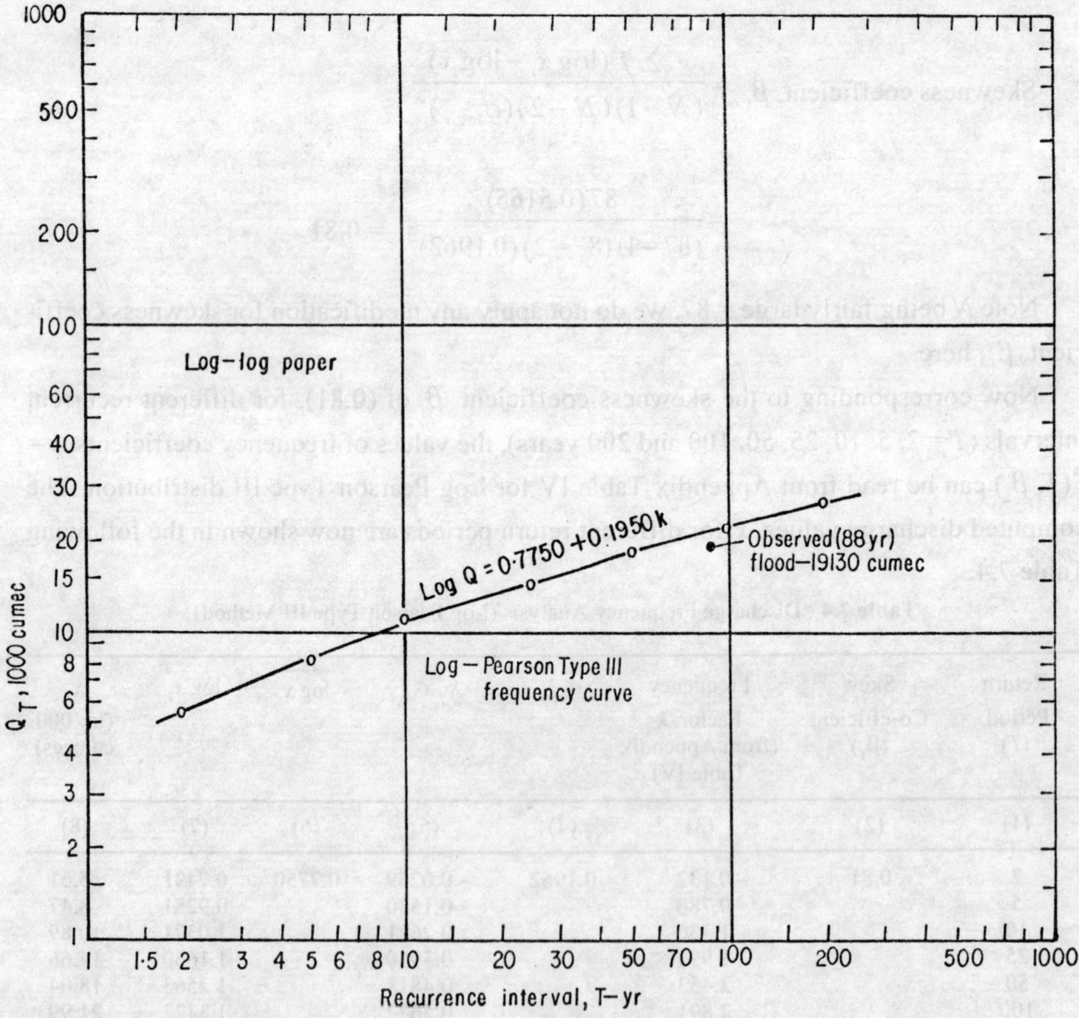

Fig. 7.3 : Log-Log Plot of Discharge Frequency (Log Pearson Type III Distribution)

LIST OF FORMULAE

Probability of exceedance of an extreme event of magnitude, m is $p = \dfrac{m}{N+1}$ = Weibull's plotting position

Recurrence interval/Return Period, T year of the event

$$T = \frac{1}{p} \text{ year} = \frac{N+1}{m}$$

General Equation for Hydrologic Frequency Analysis

$$x_T = \bar{x} + k \, . \, \sigma$$

where $\bar{x}$ = mean, σ = std. deviation, k = frequency factor and x_T = Probability of exceedance of a T-year return period.

The frequency factor, k in Gumbel distribution

$k = (0.78y - 0.45)$ where y is the reduced variate $y = -\ln\left[-\ln\left(1 - \dfrac{1}{T}\right)\right]$ with T the return period given.

General Equation for probability of exceedance of a T-year return period in log Pearson Type III Distribution is

$$\text{Log}\, x_T = \overline{\log x} + k\, \sigma_{\log x}$$

where mean, $\overline{\text{Log}\, x} = \dfrac{\sum f \, Log\, x_i}{N}$; std. deviation, $\sigma_{\log x}$

$$= \sqrt{[\sum f(Log\ x_i^2) - (\sum Log\ x_i)^2 / N^2]/(N-1)}$$

and k is the skewness coefficient

$$\beta_1 = \frac{\sum f(\log x_i - \overline{\log x})^3}{(N-1)(N-2)(\sigma_{\log x_i})^3}$$

where N is small, the correction for skewness coefficient is

$$\bar{\beta}_1 = \left(1 + \frac{8.5}{N}\right) . \beta_1$$

Frequency factor, k in Log Pearson Type III is $k = f(T, \beta_1)$ to be read from Appendix Table IV.

EXERCISES

7.1. Following is the data on annual maximum 24-hour rainfall in a period of 22 years at a station: 13, 12, 7.6, 14.3, 16.9, 6.0, 8.0, 12.5, 11.2, 8.9, 8.9, 7.8, 9.0, 10.2, 8.5, 7.5, 6.0, 8.4, 10.8, 10.6, 8.3 and 9.5 (rainfall is in cm). Plot the rainfall data over return period in years on a semi-log paper and extrapolate the following:
(a) rainfall magnitude for 13- and 50- year return period
(b) the probability $p(m)$ for rainfall = 10 cm and 2.4 years return period.

8

Spatial Statistics

(Analysis of Location Pattern Distribution and Arrangement)

8.1 ANALYSIS OF LOCATION PATTERN

The measures of central tendency, e.g. mean, median and mode summarises the informations contained in any distribution. But these measures can also be described for 'spatial data' that is the data in which location is a variable, and which can therefore be described as *spatial distribution*.

The usual way of depicting spatial distribution is by maps, and it is a commonplace that the map is the basic tool of geo-science. Geo-scientists are carefully trained to read, utilise and create maps in 2-spatial dimensions. Probably no other group of scientist is as adept at expressing and envisioning dimensional relationships. It is indeed maps are as important to earth scientists as the conventions for scales and notes to the musician for they are compact and efficient means of expressing relationships and details.

But maps by themselves are mute unless map relationships are expressed in terms of two approaches – (one) is that of the 'pen portrait' and the (other) that of the 'statistical profile.' In the former patterns are identified intuitively and characteristics with whatever verbal facility one can command. But in the latter one tries to find objective measures of various facets of the mapped distributions, the latter are always in terms of points upon the map. We are concerned with the distances between points, the density of points, degree of clustering and the values assigned to points which are all presented rather like a set of vital statistics. Again most maps are estimates of continuous functions, based on discrete observations at central points, e.g. topographic maps on which contour lines are an expression of a continuous and unbroken surface. Thus we can say that two approaches are complementary rather than mutually exclusive.

The character of continuous functions like contour lines for topographic maps, the bedding planes for the structural maps are, of course, based on the estimates made at bench mark locations (for topographic maps), boreholes for structural maps etc. Hence in the quality of the finished map reflects to great extent the density of these control points (at which measurements/observations) are made. Hence it is necessary in spatial statistics to study the effects which the control-point distributions can have on maps. But before we study these we should emphasize that the spatial data is essentially geographic data and statistical analysis of this data is concerned with indices of areal location, shifts of population, agricultural output etc. Thus when data are expressed as either of point patterns or data expressed as aggregate frequencies on a map we go by an extension of descriptive statistics but applied to data in 2-dimensional space, known as "centrographic measures", the latter are also *geostatistics*. We start this chapter by considering below the *centrographic measures*.

8.1.1 Mean Centre

Mean centre is the location on a map that can best be used to summarise the distribution of a phenomenon or event on the map. It is either expressed as point patterns or data expressed as aggregate frequencies on a map. In both cases the data must first be identified by grid coordinates.

The mean centre is taken as the centre of gravity of spatial distribution since each of them is weighted by its *bivariate measures*: location as well as the value at each point, at any particular time and thus it is counterpart to the arithmetic mean of descriptive statistics. The mean centre is the point where the sum of square of the deviations is minimised and hence it is more sensitive to the extreme value than is the median center itself, the latter is the point of intersection of two particular lines (usually N–S and E–W) each of which bisects the distribution of points.

For calculating the mean centre, a coordinate grid with equal numbers of horizontal and vertical axes are overlaid on the map showing the distributions. The limits of the grid are given by the locations of the four most extreme points in the map (Fig. 8.1). Thus starting arbitraily by drawing on X-axis and Y-axis to the south and west of the distribution of points, one can determine the X- and Y-coordinates of each point (i.e. X_i and Y_i respectively). Thus the bivariate coordinates of the mean centre ($\overline{X}_c, \overline{Y}_c$) may be calculated from the expressions:-

$$\overline{X}_c = \frac{\Sigma(X_i p_i)}{\Sigma p_i} \quad \ldots (8.1)$$

$$\text{and,} \quad \overline{Y}_c = \frac{\Sigma(Y_i p_i)}{\Sigma p_i}$$

where p_i is the value at each point

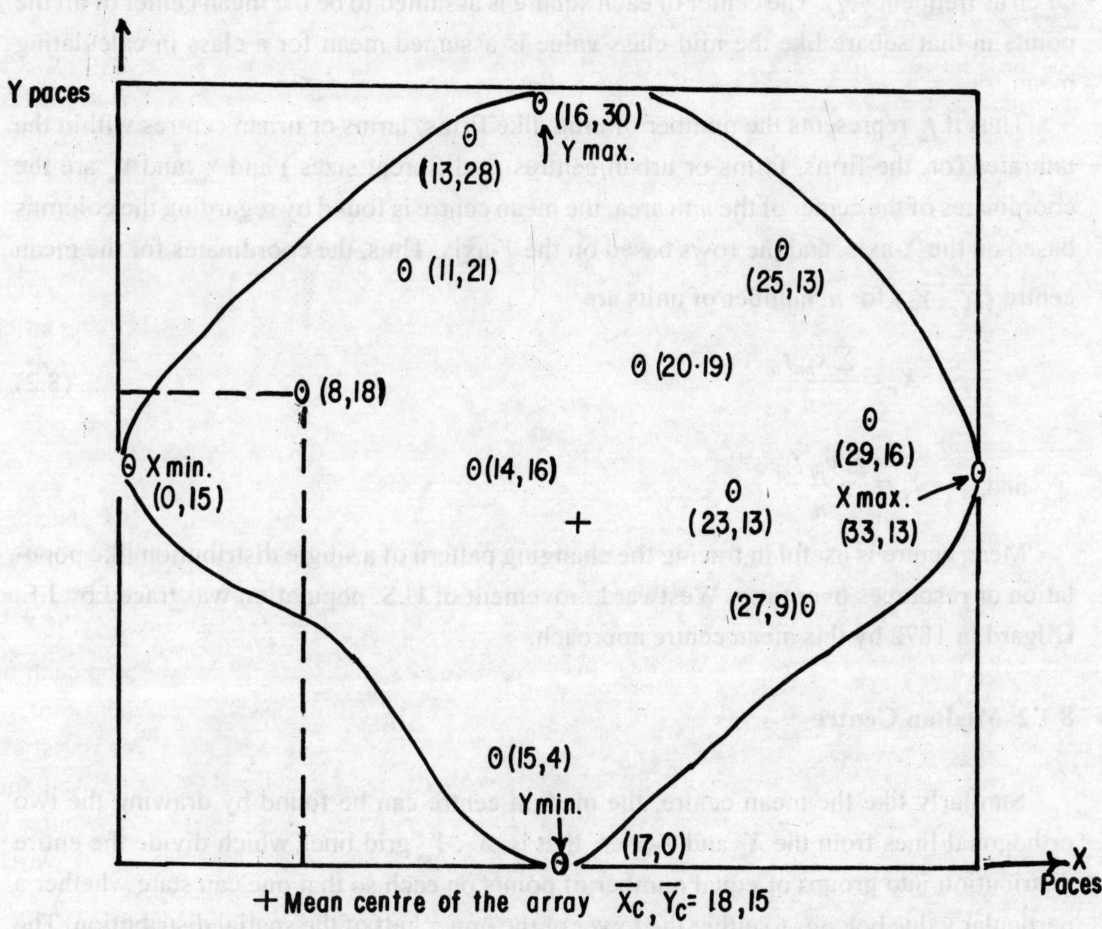

Fig. 8.1 : Mean Centre

Mean centre thus calculated serves a simplest measure for study of concentration of a particular industrial activity e.g. the concentration of different type of manufacturing in an area. Thus if we do have the information of the number of workers in each firm and the location of all the firms on the map, we can calculate the mean centre of workers in the area with the help of the Eq. (8.1). In case the specific information about each firm is not available, p_i in the Eq. (8.1) becomes equal to one and one calculate only the mean centre of the firms without weighing the means of the coordinates.

The mean centre can also be calculated from grouped data. In this case, the grid overlay should be a square giving equal number of grids column- and row-wise. The groups may be grid squares on the map within each of which the total number of units is known and are

taken as frequency, f_i. The center of each square is assumed to be the mean center of all the points in that square like the mid-class value is assumed mean for a class in calculating mean.

Thus if f_m represents the number of units like firms, farms or urban centres within the *m*th area (or, the firms, farms or urban centres of different sizes) and x_m and y_m are the coordinates of the center of the *m*th area, the mean centre is found by regarding the columns based on the *X*-axis, and the rows based on the *Y*-axis. Thus, the coordinates for the mean centre $(\overline{X}_c, \overline{Y}_c)$ for n number of units are

$$\overline{X}_c = \frac{\Sigma x_m f_m}{n} \qquad \text{... (8.2)}$$

and, $$\overline{Y}_c = \frac{\Sigma y_m f_m}{n}$$

Mean centre is useful in tracing the changing pattern of a single distribution like population or resources over time. Westward movement of U.S. population was traced by J.E. Hilgard in 1872 by this mean centre approach.

8.1.2 Median Centre

Similarly like the mean centre, the median centre can be found by drawing the two orthogonal lines from the *X*- and *Y*-axes, that is, X_m, Y_m grid lines which divide the entire distribution into groups of equal number of points on each so that one can state whether a particular value belongs to either the lower or the upper half of the spatial distribution. The first to do for this is to overlay a coordinate grid on the map, with equal numbers of horizontal and vertical axes. The limits of the grid overlay are set by the four most extreme points in the pattern (Fig. 8.2a). The intersection of the X_m and Y_m grid lines give the median centre. Larger the area of the rectangle on either side of the median centre means considerable dispersion, a small rectangle means concentration near the centre. The disadvantage of finding the median centre by this method is that mean centre can still be found using two lines still at right angles to each other, but this time going diagonally across the distribution (Fig. 8.2b).

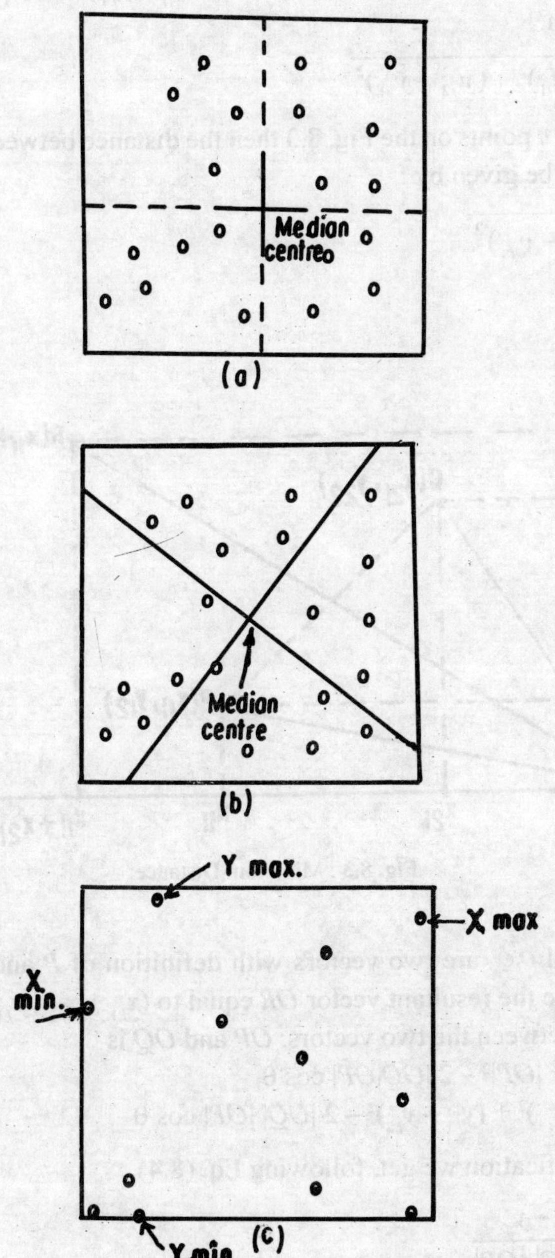

Fig. 8.2 : Three Methods of calculating the Median Centre

The *median centre* is also defined as the point in a spatial distribution at which the sum of absolute deviation of each point is minimised i.e. sum of distances between the median centre and each point are at a minimum. (Fig. 8.2c)

Thus if P and Q are two points with coordinates (x_{i1}, y_{j1}) and (x_{i2}, y_{j2}) we can write by Pythagoras' Theorem,

$$^{P}Q = \sqrt{(x_{i2} - x_{i1})^2 + (y_{j2} - y_{j1})^2} \quad \text{... (8.3)}$$

Now, if there are n points on the Fig. 8.3 then the distance between any two of them, i's and jth points would be given by :

$$d_{ij} = \sqrt{\sum_{s=1} (x_{is} - y_{js})^2} \quad \text{... (8.4)}$$

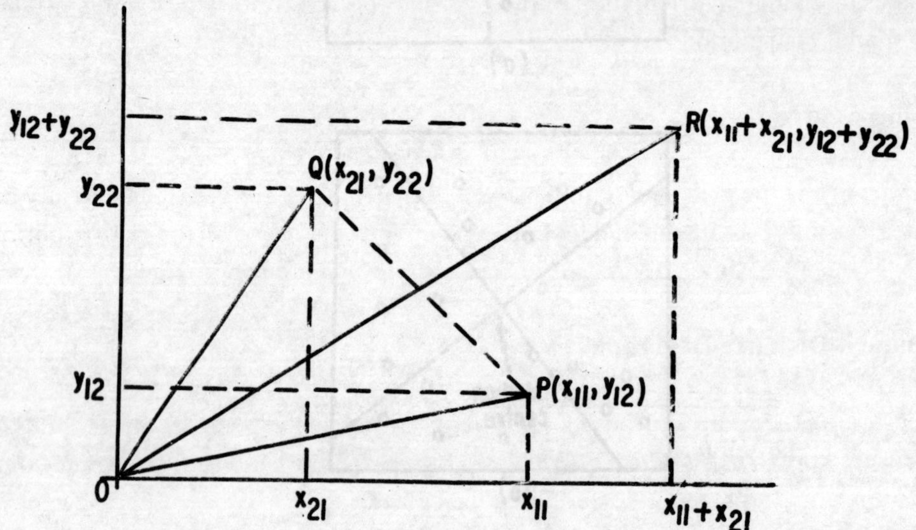

Fig. 8.3 : Minimum Distance

Again, if $\vec{OP}$ and $\vec{OQ}$ are two vectors with definition of P and Q as (x_{i1}, y_{j1}) and (x_{i2}, y_{j2}) we can write the resultant vector OR equal to $(x_{i1} + x_{i2}, y_{j1} + y_{j2})$.

Now, the angle between the two vectors, OP and OQ is

$$|QP|^2 = |OQ|^2 + |OP|^2 - 2\,|OQ|OP|\cos\theta$$

$$= (x_{i2} - x_{i1})^2 + (y_{j1} - y_{j2})^2 - 2\,|OQ|\,|OP|\cos\theta \quad \text{... (8.5)}$$

from which on simplification we get, following Eq. (8.4),

$$\cos\theta = \sum_{s=1} \frac{x_{is} - y_{js}}{2|OQ|.|OP|} \quad \text{... (8.6)}$$

This property of median centre as the point of "minimum distance travel" makes, it an important property in spatial distribution measures. For example, the median centre can be used for many locational measures because of its property of distance minimisation. Thus if a school is located in a rural area, then in theory atleast the theoretical optimum location of it can be determined regarding the distribution of students. But since there is the difficulty of defining the quadrats by orthogonal axes, this method is practically not much in use as a measure of median position. Another difficulty is that since the location of the median centre cannot be uniquely found, its use is restricted in geographical investigations. Thus the sensitive character regarding its change in location with the change in origin of the *X*-and *Y*-axes or change in orientation of the grid lines prevent it from becoming a standard index for spatial distribution.

8.1.3 Modal Centre

Modal centre is the grid square in which there are more occurrences than in any other grid square. It is thus possible to pick out modal centre easily from a gridded distribution of the values. The centre of this gives the coordinator of the modal centre (x_{mo}, y_{mo}).

8.1.4 Standard Distance Deviation

A measure of the scatter of points (or, spatial dispersion) about the mean centre is a spatial measure known as *standard distance deviation* or *root mean square distance deviation*. Since distance represents deviation of each of the points of statisticians' standard deviation. Following calculation of point to point distance, we can have the expression used for standard distance deviation as:

$$\text{Standard distance deviation, } S_D = \sqrt{\frac{\Sigma(x_i - \bar{x}_c)^2}{n} + \frac{\Sigma(y_i - \bar{y}_c)^2}{n}}$$

$$= \sqrt{\left(\frac{\sum x_i^2}{n} - \bar{x}_c^2\right) + \left(\frac{\sum y_i^2}{n} - \bar{y}_c^2\right)} \quad \text{... (8.7)}$$

where mean centre is $\bar{X}_c$ and $\bar{Y}_c$ equal to $\frac{\sum x_i p_i}{\sum p_i}$ and $\frac{\sum y_i p_i}{\sum p_i}$ respectively, x_i and y_i are the coordinates of individual units with respect to two orthogonal axes, *n* are the total number of units.

Now if the map area is divided into *k* regions, then the powerful tool of statistical analysis, the *analysis of variance* can be applied. The expression states that the sum of the squared deviations of the observation values from the grand mean for the whole area, that is, the sum of squares for total, *SST* can be partitioned into two parts: (a) the sum of the squared deviations of the values from their respective regional means (i.e. the sum of a "within regions" squared distance, *SSW*) and (b) the sum of the weighted squared deviations of their respective regional from the grand mean (i.e. a "between regions" squared distance, *SSB*). That is

$$SSD = SSW + SSB$$

$$= \sum_{i}^{k}\sum_{j}^{n} \frac{[(x_{ij}-\bar{x}_j)^2 + (y_{ij}-\bar{y}_j)^2]}{n} + \sum_{j}^{k} n_j . \frac{[(\bar{x}_j-\bar{x})^2 + (\bar{y}_j-\bar{y})^2]}{n} \quad \text{... (8.8)}$$

where (x_{ij}, y_{ij}) are the coordinates of the *i*th point in the *j*th region, $(\bar{x}_i, \bar{y}_j)$ is the regional mean centre, *k* is the number of regions, $(\bar{x}, \bar{y})$ is the mean centre of the whole distribution.

The ratio $\frac{SSB}{SST}$ provides a crude measure of significance of the regional divisions in the pattern, for if the ratio is close to 1, then something approaching maximum differentiation exists, whereas a ratio value close to zero suggests that the regions have the same mean centre as the total pattern does.

When the data set is large, they can conveniently be rearranged by one set according to the *x*-coordinates and another according to the *y*-coordinates. A *bivariate frequency distribution* is thus obtained, on which we can have the computation for the standard distance deviation.

$$S_D = \sqrt{\left(\frac{\sum f_m x_m^2}{n} - \bar{x}_c^2\right) + \left(\frac{\sum f_m y_m^2}{n} - \bar{y}_c^2\right)} \quad \text{... (8.9)}$$

where x_m and y_m are the midvalues of the grid squares and x_c, y_c are the mean centres equal to

$\frac{\sum f_m x_m}{n}$ and $\frac{\sum f_m y_m}{n}$ respectively.

Standard distance deviation appears to be a successful measure of geographical dispersion associated with any phenomenon to be studied as it's index of location. It is useful for evaluating the spread of human population over time. It is useful also for comparative analysis of the distribution of different ethnic or social groups of population. This utility of the method can be illustrated with respect to the comparative character of the spatial distribution of a number of commercial activities in a city. For example, in a city a commercial function like insurance offices are clustered in the central business district of the city whereas grocery shops are distributed throughout the city. The mean centre for both these functions need not distinguish the difference in spatial distribution pattern between them. But this difference in spatial pattern will be clearly illustrated by more standard`distance deviation of the grocery shops as compared to that of the insurance officials. Thus though standard distance deviation cannot be shown cartographically but it appears to be an important measure of geographical dispersion to be associated with the mean centre. In a comparative study, differences in shape and size of the study areas require the use of "relative standard distance measure." There are a number of ways of calculating such relative measures, depending on the phenomenon under investigation. For example, in studying the nature of dispersion of commercial functions like insurance offices/grocery shops of the cities can be related to the population distribution and the "relative dispersion" is calculated by dividing the standard distance of the commercial distribution by the standard distance of population. Again, for studying population dispersion in different countries, the relative measure can be based on the radius of the area of each country, assumed to be circular.

Example 8.1 Compute the mean centres and the standard distance deviation of population of 20 villages given in terms of rectangular coordinates below:

Serial No. of Villages	Locational Coordinates		Population		
	x	y	1951	1961	1971
1.	1.00	4.20	221	292	312
2.	1.50	3.65	328	427	492
3.	2.65	4.10	118	121	136
4.	5.25	4.35	892	1023	1521
5.	2.90	3.30	542	621	811
6.	1.15	3.20	572	623	683
7.	1.10	3.15	460	501	581
8.	0.80	2.30	747	841	891
9.	3.55	2.45	218	1292	329
10.	5.60	2.60	957	1118	1416
11.	6.90	2.80	311	518	912
12.	1.65	1.90	618	714	839
13.	3.30	1.40	551	612	781
14.	7.25	1.05	2463	3141	5128
15.	0.25	1.05	189	220	289
16.	1.35	1.05	289	373	433
17.	2.50	0.80	314	471	561
18.	3.25	0.35	171	281	392
19.	4.50	0.30	461	571	682
20.	6.80	0.40	92	217	318

Note: location coordinates are in kilometre east west and north south from an arbitrary origin.

Computation for mean centre for three different years are as follows:

S.N.	X-coordinates = $x_i p_i$			Y-coordinates = $y_i p_i$		
	1951	1961	1971	1951	1961	1971
1.	221.0	292.0	312.0	928.2	1226.4	1310.4
2.	492.0	640.5	738.0	1197.2	1558.5	1795.8
3.	312.7	320.6	360.4	483.8	496.1	557.6
4.	4683.0	5370.7	7985.2	3880.2	4450.0	6616.2
5.	1571.8	1800.9	2351.9	1788.6	2049.3	2676.3
6.	657.8	716.4	785.4	1830.4	1993.6	2185.6
7.	506.0	551.1	639.1	1449.0	1578.1	1830.1
8.	597.6	672.8	712.8	1718.1	1934.3	2049.3
9.	773.9	4586.6	4717.9	534.1	3165.4	3256.0
10.	5359.2	6260.8	7929.6	2488.2	2906.8	3681.6
11.	2145.9	3574.2	6292.8	870.8	1450.4	2553.6
12.	1019.7	1178.1	1384.3	1174.2	1356.6	1594.1
13.	1818.7	2019.6	2577.3	771.4	856.8	1093.4
14.	17856.7	22772.2	37178.0	2586.1	3298.0	5384.4
15.	47.2	55.0	72.2	198.4	231.0	303.4
16.	390.1	503.5	584.5	303.4	391.6	454.6
17.	785.0	1177.5	1402.5	251.2	376.8	448.8
18.	555.7	913.2	1274.0	59.8	98.2	137.2
19.	2074.5	2569.5	3069.0	138.3	171.3	204.6
20.	625.6	1475.6	2162.4	36.8	86.8	127.2
Total	42494.1	57450.8	82529.3	22688.2	29676.0	38260.2

Thus, mean centre for 1951 :

$$\overline{X}_c = \frac{\sum x_i\, p_i}{\sum p_i} = \frac{42494.1}{10514.0} = 4.04$$

$$\overline{Y}_c = \frac{\sum y_i p_i}{\sum p_i} = \frac{22688.2}{10514.0} = 2.16$$

Similarly, mean centre for 1961 is (4.11, 2.12) and for 1971 it is (4.46, 2.07). where Σp_is are 13977 and 17507 respectively. For standard distance deviation, S_D we have

$$S_D = \sqrt{\left(\frac{\sum x_i^2}{n} - \overline{x}_c^2\right) + \left(\frac{\sum y_i^2}{n} - \overline{y}_c^2\right)}$$

where $x_i^2 = 292.35$, $y_i^2 = 133.63$

So, S_D for 1951 is $= \sqrt{(14.62 - 16.32) + (6.67 - 4.66)} = \sqrt{0.31} = 0.56$

Similarly, S_D for 1961 and 1971 is – 0.30 and – 1.78 respectively

8.2 STANDARD DEVIATIONAL ELLIPSE

Standard distance deviation is a convenient nature of dispersion in point patterns since it summarises the spread of point in just one value. However, it takes no account of the fact that spread about the mean centre may be different in differnt directions. In Fig. 8.4 for example, standard distance deviation describes dispersion in terms of a circle about the mean centre, But from Fig. 8.4 it is apparent that dispersion is greater in a NE-SW direction than it is in a SE-NW direction.

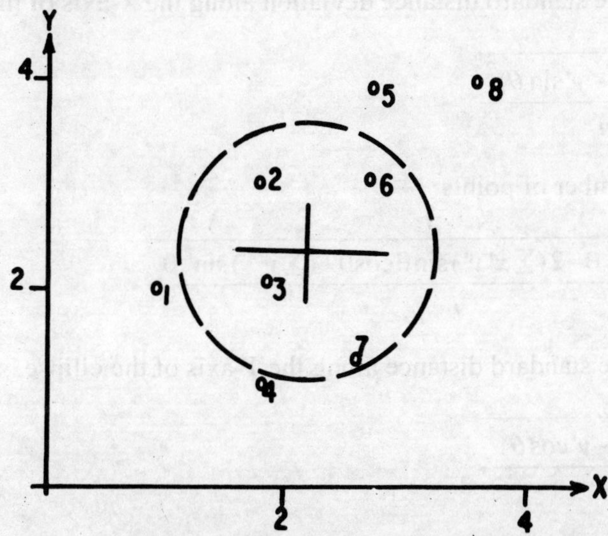

Fig. 8.4 : Standard Distance Deviation

The *standard deviational ellipse* as the name implies, is a measure which summarises dispersion in a point pattern in terms of an ellipse rather than a circle. The ellipse is centered on mean centre with its long axis in the direction of maximum dispersion and its short axis in the direction of minimum dispersion. Since mathematically, the axis of maximum dispersion in a point pattern is always at right angle (i.e. orthogonal) to the axis of minimum dispersion, the ellipse is an appropriate measure.

In order to fit an ellipse about the mean centre of a point pattern it is necessary to calculate (a) the angle of rotation of the ellipse for its orientation (b) length of standard deviation along the *x*-axis of the ellipse (i.e. the length of the long axis) and (c) the y-axis of the ellipse (i.e. the length of the short axis). The mathematical calculation for these, require first of all, the transposition of the coordinate system with reference to the mean centre, $\bar{X}_c$

and $\bar{Y}_c$ as $x' = X_i - \bar{X}_c$ and $y' = Y_i - \bar{Y}_c$. Note : x' and y' are read as x prime and y prime.

Now, (a) to calculate the angle of rotation,

$$\tan\theta = \frac{(\Sigma x'^2 - \Sigma y'^2) + \sqrt{(\Sigma x'^2 - \Sigma y'^2)^2 + 4(\Sigma x'y')^2}}{2\Sigma x'y'} \quad \text{... (8.10)}$$

In case tan θ produces a negative sign, the sign is ignored when looking up the angle in table of tangents but the angle found from the table is subtracted from 90° to get correct angle equivalent for tan θ.

(b) to calculate the standard distance deviation along the X-axis of the ellipse

$$\sigma_x = \sqrt{\frac{\Sigma(x'\cos\theta - y'\sin\theta)^2}{n}} \quad \text{... (8.11)}$$

where n is the number of points

$$\sigma_x = \sqrt{\frac{(\Sigma x'^2)\cos^2\theta - 2(\Sigma x'y')\sin\theta\cos\theta + (\Sigma y'^2)\sin^2\theta}{n}} \quad \text{... (8.11a)}$$

(c) to calculate the standard distance along the Y-axis of the ellipse, similarly,

$$\sigma_y = \sqrt{\frac{\Sigma(x'\sin\theta - y'\cos\theta)^2}{n}} \quad \text{... (8.12)}$$

$$\sigma_y = \sqrt{\frac{(\Sigma x'^2)\sin^2\theta - 2(\Sigma x'y')\sin\theta\cos\theta + (\Sigma y'^2)\cos^2\theta}{n}} \quad \text{... (8.12a)}$$

Six steps are involved in fitting a standard distance deviational ellipse to a point pattern. The Fig. 8.5 plot the location of 8 points (A to H):

A (1, 2); B (2, 3); C (2, 2); D (2, 1); E (3, 4); F (3, 3); G (3, 1) and H (4, 4).

If we have to draw the standard deviational ellipse on the point pattern of these 8 points, we first of all transpose the x, y coordinates of the points into corresponding x', y' as follows:

Points	x	y	$x' = x - \bar{x}$	$y' = y - \bar{y}$	x'^2	y'^2	$x'y'$
A	1	2	– 1.5	– 0.5	2.25	0.25	0.75
B	2	3	– 0.5	+ 0.5	0.25	0.25	– 0.25
C	2	2	– 0.5	– 0.5	0.25	0.25	0.25
D	2	1	– 0.5	– 1.5	0.25	2.25	0.75
E	3	4	+ 0.5	+ 1.5	0.25	2.25	0.75
F	3	3	+ 0.5	+ 0.5	0.25	0.25	0.25
G	3	1	+ 0.5	– 1.5	0.25	2.25	– 0.75
H	4	4	+ 1.5	+ 1.5	2.25	2.25	2.25
	Σ20	Σ20	$\Sigma x' = 0.0$	$\Sigma y'$	$\Sigma x'^2$	$\Sigma y'^2$	$\Sigma x'y'$
	$\bar{x} = 2.5$	$\bar{y} = 2.5$		= 0.0	= 6.00	= 10.00	= 4.00

Now from, Eq. (8.10),

$$\tan\theta = \frac{(\Sigma x'^2 - \Sigma y'^2) + \sqrt{(\Sigma x'^2 - \Sigma y'^2)^2 + 4(x'y')^2}}{2\Sigma x'y'}$$

$$= \frac{(6-10)+\sqrt{(6-10)^2+4(4)^2}}{2(4)} = \frac{-4+8.9443}{8}$$

$= 0.6180,$

So $\theta = 31°43'$

Now $\sin\theta = 0.5257$, $\cos\theta = 0.8507$, $\cos^2\theta = 0.7237$, $\sin^2\theta = 0.2764$ and $\sin\theta\cos\theta = 0.4472$,

again,

$$\sigma_x = \sqrt{\frac{(\Sigma x'^2)\cos^2\theta - 2(\Sigma x'y')\sin\theta\cos\theta + (\Sigma y'^2)\sin^2\theta}{n}}$$

$$= \sqrt{\frac{(6\times 0.7237) - 2(4)(0.4472) + (10)(0.2764)}{8}} = \sqrt{\frac{3.5286}{8}} = 0.6641$$

Similarly, σ_y can be calculated as equal to 1.2493. In Fig. 8.5 we show the mean centre (2.5, 2.5) and measure the angle 31° 43' clockwise at the mean centre. We get the transposed *x*- and *y*-axis. Finally, for drawing the standard deviation ellipse, we take $2\sigma_x$ = 1.3282 and $2\sigma_y$ = 2.4986 and draw it in Fig. 8.5. We further illustrate the six steps (as mentioned earlier) involved in fitting a *standard distance deviational ellipse* to a point data in Fig. 8.6.

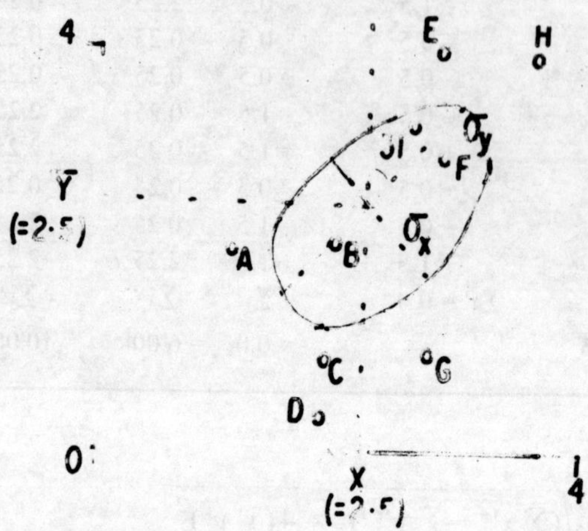

Fig. 8.5 : Standard Distance Deviational Ellipse

Example 8.2 Fit a standard deviational ellipse to the following point data:-
A (3, 4), B (3, 9), C (4, 7), D (5, 5), E (6, 3), F (7, 1) and G (7, 6)

we have $\bar{x}_c = 5$, $\bar{y}_c = 5$, $\sum x'^2 = 18$, $\sum y'^2 = 42$ **and** $\sum x'\,y' = -16$

$$\text{so, } \tan\theta = \frac{(\sum x'^2 - \sum y'^2) + \sqrt{(\sum x'^2 - \sum y'^2)^2 + 4(x'y')^2}}{2\sum x'y'}$$

$$= \frac{(18-42) + \sqrt{(18-42)^2 + 4(-16)^2}}{2(-16)}$$

$$= \frac{-24+40}{-32} = -0.5, \text{ hence } \theta = -26°\,34' = 90 - 26°34'$$

$= (90 - 26°34') = 63°26'$

Now $\sin\theta = 0.8945$, $\cos\theta = 0.4483$, $\sin^2\theta = 0.8002$, $\cos^2\theta = 0.2001$, $\sin\theta\cos\theta = 0.4008$

$$\text{Hence } \sigma_x = \sqrt{\frac{(18)(0.2001) - (2)(-16)(0.4008) + (42)(0.8002)}{7}}$$

$$= \sqrt{\frac{50.0358}{7}} = 2.6736$$

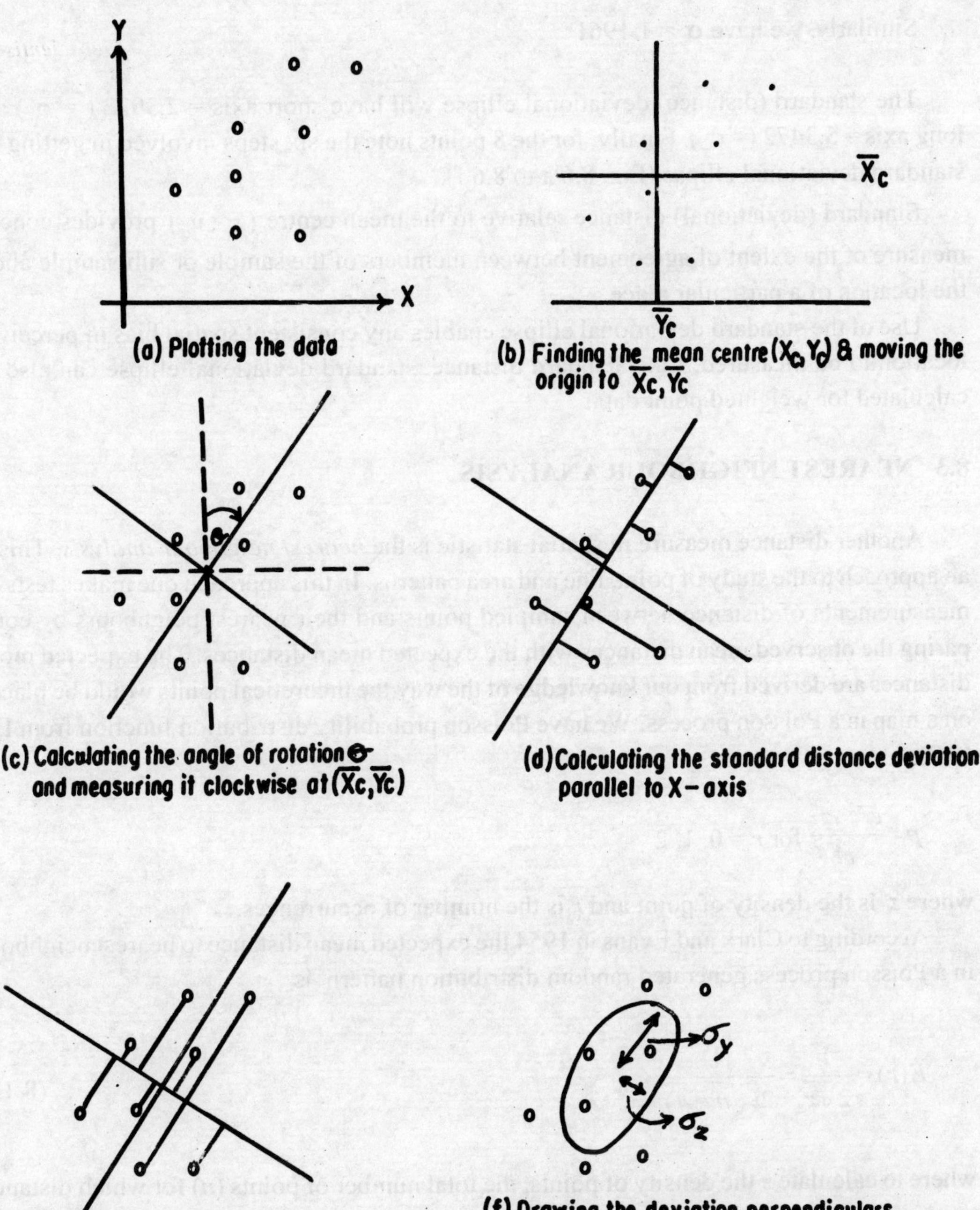

Fig 8.6 : Six steps in Fitting a Standard Distance Deviational Ellipse

Similarly, we have $\sigma_y = 1.1961$

The standard (distance) deviational ellipse will have short axis = 2.3922 (= σ_x) and long axis = 5.3472 (= σ_y). Finally, for the 8 points note the six steps involved in getting the standard deviational ellipse (Fig. 8.6 a to 8.6 f).

Standard (deviational) distance relative to the mean centre $(\bar{x}_c, \bar{y}_c)$ provides concise measure of the extent of agreement between members of the sample or sub-sample about the location of a particular place.

Use of the standard deviational ellipse enables any consistent spatial bias in perceived location to be measured. Like standard distance, standard deviational ellipse can also be calculated for weighted point data.

8.3 NEAREST NEIGHBOUR ANALYSIS

Another distance measure in spatial statistic is the *nearest neighbour analysis*. This is an approach to the study of point, line and area patterns. In this approach one makes tests on measurements of distance between sampled points and their nearest neighbours by comparing the observed mean distances with the expected mean distances. The expected mean distances are derived from our knowledge of the way the theoretical points would be placed on a map in a Poisson process. We have Poisson probability distribution function from Eq. (6.7).

$$p = \frac{e^{-z} . z^r}{r!} \text{ for } r = 0, 1, 2$$

where z is the density of point and r is the number of occurrences.

According to Clark and Evans in 1954 the expected mean distance to nearest neighbour in a Poisson process generated random distribution pattern is

$$E(r) = \frac{1}{2\sqrt{z}} = \frac{1}{2(\sqrt{n/A})} \quad \text{... (8.13)}$$

where to calculate z the density of points, the total number of points (n) for which distance measurements are made in the given area (A) are determined and hence z in nearest neighbour analysis is n/A. Now, the nearest neighbour statistic is

$$E = \frac{\bar{r}}{E(r)} \quad \text{... (8.14)}$$

where $\bar{r}$ is the mean of the distances between every point and its nearest neighbor, $E(r)$ is the expected mean distance to nearest neighbour as given in earlier. The above nearest neighbour statistic (as the ratio of the measured mean distance and expected mean distance) ranges from 0 for a distribution where all points coincide giving a clustered pattern, to 1.0 for a random distribution of points rising to a maximum value of 2.15, the latter value characterise a distribution in which the mean distance to nearest neighbour is to maximise a distribution : it has the form of an even hexagonal pattern where every point is equidistant from every other five points.

The standard error of $E(r)$ is

$$\sigma_r = \frac{0.26136}{\sqrt{nz}} \quad \text{... (8.15)}$$

where the constant 0.26136 is derived from considerations of the radius of a circle of unit area and the Poisson probability model.

It is possible to apply a significance test to nearest neighbour analysis to decide how probable is that the observed arrangement of points occurred at random i.e. by chances. The test statistic thus measures the significance of the departure from the Poisson process model as

$$z = \frac{|\bar{r} - E(r)|}{\sigma_r} \quad \text{... (8.16)}$$

where z is the standard variate of the normal curve. Note this formula is particularly applicable for samples of 100 or more

Thus the summary procedure for nearest neighbour analysis is :

(a) Calculate the area in which the points are located and count the number of points (n).

(b) Measure the distance between each points and the nearest neighbour, find the mean, r

(c) Apply the formula to find R, the nearest neighbour index

(d) Formulate H_0 and H_1 and calculate σ_r

(e) Calculate z and establish the significance of the result.

Example 8.3 For an area measuring 60 sq. km, 16 settlements are there. If the mean of the distances of each settlement and between their nearest neighbour is calculated to be 1.45 km, find out the pattern of settlement distribution and test its significance.

The accompanying Fig. 8.7 gives the location of the village settlements which appears fairly well scattered. Here area $A = 60$ sq. km, $n = 16$ so $z = \frac{n}{A} = 0.2666$. And we have

$$E(r) = \frac{1}{2\sqrt{0.2666}} = \frac{1}{1.0326} = 0.9684$$

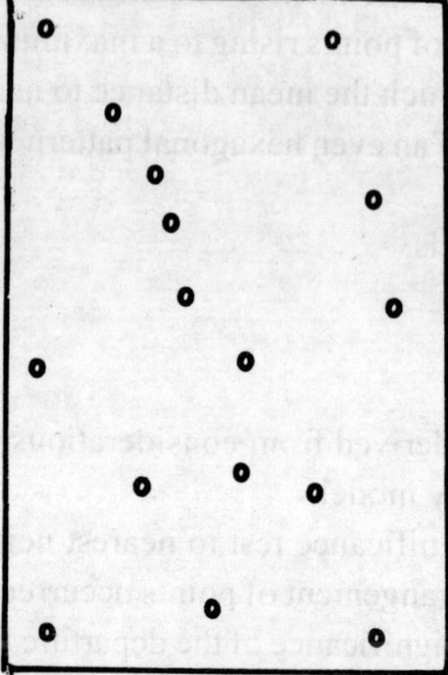

Fig. 8.7 : Location of Settlements

We have, $E = \dfrac{\bar{r}}{E(r)} = \bar{r} \cdot \dfrac{1}{E(r)} = (1.45)\dfrac{1}{(0.9684)} = 1.4973 \approx 1.50$

Now $z = \dfrac{\bar{r} - E(r)}{\sigma_r}$ where $\bar{r}$ = the mean of the nearest neighbour distance calculated, 1.45 km. Again, we calculate σ_r as

$$\sigma_r = \frac{0.26136}{\sqrt{nz}} = \frac{0.26136}{\sqrt{16 \times 0.2666}} = \frac{0.26136}{2.06533} = 0.1265$$

So, $$z = \frac{1.45 - 0.9684}{0.1265} = 3.8071$$

Now, for testing the significance of the statistic $E = 1.50$, we have:

H_0 : the pattern of villages similar to a random pattern.

H_1 : the villages distributed in a manner that is not random.

Since we have $z \geq 3.80$ so a 2-tailed test that this distribution might be expected to occur through chance variations is no less than 1 in 10,000. So, we reject H_0 and we can seek explanation for an even settlement pattern.

8.3.1 Problems in Application of Nearest Neighbour Analysis

Nearest neighbour index thus as revealed in the Example 8.3 above clearly shows that point patterns are not random, and which therefore requires explanation. So to prove that a point pattern (or, distribution) is significantly not random, we can use this analysis but it cannot be used to prove that it is. If the analysis shows randomness in the statistical sense, confirmation must lie outside measurements and figures. Like all statistical methods this method has to be applied with caution and common sense. The value of the interpretation of the results depends on the judgement and skill of the investigator. Thus though this analysis by its statistic enables more exact comparisons to be made than hitherto, this index has its certain limitations.

(1) The first is that the points are located within an infinite area and thus there is the problem of defining boundaries because as the boundary changes, the values of $\bar{r}$ get altered. Ideally, the boundary should be natural e.g. the case of the settlement pattern on an island, in which high water mark provides a precise termination of area. Similarly, the distribution of plant species over a particular rock type has natural boundary where the outcrop terminates.
When part of an area is considered, the decision as to where to place the boundary becomes much more difficult. In this case, the decision must lie with the researcher who must state a particular factor which affects it. Where no obvious boundary exists it is only possible to say that larger the area chosen the smaller will be the result of arbitrary decisions affecting the boundary.

(2) The second point is that the points are free to locate anywhere within that area. In most geographical situations these assumptions are manifestly unrealistic. In many cases where this nearest neighbour analysis is applied the delimitation of the study area is quite arbitrary. Since the size of the area influences the point density, z it also seriously affects the outcome of the analysis. If a study area is enlarged without bringing more points into it, the effect is to increase the expected values of $E(r)$ without of course, changing the value of $\bar{r}$. The nearest neighbour index, E will consequently be decreased in value, although the appearance of the observed arrangement of points and its degree of clustering or dispersion will not have changed at all. Thus there is evidence that the significance test is consistently biased towards finding significantly dispersed patterns.

(3) The third point is that it does not distinguish between a single and a multi-clustered pattern. An extreme example of this might be a series of pairs of towns on either side of a river, each pair at some distance from the next, making a linear pattern in the landscape. In these cases because each town (or, attribute) forms the nearest neighbour of the other town (or, attribute) of the pair, the index will be near zero.

(4) The fourth point is that it "averages out" sub-patterns which may exist within the area and may thereby hide contrasting sub-patterns which cancel each other out when put together to give a false impression of randomness.

(5) The fifth point is that a pattern emerges with an index approaching 1 indicates the kind of distribution which could result from the independent random selection of each location. This does not mean that locations of the observed phenomena in the real landscape are necessarily purely the result of chance.

(6) The sixth point is that the distribution of one kind of phenomenon may largely determine that of another. For example, a particular type of drift deposited by melting ice in an apparently random manner might closely affect a pattern of agricultural land use.

(7) Finally, it can be said that the selection of the nearest neighbour (or, time series) is arbitrary. The choice of the second or third nearest neighbour is possible and might produce different results.

Of these above problems, of course, it does not mean that this nearest neighbour frequently do not provide very useful descriptive measure of point pattern, particularly for qualifying the increase or decrease in dispersion or clustering of a pattern through time provided the definition of the study area remains the same. However, comparison between different areas based on this test should be treated with caution, as should the significance test. For example, in a Poisson process generated pattern the means and variances of measurements are equal. In this analysis we take distances from sampled points to their nearest neighbouring points (called, point-to-point distances, r_{pp}) or distances from sampled locations (called, location-to-point distances, r_{lp}). We square both r_{pp} and r_{lp} distances and sum them. The ratio of the two is called C

$$C = \frac{\sum r_{lp}^2}{\sum r_{pp}^2} \quad \text{... (8.17)}$$

The expected value of C is 1. This test proposed by Pielou in 1969 is based on the standardised normal variate that is,

$$z = 2\left(y - \frac{1}{2}\right)\sqrt{(2n+1)} \text{ for } n > 50 \quad \text{... (8.18)}$$

where $y = \frac{C}{1+C}$ and obviously the expected value of y would be $\frac{1}{2}$. This test is particularly accurate for $n > 50$.

8.4 SPATIAL DISTRIBUTION PATTERNING

To have a quick survey of the spatial distribution patterning of individuals occurring across a continuous community, especially where individuals are easily distinguished (e.g. village settlement in a habitat, trees in a forest etc.), we can have

(a) Quadrat sampling method
and, (b) Quadrat variance method

(a) *Quadrat sampling method* is discused above in terms of quadrats as arbitrary sampling units. In this context, the size and shape of the quadrats is having an important influence on the distribution. For example, for a clustered distribution, if the size of quadrat are doubled, the number of individuals per quadrat will be greatly influenced. On other hand, doubling the quadrat size will not be a problem when the distribution is random, since the expected number of individuals per quadrat is the same regardless of the quadrat size, in a random distribution following the Poisson series.

(b) *Quadrat variance methods* examine the changes in the mean and variance of the number of individuals per sampling units (or, quadrat) over a range of quadrat sizes. Once a given area is gridded into a belt of contiguous quadrats as that in Fig. 8.8 in which the belt data can be represented as a row vector of observations, $x = x_1, x_2, x_3, \ldots, x_n$ where n is the total number of quadrats in a row.

Now with this belt size, we can have "different block sizes" e.g. 1, 2, 3, ..., by combining the quadrats progressively in a manner known as "two-term local quadrats" (TTLQ) or "pairs of quadrats" (POQ). The TTLQ method affects the quadrat size by combining adjacent or abutting quadrats to identify the pattern intensity (the range of densities present) and the grain of the pattern (the distance between the present of individuals). The POQ method utilize change in quadrats spacing to provide this information.

In Fig. 8.8 we are having a belt transect of 8 contiguous (abutting) quadrats. Now if we have to pool the quadrats of different block sizes of spacings for the TTLQ method, at block size 1, there are 8 quadrats, at block size 2, we can have 4 quadrats, by combining the adjacent, similarly at block size 3, there will be 2 quadrats etc.

For the POQ method we keep the block size same (equal to 2) unlike the above TTLQ where the quadrats are of different sizes, and thus the spacing (or, distance) between the quadrats is the only component which can contribute to the variance calculation.

Now for the above transect of 8 contiguous quadrats in a row vector we can have the following blocks of quadrats by the above 2 methods in Fig. 8.8.

1	2	3	4	5	6	7	8

Block method	Block size	Quadrats
TTLQ	1	(1); (2); (3); (4); (5); (6); (7); (8)
	2	(1, 2); (3, 4); (5, 6); (7, 8)
	3	(1, 2, 3); (4, 5, 6)
	4	(1, 2, 3, 4); (5, 6, 7, 8)
POQ	Spacing	Quadrats
	1	(1); (2); (3); (4); (5); (6); (7); (8)
	2	(1,2); (2,3); (3,4); (4,5); (5,6); (6,7); (7,8)
	3	(1,3); (2,4); (3,5); (4,6); (5,7); (6,8)
	4	(1,4); (2,5); (3,6) (4,7); (5,8)
	5	(1,5); (2,6); (3,7); (4,8)
	6	(1,7); (2,8); (1, 6); (2, 7); (3, 8)

Fig. 8.8 : Belt Transect of 8 Contiguous Quadrats

Now, the variance for the TTLQ at block size 1 is

$$\text{Var}(x)\,1 = \frac{1}{(n-1)}\left[\frac{1}{2}\sum_{i=2}^{n}\left(x_{i-1} - x_i\right)^2\right] \quad \text{... (8.19)}$$

for block size 2 it is

$$\text{Var}(x)2 = \frac{1}{n-3}\left[\left(\frac{1}{4}\right)\sum_{i=4}^{n}(x_{i-3} + x_{i-2} - x_{i-1} - x_i)^2\right] \quad \text{... (8.20)}$$

for block size 3 it is

$$\text{Var}(x)3 = \frac{1}{n-5}\left[\left(\frac{1}{6}\right)\sum_{i=4}^{n}(x_{i-3} + x_{i-2} + x_{i-1} - x_i - x_{i+1} - x_{i+2})^2\right] \quad \text{... (8.21)}$$

for block size 4 it is

$$\text{Var}(x)\,4 = \frac{1}{n-7}\left[\left(\frac{1}{8}\right)\sum_{i=4}^{n}(x_{i-4} + x_{i-3} + x_{i-2} + x_{i-1} - x_i - x_{i+1} - x_{i+2} - x_{i+3})^2\right] \quad \text{... (8.22)}$$

Now for the variance for the POQ, since the quadrat sizes remain the same but spacing is different, we can see now the difference in quadrat spacing only can contribute to the difference in variance. Hence, the variance at a spacing of 1 by the POQ method is:

$$\text{Var }(x)1 = \frac{1}{n-1}\left(\frac{1}{2}\right)\sum_{i=2}^{n}(x_{i-1}-x_i)^2 \quad \text{... (8.23)}$$

Again, the variance at a spacing of 2 is

$$\text{Var }(x)2 = \frac{1}{n-2}\left(\frac{1}{2}\right)\sum_{i=3}^{n}(x_{i-2}-x_i)^2 \quad \text{... (8.24)}$$

Similarly, the variances at spacing 3 and 4 are respectively,

$$\text{Var }(x)3 = \frac{1}{n-3}\left[\left(\frac{1}{2}\right)\sum_{i=4}^{n}(x_{i-3}-x_i)^2\right] \quad \text{... (8.25)}$$

and, $$\text{Var }(x)4 = \frac{1}{n-4}\left[\left(\frac{1}{2}\right)\sum_{i=5}^{n}(x_{i-4}-x_i)^2\right] \quad \text{... (8.26)}$$

It can be seen from Fig. 8.8 the largest block size for which a variance estimate would be obtained with respect to n= 8 is 4 by the TTLQ method and 7 by the POQ method.

Example 8.4 For 8 quadrats given in contiguity for a belt transect, the number of individuals are

$x = (x_1, x_2, x_3, \ldots, x_8)$

$= (1, 0, 3, 5, 2, 1, 4, 6)$

Compute the variances up to the block-size or block-spacing by the two term local quadrat (TTLQ) and pair of quadrat (POQ) methods. Plot the variances by these methods and interpret.

Variances by TTLQ method

At block size 1 it is

for var (x) 1 $$= \frac{1}{n-1}\left[\left(\frac{1}{2}\right)\sum_{i=2}^{n}(x_{i-1}-x_i)^2\right]$$

$$= \frac{1}{(8-1)}\frac{1}{2}\left[(x_1-x_2)^2+(x_2-x_3)^2+\ldots(x_7-x_8)^2\right]$$

$$=\left(\frac{1}{7}\right)\left(\frac{1}{2}\right)\ [(1-0)^2+(0-3)^2+(3-5)^2+(5-2)^2+\ldots(4-6)^2]$$

$$=\left(\frac{1}{7}\right)(0.5)\ [1+9+49+\ldots+4]=\frac{1}{7}\ (18.5)=2.64$$

$$\text{for var }(x)2=\frac{1}{(8-3)}\{\frac{1}{4}(1+0-3-5)^2+\frac{1}{4}(0+3-5-2)^2$$

$$+\ldots+\frac{1}{4}\ (2+1-4-6)^2\}$$

$$=7.15$$

Similarly, for var (x) 3 = 1.00 and for var (x) 4 = 2.00

Variances by POQ method

At block spacing of 1, by the pairs of quadrat method, the variance is : var (x) 1

$$=\frac{1}{n-1}\left[\left(\frac{1}{2}\right)\sum_{i=2}^{n}(x_{i-1}-x_i)^2\right]$$

$$=\frac{1}{(8-1)}\left[\left(\frac{1}{2}\right)\{(1-0)^2+(0-3)^2+\ldots+(4-6)^2\}\right]$$

$$=\frac{1}{7}\ (0.5+4.5+\ldots+2.0)=2.64$$

Now, for variance at a spacing of 2, var (x)2

$$=\frac{1}{n-2}\left[\frac{1}{2}\sum_{i=3}^{n}(x_{i-2}-x_i)^2\right]$$

$$=\frac{1}{6}\left[\frac{1}{2}\{(1-3)^2+(0-5)^2+\ldots+(1-6)^2\}\right]$$

$$=\frac{1}{6}\ (2.0+12.5+\ldots+12.5)=6.25$$

Similarly for variances at a spacing of 3 and 4 by the same POV method are 4.10 and 0.5 respectively.

Now the above variances obtained by using the TTLQ and POQ methods are plotted against block sizes in Fig. 8.9 which shows clustering at block size/spacing both at 2 quadrat sizes.

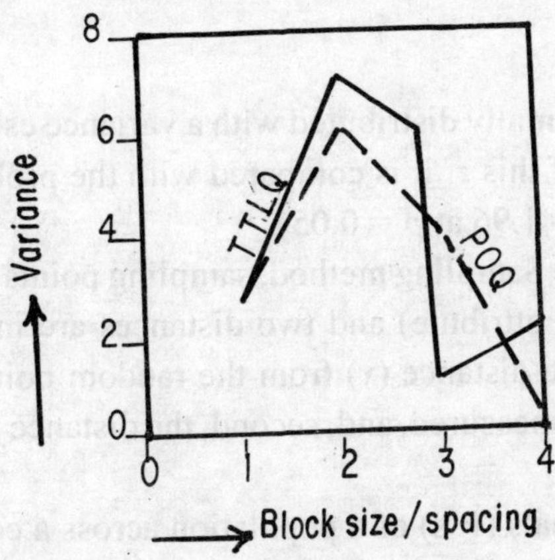

Fig. 8.9 : Clusterng at Block Size (SU)

8.4.1 Distance-sampling Method

Spatial pattern analysis (SPA) in the previous quadrat variance method is for sample data obtained from contiguous quadrats, that is, observations from a series of abutting sampling plots (quadrats) across a continuous surface. An alternative to this method is to sample using "plotless" methods. This is known as "distance sampling" methods which describe basic distance sampling procedures, known as the "nearest neighbor sampling." The "*T*-Square Distance Sampling" is a well-known method for indexing the distribution pattern. The other method is the "Distance Index of Dispersion," *I*. We discuss below both these methods which give equally comparable powerful tests for spatial patterning.

8.4.1.1 *T*-Square Distance Sampling

The *T*-square distance can be derived as a ratio of squared point to individual distances, x_i and squared individual-to-nearest neighbour distances, y_i as

$$C = \frac{\sum_{i=1}^{n}\left[x_i^2 / (x_i^2 + \frac{1}{2} y_i^2)\right]}{n} \qquad \text{... (8.27)}$$

where n = total number of sample points

This value of C is approximately 1/2 for random patterns, significantly less than 1/2 for uniform patterns and greater than 1/2 for clustered patterns.

To test the significance of any departure from 1/2, we compute a value for z as

$$z = \frac{C - 0.5}{\sqrt{1/(12n)}} \quad \text{... (8.28)}$$

since C is approximately normally distributed with a variance estimated by $1/(12n)$. Hence for statistical significance of this z, it is compared with the probability table for standard normal distribution (with $z = 1.96$ at $p = 0.05$).

In the T-Square Distance Sampling method, sampling points are selected at random (or on a regular grid within the attribute) and two distances are measured at each point, as shown in Fig. 8.10. First, the distance (x) from the random point (O) to the nearest individual (P) in any direction is measured, and, second, the distance y from individual (P) to its nearest neighbour (Q)

If the pattern of individuals (P's) of a population across a continuous surface is completely random, then with distance sampling the expected square of point to individual distances will be approximately equal to one-half the expected square of T-square nearest neighbour distances, i.e. $E(x^2) = \frac{1}{2} E(y^2)$. For clustered distribution, $E(x^2) > \frac{1}{2} E(y^2)$ and for uniform pattern, $E(x^2) < \frac{1}{2} E(y^2)$

8.4.1.2 Distance Index of Dispersion

This method is also a new distance index of spatial dispersion, I based on point-to-individual distances. Given a sample of n points with distances x_i from the ith point to the nearest individual then this index of dispersion (I) is defined as the ratio of the sums of squares of squared distances to the square of the sum of squares of the distances. This index thus involves a lot of "squaring" and in equation form it is given as:

$$I = (n+1) \frac{\sum_{i-1}^{n} (x_i^2)^2}{\left[\sum_{i-1}^{n} (x_i^2)\right]^2} \quad \text{... (8.29)}$$

Now, from the T-square sampling distances obtained, an index of spatial pattern can be obtained since the above T-square Index also converges to a normal distribution with

$$z = \frac{I - 2}{\sqrt{4(n-1)/(n+2)(n+3)}} \quad \text{... (8.30)}$$

For getting information on spatial pattern of a distribution, both these indexes of distance methods should be applied together.

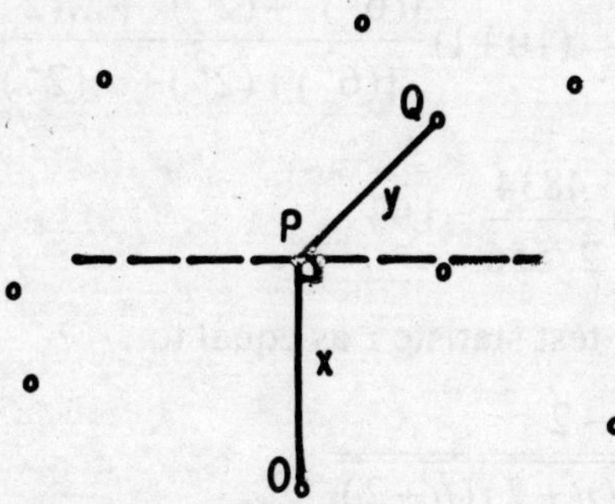

Fig. 8.10 : T-distances

Example 8.5 Following is a sample set of 10 point-to-individual (X_i) and individual-to nearest neighbour y_i distance pairs in metres

S. P.	1	2	3	4	5	6	7	8	9	10
x_i	6	2	6	5	6	2	1	2	4	2
y_i	4	9	3	7	1	8	3	5	6	5

Compute the *T*-square Index (*C*) and Distance Index (*I*) of the settlement and comment on the settlement pattern.

For *T*-Square Index of spatial patterns we compute *C* as

$$C = \frac{\sum_{i=1}^{n}\left[x_i^2 / (x_i^2 + \frac{1}{2} y_i^2)\right]}{n}$$

$$= \frac{1}{10}\left[\frac{6^2}{6^2 + 4^2/2} + \frac{2^2}{2^2 + 9^2/2} + \ldots + \frac{2^2}{2^2 + 5^2/2}\right]$$

$$= \frac{1}{10}(0.82 + 0.09 + \ldots + 0.24) = 0.454$$

Now for significance test of computed *C*, we have

$$z = \frac{C - 0.5}{\sqrt{1/[(12)(10)]}} = \frac{-0.046}{0.091} = -0.51$$

Since the computed value of *z* for $C = 0.454$ is less than the critical *z* values of 1.96 at $p = 0.05$, so we accept the null hypothesis that the distances were measured between point and individuals within an underlying random pattern.

Again, for the Distance Index of Dispersion (I) we have,

$$I = (n+1)\frac{\Sigma(x_i^2)^2}{[\Sigma(x_i^2)]^2} = (10+1)\frac{[(6^2)^2+(2^2)^2+...(2^2)^2]}{[(6^2)+(2^2)+...(2^2)]^2}$$

$$=(11)\frac{4834}{(166)^2} = (11)\frac{4834}{27556} = 1.93$$

Now, we compute the test statistc z as equal to

$$\frac{I-2}{\sqrt{4(n-1)/(n+2)(n+2)}} \text{ as}$$

$$z = \frac{1.93-2}{\sqrt{[4(10-1)]/[(10+2)(10+3)]}} = \frac{-0.07}{\sqrt{36/[(12)(13)]}} = -\frac{0.07}{\sqrt{0.23}} = -0.15$$

Since this value of z is less than its critical of 1.96 at $p = 0.05$ so the random pattern of distribution of the villages is accepted.

We discuss the various measures of spatial statistics elsewhere in the book like the Poisson distribution in a geographic area context earlier and multicollinearity and autocorrelation separately in a chapter on regression. Finally, we have concluded the last chapter of the book with trend surface mapping. These all come under the purview of spatial statistics.

LIST OF FORMULAE

Mean centre for listed data : $\bar{X}_c = \frac{\Sigma(X_i p_i)}{\Sigma p_i}, \bar{Y}_c = \frac{\Sigma(Y_i p_i)}{\Sigma p_i}$

for grouped data : $\bar{X}_c = \frac{\Sigma(x_m f_m)}{n}, \bar{Y}_c = \frac{\Sigma(y_m f_m)}{n}$

Standard Distance Deviation, (for listed data)

$$S_D = \sqrt{\left(\frac{\Sigma x_i^2}{n} - \bar{x}_c^2\right) + \left(\frac{\Sigma y_i^2}{n} - \bar{y}_c^2\right)}$$

for grouped data, $S_D = \sqrt{\left(\frac{\Sigma f_m \cdot x_m^2}{n} - \bar{x}_c^2\right) + \left(\frac{\Sigma f_m y_m^2}{n} - \bar{y}_c^2\right)}$

Standard Deviational Ellpse :

S_D along the *x*-axis, $\sigma_x = \sqrt{\dfrac{\Sigma(x'\cos\theta - y'\sin\theta)^2}{n}}$

S_D along the *y*-axis, $\sigma_y = \sqrt{\dfrac{\Sigma(x'\sin\theta - y'\cos\theta)^2}{n}}$

Angle of rotation, $\tan\theta = \dfrac{(\Sigma x'^2 - \Sigma y'^2) + \sqrt{(\Sigma x'^2 - \Sigma y'^2)^2 + 4(x'y')^2}}{2\Sigma x'y'}$

Nearest Neighbour Statistic, $E = \dfrac{\bar{r}}{E(r)}$ where $E(r) = \dfrac{1}{2(\sqrt{n/A})}$

and $\bar{r}$ = mean of the distance between every point and its nearest neighbour, n = total number of points and A = area

Standard error of $E(r)$, the expected mean distance to nearest neighbour is $\sigma_r = \dfrac{0.26136}{\sqrt{nz}}$

where z = mean density of point locations; The test statistic for nearest neibhour statistic, E

is $z = \dfrac{|\bar{r} - E(r)|}{\sigma_r}$

Distance Index of Dispersion, $I = (n+1)\ \dfrac{\Sigma(x_i^2)^2}{[\Sigma(x_i^2)]^2}$; For $I = 2$ pattern is random, < 2 for uniform and > 2 for clustered pattern.

EXERCISES

8.1. In a residential survey of an area, dwelling units are located in terms of eastings and northings from a map origin. The coordinates in terms of eastings and northings of 25 dwelling units are given below:

1.0, 0.9; 1.3, 1.2; 3.7, 5.4; 0.8, 4.2; 9.9, 0.1; 1.2, 8.0; 6.6, 0.6; 2.5, 0.8; 3.1, 0.6; 8.5, 2.6; 6.3, 2.7; 7.9, 9.8; 5.2, 1.1; 8.1, 4.5; 9.9, 5.9; 6.5, 4.8; 8.0, 1.2; 7.4, 3.5; 6.9, 9.1; 2.7, 7.3; 0.9, 8.9; 6.3, 6.6; 1.2, 5.5; 4.4, 0.8; 8.0, 3.3

Compute the mean, median and modal centres of the dwelling units. Calculate the standard distance deviation also.

8.2. Group the data given in Exercise 8.1 above into five classes and calculate the standard distance deviation.

8.3. Within 2 kilometres of a city centre, a map on a scale of 3/4 cm. to 1 km, provides distribution of 27 gas stations. Assuming that the gas stations do not exist outside this city boundary, calculate the mean distance of the gas stations in a random distribution.

8.4. It R in a nearest neighbourhood analysis is 1.10 and the z value of the points is 0.937, what pattern of the distribution will be revealed?

8.5. An area of 25 sq. kms divided into 25 square quadrats is having 50 location points of an attribute. Now out of these 50 points covering the area, if 41 points are sampled and observed mean distance to their nearest neighbours are found to be 0.3061 km, test whether the points will have a random distribution. Again for this data of 41 sampled points if we do have the calculations for $\sum r_{lp}^2 = 5.915$ and $\sum r_{pp}^2 = 4.940$ can we conclude that the point pattern was generated by a Poisson spatial process?

9

Sampling and Sampling Plan in Geo-science Systems

9.1 THEORETICAL BASIS AND SAMPLING

Generally, a set of data represents only a small fraction of the total number of observations that might in principle be available about the subject of interest. The total set of potential observations or the aggregate of objects is named a *population*, and the set of observations actually selected from this population is named a *sample*. The selection procedure by which the sample is selected is known as "sampling".

The *concept of population* comprising the total number of possible observations of a specified physical object or phenomenon or a survey is known as "census". *Census* is a complete approach which may not be possible in data analysis because of the stupendous task involved in terms of cost, man-power and time. Often they are hypothetical. Again, the statistical definition of a sample as a group of observations does not agree with the set of observations at hand of a geo-scientist who likes to treat it as a sample and a representative of the total information of interest. Hence two kinds of populations can be distinguished in most geo-science systems, the (i) one from which the sample is actually drawn which is named the *sampled population* and (ii) the population of interest that is the *target population* about which we wish to make inferences or draw conclusions. Target population is thus the population which we are trying to study and we come to know about it when the behaviour of all possible samples from a population is analysed. The sampled population is an important concept because by statistical theory we can make quantitative inferential statements, with known chances of error, from sample to sampled population. It must be carefully distinguished from the target population, the population of interest, about which we are tempted to make similar inferential statements. For example, the target population in landuse studies might be conceptualised as the kaleidoscope of patterns that exist from month to month and year to year. The sampled population in such dynamic issues like landuse is the pattern which exists at the particular time a survey is taken.

On the basis of sample study we can predict and generalise the behaviour of mass phenomena. This is possible because there is "no statistical population" whose elements would vary from each other within limit. For example, flow in any river varies from day to day and no two flows are alike but they can always be identified as discharge. Thus we find that although diversity is universal quality of mass data, every population has characteristic properties with limited variation. This makes possible to select a relatively small unbiased random sample and a representative cross section of all cases or of the whole area being studied (i.e. target population). Hence if the target population is a well-defined set (finite or infinite) of elements, a sampled population is a subset of elements taken from a population so that the samples are (i) adequate by size, (ii) independent of each other, (iii) representative and (iv) homogeneous for representing the target population. The scientific methods of selecting samples and the relations between the sample and the population form the subject matter of the theory of sampling.

Thus sampling could be used in studies such as nature of variations in landuse of an area, innovation adoption of farmers within a group of villages or the collection of data about slope characteristics along a river. In all these cases making use of the total information available will raise problems of time, effort and money which can be avoided if we can first define the target population explicitly and precisely and then adopt a reliable method for sampled population. This involves two things: (i) the samples must first be collected by the use of an appropriate "sampling plan" and (ii) the samples must first be related by the use of an appropriate "degree of confidence" for estimating the size of samples. The problem of the sampling plan, then, is to determine how to select the sample and how large a sample to select. In actual practice since a sample is not studied for its own sake these sampling procedures cannot be considered independently of the method of deriving an estimate from the sample, i.e. the "sampling estimation" procedure. The values obtained from the study of a sample, such as the mean and variance, are known as "statistics" and these values for the population are called "parameters". In practice, the *parameter values* are not known and their estimates based on the sample values are generally used. Thus *statistic* which may be regarded as an estimate of the parameter obtained from the sample is a function of the sampling plan out of which the sample values are drawn. Certain symbols traditionally are assigned to measures of distribution curves. Generally, the symbols for population distributions are Greek letters, and those for sample distributions are Roman. The population mean, for example is μ (mu), population standard deviation is σ (sigma) and the sample mean is designated as $\bar{x}$ and sample standard deviation, s. For purposes of exposition we will concentrate on the "sampling plan" in this chapter. The "sampling estimate" which is the second objective of sampling will be discussed in the next chapter.

9.2 SAMPLING PLAN

The first element of a sampling plan is the "sampling frame" which refers to a list, map or other specifications covering all the sampling units or observation about which is of interest in the sampling plan. Thus a *sampling frame* consists of sampling units which, if they are all covered in the data collection survey, will constitute a 100% sample. It can be said that without a sampling frame there can neither be any complete coverage nor there can be any probability sample. Therefore if the sampling frame does not have hundred percent coverage of the sampling units, the result of that sampling survey would not be correct from the point of view of providing desired information about the population it intends to study. The sampling frame used for enumeration of the population in a census is often a list of bounded areas referred to by census authorities as "zones". In an urban context, a zone is a small group of city "blocks" or "circles" each with a population say, between 3,000 to 5,000. The census publishes the statistics for the blocks/circles. In the frame of the circles all households bear a serial number.

Thus maximum representation of the population by a sample data may be achieved by drawing a sampling plan. The different methods of sampling which give a better representation under different contexts are basically two–(i) simple random or probability sampling in which every item in the population or the universe has a chance to occur and (ii) restricted random sampling, the latter can be further divided into three like (a) systematic or grid sampling, (b) stratified sampling and (c) cluster sampling.

9.3 SIMPLE RANDOM SAMPLING

In this method of drawing sample, there is the equal chance or probability for every member or every combination of individuals of population in the sampling frame to be selected as sample. "Simple random sampling" is a scientific technique based on the essential condition that every individual unit in the population is having an equal and independent chance or probability of being selected as sample. Thus in simple random sampling, the probability of drawing any unit from a population of n units is $1/n$, the probability of drawing any unit in the second draw from among the available $(n-1)$ units is $\left(\frac{1}{n-1}\right)$ and so on. Now if the event E is selected at the rth draw, then the probability of the event not selected at any previous $(r-1)$ draw and then selected at rth draw is the compound probability and it is

$$f(r) = \sum_{i=1}^{r=1}\left[1-\frac{1}{n-(i-1)}\right]\times\left[\frac{1}{n-(r-1)}\right] = \sum_{i=1}^{r=1}\left(\frac{n-i}{n-i+1}\right)\left(\frac{1}{n-r+1}\right)$$

$$= \left[\frac{n-1}{n}\times\frac{n-2}{n-1}\times\ldots\frac{n-r+1}{n-r+2}\right]\times\frac{1}{n-r+1} = \frac{1}{n} \quad \ldots (9.1)$$

Thus the probability of selecting a specific unit of the population at any given draw is the probability of its being selected at the first draw. In order to take a random sample from a population, unambiguous rules must be thoughtfully made and carefully followed. One useful device is to assign to each item in the population a unique serial number. For example, everyone eligible to receive a social security payment in the U.S.A. has a social security number. The social security numbers are serial numbers, and the monthly payments by the individuals, including the zero ones, are "observations". Observations are the same if many individuals are eligible to receive the same payment each month, but the individuals are differentiated by the serial numbers. In order to take a random sample from this population, it is only necessary to find a method of specifying the serial number for the individuals to be selected.

The serial numbers may be specified and the random samples are then selected by methods of chance like tossing a coin, drawing a card (from pack of cards). In practice, it is more convenient to use a "table of random numbers". *Random numbers* are generated by random processes, which are devised so that each digit from 0 to 9, has an equal probability (or chance) of occurring independently of each other. Because there are ten digits in the range 0 to 9, the chance or probability of one occurring each time is 0.1. The mathematical process is such that the digits are generated with statistical independence, just as the toss of a coin does not depend on the result of the previous toss.

Tables of random numbers are available. In the Appendix Table V, the random numbers are arranged in pairs with 18 two-digit columns of random numbers to the page. To use the table it makes no difference which column or row or diagonal we start with in picking up the random numbers. In a list of random numbers taken, there is no guarantee that the same numbers will not repeat, i.e. 5 can follow 5 more frequently than any other number. However, in a long list of random numbers, any pattern is highly unlikely. The use of random sampling is illustrated below with point random sampling. It is to be noted that random sampling applies to sampling without replacement, that is, once an item by a random number is selected in the sample it cannot appear in the sample again.

The units of observation may differ being either points, areas or lines. Thus depending on the selection of a single number, an area or a strip, the simple random sampling is known as point, areal or line random sampling respectively.

9.3.1 Point Sampling

From a list of values, single numbers locate a *point sample*. In a practical exercise of random sampling, the members of the population are first put in a list and every number is assigned a serial number. Then for selecting the samples from the serial numbers of the members in the sampling frame, the set of random numbers are chosen. For example, if the sampling frame contains 88 observations, two digits in Appendix Table V have to be taken and any random number more than 88 have to be discarded. Similarly, if the sampling frame contains 3,000 items, digits in four must be used for the random numbers. In this case if the random numbers between 3,001 and 9,999 are rejected, there will be a very high rejection rate. So it will be desirable to rephrase the numbers above 3,000 series, i.e. numbers 3,001 to 6,000 and 6,001 to 9000 can each be taken as a fresh series of values of 0001 to 3,000. Thus all the numbers are used and much time is saved. Example 9.1 is a specific example in this respect.

Example 9.1 Select 4 days at random from the days of a year numbered 1 to 365.

In this case, we read the digits in threes from the random numbers of the Appendix Table V which are:

201, 744, 947, 221, 932, 450, 449, 162, 045, and so on.

From the above reading, accepting the numbers between 1 to 365, we have 4 random numbers as follows:

201, 221, 162 and 045

The rejection in the above reading of the random numbers can be reduced by using the equivalence from the next series of 366 to 730 as: 001/366, 002/367, ..., 365/730, we then only reject 000 and numbers greater than 730 to have the random numbers as follows:

201, 221, 450 – 365 = 085 and 449 – 365 = 084

Another occasion when the listing of the data and its serial numberings are required if the data are available in a series of groups. In this case if it is desired to obtain an overall sample rather than a sample of each group, the serial numbers have to be made to run consecutively through the whole population before the samples are selected from the data. Example 9.2 below provides the selection of random samples from a frequency table.

Example 9.2 In Table 9.1 a frequency distribution of heights of 4,985 persons are given. Select 10 samples and compare the sample mean height with the population mean height.

Table 9.1 : Frequency distribution of height of persons

Class interval of height of persons (in cms)	Frequency (in numbers)	Range of random number (in serial order)
138.1–143.0	07	0001–0007
143.1–148.0	07	0008–0014
148.1–153.0	38	0015–0052
153.1–158.0	165	0053–0217
158.1–163.0	544	0218–0761
163.1–168.0	1174	0762–1935
168.1–173.0	1446	1936–3381
173.1–178.0	1032	3382–4413
178.1–183.0	441	4414–4854
183.1–188.0	107	4855–4961
188.1–193.0	18	4962–4979
193.1–198.0	04	4980–4983
198.1–203.0	02	4984–4985

In Table 9.1 the range of the random numbers extend from 0001 to 4985 inclusive – this is because though 0000 is a random number it does not occur in the random numbers of the Appendix Table V and hence it is excluded from the range.

Now to draw a random sample of 10 from this frequency distribution of height of persons, we use the Appendix Table V as the random number table and proceed down the column beginning on the left, ignoring any four-digit number larger than 4985. The first 10 random numbers and the class interval of height in which they fall are as follows:

Random number	Midpoint of the class
2017, 2215, 3270	170.55
0450, 0364	160.55
0172	155.55
2756, 2026	170.55
4905	185.55
4974	190.55

After rearranging the previous columns we get the frequency distribution of this 10 simple random sample as follows.

Midpoint	Frequency	(Frequency) (mid-point)
155.55	1	155.55
160.55	2	321.10
170.55	5	852.75
185.55	1	185.55
190.55	1	190.55
	$\Sigma 10$	$\Sigma 1705.50$

Hence the arithmetic mean of the 10 samples is $\bar{x}$ = 170.55 cm which compares favourably with the population mean of the height of the persons, μ = 169.90 cm.

9.3.2 Area Sampling

Area sampling by random numbers requires that firstly, the area under study should be gridded – if a map of the area is available it can be placed within the two axis at right angle. The grid can then be numbered with the zero in its southwest corner increasing steadily eastwards and northwards to generate a population of coordinate points. If the grid numbers are in two digits, four digits of random numbers have to be taken from the random number table, of which two conscutive digits can be assigned to the x coordinate. The intersections of row and column will give the samples of the data/observations in the area.

Thus when the units of observation are points, they are simply located by a series of random numbers as mentioned above. The probabilities of any point to be selected in this irregular pattern can also be known. For example, if an area or a map contains a range of coordinates from (00,00) to (99,99), the probability of any point to be chosen in one single chance is 1/(100 × 100) or 1/10,000 or 0.0001. It should be mentioned that the observational points need not coincide with the chosen grids. In such cases one can take the nearest observational point as the member of the sample and count the number of points falling within each grid. When choosing a series of the samples, areas are preferred since the numbering of the grid can apply to the space between the grid lines.

The most important step in area random sampling is to decide on the size of the grid area (better termed as the 'quadrat',which is a sampling area of a consistent size and shape, mostly square or circular), that can minimize the variances of the observed set of frequencies. In the case of large quadrats, *variance* will be high because of many points in same quadrats and a few in others. On the other hand, if the sample quadrats are too small in size they will cut through a clump of points and will be too large to contain atmost one point, thus leading to binomial probability distribution even for a clustered pattern on the map so that the number of points per quadrat will be either equal to 0 or 1. This latter situation is also undesirable from the point of view of the size of the quadrat in the area sampling. Thus there is an apparent indeterminacy for the data describing a spatial process and some compromise must be found between these two extremes. Since the size of the grid/quadrat is very important in determining the number of points in the area sampling method for using a probability distribution theory to model the frequency distribution of the observed point pattern (that is the density of points varying over the study area), it is suggested that the observed set of frequencies should be reasonably symmetrical. This can be achieved by having the mean number of points per quadrat large enough to allow about as many quadrats with fewer points per quadrat than the mean as there are quadrats with more points per quadrat than the mean. Thus the rule of thumb is that the proper size of a quadrat can be approximated as twice the size of the mean area per point. But one should first select a number of quadrat size and repeat the tests before any rule of thumb is applied. If results differ quadrats can be combined or rotated.

The decision of any area random sampling by grid or quadrat also begins with defining the boundaries of the study area. The guidelines for designing quadrat is that firstly, every part of the study area should be a possible location for a point. For example, if we study the frequency distribution of karst depressions, the study area should be exclusively a limestone region. Proper delimitation of the study area boundary automatically fixes the scale of resolution of the problem and it helps in predetermining the size of quadrats over the map of area delineated.

In the case of a definite clustering or agglomerating component of a particular distribution, the *sample variance* of the number of points per quadrats is likely to increase because there will be more empty quadrats and more quadrats containing a relatively large number of points, thus resulting in a larger spread about the mean. In contrast, for a pattern leading to a regularity or dispersion pattern in the map will produce relatively few empty quadrats and also relatively few quadrats with more points leading to a reduction in the sample variance. Thus it can be said that for a dispersed (or uniform) pattern of points the variance will be the least and it will be maximum for a clustered pattern with the random one occupying a medial position. The probability models for spatial point patterns take their assumptions from some qualitative ideas about the geographical processes which control the evolution of spatial pattern through space and time. Fig. 9.1 illustrates six patterns as the result of the continuum of possible point pattern which ranges from the totally clustered pattern (Fig. 9.1a) to the totally uniform pattern (9.1f) through the medial random position (Fig. 9.1c). Clustered patterns are usually controlled by contagious processes whereby the location of a point in space and time attracts subsequent point locations to its immediate vicinity (Fig. 9.1c). In fact many seemingly dissimilar spatial entities are found to exhibit clustering when mapped as point patterns, for example, the distribution of karst depressions in limestone terrain show a tendency of clustering. The extreme case of a clustered pattern is where all the points are attracted to the same location. This theoretical limiting case of a clustered point pattern is assumed to occur when the contagious process controlling the clustering changes from probabilities to a deterministic nature (Fig. 9.1a).

Random patterns are the result of an independent probabilistic process where each point is located in the plane independent of existing or subsequent point locations. The operation of an independent, spatial process usually results in the haphazard distribution of points as in Fig. 9.1c where visually there is no underlying structure to the pattern. Observed random point patterns are apparently of no interest because independent process is chance like and so they are not explainable in spatial context. However, theoretical random processes are simple to represent statistically by the binomial and Poisson distributions.

Uniform patterns are created by competitive processes where, during the evolution of the pattern, the points compete for space in the plane (Fig 9.1d and 9.1f). The distribution of

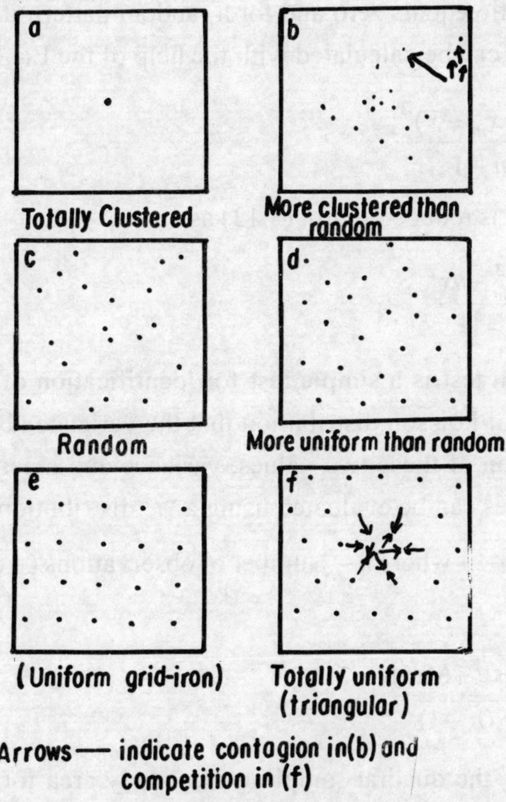

Fig. 9.1 : The Continuum of Spatial Point Pattern

settlements in predominantly agricultural regions often tend to uniformity because the settlements compete with one another for market areas (Fig. 9.1f). Therefore, in a competitive probabilistic process, the location of one point will tend to repel the location of other points in its immediate vicinity. The upper limit of uniform patterns is where the points are located on the vertices of an imaginary grid of equilateral triangles (Fig. 9.1f). This limiting uniform pattern is created by a deterministic competitive process because as soon as one point has been located and the distance between neighbouring points is established, the location of all other points is immediately pre-determined. For the same reason, the grid-iron pattern where the points are located at the vertices of a square lattice is also caused by deterministic competitive process. But the triangular pattern is considered the more uniform one because when the distance between neighbouring points is identical, more points can be packed into the same area with an equilateral grid than with a square grid.

The simple random models such as binomial and Poisson distributions are used to classify observed patterns into one of the three pattern types: clustered, random or uniform. Depending on the variance and the variance–mean ratio test, described earlier we can say the pattern of point distribution in an area is clustered when the ratio is infinite, for an uniform pattern this ratio equals zero and for a random pattern it equals one. The sample variance in these cases can be calculated with the help of the Eq. (3.11) as

$$s^2 = \frac{\Sigma f\,(x_m - \bar{x})^2}{n-1}$$

which for computation is reduced in Eq. (3.13) as

$$s^2 = \frac{\Sigma f\, x_m^2 - n\bar{x}^2}{n-1}$$

This variance mean test is a simple test for identification of spatial pattern and it is based on the property of Poisson distribution that the variance about the mean is equal to the mean. A comparison of these two values, variance and mean, obtained from the observed set of frequencies can be evaluated using a "*t*" distribution test with standard error $\sqrt{2/(n-1)}$ and $d.f. = n - 1$ where n = Number of observations (= quadrats). The t distribution test is

$$t = \frac{\text{Variance} - \text{Mean}}{\sqrt{2/(n-1)}} \qquad \text{... (9.2)}$$

Thus, for example, the quadrat sampling of a study area for an event reveal that 34 quadrats contain no point for the event, 8 quadrats contain one point, 5 quadrats two points and 1 quadrat three points. Now if we have to identify the spatial pattern of the distribution by variance/mean ratio test, we calculate the mean number of points per quadrat equal to 0.44 and the variance equal to 0.58. Hence the ratio 1.32 which is close to 1.0 identifying the random pattern of the distribution of the events in the area. The use of the above method of area sampling by quadrats is illustrated by Example 9.3 below for identifying a spatial distribution pattern.

Example 9.3. Figure 9.2 shows 50 points in an area divided into a lattice of 25 square quadrats. The frequency distribution of these points is shown in Table 9.2.

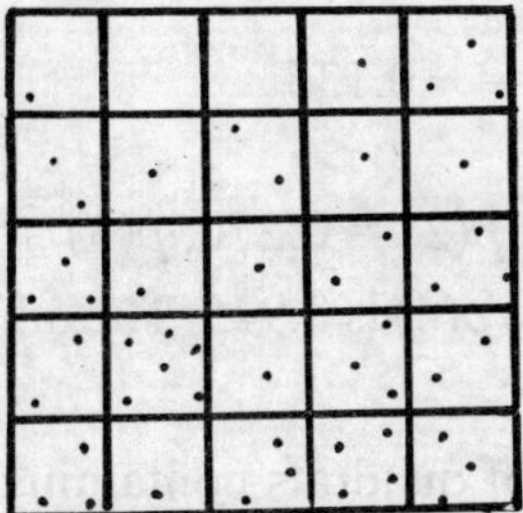

Fig. 9.2 : Distribution of Points in a Quadrat Lattice

Table 9.2 : Frequency of point

No. of points per quadrat	Observed frequency
0	2
1	9
2	5
3	7
4	1
5	0
6	1

Based on the above distribution of points in quadrats can it be decided that the points give a clustered pattern which are created by the Poisson distribution?

It can be seen that the mean number of points per quadrat is 2 and the variance 2.68. Now since the mean and the variance values are close so we can evaluate these two values using a *t*-distribution test by the Eq. (9.1) to find

$$t_{(d.f.=n-1)} = \frac{\text{Variance} - \text{Mean}}{\sqrt{2/(n-1)}}$$

$$= \frac{2.68 - 2.00}{\sqrt{2/24}} = \frac{0.68}{0.29}$$

$$= 2.36$$

t-value is such that at higher significance beyond 98% there is no significant difference between the mean and variance and so the distribution of points by the quadrats similars a Poisson probability distribution.

Now assuming that the size of quadrats can be infinitely small, capable of containing one or no random point, we can have the Poisson probability distribution. Thus the probability that any given quadrat will contain no point as in Fig. 9.2 is

$$p(0) = \frac{2.71828^{-2.0}.(2.0)^0}{0!} = 0.135$$

Similarly, we get, $p(1) = 0.270$, $p(2) = 0.270$, $p(3) = 0.180$, $p(4) = 0.090$ and so on. Thus the sum of the probabilities up to $p(4)$ is 0.945. The other higher probabilities altogether will be low, being equal to 0.055.

The long run expected number of quadrats containing the given number of points are obtained by multiplying the appropriate probability with the total number of quadrats. Thus in the above case, the expected number of quadrats containing 0,1,2,3, and 4 points will be 3.37, 6.75, 6.75, 4.50, and 2.25 respectively. Now whether these expected frequencies tally with the observed frequency in Table 9.2 so that we can say that the latter correspond to the Poisson distribution will be taken up in the "goodness-of-fit" test of the "chi square distribution" in Chapter 13. In case the Poisson character of the distribution of points is verified by the above test, the point distribution in this quadrat sampling can be either clustered or dispersed.

9.3.3 Line Sampling

A further possibility of area sampling is by 'line sampling' in which only eastings or northings are needed and the cross-sectional transect or the grid line from either of these two points form the sample line. Along the transect or the grid line the distances possessing given characteristics are measured and random selection from these values provide the sample data. The line random sampling is context dependent, e.g. it is of interest in forestry where orientation is not important and a number of parallel traverses provide efficient results.

9.3.4 Comparison of Random Sampling

The Fig. 9.3 illustrates the three methods of random sampling for a spatial distribution. It can be seen that these three random sampling methods – point, area and line differ little from each other, thus making possible 'sampling plans' for any population (Fig. 9.3). Point sampling is easier to work and the returns are high; areal sampling involves more work in plotting and calculation. Line sampling is apparently simple but here again the labour is more. Line sampling is useful with binomial distribution where we are concerned with the presence or absence of only one attribute.

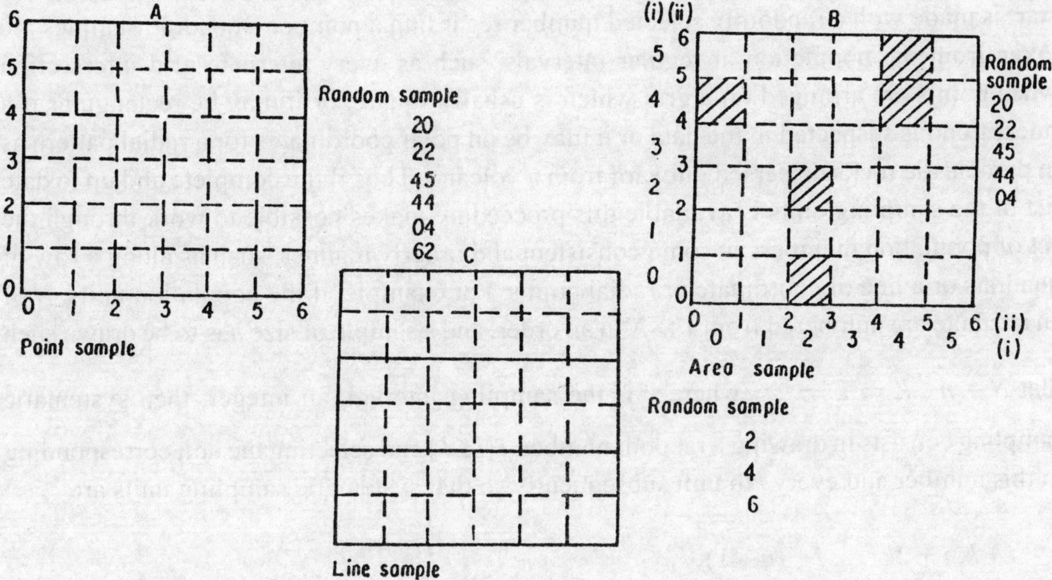

Fig. 9.3 : Methods of Random Sampling for Spatial Distribution

9.4 RESTRICTED RANDOM SAMPLING

In random sampling there is an important assumption that the sample observations are independent and have equal probabilities of being chosen. But on the more practical side, however, there is one disadvantage also: the procedure in random sampling does, not always ensure an adequate areal coverage of a study area. Some clustering of the sample points often result and as a consequence, important spatial properties may be overlooked. Again in geo-science systems, the larger is the area, the larger is the variance. Hence another sampling plan is involved. This is the *restricted random sampling* in which the use of the random number table is restricted and the most thorough coverage of the area of interest is emphasized. Restricted random sampling is used where evenly spaced areal coverage is needed for setting up the experimental design for the major factors of an investigation. As mentioned earlier, restricted random sampling has three types : systematic (or grid), stratified and cluster sampling.

9.4.1 Systematic or Grid Sampling

To avoid the shortcoming of random sampling regarding the coverage of an area this is another type of sampling plan, quite frequently used often interchangeably with simple random sampling. "Systematic" or "grid sampling" starts simply with a randomly selected number. Thus instead of drawing the samples from the table of random numbers once the start is made with a randomly selected number regarding a point or a quadrat, samples are taken from the population at regular intervals, such as every alternate grid intersection when points are arranged on a grid which is usually square, or it may be rectangular if a linear trend is suspected in the data or it may be on polar coordinates for a radial pattern as in pyroclastic rocks dispersed outward from a volcano. Thus if the complete and up to date list of the sampling units is available this proceedure makes possible to work through the set of population members in some consistent and orderly manner, such as along a row of quadrats or a line of coordinate or radial points. For example, if the population units, N of an attribute are numbered from 1 to N in an order, and a sample of size n is to be drawn such that $N = n \,.\, k \Rightarrow k \Rightarrow \frac{N}{n}$ where k is the sampling interval, an integer, then systematic sampling consists in drawing a random number, $i\,(i \leq k)$ and selecting the unit corresponding to this number and every kth unit subsequently so that systematic sampling units are

$$i + k,\ i + 2k, \ldots\ i + (n - 1)\,k$$

Two basic designs are considered in systematic sampling:(one) has points 'aligned' in a checkerboard fashion, in the (other) the points are 'unaligned'. For the first of these designs, the study area is divided into a checkerboard fashion by even spread of points with regular spacing on both abscissa and ordinate. One point is chosen randomly in the first cell, the corresponding point locations in the other cells are selected then as the remaining sample members. Thus for a point pattern if we want a sample size of 300 from a list of a large population, say 10,500, we can take a random number within 1 to 300 and simply add 300 to the previous random number. For example, if we take our random number to start with as 201, the subsequent sample numbers will have the following sequence:

201 (to start with), 501, 801,......., 9801, 10101 and 10401

Thus in this systematic grid sampling, there is no requirement of using the random number table and it is preferable to random sampling because of its simplicity and also in the case when a list of the population is extremely long or when the size of the sample drawn is large. Adopting the method of systematic sampling, it is easier for an interviewer to go to every third house following a complete listing of the houses in an area for a primary survey than to sample houses following the table of random numbers. Thus, in practice, systematic sampling is widely employed in sample surveys because of its ease and convenience. Also, since systematic sampling gives uniform cover of the population it is more useful than random sampling, particularly where the concerned population includes clusters and it itself is randomly taken.

However, since the systematic sampling method does not really give a random sample of individuals as such but a random sample of groups determined by the random starting point, it can get biased if there are regularities or periodicities in the population itself. For example, if the sampling frame listed men and women alternately, the systematic sampling would cover either all men or women in the sample. Again, in a locality having rectangular grid pattern, a systematic sampling can result in samples derived from street-corner locations. Thus one has to pay a price for selecting systematic sampling for simplifying the data collection procedure.

To avoid the above type of unexpected periodicity in the data which may wreak havoc with conclusions in the systematic sampling, unaligned point design can be suggested. With this design the first point is chosen in the same way as the aligned one. The x-coordinate of this point in the first cell is then held constant for all the other grid squares across the top row and a point in each of these squares is chosen by randomly selecting new y-coordinates. Similarly for the next row the y-coordinate is held constant for all grid squares while the x-coordinate is randomized. This procedure is iterated until all the grid cells in the area get the coordinates for the sample elements. It can be noted in the unaligned design, the sample point in the first grid of every row in a gridded area is selected by randomizing either the x or the y-coordinate and hence the problem of periodicities in the data is alleviated in this unaligned designing of systematic sampling.

9.4.2 Stratified Sampling

So long as there are not marked differences in the population, random sampling is good. But a problem arises when the population is composed of highly-varied observations. In this case a balanced representation can be achieved by dividing the population into groups or sub-populations called 'strata' or layers to ensure the representation of each sample and drawing separate samples from each strata. The strata is defined in such a way that each individual appears in one and only one stratum and the number of strata are clearly related to how effective the attributes are in reducing the variances in the observation for which estimates are made. This refinement in the sampling procedure is known as 'stratified sampling'. Thus if a population of size N is divided into k relatively homogeneous subgroups called *strata*, of size $N_1, N_2, \ldots, N_K$ such that $N = \sum_{i=1}^{k} N_i$ and simple random sample of size n_i ($i = 1, 2, \ldots, k$) is drawn from each of the stratum respectively such that $n = \sum_{i=1}^{k} n_i$, the sample is termed as *stratified random sample.*

Thus a stratified sampling is restricted to the selection of the sampling units from each stratum until all the strata covering the population are included. The number of units drawn at random from each stratum need not be the same. When the sampling fraction for each stratum is equal, it is called *proportional stratified sampling*, on the other hand, when an equal number of sampling units are selected from each stratum of the population, it is called *disproportional stratified sampling*. For example, suppose in an innovation adoption survey, it is desired to take a sample size of 1000 farmers from a population consisting of 600 jotedars (i.e. big farmers), 300 medium-sized farmers and 100 small farmers. In disproportional stratified sampling an equal number, say 33, will be selected by sampling from each of the three groups or stratum. On the other hand, in proportional stratified sampling, 10 percent of the samples will include 60 jotedars, 30 medium-sized farmers and 10 small farmers.

In statistical analysis many population distributions are highly skewed so that the highest values of the distribution are hundred or even thousand times as large as the lowest values. Under such conditions, an estimate based on a sample drawn at random from the entire population is itself subject to considerable variations. For reducing the variability of estimations made from distribution of wide ranges, stratified random sampling is an effective method. This is illustrated below in Example 9.4.

Example 9.4 The following is an empirical population: 1, 2, 3, 10, 11, 12, 50, 51, 52, 100, 101 and 102. Divide this population into four strata and by drawing one sample at random from each stratum prove that the range between the lowest and highest possible estimate of these sample of four will be lower than the range between the lowest and highest possible estimates of the 11 samples drawn at random.

The above population of 12 has been divided into 4 strata as follows:

A	*B*	*C*	*D*
1	10	50	100
2	11	51	101
3	12	52	102

Now if we draw at random a sample of one from each stratum to get total of four stratified samples we can have the sample mean as low as 40.25 $\left(=\dfrac{1+10+50+100}{4}\right)$ or as high as 42.25 $\left(=\dfrac{2+12+52+102}{4}\right)$. So the range in the sample mean by systematic stratified sampling is $42.25-40.25=2.00$ only. As a matter of fact, we would not get as good an estimate of an average if we draw sample of 11 at random from the above unstratified population of 12. In the latter case, the lowest sample mean is 35.4 whereas the highest sample mean is 44.9. The position becomes worse if a sample set of four are drawn from the above unstratified population. In this case the lowest sample mean is 4.00 $\left(=\dfrac{1+2+3+10}{4}\right)$ and the highest sample mean is 88.75 $\left(=\dfrac{52+100+101+102}{4}\right)$.

Thus summarizing the above three sets of conditions, we say that the minimum range between the lowest and highest sample means is attained when the stratified sampling system is adopted for the correct estimate of the population mean.

	Lowest possible estimate of mean	Highest posible estimate of mean	Range	Population mean
Unstratified sample of 4	4.00	88.75	84.75	41.25
Unstratified sample of 11	35.40	44.90	9.50	41.25
Stratified sample of 4	40.25	42.25	2.00	41.25

9.4.3 Cluster Sampling

In stratified sampling, we divide the population into categories or groups called *strata* and then take a sample from every stratum. But for some population it may be difficult or impossible to list each member of the population; yet it is advantageous to divide the population into large neighbourhood groups called "cluster". Thus "cluster sampling" refers to sampling taken in local groups or clusters. *Clustering* is done by dividing the population according to some criteria, such as geographical proximity and, thus, a *cluster* is a geographical or social unit for which a complete count is made and from which samples are selected on a random basis. Again, unlike stratified sampling, the resulting clusters are based on heterogeneity because each cluster or neighbourhood covers the whole range of properties exhibited by the total population itself. Hence for cluster sampling the population is first divided into clusters or neighbourhood blocks, census tracts etc. The clusters are selected from the population on the basis of simple random sampling and then every individual of each random number is listed for subsampling within the clusters. Examples of clusters are landform units in a landform pattern, dwelling units in a city block, etc. Thus the cluster sampling is basically a "two-stage sampling", clusters being termed as "primary units" and the sampling units within the clusters as "secondary units". This technique, as explained here afterwords, may be generalised to what is called "multistage sampling".

The aim in cluster sampling is to select heterogenous clusters. The accuracy of selection depends entirely on how well the clusters represent the variability of the total population. When the clusters are areal units, cluster sampling is known as "nested sampling". *Cluster sampling* is a statistically efficient sampling technique when the local and regional variability are of the same size.

In comparison to other sampling methods, random or stratified cluster sampling is always less efficient; the only justification for this sampling is that it is economic, i.e. they are always less expensive and quick to carry out. For example, if we have to collect 100 responses to particular enquiry in a city population of 100,000 by simple random sampling, we have to construct a list of the entire city population (from where the random sampling will be taken) which itself can be quite expensive and time consuming. Secondly, once the simple random sampling is done it will be found that the chosen respondents are scattered throughout the city and collecting the responses requires considerable travel time, energy and consequently the entire procedure becomes a big botheration. But if instead of random sampling, the cluster sampling method is undertaken it will involve door to door canvassing for only a few selected number of neighbourhoods in the city and collection of responses, also become easier and economical. In cluster sampling this is the simplest design because sampling occurs once only for selecting the cluster sample – this sampling is known as "single stage cluster sampling". On other hand, the design may be more complicated. For example, when for the city survey, the city is divided into a number of blocks called "primary sampling units", we take a simple random sample of blocks (smaller clusters) within each cluster and the interviewer is asked to select every third household within the selecting blocks and to interview every second adult within each of these third households, then in this case, sampling procedures enter the selection process at a number of points and we call this type of clusters within a cluster design as *multistage cluster sampling*.

Multistage cluster sampling is desirable when the primary sampling units are large and heterogenous. This sampling is also economical and convenient to work out because framing of the sampling units is really involved at the third level or household level. Within the selected blocks the sampling at the higher level is supposed to be available. Finally, because of the potential gains, one should always try to apportion sampling so that an appropriate amount is done at individual locations as opposed to occupying other locations.

Note: All the statistical tests we will consider now onwards throughout this book will require the assumption that the sample informations have been obtained on the basis of simple random sampling procedure. The other three sampling plans require more complicated methodology which is beyond the scope of this book.

9.5 DESIGNING AND CONDUCTING A SAMPLING SURVEY

Sampling is required for convenience in estimating the population from which the sampling itself is derived. Without the sampling, there would be no need for statistical theory. As mentioned earlier, samples should be drawn in such a way that meaningful conclusions can be drawn regarding the entire population This sample designing and sample survey for estimation of population consists of three stages: (1) planning, (2) data collection, (3) data analysis and conclusions.

9.5.1 Planning A Sample Survey

Planning a sample survey is the most important step for ensuring that the sampling plan achieves the best result. Poor planning can result in the eventual invalidation of the entire survey and gives a strong counter productive result. Planning begins with the identification of a population which is oriented to the goal of study. This is followed by an observation procedure that can include the design of a questionnaire for the choice of a particular source of data or for tests for field sample collection in geo-science system surveys or in social sciences for various primary surveys such as rural or urban surveys, etc. The population, as the aggregate of objects should be defined in clear and unambiguous terms. It must be capable of division into what are called sampling units for purposes of sampling. The sampling units together must cover the entire population so that every element of the population belongs to one and only one sampling unit. Hence a sampling plan is required to be prepared very precisely. For this purpose, in the planning stage understanding the target and sampled populations is of great importance because it gives the choice of the sampling plan itself. This sampling plan helps the listing of all those elements of the population that are very essential. We have seen the ranges within the sampling plans from the most "scientific" random sample downward to the "convenience" sample (e.g. nested sample). The random sample is the most important because, as mentioned earlier, statistical inferences apply only to this sample. A defective sampling plan gives to faulty selection of sample leading to sampling errors in data collection.

Choosing an appropriate sample size and deciding on a statistical test for estimation of population both involve a heavy dose of inferential statistics which will be discussed from next chapter onwards.

Data Collection : The obtaining of data is the second phase of the sampling survey. This requires the greatest amount of time and effort. Adequate controls must be provided to prevent the sampling units from becoming biased when they are collected either from measurements or from questionnaires, the latter by either the interview or the mailed method. An improper collection and a poorly asked question both lead to incorrect results. Hence the data should be collected keeping in view the objectives of the survey.

Data Analysis and Conclusions : This last stage of sample survey is extremely important. Here we can communicate the results in terms of computation or measures of descriptive statistics like the central tendency measures of mean, median, mode and dispersions etc. or we can make use of inferential statistics by testing the distributional forms of the statistics, e.g. normal, log normal, etc. and estimating the probabilities of population parameters. The choice of which of the various tests are available to use will depend upon the goals of the study and the resources available.

The phenomena related with geo-science systems are highly erratic, complex and random in character and hence they can be interpreted only in a probabilistic sense. One of the important problems in geo-science systems like in hydrology or in climatology is concerned with interpreting a past record of hydrologic or climatologic events in terms of future probabilities of occurrence. Again there are other geo-science systems like forestry or landuse which are concerned with problems of sampling spatial distributions. Many of the sampling designs used in these fields can behoove a spatial scientist to be aware of them. Again frequently the data set in geo-science systems is based on historical data which are measured and collected from natural phenomena that can be observed only once and then will not occur again. This historical data may be of small duration or it may be fragmentary for any purposeful sampling towards estimation of the population data. Despite these difficulties in geo-science systems, the available material and the purpose govern the plan of work, the plan in turn governs the sampling and sampling provides the data from which valid conclusions can be inferred. These are all incorporated in the report which highlights the technical aspect of the survey, namely, the types of the estimators along with the amount of error to be expected in the most important estimate.

Summary Table : Types of Sampling Designs

Sampling Plans	*Characteristics*
1. Simple Random	1. Assign to each element a serial number. Select sample units by random number table.
2. Systematic	2. Select the first sample unit randomly and the rest according to a sampling interval, e.g. for every 10th........
3. Stratified	3. Dividing the population into strata and selecting random sample for each stratum by analytical considerations.
4. Cluster	4. Dividing the population into a number of levels of clusters or neighbourhoods and then selecting each level of cluster randomly; suited for quick economical surveys.

LIST OF FORMULAE

Probability of an event selected at rth draw, $f(r) = \frac{1}{n}$

Variance mean ratio test, $t = \dfrac{\text{Variance} - \text{Mean}}{\sqrt{2/(n-1)}}$ for $d.f. = n - 1$, for uniform pattern this ratio is zero, for random pattern it is 1.

EXERCISES

9.1. Draw a sample of 100 one digit random numbers from the table of random numbers by taking consecutive digits from as many columns needed. Classify these numbers into a frequency table and test whether the frequency distribution is having a normal probability character.

9.2. Ten random numbers : 33, 43, 61, 87, 49, 78, 50, 02, 27,and 15 are provided to a field investigator, If he has to select 5 plots randomly from 40 plots numbered from 1 to 40, how he should use the random numbers given to him for selection of 5 sample plots.

9.3. Consider a population of 6 units with values 1 to 6 in sequence. Write down all possible samples of 2 without replacement from this population and justify that sample mean is an unbiased estimate of population mean.

[Hint: There will be 15 sets of samples of 2, total of whose sample means is 52.5, this divided by 15 is 3.5 equal to the population mean].

9.4. Draw 40 sets of samples of two digits for two at a time from the random number table. Set out the results by combination as shown below:

No. of samples	Sample pairs	Mean (sample of 2)	Mean (sample of 4)	Mean (sample of 5)
1.	10, 37	23.5		
			38.5	
2.	08, 99	53.5		
3.	12, 66	39.0		43.5
			48.5	
4.	31, 85	58.0		
.				
.				
.				
40.				

Treat the above three columns of means as three frequency distributions. Are the distributions grouped around the population mean of 44.8? Do they become more compressed about it as sample size increases?

9.5 The following is an empirical data arranged in row by column mode:

	a	b	c	Total
p	2000	240	50	2290
q	3000	220	80	3300
r	1500	170	30	1700
s	880	80	40	1000

Work out how many a, b, and c's will be included from each category if 10% of the total is selected as samples by (i) stratified sampling and (ii) if in the 10% of sampling if the a, b, and c's are to be in the ratio of 5 : 3 : 2 and weightage of p, q, r and s is to be in the ratio of 4 : 3 : 2 : 1 .

10

Parametric Statistics in Geo-science Systems I

(Sampling Estimates for Large-sized Samples)

10.1 SAMPLING THEORY AND PARAMETRIC STATISTICS

When data have been collected by sampling, it is necessary to formulate hypotheses concerning the data. In statistical analysis interest is not limited to organising and classifying the data supplied by the observations (Chapters 2 to 3) nor in calculating the value of their probabilities (Chapters 4 to 7) nor in drawing the set of observations from its total population as sampling units in terms of a sampling plan (Chapter 9). More often, a comparison is suggested between one set of values with one or more other sets to assess the degree of similarity or difference between the sets of values. In fact, it is this very capability of comparing the degree of similarity or dissimilarity between different sets of data collected as samples which make statistical techniques a useful and objective means of analysis. This takes us to "sampling theory" which is a study of relationships existing between a population and samples drawn from the population. This *sampling theory* is applicable only to random samples. In context of sampling theory it is needed to explain the two terms, viz "parameter" and "statistic". All the statistical measures based on all items of the universe (which may be finite or infinite, hypothetical or existent, like tossing of a coin which is hypothetical) like population mean or population standard deviation are termed as *parameters* whereas the statistical measures worked out on the basis of sample statistics like "sample mean or a sample standard deviation" is an example of a *statistic*. The main problem of sampling theory is the problem of relationship between a parameter and a statistic and is thus concerned with estimating the properties of population from those of the sample and also with gauzing the precision of the estimate. This sort of movement from particular (sample) towards general (universe) is what known as "statistical induction" or "statistical inference" and in doing so we must follow a deductive argument which is that we imagine a *population* or *universe* and investigate the behaviour of the samples drawn from the universe applying the laws of probability.

The statistical methods that can be used for the purpose of comparison are many and varied depending on the scale of measurement involved in collecting the set of data. Actually a basic distinction is drawn between those tests which make use of the actual measurement in the interval or ratio scale and those which make use of the data collected in the nominal or ordinal scale. The former are known as parametric tests and the latter non-parametric tests. "Parametric statistics" include all classical statistics and by the word "parametric" they refer to the characteristics of population. The characteristics of the background population of the observations in the context of parametric statistics are essentially that:

1. The population should be normally distributed so that the standard deviation of the observations are used predictively;
2. The observations are independent of each other so that the value of any one observation does not affect that of others; and
3. The measurement of the observations are available on the interval or ratio scale.

Thus the *parametric tests* are based on the mean and standard deviation and they are essentially tied to the normal frequency distribution. The three most commonly used parametric tests are the normal z test, Student's t test and Fisher's F test. In this chapter, the z test will be discussed for estimating the population for large-sized samples; the t test and F test will be discussed in Chapters 11 and 12 respectively. The non-parametric tests which do not require the above assumptions of parametric statistics (except the independence of observations) will be introduced from Chapter 13 and will continue up to Chapter 19.

At this stage it will be useful to go back to the concept of sampling theory which is to attain the following objectives:—

(a) Null hypothesis and significance level – this enables one to decide whether to accept or reject hypotheses or to determine whether observed samples differ significantly from expected results.

(b) One- and two-tailed tests – the significance level identifies two regions for critical values; and,

(c) Sampling distribution

10.1.1 Null Hypothesis and Significance Level

The logic of the theory of sampling is the logic of induction, i.e. we pass from particular to general and hence all the results will have to be expressed in terms of probability. Thus if we have to test whether a sample is drawn from a population or not and whether the apparent difference between the two sets of samples can be accepted as real or not, we require to start by setting an hypothesis that there is, in fact, no difference between the sets of sample data being compared, even though the sample values might suggest such a difference. Such a difference, one hypothesis assumes, results merely from the fact that they are samples, and if additional samples are taken, then there would be no such difference or even a difference in the opposite direction. Thus this hypothesis which assumes no real difference between the data is named as 'null hypothesis,' and is usually written as 'H_0'. The *null hypothesis* is thus akin to the legal principle that a man is innocent until he is proved guilty. Statistical tests then consist of deciding whether one should accept or reject the null hypothesis. If the null hypothesis is rejected, then the inverse of the null hypothesis is accepted that is there is a real difference between the set of data, as suggested by the samples. This latter hypothesis is known as the "research hypothesis" and it is written as 'H_1'. The null hypothesis is a statement only, but the explanation of their acceptance or rejection which form the *research hypothesis*, is the essence of sampling theory.

Thus, a climatologist who wishes to find whether the rainfall in a particular year is higher than the mean annual rainfall of the place which is 60 cm, he might establish the following null and research hypotheses:

$H_0 : \mu = 60$ cm (null hypothesis)
$H_1 : \mu \neq 60$ cm (research/alternative hypothesis)

or, if the climatologist is interested in testing the differences in the mean annual rainfall of two different periods, he may like to establish the null hypothesis test that the two periods have equal means ($\mu_1 - \mu_2 = 0$) and the research hypothesis that their means are not equal ($\mu_1 - \mu_2 \neq 0$);

$H_0 : \mu_1 - \mu_2 = 0$ (null hypothesis)
$H_1 : \mu_1 - \mu_2 \neq 0$ (research hypothesis)

Once a hypothesis is expressed, we can make a decision to either accept or reject it on the basis of our statistical test. There are also two possible states of the hypothesis: it may be 'true' or 'false'. This combination produces four possible outcomes, of which two are correct and two incorrect. The possibilities can be graphically illustrated:

Decision	Hypothesis is correct	Hypothesis is incorrect
Hypothesis is accepted	Correct decision	Type II error, β
Hypothesis is rejected	Type I Error, α	Correct decision

Either acceptance of a true hypothesis or rejection of a false hypothesis will result in a correct decision. If a null hypothesis is rejected when it is in fact true, a type I error has been committed. Conversely, if an erroneous hypothesis is accepted, a type II error is committed.

In standard statistical procedures, the validity of either of the above hypotheses is tested at a certain significance level or at confidence limits in terms of probability because in statistical tests on analyses we, in fact, do not make an absolute acceptance or rejection of a hypothesis. The "confidence limits" give a measure of the accuracy with which a population mean is known in relation to the sample mean. Significance level nearer 100 per cent gives less scope for the result to occur by chance. At 99 per cent probability a result can be considered to be highly significant, but below 95 per cent the result is only suggestive. Thus the probability of our citing a particular hypothesis being in error is called the "significance level" of the test and it is usually taken at 1 per cent or 5 per cent. So the *level of significance* can be defined as the probability of rejecting a hypothesis.

The level of significance is denoted by the Greek letter α. It is an integral part of the null hypothesis decision procedure and it has to be specified before a test is conducted. Thus $\alpha = 0.05$ or 0.01 meaning that the probability of rejecting the null hypothesis is 5 times in 100 (or 1 in 20) and 1 in 100 times respectively. The term "level of significance" is used interchangeably with the term "confidence level", thus 95% confidence level is the same as 0.05 significance level, and so on. In order to minimize the possibility of committing a type II error, we write the null hypothesis with the intention that it will be rejected. If the null hypothesis is rejected there is no chance for a type II error, and the probability of type I error is known because it was specified. If, however, the test fails and does not reject the null hypothesis, the probability of a mistake (called β) in the form of a type II error remains. This probability, β is not known. Thus if H_0 is rejected (= hypothesis of equality), we can state that the two parent populations have different means. The probability that this statement is correct and our decision is an error is equal to α. On other hand, if H_0 is not rejected, the statement H_1 is true is accompanied by an error, equal to β.

10.1.2 One-tailed and Two-tailed Tests

The level of significance of a sample is plotted in the frequency curve of the sampling distribution which like the standardised normal distribution curve corresponds to a probability of 1.0. Within this total area there is a "critical region" which is a region in which one is critical of the null hypothesis. The boundary of the *critical region* is defined by the significance level adopted in carrying out the test. The significance level identifies two regions for the critical value: the "acceptance region" in which the null hypothesis is accepted and the "rejection region" in which it is rejected (Fig 10.1). For example, if we specify that we are willing to reject the null hypothesis (H_0, the hypothesis of equality) when they are actually equal one time out of twenty, i.e. we will accept a 5 per cent risk of a type I error. On the standardized normal distribution curve we determine the extreme regions (which will be two subregions in the extremes each equal to 2.5 per cent if we are

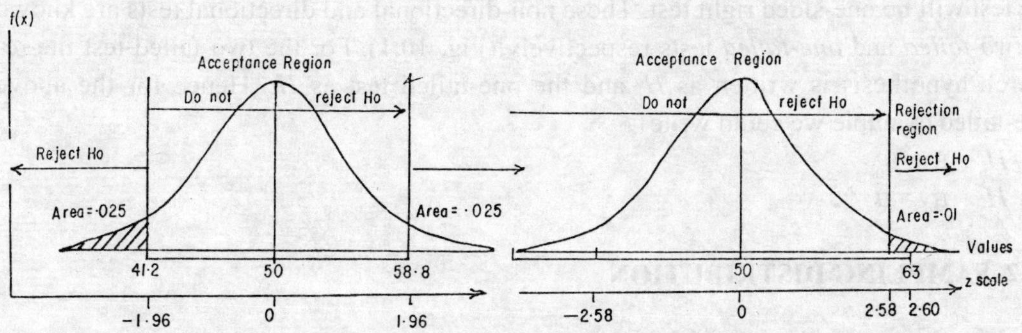

Fig. 10.1 : One-tailed and Two-tailed Tests

concerned with simple inequality) which contain 5 per cent of the area of the curve. This part of the probability curve is called the "area of rejection" or the *Critical Region*.

In comparing one sample group with the other or in estimating the population mean from the sample mean, we can phrase the hypothesis inverse (or alternative) to the null hypothesis in either of the following two ways: we can phrase it either as "one group is significantly different from the other" or, "one group is significantly greater than the other". The former merely states a difference but the second phrasing produces a "direction of difference" test of the hypothesis. Thus when a statistical test is designed to distinguish between two groups, then this is called a "two-tailed test". If we know, before we compute the statistic we are going to test, that we are interested in finding the greater or lesser between the two groups then we call it a "one-tailed test". In this case, we take the critical value of z in one tail only, whichever is specified by H_1. For example, if the significance level is $\alpha = 0.05$, then the critical region occupies 5 per cent of the total area under the curve describing the sampling distribution. In the two-tailed test there is a critical region in both tails, and is divided equally. Between the two critical regions, this means that when

$\alpha = 0.05$, the upper and lower critical regions, both occupy 2.5% of the area under the curve. In the case of directional one-tailed test, the critical region will be at either the upper or lower tail of the distribution, depending upon the research hypothesis formulated. For example, if we are interested in testing a hypothesis that the mean annual rainfall at a place is more than 60 cm against the research hypothesis that the mean annual rainfall is 50 cm we will place our alpha (α) risk on the right side of the theoretical sampling distribution and the test will be one-sided right test. These non-directional and directional tests are known as *two-tailed* and *one-tailed* tests respectively (Fig. 10.1). For the two-tailed test the research hypothesis is written as H_1 and the one-tailed test as H_2. Hence for the above one-tailed example we could write

$$H_0 : \mu = \mu_0$$
$$H_2 : \mu_0 > \mu$$

10.2 SAMPLING DISTRIBUTION

If a large number of samples of equal size n are to be repeatedly drawn from a population and if $\bar{x}_1, \bar{x}_2, \bar{x}_3, \ldots$ be their possible random sample means, then the frequency distribution of these sample means is known as "sampling distribution" of means. For example, let us assume that we have a 200-year long annual peak discharge data and we make its 20 series, each with a 10-year long data. On computing the mean of each series we get 20 different values of the mean, say $\bar{x}_1, \bar{x}_2, \ldots \bar{x}_{20}$ to give us the *sampling distribution* of the mean. The values used to form the sampling distribution are now observations that together form a new population.

One of the most important tools in statistical analysis is the "central limit theorem" which is a set of theorems in mathematical statistics. This theorem states that if repeated random samples of size n are drawn from a normal population with mean μ and standard deviation σ the sampling distribution of the mean, $\bar{x}_i$ approaches a general normal distribution with a mean μ and standard deviation $\sigma_{\bar{x}}$ irrespective of the parent population as sample size increases. The more skewed the parent population is the larger the sample size has to be to ensure approximate normality. However, it may not be true in the case of small size of samples. The smaller the sample, the more skewed is the form of the distribution of the means. This relationship between a population distribution and a distribution of sample mean is critical for drawing inferences about parameters. Reliability, therefore, is almost completely a function of sample size.

In summary, the means and standard deviations of three kinds of distributions are as follows :

	Mean	Standard deviation
Population	μ	$\sigma_p = \sqrt{\Sigma(x_i - \mu)^2 / N}$
Sample	$\bar{x}$	$s = \sqrt{\Sigma(x_i - \bar{x})^2 / (n-1)}$
Sampling distribution	$\bar{x}_i$	$\sigma_{\bar{x}} = \sigma_p / \sqrt{n}$

Before we go to discuss the usefulness of sampling distribution in statistical inferences, we will like to know how the statistic $\bar{x}_i$ behaves. It is obvious that the $\bar{x}_i$ value will depend upon the particular sample we choose. Each possible sample of size n would have it's own value of $\bar{x}_i$. For example, we take a population: 1, 4, 7, 10 and 13. If we plan to choose a sample of size 3 from this population for calculating $\bar{x}_i$, we have samples in

$$\frac{5!}{(5-3)!3!} = 10 \text{ possibilities}$$

Table 10.1 lists these and other samples with other possible sizes.

Table 10.1 : Listing of Samples and their Sample Means from a Population of Size 5 : 1,4,7,10 and 13

Sample of size 1	$\bar{x}_i$	Sample of size 2	$\bar{x}_i$	Sample of size 3	$\bar{x}_i$	Sample of size 4	$\bar{x}_i$	Population mean μ
1	1	1.4	2.5	1. 4. 7	4.0	1. 4. 7. 10	5.50	
4	4	1. 7	4.0	1. 4. 10	5.0	1. 4. 7. 13	6.25	
7	7	1. 10	5.5	1. 7. 10	6.0	4. 7. 10. 13	8.50	
10	10	1. 13	7.0	1. 10. 1	4.0	1. 10. 7. 13	7.75	7.0
13	13	4. 7	5.5	1. 13. 7	7.0	1. 10. 13. 4	7.00.	
		4. 10	7.0	1. 13. 4	6.0			
		4. 13	8.5	4. 7. 13	8.0			
		7. 10	8.5	4. 10. 13	9.0			
		7. 13	10.0	7. 10. 13	10.0			
		10. 13	11.5	4. 10. 7	7.0			

From the above table we can calculate the standard deviation of these sample means with the help of Eq. (3.10) which is

$$s = \sqrt{\frac{\Sigma(x_i - \bar{x})^2}{n-1}}$$

where $(n-1)$ in the denominator is one less from n, the number of samples in the data. This introduces the concept of the "degree of freedom" which is always the number of independent observations in the data less the number of constraints, which is 1 here, the mean.

But the above method of taking a number of samples to calculate the sample mean and sample standard deviation can be tedious and not always possible. Fortunately, we can calculate the sample standard deviation from a single sample. The sample mean $\bar{x}$ will be based on sample measurements and the standard deviation of this sampling distribution is dependent on the number of items size in the sample, n and is known as the "standard error of the sample mean". Thus the *standard error of the sampling distribution* for an infinite population is related to the standard deviation of the individual values of the population by

$$\sigma_{\bar{x}} = \frac{\sigma_p}{\sqrt{n}} \qquad \text{... (10.1)}$$

where $\sigma_{\bar{x}}$ is the estimated standard error of the sample mean when the standard deviation σ_p of the population is known. We call the standard deviation $\sigma_{\bar{x}}$ of a sample as the *standard error* because owing to sampling fluctuations (as shown in example above), there is a certain error in considering the sample means as estimates of the population mean.

This expression $\sigma_{\bar{x}} = \frac{\sigma_p}{\sqrt{n}}$ shows the relationship between the dispersion of a population distribution and that of the sample mean. It indicates the spread in the distribution of all possible sample means which will be smaller than population standard deviation whenever n exceeds 1. This simply reflects that the values of the sample mean $\bar{x}$ are more alike (i.e. variability is much less) than the individual population values. Again, since the standard error of the sample mean is inversely proportional to the square root of the sample size, the larger the sample, the smaller will be the standard error and the possible values of $\bar{x}_i$ will cluster more tightly about the population mean, μ. This is reflected in the shape of the normal curve, which is more skewed when $\sigma_{\bar{x}}$ is small (Fig. 10.2). We can thus conclude that the smaller the value of $\sigma_{\bar{x}}$, the more accurate would be the sample result. Again more varied is the population, wider will be the size of the samples to cover the range. If a population has no variation, i.e. all numbers are identical, a sample of one will provide the desired information. Thus the term 'standard error' implies that the difference between a population standard deviation and a sample standard deviation is due to an error caused by the samples. In fact the standard error, $\sigma_{\bar{x}}$ is an index of the difference between sample and true means.

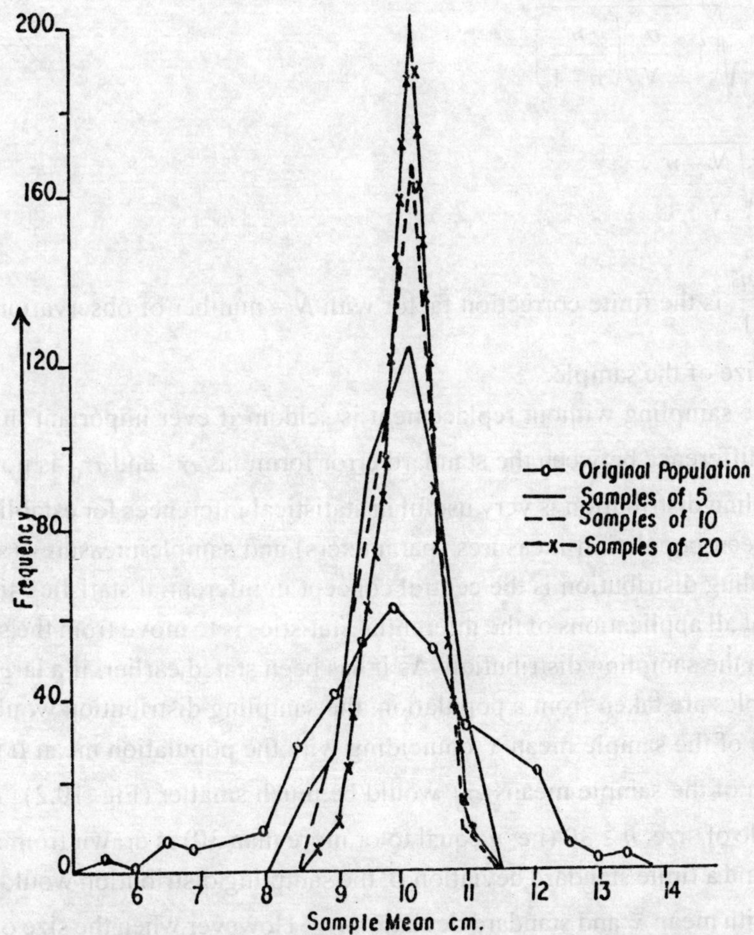

Fig. 10.2 : Sampling Distributions (with samples of 5, 10 and 20)

The *central limit theorem* is quoted in this context to assume that sampling is done "with replacement" corresponding to statistical independence. Now if the sampling is done "without replacement" and the sample is large relative to the population (for example, when population is having 5 observations the number of samples of size 2 will be 10 for samples selected without replacement and it is 4 for samples selected with replacement), a modification must be made in calculating standard error, $\sigma_{\bar{x}}$. If n is greater than 5% of the population size and sampling is performed without replacement we have :

$$\sigma_{\overline{w}} = \frac{\sigma_p}{\sqrt{n}} \cdot \sqrt{\left(1 - \frac{n}{N}\right)\left(\frac{n}{n-1}\right)}$$

$$= \frac{\sigma_p}{\sqrt{n}} \cdot \sqrt{\frac{N-n}{N-1}} \qquad \text{... (10.2)}$$

where $\sqrt{\frac{N-n}{N-1}}$ is the finite correction factor with N = number of observations in population and n = size of the sample.

Note since sampling without replacement is seldom if ever important in geo-science systems, the difference between the standard error formulas $\sigma_{\bar{x}}$ and $\sigma_{\overline{w}}$ is not discussed.

The sampling distribution is very useful in statistical inferences for establishing a relationship between population measures (parameters) and sample measures (statistics). In fact, the sampling distribution is the central concept in inferential statistics since the general strategy of all applications of the inferential statistics is to move from the sample to the population via the sampling distribution. As it has been stated earlier, if a large number of different samples are taken from a population, the sampling distribution would be normal with the mean of the sample mean $\bar{x}$ coinciding with the population mean μ but the standard deviation of the sample means $\sigma_{\bar{x}}$ would be much smaller (Fig. 10.2). Thus when a random sample of size, $n \geq 30$ (i.e. n equal to or more than 30) is drawn from a population with mean μ and a finite standard deviation σ, the sampling distribution would be a normal distribution with mean $\bar{x}$ and standard deviation $\sigma_{\bar{x}}$. However when the size of a sample is less than 30 ($n < 30$), the sampling distribution of the mean would correspond to a good estimate of the *Student's t distribution*. This distribution curve would be very close to the normal curve, the only difference being that the former would be much flatter than the latter. It also appears that the distribution of t depends only on the size of the sample. For this reason, in the case of the sampling distribution of the mean, a sample of size 30 and more is considered to be a "large sample". Again, the sample is not necessarily the "numbers of cases", rather it is the number of sampling units selected at random from the population. If the sampling unit is an area or group, the size of the sample is the number of such areas or groups which are selected at random, and not the number of individual elements. Thus, if we select at random in a given state a sample of 40 villages for the purpose of finding the effectiveness of a high-yielding crop, the size of the sample is 40 villages and not the total numbers of farmers in these villages.

10.3 POINT AND INTERVAL ESTIMATES IN SAMPLES

There are two types of estimates in parameters in statistics – the *point estimate* and the *interval estimate*. The 'point estimate' is basically a single number calculated from the sample values that serves as an approximation to the parameter being estimated. For example, the sample mean $\bar{x}$, sample proportion $\bar{p}$ or sample variance $\sigma_{\bar{x}}^2$ is a point estimate of the population mean, population proportion or population variance, as the case may be. For example, it has been proved earlier that the mean of the theoretical sampling distribution of $\bar{x}_i$ is the population μ, and therefore, $\bar{x}_i$ is an unbiased estimator of μ. Thus we call a statistic unbiased if on the average its values can be expected to equal the parameter it is supposed to estimate e.g. the sample mean is an unbiased estimate of the population mean, but the sample standard deviation is a biased estimate of the population standard deviation. The bias δ is equal to difference between the average value of the statistic and the value of the parameter being estimated. Thus $\sigma = \bar{x} - \mu = 0$ but $\sigma = \sigma_{\bar{x}} - \sigma \neq 0$. Again, since there can be more than one unbiased estimate, the sampling distribution whose standard error is less than that of the other unbiased estimate is taken as the more efficient unbiased estimate.

The "interval estimate" is an interval for a parameter determined by two numbers obtained from computations on the sample values that is expected to contain the value of the parameter in its interior. The interval estimate is usually constructed in such a manner that the probability of the interval containing the parameter can be specified. Since the point estimate of a sample mean $\bar{x}$ cannot be made exactly equal to the population mean μ, it is desirable to have an interval estimate of the sample mean. The advantage of the interval estimate is that it shows how accurately the parameter is being estimated. If the length of the interval is very small, it shows that high accuracy has been achieved. The upper and lower boundaries of such an interval estimate is called the "confidence interval". The concept of the interval estimate is shown in Fig. 10.3.

The distinction between the point and interval estimates is simple. In the point estimate, we compute one number and infer that the parameter is that number. In the interval estimate we compute two numbers and infer that the parameter lies in that interval between them. In this chapter both point and interval estimates are determined for normal and binomial distribution parameters.

10.3.1 Specification of Sample Mean and Sample Size

As mentioned earlier, while estimating the mean by a random sample, we do not get the exact value for the population mean, i.e. $\bar{x} = \mu$. However, a random sample gives us the class interval within which the population mean falls with some confidence limit. These are given below.

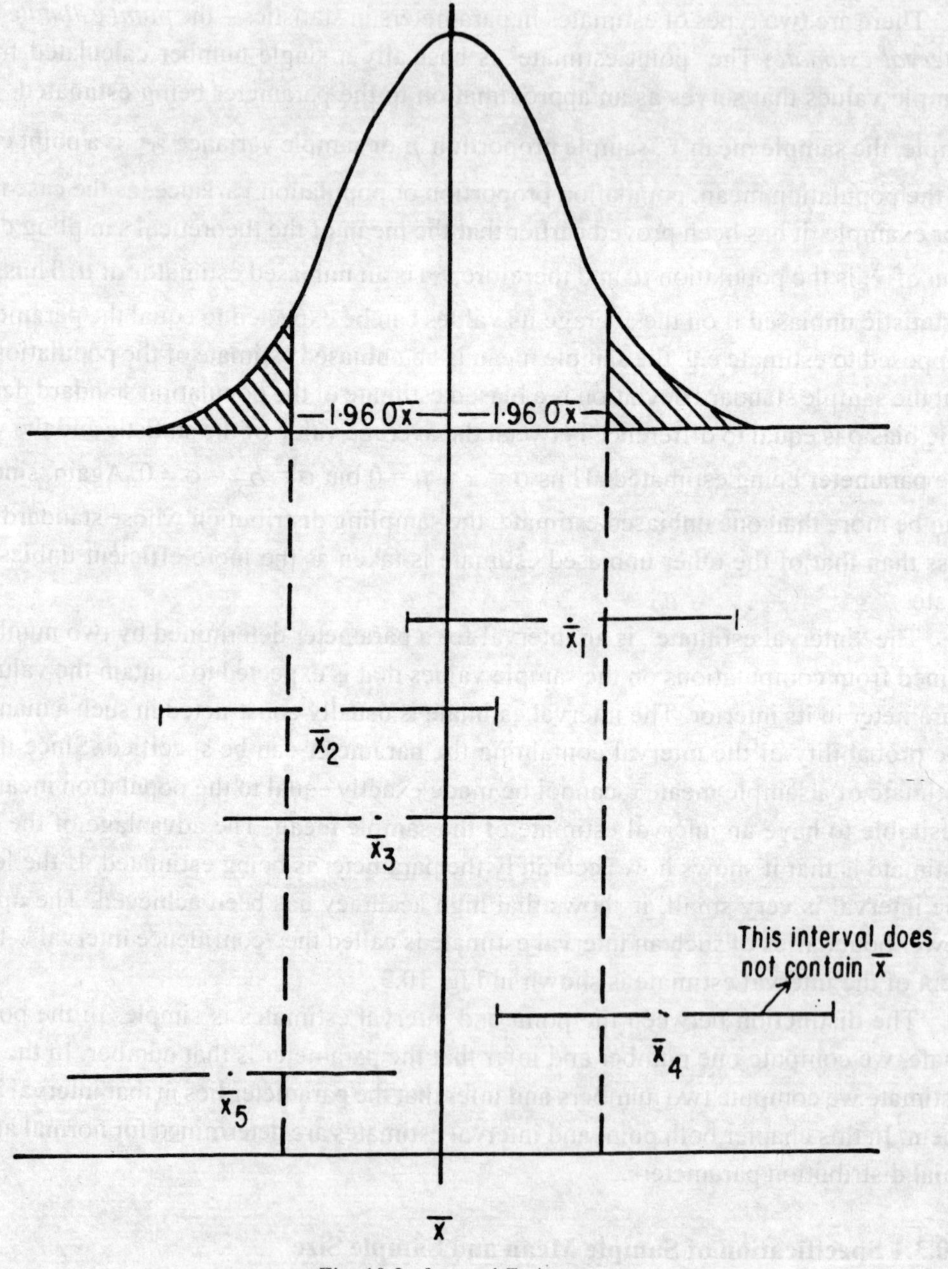

Fig. 10.3 : Interval Estimate

The mean of the population will be:

For 99% confidence limit : $\bar{x} - 2.58\,\sigma_{\bar{x}}$ to $\bar{x} + 2.58\,\sigma_{\bar{x}}$

i.e. $$\bar{x} - 2.58\,\sigma_{\bar{x}} < \mu < \bar{x} + 2.58\,\sigma_{\bar{x}} \quad \text{... (10.3)}$$

For 95% confidence limit : $\bar{x} - 1.96\,\sigma_{\bar{x}}$ to $\bar{x} + 1.96\,\sigma_{\bar{x}}$

i.e. $$\bar{x} - 1.96\,\sigma_{\bar{x}} < \mu < \bar{x} + 1.96\,\sigma_{\bar{x}} \quad \text{... (10.4)}$$

where μ is the population mean, $\bar{x}$ and $\sigma_{\bar{x}}$ are the mean and standard error of the random sample respectively.

Thus this interval is centered at $\bar{x}$ and the end points are partly determined by the value chosen for the confidence limits, e.g. $2.58\,\sigma_{\bar{x}}$ or $1.96\,\sigma_{\bar{x}}$. These two values correspond to the critical regions for the normal curve using the 99% and 95% confidence limits (i.e. 0.01 and 0.05 levels of significance) respectively.

Example 10.1 Calculate the estimate for the population mean at 99% confidence interval when the population standard deviation is known and equal to 5.0. The sample size is 100 and the sample mean is 15. Calculate also the sample size required to keep the estimate of the sample mean within ± 1.

For the above given conditions, the population mean of 99% confidence would be

$$= 15 \pm 2.58\,\frac{5}{\sqrt{100}} = 15 \pm 1.29$$

So the interval for the population mean at 99% confidence interval would range from 13.71 to 16.29.

For the second part, since we wish to estimate the sample mean to within an accuracy of ± 1, we make use of a 99% confidence interval. Then the given range of the sample estimate (at 99% confidence limit) would be equal to $5.16\,\sigma_{\bar{x}}$. From this, n, the desired size of the sample can be calculated as follows:

$$1 = 5.16\,.\,\frac{\sigma}{\sqrt{n}}$$

If σ is 5 then $\sqrt{n} = 5 \times 5.16 = 25.80$. Therefore the sample size will be 665.64 or 666.

10.3.2 Selecting The Level: Type I And Type II Errors

In using a sample to draw conclusions about a population there is a risk that an incorrect conclusion will be reached. Indeed, as it has been said earlier two different errors can occur in applying the hypothesis-testing approach to decision making concerning population parameters. The first error is called the Type I error.

"Type I" (also called the "alpha") error can be defined as the rejection of a null hypothesis, when it is true. To minimise this type of error, very small alphas should be used because as alpha α goes down the critical region moves further away from the mean of the sampling distribution (i.e. at $\alpha = 0.05$, the critical region starts at $z = \pm 1.96$, at $\alpha = 0.01$, it starts at $z = \pm 2.58$). Thus with decreasing α, the critical region also decreases and thus allows larger acceptance region.

But with a larger acceptance region, following a decreasing α, there is also a possibility that all other things being equal, there occurs the possibility of a second type of incorrect decision, called "Type II" (also called the "Beta") error which can be defined as failing to reject a null hypothesis that is in fact false. For example, in a criminal trial, if jury declares an innocent person as guilty, a Type I ("Alpha") error is made because a true hypothesis, H_0 is rejected. A Type II error occurs when a guilty person is acquitted.

Thus when the probability of Type I error decreases with lower levels of α, the probability of Type II error increases, and *vice versa.* Hence the two types of errors are inversely related and it is not possible to minimise simultaneously both in the same test. Normally, however, Type I error is minimised and the lower α-levels are used. Conventionally α is set at 0.05 or somewhat less frequently at 0.10 or higher level at 0.01 or 0.001.

One way in which a researcher can control the probability of these errors is to increase the sample size. Table 10.2 illustrates the two types of errors which is shown also earlier in 10.1.1.

Table 10.2: Hypothesis Testing Outcomes

Statistical Decision of the investigator	Actual Situation (H_0 true; H_1 false)	(H_0 false; H_1 true)
Do not reject H_0	Correct decision	Type II or β Error
Reject H_0	Type I or α Error	Correct decision

Example 10.2 Given the null and research hypotheses that $H_0 : \mu \geq 1200$ and $H_1 < 1200$ and the constraints : (1) the probability of a Type I error must be ≤ 0.02 significance and (2) if the true mean μ is 1180, the chances of a Type II error (or, a false hypothesis) must not exceed 0.05; find the minimum sample size to hold both Type I and Type II errors to the levels specified. The standard deviation of population, α is known equal to 40.

The first step is to solve for critical value, we can do this in two ways : (a) for the hypothetical distribution,

$$z = \frac{x - \mu_i}{\sigma_{\bar{x}}} \text{ where } \sigma_{\bar{x}} = \frac{40}{\sqrt{n}}$$

or $x - 1200 = [40/\sqrt{n}].z$ for $\alpha = 0.02$ we have

$z = -2.05$ and we have $x = 1200 - 2.05\ [40/\sqrt{n}]$.

Now for the assumed true distribution

$$z = \frac{x - \mu}{\sigma_{\bar{x}}} \text{ or } 1.65 = \frac{x - 1180}{40/\sqrt{n}} \text{ or } x = 1180 + 1.65\ (40/\sqrt{n}).$$

Again, if we make one set equal to the other, we can write

$$x = 1200 - 2.05 \cdot \frac{40}{\sqrt{n}} = 1180 + 1.65 \cdot \frac{40}{\sqrt{n}}$$

or, $20\sqrt{n} - 82.0 = 66.0$ or $20\sqrt{n} = 148.0$ or $n = 55$

Thus the sample size should be 55.

10.4 MODEL FOR HYPOTHESIS TESTING

In order to make use of the central limit theorem this is a five-step model for organizing hypothesis testing. These five steps are.

Step 1 : Making Assumption : This is usually stated as :

Model : Random samples

Level of Measurement : State the level of measurement of data.

Sampling Distribution : State the name of the probability distribution

Step 2 : Stating the null hypothesis : $H_0 : \bar{x} = \mu$ against the research hypothesis $H_1 : \bar{x} > \mu$

State the one- or two-tailed test

Step 3 : Selecting the sampling distribution and establishing the critical region, by stating the significance level α

Step 4 : Computing the test statistic and finally,

Step 5 : Making a decision.

10.4.1 Test of Significance between Sample Mean and Population Mean (when Population Standard Deviation is Known or Unknown)

Given a sample of size n drawn from a population with its mean and standard deviation known, we can test the null hypothesis that the mean value of a given sample could have been drawn randomly from a population. Symbolically, we ask, can we write : $\bar{x} = \mu$?

The procedure for this test is to use the normal curve with the mean μ to represent the sampling distribution of $\bar{x}$. In employing the normal curve any value for $\bar{x}$ corresponds to a particular normal deviate given by the relation

$$z = \frac{\bar{x} - \mu}{\sigma_{\bar{x}}} \quad \text{... (10.5)}$$

Therefore, the normal deviate $\bar{x}$ is given by

$$\bar{x} = \mu + z.\sigma_{\bar{x}} \quad \text{... (10.6)}$$

Now, the test is to compare the absolute difference between the sample mean $\bar{x}$ and the population mean μ with the standard error of the sample mean $\sigma_{\bar{x}}$ at a particular level of significance. If the difference of $\bar{x} - \mu$ is less than 2.58 $\sigma_{\bar{x}}$, it is said to be insignificant at 1% level of significance and the sample mean can be considered equal to population mean μ (i.e. the H_0 can be accepted). On other hand, if $\bar{x} - \mu$ is greater than or equal to $2.58\sigma_{\bar{x}}$, is rejected and the sample mean cannot be considered equal to the population mean.

Example 10.3 Set up the probable limits of μ with 99.73% of confidence level for $\bar{x} = 64$ inches, $n = 100$ and $\sigma = 3$ inches.

We have $\sigma_{\bar{x}} = \dfrac{\sigma}{\sqrt{n}} = \dfrac{3}{10} = 0.3$

So, probable limits of $\mu = 64 \pm 3\ (0.3) = 64 \pm 0.9$ inches.

Note : For tests of significance between sample mean and population mean when σ_p is not known, we require to use the *t*-ratio score which is discussed in the next chapter.

10.4.2 Test of Significance between the Means of Two Samples and Standard Error of Difference

A comparison between two samples is frequently of greater interest than a comparison between a sample mean and a hypothetical value of population mean.

Two samples are drawn from two populations, μ_1 and μ_2 with sample sizes n_1 and n_2, sample mean $\bar{x}_1$ and $\bar{x}_2$ and standard error $\sigma_{\bar{x}_1}$ and $\sigma_{\bar{x}_2}$. Now the standard error of difference of two samples is actually an estimate of the standard deviation of the distribution of differences between two samples. For this test, the research worker has to draw a large number of sample pairs. Yet this particular distribution plays an important role in the method of testing hypotheses about means differences which cannot be ignored. The logic is that when the two samples x_1 and x_2 possess normal distribution (being drawn from populations with normal distributions), then it can be shown that the difference (or, sum) of the two samples x_1 and x_2 will also approach a normal distribution. Thus, the standard error of differences tells us how much on an average a given $(\bar{x}_1 - \bar{x}_2)$ is likely to differ from the central point of the distribution of differences. Fortunately, this standard error of the difference is calculated as the square root of the sum of squares of their standard errors. Symbolically this can be written as

$$\sigma_{\text{diff}} = \sqrt{\sigma_{\bar{x}_1^2} + \sigma_{\bar{x}_2^2}} \qquad \text{... (10.7)}$$

or, more precisely

$$\sigma_{\text{diff}} = \sqrt{\frac{\sigma_{\bar{x}_1^2}}{n_1 - 1} + \frac{\sigma_{\bar{x}_2^2}}{n_2 - 1}} \qquad \text{... (10.8)}$$

Now, if we decide to have 0.01 level of significance for testing the null hypothesis that the means of the two populations from which the two samples have been drawn are equal (i.e. symbolically, $H_0 : \mu_1 - \mu_2 = 0$), the absolute difference of the sample means, $\bar{x}_1 - \bar{x}_2$ should not exceed 2.58 times the standard error of difference of the samples, σ_{diff}.

Note: For tests of significance between small samples (for $n < 30$), instead of the above z score, a t ratio is required. Like a z score, the t ratio score can be used to translate a sample mean difference $\bar{x}_1 - \bar{x}_2$ into units of standard error of the difference of the samples. This is discussed in the next chapter.

Example 10.4 To test the adoption of innovation measures by the farmers for a drive in increasing agricultural production, 50 sample villages were selected in two districts each. If the sample mean and standard error of the farmers adopting the innovation measures are found to be 7.0 and 0.29 for district A and 6.0 and 0.21 for district B, test the difference in the sample means between the two districts.

We have here $\sigma_{\text{diff}} = \sqrt{\sigma_{\bar{x}_1^2} + \sigma_{\bar{x}_2^2}} = \sqrt{.29 + .21} = 0.35.$

Now since the $\bar{x}_1 - \bar{x}_2$ is $7.0 - 6.0 = 1.0$ which is more than 2.58 times σ_{diff} at 0.01 level of significance, hence the null hypothesis that the sample means are equal belonging to the same population is rejected.

Example 10.5 Mean produce of wheat of a sample of 100 plots is 1230 kgs/hectare, $\sigma_{\bar{x}_1}$ = 1·0 kg, another sample of 150 plots give the mean 1350 kg/hectare with $\sigma_{\bar{x}_2}$ = 12 kg. Assuming the standard deviation of the mean yields is 11 kg/hectare for the universe, find if the results are consistent.

Now since the population standard deviation α_p is given, $\sigma_{\text{diff}} = \sqrt{\frac{(\sigma_{p_1})^2}{n_1} + \frac{(\sigma_{p_2})^2}{n_2}}$

$= \sqrt{(10)^2/100 + (12)^2/150} = \sqrt{1.00 + 0.96} = 1.40$ But since the observed mean difference in yield is 120 kg which is many times more than σ_{diff}, so the difference is significant.

10.4.3 Test of Significance between Sample Standard Deviation and Population Standard Deviation

The standard error of a standard deviation also records fluctuations from sample to sample which can also be estimated and used to find the limits within which population standard deviation, σ_p will fall. The necessary formula is

$$\sigma_{\bar{\sigma}} = \frac{\sigma_p}{\sqrt{2n}} \qquad \text{... (10.9)}$$

For small samples with $n < 100$, the sampling distribution of standard deviation of samples is somewhat skewed but it approaches normality as n increases. Hence inferences based on normal distribution can be drawn. Compared to the formula for standard error of the mean in Eq. (10.1), the denominator in the Eq. (10.9) here is about 40% less and hence the $\sigma_{\bar{x}}$ is less stable than $\sigma_{\bar{\sigma}}$.

Example 10.6 Variability in percentage of sand content in soil samples (measured as standard deviation) is found to be 0.80%. The sample size was 350 but how reliable is this measure of variability?

We have $\sigma_{\bar{\sigma}} = \frac{0.80}{\sqrt{2 \times 350}} = 0.030$; now the 95% confidence limit of $\sigma_{\bar{\sigma}} = 0.80 \pm 1.96\ \sigma_{\bar{\sigma}}$

$= 0.80 \pm (1.96)(0.030) = 0.80 \pm 0.059$. Thus at 95 per cent confidence level, the population standard deviation of percentage of sand content lies between 0.741 and 0.859 per cent.

10.4.4 Test of Significance between Sample Variance and Population Variance

The necessary formula for sample variance, $\sigma^2_{\bar{\sigma}}$ is

$$\sigma^2_{\bar{\sigma}} = \sigma^2_p \cdot \sqrt{\frac{2}{n}} \qquad \text{... (10.10)}$$

where σ_p^2 is the population variance.

10.4.5 Test of Significance between Standard Deviation of Two Samples

There can be two contexts: (i) when population standard deviations, σ_p s are known, for which the standard error of differences of the standard deviation of the samples will be

$$\sigma_{\sigma_1-\sigma_2} = \sqrt{\frac{(\sigma_{p_1})^2}{2n_1} + \frac{(\sigma_{p_2})^2}{2n_2}} \qquad \text{... (10.11)}$$

where σ_{p_1} and σ_{p_2} are the population standard deviations, n_1 and n_2 are size of the samples.

When the population standard deviations are not known, we can use the sample standard deviation, $\sigma_{\bar{\sigma}}$ in the above Eq. 10.11. Thus we have

$$\sigma_{\sigma_1-\sigma_2} = \sqrt{\frac{(\sigma_{\bar{\sigma}_1})^2}{2n_1} + \frac{(\sigma_{\bar{\sigma}_2})^2}{2n_2}} \qquad \text{... (10.12)}$$

Example 10.7 For a manufactured item 100 samples of make *x* has a mean life of 1300 hours with a standard deviation of 8 hours, while the same item of make *y* for 100 samples has a mean life of 1250 hours with a standard deviation of 90 hours respectively. Test the difference.

We have $$\sigma_{\sigma_1-\sigma_2} = \sqrt{\frac{(\sigma_{\bar{\sigma}_1})^2}{2n_1} + \frac{(\sigma_{\bar{\sigma}_2})^2}{2n_2}}$$

$$= \sqrt{\frac{(80)^2}{200} + \frac{(90)^2}{200}} = 8.51$$

Since the observed difference in the sample means is much more than the above difference in standard deviation of the samples we find the difference significant.

10.4.6 Test of Significance between Sample Proportion and Population Proportion

So far we have focused on the distribution of the mean of a quantitative variable. In many cases we are interested in behavioural research and qualitative (or categorical) characteristics such as "number of successes". We know a random variable like number of successes (or failures) follows a binomial distribution where the mean number of successes, is equal to np and the standard deviation is equal to $\sqrt{np(1-p)}$ with p as the probability of successes. Now instead of expressing the variable in terms of the probability of successes we can convert to the proportion of successes by dividing the number of successes by sample size so that the average proportion of successes is equal to

$$\frac{\text{Number of successes}}{\text{Sample size}} = \frac{np}{n} = p \qquad \text{... (10.13)}$$

while the standard deviation of the proportion of successes $\sigma_{\bar{p}}$ is equal to

$$\frac{\text{Standard deviation of successes}}{\text{Sample size}} = \frac{\sqrt{np.(1-p)}}{n} = \sqrt{\frac{pq}{n}} \qquad \text{... (10.14)}$$

where $q = 1 - p$ and $n \geq 30$, n is the sample size.

We know that when the sample size is large ($n > 30$), the binomial distribution can be approximated by the normal distribution. The rule of thumb is that if np and nq each are at least 5, the normal distribution provides a good approximation to the binomial distribution. In most cases in which inferences are to be made about the proportion, the sample size is substantial enough to meet the conditions for using the normal approximation. Thus, in many instances, we may use the normal distribution to evaluate the sampling distribution of the proportion.

We may develop this idea by referring to a hypothetical situation of two-candidate election to make what proportion of voters would vote for a specific candidate. In this case the only information available is the percentage of votes polled by the candidates and this sample data is arranged in the binomial form with the expression $(p + q)^n$ where n is the total number of sample observations, p is the proportion of an attribute with respect to n and q is the proportion not having that attribute. Now assuming a two-candidate, a political pollster assess that if in a sample of 100 voters, a specific candidate receives at least 55% (i.e. $p_s \geq 0.55$) of the vote, then that candidate will be winning the election. Now if the candidate actually got 48%, we may like to determine the probability that a candidate will be projected as winner. Since the sampling distribution of the proportion can be assumed to be approximately normally distributed, we have

$$z = \frac{\mu - \bar{x}}{\sigma_{p-\bar{p}}} \quad \text{... (10.15)}$$

and, because we are dealing with a sample proportion (not a sample mean) we can substitute p_s for μ, p for $\bar{x}$ and the Eq. (10.14), $\sqrt{pq/n}$ for $\sigma_{p-\bar{p}}$ and we have

$$z = \frac{p_s - p}{\sqrt{\dfrac{pq}{n}}} \quad \text{... (10.16)}$$

which is $z = \dfrac{0.55 - 0.48}{\sqrt{\dfrac{0.48 \times 0.52}{100}}} = \dfrac{0.07}{0.04996} = +\,1.40$

Using Appendix Table I(a), the area under the normal curve from $z = 0$ to $z = 1.40$ is 0.4192. Since we need to find the probability of obtaining a sample proportion above 0.55 (i.e. 55%), we need to subtract 0.4192 from 0.5000 to obtain the desired value of 0.0808. Thus, if the true population proportion, p is 0.48, there is an 8.08% chance of obtaining a sample proportion of 0.55 or above (and thereby incorrectly projecting the winner) if a sample of 100 voters is selected.

Proportions are widely used in analysing questionnaire surveys in human geography. For example, in the report of a census survey, a 10 per cent sample is generally published to give an impression of the trends since the previous census. One 1991 10% sample census report for an area showed that out of the 90,000 sample population surveyed, 4050 are unemployed. Find out how reliable this proportion of unemployed person at .05% significance level?

In this two-trait survey of classifying whether population of the area is employed or unemployed, the only information available is the proportion of the people unemployed which is p = 4050/90,000 = 0.045. Now $\sigma_{\bar{p}} = \sqrt{pq/n} = \sqrt{(0.045 \times 0.955)/(90{,}000)}$ = 0.00063. For a 0.05 significance level, the proportion of unemployed in the area at the time of the census was within $0.045 \pm 1.96 \times 0.00063 = 0.045 \pm 0.00123$ that is within 4.38 to 4.62 percentage.

Example 10.8 In a sample survey of 3200 births 1680 are boys and 1520 are girls. Do this figure conform to the hypothesis that the sex ratio is 50 : 50?

If we have the 99% probability for testing the hypothesis, for H_0 we can write

$|p_s - p| < 2.58 \cdot \sqrt{\frac{pq}{n}}$ as proportion of boy child birth, p_s is $\frac{1680}{3200} = 0.5250$

or, $|0.5250 - 0.5000| < 2.58 \cdot \sqrt{\frac{0.525 \times 0.475}{3200}}$

or $0.025 < 2.58 \times 0.008626 < 0.0228$

So we reject H_0 that the sex ratio cannot be taken 50 : 50.

10.4.7 Test of Significance of the Difference between Proportions of Two Samples

The problem of determining whether two populations differ with respect to a certain attribute is of much importance in statistical work and it is widely used in social sciences. A familiar example is whether there is any significant difference between percentage of smokers and non-smokers who have heart ailments?

The problem of this type can be solved in the following way: if n_1 and n_2 denote the size of the samples taken and p_1 and p_2 the sample proportions obtained, then the variable to be used in this problem is the z statistic on the assumption that we are testing a null hypothesis, i.e. there is an unbiased difference between the sample proportions, $(p_1 - p_2)$. This corresponds to using the standard error of the difference of the two samples in testing the null hypothesis that $\bar{x}_1 = \bar{x}_2$. Thus the difference in the sample proportions, $(p_1 - p_2)$ can be considered as being approximately normally distributed with the mean, $(\bar{p}_1 - \bar{p}_2)$ and standard deviation $\sigma_{(\bar{p}_1 - \bar{p}_2)}$ to use a z statistic as follows:

$$z = \frac{\bar{p}_1 - \bar{p}_2}{\sqrt{\frac{p_1 q_1}{n_1} + \frac{p_2 q_2}{n_2}}} \quad \text{... (10.17)}$$

or

$$z = \frac{\bar{p}_1 - \bar{p}_2}{\sqrt{\sigma_{\bar{p}_1}^2 + \sigma_{\bar{p}_2}^2}} \quad \text{... (10.18)}$$

Alternatively, when testing the null hypothesis 'H_0': $p_1 = p_2 = p_{est}$, the mean distribution of $(p_1 - p_2)$ will, of course, be equal to zero. Therefore, the estimated standard error of $(p_1 - p_2)$ is

$$\sigma_{(\bar{p}_1 - \bar{p}_2)} = \sqrt{p_{est}\, q_{est}\left(\frac{1}{n_1} + \frac{1}{n_2}\right)} \quad \text{... (10.19)}$$

where p_{est} is a pooled estimate disregarding the group membership and it is given by

$$p_{est} = \frac{x_1 + x_2}{n_1 + n_2} \text{ or } \left(= \frac{n_1 p_1 + n_2 p_2}{n_1 + n_2} \right) \qquad \text{... (10.20)}$$

where x_1 and x_2 are the numbers in the two groups out of the respective sample sizes.

Finally, if $z > 1.96$ or 2.58, we reject the null hypothesis at 95% or 99% confidence level.

At times we may be interested in comparing the proportion of the possessing an attribute in a sample with the proportions given by the population. In such a case we can have:

$$\sigma_{(p_0 - p_1)} = \sqrt{p_0 q_0 \times \frac{n_2}{n_1 (N)}} \qquad \text{... (10.21)}$$

where p_1 = sample proportion, p_0 = population proportion, $q_0 = 1 - p_0$, n_1 = number of observations in the sample, N = size of population and $n_2 = N - n_1$.

Example 10.9 Four hundred farmers were divided into two equal groups for carrying out a field survey for testing the improvement in crop production by using a high-yielding variety. The numbers who were found successful in these two groups of farmers selected randomly were 152 and 132. Can we conclude that there is no real difference in the effectiveness among high-yielding variety?

Here $n_1 = n_2 = 200$

Therefore, $p_1 = 152/200 = 0.76$ and $p_2 = 132/200 = 0.66$

$$\text{Now, in } z = \frac{p_1 - p_2}{\sqrt{\frac{p_1 q_1}{n_1} + \frac{p_2 q_2}{n_2}}} = \frac{0.76 - 0.66}{\sqrt{\frac{(0.76)(0.24)}{200} + \frac{(0.66)(0.34)}{200}}}$$

$$= \frac{0.100}{\sqrt{0.000912 + 0.001122}} = \frac{0.100}{0.045} = 2.22$$

Since z is 2.22 it is itself larger than the critical value (= 1.96) at 95% confidence level. But since the above z value is smaller than the critical value (= 2.58) at 99% confidence level, we conclude that the null hypothesis cannot be rejected finally and there is no significant difference in the two high-yielding varieties.

Note the same result can also be obtained if we could calculate the probability of successful farmers in the population (which is not known) by a pooled estimate of overall averages, disregarding the group membership as in Eq. (10.20)

$$p_{est} = \frac{x_1 + x_2}{n_1 + n_2} = \frac{152+132}{200+200} = \frac{284}{400} = 0.71$$

and using it in Eq. (10.19) for the estimated standard error of differences in the proportion of the groups, $p_1 - p_2$ as

$$\sigma_{(\bar{P}_1 - \bar{P}_2)} = \sqrt{p_{est}\, q_{est}\left(\frac{1}{n_1} + \frac{1}{n_2}\right)}$$

$$= \sqrt{(0.71)(0.29)\left(\frac{1}{200} + \frac{1}{200}\right)}$$

$$= \sqrt{(0.71)(0.29)(0.01)} = 0.045$$

and, hence z value will be the same (= 2.22) as in before.

Example 10.10 There are 1000 households in a village and 20,000 households in the district of the village. Two hundred households in the village are found having tractors and 1000 in the whole district. Can we say that there is a significant difference between the proportion of tractor owners in the village and the district?

Proportion of households having tractors in the village p_1 is 0.20 and proportion of households having tractors in the district of the village, p_0 is $\frac{1000}{20000}$ = 0.05. Difference between p_0 and p_1 = 0.20 – 0.05 = 0.15. Now $q_0 = 1 - p_0 = 0.95$; $n_1 = 1000$, $n = 20{,}000$.

$$\text{Hence } \sigma_{(\bar{p}_0 - \bar{p}_1)} = \sqrt{p_0 q_0 \times \frac{n_2}{n_1 (N)}}$$

where $n_1 = 1000$, $N = 20{,}000$ and $n_2 = 19{,}000$

$$\text{So, } \sigma_{(\bar{p}_0 - \bar{p}_1)} = \sqrt{0.05 \times 0.95\left(\frac{19{,}000}{1{,}000 \times 20{,}000}\right)} = \sqrt{0.000045} = 0.0067$$

So, $z = \frac{0.15}{0.0067} = 22.39$ which is > 2.58. Hence H_1 is accepted, that is, there is a significant difference between the proportion of households having tractors in the village and the district, the proportion being significantly higher in the village.

10.4.8 Test of Significance of the Difference between Coefficient of Variation of Two Samples

In climatology, maps are often prepared based on the coefficient of variation and then comparisons are made between places with different values of this coefficient. Whether these differences in values are statistically significant or not (though this is looked into rarely), can be estimated from the standard error of the coefficient of variations, $\sigma_{(v_1-v_2)}$. This can be written as

$$\sigma_{(v_1-v_2)} = z_\alpha \cdot \sqrt{\frac{v_1^2}{2n_1} + \frac{v_{\hat{s}}^2}{2n_2}} \qquad \text{... (10.22)}$$

where z_α is the critical value of the score.

[*Note*: if σ is taken instead of the coefficient of variation, the formula on the right side would be $z_\alpha \cdot \sqrt{\frac{\sigma_1^2}{2n_1} + \frac{\sigma_2^2}{2n_2}}$ as in Eq. (10.22).]

Now, following the logic given in the test of significance between the means of two samples, it can be said that for the null hypothesis $H_0 : v_1 = v_2$ to be retained, the absolute difference between the coefficient of variation of two samples should not be more than the critical value at the significance level multiplied by the standard error of the coefficient of variations.

Example 10.11 A comparison is made between two areas each with a 30-year record, but one area has a coefficient of variation as 13% and another area has 10% coefficient of variation. Test whether the difference in the coefficient of variation is significant or not.

The absolute difference in the coefficient of variation of the two areas is 13 – 10 = 3. Now the standard error of the coefficient of variations is

$$\sigma_{(v_1-v_2)} = \sqrt{\frac{13^2}{60} + \frac{10^2}{60}} = 2.11$$

Thus absolute difference in the coefficient of variation (equal to 3) is less than 1.96 times the standard error of the coefficient of variations at 0.05 significant level. Hence we can conclude that the difference in the coefficient of variation is not at all significant.

10.5 LIMITATIONS OF TESTING OF HYPOTHESIS

Finally we can discuss the limitations of testing of hypothesis. Given that we are usually interested in rejecting the null hypothesis, the probability of rejecting the null hypothesis (and accepting the research hypothesis) is a function of four independent factors:

1. *The size of sample* : For all tests of significance, the probabilities for rejecting H_0 with larger samples are high. But this relationship states that larger samples are, after all, better approximations of the populations they represent. Thus, decisions based on larger samples are more trustworthy than decisions based on small samples.
2. *The size of the observed difference(s)* : This in terms of difference between the sample outcome and the population or between two sample outcomes is a function in part of the testing procedures (i.e. how variables are measured) but for the most part should reflect the underlying realities we are trying to probe.
3. *The alpha* (α) *level* : The relationship between this and the probability of rejection is straightforward. The higher the α level, the larger the critical region and thus greater is the probability of rejection.
4. *The use of one- or two-tailed tests* : The higher α level in the previous one will lead to more frequent Type-I errors and we might find ourselves declaring rather small differences to be statistically significant. In similar fashion, use of the one-tailed test increases the probability of rejection (assuming that the proper direction of the difference has been predicted).

LIST OF FORMULAE

I. *Standard Error of parameters of Sampling Distribution*

1. Mean, $\sigma_{\bar{x}} = \sigma_p / \sqrt{n}$ where $\sigma_p = \sqrt{\Sigma(x_i - \mu)^2 / N}$

Standard error or Standard deviation, $\sigma_{\bar{\sigma}} = \sigma_p / \sqrt{2n}$

Note : for sampling 'without replacement' the above formula should be multiplied by $\sqrt{(N-n)/(N-1)}$, where N is number of observations in population and n is the sample size

2. Specification of Sample Mean : $\bar{x} - \alpha.\sigma_{\bar{x}} < \mu < \bar{x} + \alpha.\sigma_{\bar{x}}$

3. Desired Size of Sample for α significance level within an accuracy of $\pm$ 1,

$$1 = 2\,\alpha\,.\,\frac{\sigma_p}{\sqrt{n}}$$

II. ***Table of Test Statistics and Limits for a variety of different parametric tests for large samples***

Name of the Test of Significance between (with what is known/given)	Test statistic	One-or two-tailed test	Limit(s)
1	2	3	4
(a) Sample mean and population mean (mean and standard deviation for population are known)	$z = \frac{\bar{x} - \mu}{\sigma_{\bar{x}}}$ in case σ_p is not known we use $s = \sqrt{\frac{\Sigma(x_i - \bar{x})^2}{n-1}}$	Two	Upper $= z.\left(1 - \frac{\alpha}{2}\right)$ Lower $= z\,.\,\frac{\alpha}{2}$
(b) the means of two samples and standard deviation/variances are known	$z = \frac{\bar{x}_1 - \bar{x}_2}{\sigma\sqrt{\frac{1}{n_1} + \frac{1}{n_2}}}$ when σ, s are equal,	Two	
	or, $z = \frac{\bar{x}_1 - \bar{x}_2}{\sqrt{\frac{\sigma_1^2}{n_1} + \frac{\sigma_2^2}{n_2}}}$ when σ, s are not equal	One $(\bar{x}_1 > \bar{x}_2)$	$z_{1-\alpha}$
(c) Sample standard deviation and population standard deviation : for specification of sample standard deviation (population standard deviation being known)	$\sigma_{\bar{p}} - \sigma_{\bar{\sigma}} < \mu < \sigma_{\bar{p}} + \sigma_{\bar{\sigma}}\,.\,\alpha$ where $\sigma_{\bar{\sigma}} = \sigma_p \,/\, \sqrt{2n}$		

1	2	3	4
(d) Standard deviation of two samples	$z = \dfrac{\sigma_{\sigma_1} - \sigma_{\sigma_1}}{\sqrt{\dfrac{(\sigma_{p_1})^2}{2n_1} + \dfrac{(\sigma_{p_2})^2}{2n_2}}}$ in case σ_p s are not known, we use $\sigma_{\sigma_1 - \sigma_2} = \sqrt{\dfrac{(\sigma_{\bar{\sigma}_1})^2}{2n_1} + \dfrac{(\sigma_{\bar{\sigma}_2})^2}{2n_2}}$		
(e) Proportions for repeated independent trials (sample size large so that normal approximation of binomial distribution is possible), population proportion, p is known	$z = \dfrac{p_S - p}{\sqrt{pq / n}}$		
(f) Proportions for two samples	$z = \dfrac{p_1 - p_2}{\sqrt{\dfrac{p_1 q_1}{n_1} + \dfrac{p_2 q_2}{n_2}}}$ where $\dfrac{p_1 q_1}{n_1}, \dfrac{p_2 q_2}{n_2}$ are $\sigma_{\bar{p}_1}^2$ and $\sigma_{\bar{p}_2}^2$ or . $z = \dfrac{p_1 - p_2}{\sqrt{p_{est}\, q_{est} \left(\dfrac{1}{n_1} + \dfrac{1}{n_2} \right)}}$ where two population are similar, p_{est} is the pooled estimate of the two sample proportions equal to $\dfrac{x_1 + x_2}{n_1 + n_2}$ or $\dfrac{n_1\, p_1 + n_2\, p_2}{n_1 + n_2}$		
(g) Coefficient of variation of two samples	$z = \dfrac{v_1 - v_2}{\sqrt{\dfrac{v_1^2}{2n_1} + \dfrac{v_2^2}{2n_2}}}$		

EXERCISES

10.1 Draw 50 one-digit random samples of size $n = 10$ from random number table. Work out a sampling of the mean and draw its histogram.

10.2 A random sample of 100 boring points has been selected from an area and the depth of the groundwater level is recorded. The mean and standard deviation of the boring depths are 900 cm and 480 cm respectively. Find out the confidence limits for the mean depth of groundwater in the whole area.

10.3 A random sample of 100 people in rural areas yields a 64% chance of timely monsoon in one year. For the same year a sample of 120 people in a city gives a 40% chance. Does the evidence depict the difference in the expectation of people on timely monsoon in the areas ?

10.4 A random sample of 125 upper-income families and a sample of 145 from lower-income families revealed a mean and standard deviation of 3.3, 0.9 and 2.8, 1.2 respectively. Can we conclude that both the sets of samples are drawn from the same population?

10.5 In a hilly area, 30 out of 50 fields with a southerly aspect and 16 out of 40 with a northerly aspect are found to have areas of less than 2 hectares. Do these figures support that field size and aspect are related in the area studied?

10.6 In an innovation adoption survey in a region it was found that tractors are widely used in 125 villages out of 400 villages, but they are adopted in only 115 out of 300 villages in another region. Determine whether the adoption of innovation in terms of tractors are the same irrespective of the regions.

10.7 From a random sample, 60 farms are found to have a mean area of 4.5 hectares with a standard deviation of 1.2. Find the 95% and 99% confidence limits for the mean area. If the size of sample is increased to 144, what effect would there be on the confidence limits?

11

Parametric Statistics In Geo-science Systems II (Sampling Estimates for Small-sized Samples)

11.1 STUDENT'S *t* TEST

A small-sized sample is usually referred to as one which has fewer then 30 sampling units. Usually, the number 30 is based upon random selection from an infinite population but this does not mean that a sample of 40 can be considered to be a large sample at all circumstances. The value of the distinction between small and large sample lies in the fact that although certain statistical operations can be applied to large samples, they cannot be applied to small samples. For example, the approximation that the sampling distributions of the mean are approximately normal does not hold good for small samples (of size $n < 30$). Hence we require a new probability of the sampling distribution of statistics for small samples to serve the purpose of a test of significance in terms of the t test.

We have seen in Chapter 10 that the difference between the mean of a sample $\bar{x}$ and the mean of the population from which the sample is drawn can be "standardised" by expressing it in terms of the standard deviation of the population. This is done in normal distribution by calculating the z statistic as given in Eq. (4.13.) The t distribution score is also very much like the z score which is discussed in this chapter.

In Chapter 10, an assumption was made that in sampling estimates, we need to know the standard deviation of the population, σ. However, when working with samples, in most practical situations, the standard deviation of the population is unlikely to be known. Therefore, σ must be replaced by its estimate, the standard deviation of the sample scores and we have the statistic $s_{\bar{x}}$ as the standard error of the sample means as

$$s_{\bar{x}} = \frac{\text{Standard deviation of the sample raw scores}}{\sqrt{\text{No. of observations in the sample}}}$$

$$= \frac{\sqrt{\Sigma(x-\bar{x})^2 / n}}{\sqrt{n}} = \frac{s}{\sqrt{n}} \quad \text{... (11.1)}$$

where $s = \sqrt{\Sigma(x-\bar{x})^2 / n}$ that is, we replace σ with s, the sample standard deviation and s could be computed directly from sample data.

Where n/N is approximately 5 per cent, for a best estimate of $s_{\bar{x}}$ the Eq. (11.1) should be multipled by the Bessel's Correction $\sqrt{\frac{n}{(n-1)}}$ Hence

$$s_{\bar{x}} = s.\sqrt{\frac{n}{(n-1)}} = \sqrt{\frac{\Sigma(x-\bar{x})^2}{n}}\sqrt{\frac{n}{(n-1)}} = \sqrt{\frac{\Sigma(x-\bar{x})^2}{n-1}} \quad \text{... (11.2)}$$

William S. Gosett in 1908, an employee of Guinness Breweries in Ireland was interested in making inferences about the sample mean when standard deviation of population, σ is not known and it had to be estimated. Since his employer did not allow him to publish research work under his name, so Gosett studied this "standardised difference" under the nom de plume "Student" and the test is known as "Student's t test". He expressed this standard error of the means, $s_{\bar{x}}$ in terms of a ratio as in the z score [Eq. (4.13)]. Thus the test is carried out as in the large sample estimate, but instead of stating a value of z which must be exceeded (for rejecting the null hypothesis), we state a value of t. Hence the *Student's t statistic* is a ratio which can be expressed as

$$t = \frac{\text{Observed sample mean} - \text{Population mean}}{\text{Standard error of the means}}$$

$$= \frac{\bar{x} - \mu}{s_{\bar{x}}} \quad \text{... (11.3)}$$

which is analogous to the z statistic given by $z = \frac{\bar{x} - \mu}{\sigma}$. When this t statistic is calculated for all samples of size n, it has a probability distribution which is known as "Student's t distribution". Like normal distribution the t distribution is symmetrical with the mean zero, but it has a wider 'spread' than the normal distribution and its standard deviation depends on a specific parameter known as *degrees of freedom*. The *degrees of freedom* is a number and it is defined as the number of independent observations less the number of the fixed values imposed on the determination. For example, we have a sample of five values which had a mean of 10 and we raise the question the number of individual values we require to know before we could obtain the remainder, the answer will be $n - 1 = 5 - 1 = 4$ values which itself is the degrees of freedom (henceforth expressed as *d.f.*). Thus in this example, if we have the four means as 8, 14, 9 and 6, the fifth value can only be 13 since the sum of the means is $\sum x = 50$ and total of the four values is 37. Thus the *d.f.* for one fixed value, the mean $\bar{x}$ in a set of observation n is $n - 1$ i.e. the sample size less one. Appendix Table VI gives the value of the t distribution corresponding to the number of *d.f.* denoted by the Greek letter ν (pronounced as nu) and the probabilities. Thus the t distribution table similars the cumulative standard normal distribution except that in the t distribution table there are two entries: (i) the level of significance and (ii) degrees of freedom ν. For example, Appendix Table VI shows that for a two-tailed test at 5% significance level and 8 *d.f.* the t value must exceed 2.31. Thus the t distribution is not a single distribution but a family of distributions, each number of which depends upon *d.f.* Hence in contrast to standarised normal distribution, which is one curve, the t distribution is a family of curves one for each number of *d.f.* Since the t test is based on samples, we require to estimate the population parameters to calculate the test statstic. So to estimate the parameters and to perform the test from the same set of data without somehow compensating for the double use of observations, it is fair that we consider the *d.f.* defined as the number of observations in a sample minus the number of parameters estimated from the sample. In other words, the *d.f.* are the number of observations in excess of those necessary to estimate the parameters of the distribution. From Appendix Table VI it can also be seen that since t is a function of *d.f.*, when the sample size is large (and cons quently the *d.f.*), the Student's t distribution closely approximates the z value in the normal distribution, till the limiting case (i.e. at *d.f.* equal to infinity) where they become identical. Thus at *d.f.* = 120, the t value at 0.05 and 0.01 significance levels are 1.98 and 2.62 respectively but at *d.f.* = ∞ values at all the significance levels become identical with the corresponding z values. On the other hand when there are only a few *d.f.* the t distribution curve is flatter and more spread out (i.e. dispersion) than the standard normal curve and hence it is platykurtic containing more area in the tails and less in the centre than does the normal distribution. Figure 11.1 illustrates the comparison between a normal and a t distribution curve. The basic difference between the two distributions is in spread of the two distributions. The standard deviation of the standard normal distribution is 1.0 whereas the variance of the t distribution is always greater than 1.0. Otherwise both the distributions are symmetrical, both range from $-\infty$ to $+\infty$ and both standardised distributions have a mean of zero.

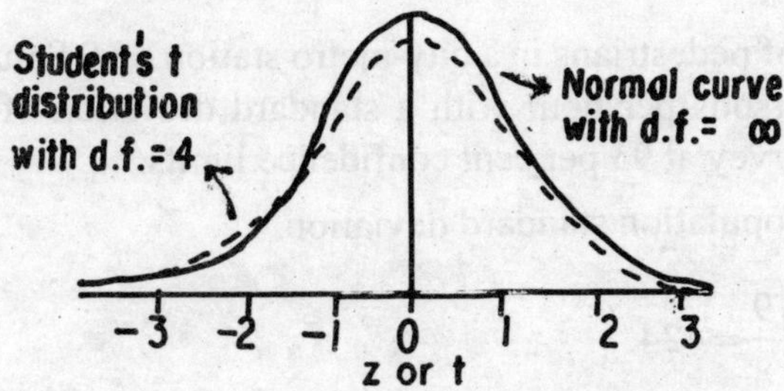

Fig. 11.1 : Comparison between Standard Normal and Standardized Student's *t* distribution curves

11.2 IMPORTANCE OF *t* STATISTIC

In many statistical tests, the sizes of the samples are severely limited because of economy of time and inaccessibility (as well as of infinity) of many populations. In these cases the availability of the Student's *t* distribution makes the interval estimates possible. Again since *t* does not require a knowledge of σ, as in the case with *z*, its value can be computed from the sample data, whereas the value of *z* cannot be computed unless σ is known. This is the reason why *t* can be used to solve problems without the necessity of introducing approximation to population parameters.

Like large sample estimates, in the previous chapter, we discuss below the point and the interval estimates for small samples ($n < 30$) by the Student's *t* test.

11.2.1 Interval Estimates by *t* Test

The confidence interval estimates for the population mean μ with the σ unknown from samples ($n < 30$) as given by [from Eqs. (11.1), (11.2) and (11.3)] are

$$\mu = \bar{x} \pm t_{\alpha} \cdot \frac{s}{\sqrt{n}} \qquad \text{... (11.4)}$$

where the value of $\pm\, t_{\alpha}$ are critical values of *t* at significance levels of 1 or 5 percent at *d.f.* $= (n - 1)$.

Equation (11.4) can also be written as

$$\bar{x}-t_{\alpha}.\frac{s}{\sqrt{n}} \leq \mu \leq \bar{x}+t_{\alpha}.\frac{s}{\sqrt{n}} \qquad \text{... (11.5)}$$

Example 11.1 A survey of pedestrians in a city-metro station on 9 Saturday mornings give a mean flow of 2500 persons per hour with a standard deviation of 400. Find the best estimate of the sample survey at 95 per cent confidence limits.

The best estimate of population standard deviation,

$$\sigma = s_{\bar{x}} . \sqrt{\frac{n}{n-1}} = 400\sqrt{\frac{9}{8}} = 424$$

Now the standard error of the sampling distribution is

$$s_{\bar{x}} = \frac{\sigma}{\sqrt{n}} = \frac{424}{\sqrt{n}} = 141$$

The 95 per cent confidence level indicates that there would be an 0.025 proportion in each tail of the distribution. Thus, the crtical value of t at $\alpha = 0.025$ and $d.f. = n - 1 = 8$ is 2.31 and hence we can have the best estimate of the sample survey at 95 per cent confidence limits as 2500 ± (2.31) (141) = 2500 ± 325, that is between 2175 and 2825.

11.2.2 Point Estimates by *t* Test

Now that the interval estimates obtained from the mean in the above example is too wide to be satisfied. Thus it may be needed to determine the sample size for a desired confidence level.

To determine the sample size, we begin by considering the sampling error, that is, the difference between the observed sampling mean and the population mean, $\bar{x} - \mu$, we recall from Eq. (11.3) at a significance level α

$$t = \frac{\bar{x}-\mu}{s_{\bar{x}}} = \frac{\bar{x}-\mu}{s/\sqrt{n}}$$

Now, multiplying each side by $s/\sqrt{n}$, we have

$$t \times s/\sqrt{n} = \bar{x} - \mu$$

Again, solving for sample size, n on squaring and on simplification, we obtain

$$n = \frac{t^2 s^2}{(\bar{x}-\mu)^2} \qquad \text{... (11.6)}$$

In the real world often a research worker is required to carry on experiments in the field and this is particularly true for agricultural research where one has to try out new varieties of cultivation practises or methods of "seed treatment" in the field plots repeatedly. These objects of comparison in his trials/plots is known as *treatments*. For getting good results, the field plots are divided into small units to provide uniform treatments. But even with this apparent uniformity in plots, there can be an appreciable variation in fertility which does not follow any regular pattern in any direction except the fact that relatively small areas are homogeneous. Such variation from plot to plot as caused by uncontrolled factors is taken as *experimental error*. Hence a research worker has to evaluate the magnitude of the experimental error and to compare with it the observed variation between treatments to find out whether the experiment indicates any real differences in the effects of the treatment. In this process, he repeats the treatments under investgation and this is known as "replication". Since the variation in fertility cannot be allowed for directly owing to its unpredictable nature, the research worker has to average out its influence over the different treatments by replication. Thus replication amounts to sampling. If we repeat a single treatment "n" number of times, the mean of these replications will be subject to a standard error of $s/\sqrt{n}$ where s is the standard deviation of individual plots in the experiment.

The standard error of difference of treatment means based on n replications expressed as percentage of the common mean is

$$C_v . \sqrt{\frac{2}{n}} \quad \text{... (11.7a)}$$

On this basis we can calculate the number of replications (or, samples) required to enable us to infer the difference between two treatments as significant at a given level, when the true value of coefficient of variation, C_v is known. This t ratio is

$$t = \frac{\text{Observed difference of treatment means}}{\text{Standard error of difference}}$$

$$= \frac{\bar{x}_1 - \bar{x}_2}{C_v . \sqrt{\frac{2}{n}}} \quad \text{... (11.7)}$$

when this t ratio in Eq.(11.7) exceeds the value of 1.96, the difference between the treatment means based on n replications will be taken significant at $\alpha = 0.05$ level. We illustrate this estimation of the minimum number of replications at $\alpha = 0.05$ level of significance by an example below.

Example 11.2 Esimate the minimum number of replications required for an observed difference of 15 per cent of the mean, significant at 5 per cent level, the coefficient of variation of the individual plot values is known being 15 per cent.

Now, from Eq. (11.7) at $\alpha = 0.05$ level, we can write

$$1.96 = \frac{15}{15.\left(\sqrt{\frac{2}{n}}\right)} \text{ or } 1.96\left(\sqrt{\frac{2}{n}}\right) = 1 \text{ or } \frac{7.6832}{n} = 1 \text{ or } n \approx 8 \text{ replications}$$

11.3 TEST OF SIGNIFICANCE BETWEEN SAMPLE MEAN AND POPULATION MEAN WHEN STANDARD DEVIATION IS UNKNOWN

Under the null hypothesis for a given sample mean, $\bar{x}$ the calculated value of t should not exceed the critical value of t given in Appendix Table VI, for $d.f. = n - 1$ at a given significance level. For example, with $n = 31$, $\bar{x} = 5.7$, $\mu = 4.0$ and $s = 2.1$, if we select a 0.10 level of significance for a one-tailed test the critical value of t with $d.f. = 31 - 1 = 30$ can be obtained from the Appendix Table VI as 1.70 (the rejection region of 0.10 being divided into two equal parts of 0.05 each for the one-tailed test).

Thus, we frame :

Reject H_0 if $t > t_{30} = +1.70$
or if $t < -t_{30} = -1.70$
Otherwise do not reject H_0

Example 11.3 If in a sample of $n = 15$, the mean of the sample is 19 and standard deviation of the sample is 3.50, can we conclude that the population mean of the sample is not 17.50?

We have, $t = \dfrac{\bar{x} - \mu}{s_{\bar{x}}} = \dfrac{19 - 17.50}{3.50/\sqrt{15}} = \dfrac{1.50}{0.90} = 1.66$

This calculated value of t is less than the critical value of t at $d.f. = 14$ which is $t_{\alpha} = 2.15$ at $\alpha = 0.05$. Hence the null hypothesis, H_0 is retained and μ is 17.50.

11.3.1 Test of Significance between the Means of Two Independent Samples

This is an extension of the single sample mean test to a test in which a comparison can be made between the means of two sets of random samples from each population. But in testing the sampling difference between two samples, with the population standard deviation, σ not known there can be three types of situations as the

(a) two sample sizes n_1 and n_2 are unequal but their population variances are equal
(b) two samples are of equal sizes, and
(c) two samples are having unequal size and heterogeneous variance

For (a) above when the sample sizes are unequal ($n_1 \neq n_2$) but their population variances are equal ($\sigma_1^2 = \sigma_2^2$) and we know only the sample means and the sample standard deviations $(\bar{x}_1, \bar{x}_2; s_{\bar{x}_1}, s_{\bar{x}_2})$, the t distribution with *d.f.* equal to $n_1 + n_2 - 2$ can be used to test for the sampling difference between the means of two samples $\bar{x}_1$ and $\bar{x}_2$.

Since we are assuming equal variance in the two populations, the variances of the two samples can be pooled together (i.e. combined) to form a pooled variance estimate, $s_{\bar{x}'}^2$. So the test statistic

$$t = \frac{\bar{x}_1 - \bar{x}_2}{s_{\bar{x}_1 - \bar{x}_2}} \text{ will be}$$

$$t = \frac{\bar{x}_1 - \bar{x}_2}{\sqrt{s_{\bar{x}'}^2\left(\frac{1}{n_1} + \frac{1}{n_2}\right)}} = \frac{\bar{x}_1 - \bar{x}_2}{s_{\bar{x}'}} \cdot \sqrt{\frac{n_1 n_2}{n_1 + n_2}} \quad \text{... (11.8)}$$

Since the population variance is unknown, the pooled variance $s_{\bar{x}}^2{}'$ can now be estimated by obtaining a combined estimate from both samples. Again, since the sample sizes are unequal, we can obtain an estimate of the variance by aggregating the sample sum of squares equal to

$$\Sigma(x_{1i} - \bar{x}_1)^2 + \Sigma(x_{2i} - \bar{x}_2)^2$$

and then this is divided by the proper *d.f.* which is equal to $(n_1 - 1) + (n_2 - 1)$ i.e. $(n_1 + n_2 - 2)$ for the two samples taken together in order to obtain the unbiased estimate for the variance. Taking the square root, we have an estimate of $s_{x'}$ as

$$s_{\bar{x}'} = \sqrt{\frac{\Sigma(x_{1i} - \bar{x}_1)^2 + \Sigma(x_{2i} - \bar{x}_2)^2}{n_1 + n_2 - 2}} \quad \text{... (11.9)}$$

Now, since $s_{\bar{x}_1}^2 = \frac{\Sigma(x_{1i} - \bar{x}_1)^2}{n_1}$ we can write $n_1 s_{\bar{x}_1}^2 = \Sigma(x_{1i} - \bar{x}_1)^2$

Similarly, $n_2 . s_{\bar{x}_2}^2 = \Sigma(x_{2i} - \bar{x}_2)^2$

Hence the Eq. (11.9) can be rewritten as

$$s_{\bar{x}'} = \sqrt{(n_1\, s_{\bar{x}_1}^2 + n_2\, s_{\bar{x}_2}^2)/(n_1 + n_2 - 2)} \qquad \text{... (11.10)}$$

Note for a best estimate of $s_{x'}$ we can have the Eq. (11.8) multiplied by a "finite population correction factor" as

$\sqrt{\left(\dfrac{n_1 + n_2}{n_1 . n_2}\right)}$ so that $s_{\bar{x}}$

$$= \sqrt{(n_1\, s_{\bar{x}_1}^2 + n_2\, s_{\bar{x}_2}^2)/(n_1 + n_2 - 2)} \,.\, \sqrt{(n_1 + n_2)/(n_1 n_2)} \qquad \text{... (11.11)}$$

If the calculated value of t is greater than the critical value at *d.f.* $n_1 + n_2 - 2$ at the level of significance, the difference of sample means is significant and the null hypothesis is rejected.

Example 11.4 The runoff data (in cm/year) are measured at two places A and B for $n = 21$ years. If the mean and the variance of the runoff at A and B are calculated as 80.33 cm and 1079.51 cm for A and 23.97 cm and 109.51 cm for B can we conclude that the sampling difference for runoff between A and B is significant.

We can have the test statistic t for *d.f.* $= n_A - 1 + n_B - 1 = 40$ as

$$t = \frac{\bar{x}_1 - \bar{x}_2}{\sqrt{\dfrac{s_A^2}{n_A - 1} + \dfrac{s_B^2}{n_B - 1}}} = \frac{80.33 - 23.97}{\sqrt{\dfrac{1079.51}{21-1} + \dfrac{109.51}{20}}}$$

$$= \frac{56.36}{7.71} = 7.31$$

Since this value is higher than the critical value of t at any level of significance so we can conclude that the two samples are drawn from different populations.

Note : Assuming equal variance in the two populations, if we calculate the pooled standard deviation, $s_{\bar{x}'}$, as in the Eq. (11.11) we will get $s_{\bar{x}'} = 7.71$ to read the same conclusion as before.

For (b) samples of equal sizes (that is for matched or paired samples) say n, the test statistic, t in Eq. (11.8) changes to

$$t = \frac{\bar{x}_1 - \bar{x}_2}{\sqrt{s_{x'}^2}} . \sqrt{\frac{n}{2}} \qquad \text{...(11.12)}$$

and the value of t will be entered in the table of t statistic with $d.f.=2\,(n-1)$ instead of $d.f. = n_1 + n_2 - 2$ for unequal samples.

Example 11.5 Following is the yield of cotton in kg. per plot in a randomized experiment with two treatments A and B :

Treatment A : 14.05, 13.82, 12.44, 12.71, 12.09, 10.33, 11.55, 10.98, 11.78 and 11.65

Treatment B : 15.33, 14.47, 14.00, 13.75, 10.68, 10.42, 11.28, 11.79, 10.67 and 11.91

Test the significance of the difference between A and B.

We have the mean for A = 12.14 kg and for B = 12.43 kg and we have their sum of squares of differences, $\sum (x_i - \bar{x})^2$ for A as 12.2274 and for B as 28.9816.

So the variances are : $s_A^2 = 12.2274$ and $s_B^2 = 28.9816$. Now we have

$$s_{\overline{A}-\overline{B}} = \sqrt{\frac{12.2274}{9} + \frac{28.9816}{9}}$$

$$= \sqrt{4.5788} = 2.14$$

$$\text{Hence } t = \frac{\overline{A}-\overline{B}}{s_{\overline{A}-\overline{B}}}\cdot\sqrt{\frac{n}{2}} = \frac{12.14-12.43}{2.14}\cdot\sqrt{\frac{10}{2}}$$

$= (-0.1355)\,(2.2360) = -0.30$, a non-significant value and so we conclude that there is no significant difference between A and B.

Again for (c) tests of significance of difference of means with heterogeneous variance, we are having two samples drawn from populations with the same means $\mu_1 = \mu_2$ but $\sigma_1 \neq \sigma_2$. In this case we work out the variance of the sampling differences, $s_{\bar{x}_1 - \bar{x}_2}$ and compute the t statistic as in testing the difference between two means:

$$t = \frac{\bar{x}_1 - \bar{x}_2}{s_{\bar{x}_1 - \bar{x}_2}}$$

where $s_{\bar{x}_1 - \bar{x}_2}$ is equal to $\sqrt{(s_{\bar{x}_1})^2 / n_1 + (s_{\bar{x}_2})^2 / n_2}$

Now for the chosen level of significance $\alpha = 0.05$ we compute the value of t' for both the samples as

$$t' = \frac{[(s_{\bar{x}_1})^2 / n_1]t_1 + [(s_{\bar{x}_2})^2 / n_2]t_2}{\dfrac{s_{\bar{x}_1}^2}{n_1} + \dfrac{s_{\bar{x}2}^2}{n_2}} \quad \text{... (11.13)}$$

where t_1 and t_2 are critical values of t at $d.f. = n_1 - 1$ for sample one and at $d.f. = n_2 - 1$ for sample two respectively.

Finally, in comparing t' with t, if $t' \geq t$, reject the null hypothesis but if $t \geq t'$, the null hypothesis is accepted. It can be observed that it is the high variation of the second sample which makes us to suspect that the two samples belong to populations with heterogeneous variances.

Example 11.6 Two samples of sizes 25 and 12 give $\bar{x}_1 = 1500$, $s^2_{\bar{x}_1} = 6000$ for the sample one and $\bar{x}_2 = 600$, $s^2_{\bar{x}_2} = 24000$ for the sample two. Now if these two sets of samples are drawn from populations with $\mu_1 = \mu_2$ but $\sigma_1 \neq \sigma_2$ we can compute

$$t = \frac{\bar{x}_1 - \bar{x}_2}{\sqrt{\dfrac{s^2_{\bar{x}_1}}{n_1} + \dfrac{s^2_{\bar{x}_2}}{n_2}}} = \frac{1500 - 600}{\sqrt{\dfrac{6000}{25} + \dfrac{24000}{12}}} = \frac{900}{\sqrt{2240}} = \frac{900}{47.33} = 19.02$$

Now for the chosen $\alpha = 0.05$, t_1 for $d.f. = 24$ is 2.06 and for $d.f. = 11$, t_2 is 2.20. So we compute

$$t' = \frac{\left(\dfrac{s^2_{\bar{x}_1}}{n_1}\right).t_1 + \left(\dfrac{s^2_{\bar{x}_2}}{n_2}\right).t_2}{\left(\dfrac{s_{\bar{x}_1^2}}{n_1} + \dfrac{s_{\bar{x}_2^2}}{n_2}\right)} = \frac{(240)\ (2.06) + (2000)\ (2.20)}{(240 + 2000)}$$

$$= \frac{494.4 + 4400}{2240} = 2.19$$

Since $t' < t$ so we cannot reject H_0 that the two samples are not drawn from populations with the same mean.

11.4 CONCLUSION

It should be noted that the t test is appropriate when the variance (or, standard deviation) of the population concerned is estimated from the set of values of the samples. And in this estimation, the samples should be independent. In certain cases, for analysing the treatment of rainfall data of two places or treatment of two sets of samples of agricultural plots, the interdependence of the values at the two places (like rainfall in the same season) is often ignored. But because of interdependence, there would be parallel changes (for example, of rainfall) at the two places and the differences will be far less variable than what they would have been if the places had been independent. Hence while applying the *t* test for finding the significance of difference between two sets of samples, one should bear in mind about the requirement of independence of the two samples and guard himself against an erroneous application of the test. In case there is any kind of correspondence between the individual values in the two samples, they should be paired and their differences taken for analysis directly. We illustrate this interdependence of data in the samples and a solution for treating the data independent from the point of view of testing the significance between the means of the samples.

Example 11.7 Following is the annual rainfall data (in cm) at two places A and B for 24 years in a sequence from the initial year to the 24th year

A: 39.59, 19.93, 23.91, 29.38, 43.09, 25.34, 49.35, 39.62, 42.90, 53.35, 57.66, 37.05, 34.14, 38.01, 52.40, 32.20, 47.81, 33.98, 39.46, 37.78, 63.24, 39.04, 60.51 and 38.08

B: 39.48, 17.81, 24.47, 24.32, 42.18, 23.41, 45.13, 42.83, 46.94, 51.51, 57.50, 34.35, 34.29, 38.65, 50.32, 29.94, 45.24, 34.13, 40.68, 35.54, 57.24, 42.05, 55.33 and 37.45

Assuming that the above sequential data for 24 years constitute a representative sample of the rainfall at the two places, can we consider the two places to have the same mean annual rainfall?

We treat the places A and B as independent samples of rainfall at two places. From the above data $\bar{x}_A = 40.74$, $\bar{x}_B = 39.57$; the sum of the squares of the differences, $\sum(x_A - \bar{x}_A)^2 = 2901.38$ and for B it is $\sum(x_B - \bar{x}_B)^2 = 2672.46$. Now from Eq. (11.9) the pooled estimate of standard deviation, $s_{\bar{x}'}$ equal to 11.01.

Again since the samples are paired, so from Eq. (11.12) we have

$$t = \frac{\bar{x}_1 - \bar{x}_2}{s_{\bar{x}'}} \cdot \sqrt{\frac{n}{2}} = \frac{40.74 - 39.57}{11.01} \cdot \sqrt{\frac{24}{2}} = \frac{4.05}{11.01} = 0.37$$

This calculated value of *t* with $d.f. = 46$ at $\alpha = 0.05$ is much lower than the critical value and hence we would conclude that there is no significant difference in the rainfall at places A and B.

But the above treatment of rainfall data may be erroneous because of 'spatial autocorrelation' of the places giving a probable interdependence of the rainfall at the two places in the same season. The appropriate method is to take the differences in the corresponding values of the variate and the differences being taken in the same direction with due regard to sign. This (A–B) set is as follows :

(A–B) or difference set: 0.11, 2.12, –0.56, 5.06, 0.91, 1.93, 4.22, –3.21, –4.04, 1.84, 0.16, 2.70, –0.15, –0.64, 2.08, 2.26, 2.57, –0.15, –1.22, 2.24, 6.00, –3.01, 5.18 and 0.63

Now we analyse these (A–B) set values as the values of a separate variate. If there is no difference in the mean rainfall at the two places, the expected sampling difference would be zero. We can calculate directly the standard deviation of the difference and test the significance of the deviation of the mean difference from the hypothetical value, zero.

From the above set of (A–B) = d values, the mean $\bar{d}$ is calculated as 1.13 (taking into consideration the sign of the differences in the (A–B) values) and the variance of the difference is $\frac{1}{n-1}\Sigma(d-\bar{d})^2$ where $d = A$–B, $\bar{d} = \bar{A} - \bar{B}$ and it is found to be 6.66 i.e. $s_d = 2.58$. Now since there are 24 differences, the estimate of their standard deviation is based on 23 *d.f.* for which at 0.05 significance level, the critical value of t is 2.07. Now for $n = 24$ the computed value of t is $t = \frac{\bar{d}}{s_{\bar{d}}}$ (where $s_{\bar{d}} = \frac{s_{\bar{d}}}{\sqrt{n}}$)

$$= \frac{1.21}{2.58}.\sqrt{24}$$

$$= 2.30$$

We find that this t value is larger than t_c so we conclude that the difference between the mean rainfall at the two places is significant and thus we obtain a result different from the previous one.

It is very rare for proportions and coefficient of variations to be based on small size samples, $n < 30$. So no test of significance for them is presented here as these are done for large size samples in the previous chapter.

LIST OF FORMULAE

I. *Unbiased Estimates of standard error of the means of samples happened to be small*

$$s_{\bar{x}} = \sqrt{\frac{\Sigma(x-\bar{x})^2/n}{n}} = \frac{s}{\sqrt{n}}$$

where s is the sample standard deviation = $\sqrt{\Sigma(x-\bar{x})^2/n}$

For best estimate of $s_{\bar{x}}$, $s_{\bar{x}} = s.\sqrt{\frac{n}{n-1}}$ or $\sqrt{\frac{\Sigma(x-\bar{x})^2}{n-1}}$

II. ***Student's t statistic:*** $t = \dfrac{\bar{x} - \mu}{s_{\bar{x}}}$

Interval estimate for small samples $\bar{x} = \mu \mp t_{\alpha} \cdot \dfrac{s}{\sqrt{n}}$

Point estimate of small samples, $t = \dfrac{\bar{x}_1 - \bar{x}_2}{C_v \cdot \sqrt{\dfrac{2}{n}}}$ **with** C_v **as coefficient of variation**

III. ***Table of test statistics and limits for a variety of different parametric tests for small samples***

Name of the test of significance	Test statistic	One-or two-tailed test	Limit(s)
(a) Population normal but sample size small (n < 30) and population variance unknown	t test and the test statistic $t = \dfrac{\bar{x} - \mu}{s_{\bar{x}}}$ with $d.f. = n - 1$ where $s_{\bar{x}} = \dfrac{s}{\sqrt{n}}$	Two	$d.f. = n - 1$
(b) Interval estimate for one population mean	$\mu = \bar{x} \pm t_{\alpha} \cdot \dfrac{s}{\sqrt{n}}$ where t_{α} is the t statistic at α significance level	Two	$d.f. = n - 1$
(c) Difference between sample mean and population mean	Same as in (a) above		
(d) Difference between two sample means with (i) population standard deviation or variances are not known but assumed to be equal	(i) $t = \dfrac{\bar{x}_1 - \bar{x}_2}{\sqrt{s_{\bar{x}'}^2 \left(\dfrac{1}{n_1} + \dfrac{1}{n_2}\right)}}$ where $s_{\bar{x}'}^2 = \dfrac{\Sigma(x_{1i} - \bar{x}_1)^2 + \Sigma(x_{2i} - \bar{x}_2)^2}{n_1 + n_2 - 2}$ or		

$$s_{\bar{x}}^{2}{}' = \frac{n_1 s_{\bar{x}_1}^2 + n_2 s_{\bar{x}_2}^2}{n_1 + n_2 - 2}$$

(ii) for paired samples or sample of equal sizes, $n_1 = n_2$ $\quad t = \dfrac{\bar{x}_1 - \bar{x}_2}{\sqrt{s_{\bar{x}}^2{}'}} \cdot \sqrt{\dfrac{n}{2}}$ with $s_{\bar{x}}^2{}'$ as in (i) above

(iii) for samples with $\mu_1 = \mu_2$ but variances not equal $\sigma_1^2 \neq \sigma_2^2$ $\quad t = \dfrac{\bar{x}_1 - \bar{x}_2}{\sqrt{w_1 + w_2}}$ with pooled estimate, $w_1 + w_2 = \dfrac{s_{\bar{x}_1}^2}{n_1} + \dfrac{s_{\bar{x}_2}^2}{n_2}$

and $t' = \dfrac{w_1 . t_1 + w_2 . t_2}{w_1 + w_2}$

For $t' > t$, reject H_0.

EXERCISES

11.1 From a lot of rain-gauge stations in a plain area, 16 stations were taken in sample and their mean annual rainfall was found to be 120 cm with a standard deviation of 10 cm. Find the 95% confidence limit of the annual rainfall in the area.

11.2 Given that x_i is normally distributed and $\bar{x} = 42$, $s_{\bar{x}} = 5$, $n = 20$, test the hypothesis that $\mu = 44$ using the t distribution.

11.3 The runoff observations for two soil samples: clay and sandy soils were examined. If the data for these two sets of soils are: $\bar{x}_1 = 40$ cm/years, $\bar{x}_2 = 35$ cm/year, $s_{\bar{x}_1} = 3.5$ cm/year, $s_{\bar{x}_2} = 3.3$ cm/year, $n_1 = 15$ and $n_2 = 12$, test the significance of differences between the two sets of soil samples.

11.4 Following is a paired set of samples:

x	10	8	7	6	6	5
y	8	7	7	6	5	3

Test the hypothesis that these 2 paired sets of samples are not different. Use $t = \dfrac{d}{s_{\bar{d}}}$ with $s_{\bar{d}} = \dfrac{\sigma_d}{\sqrt{n}}$ and

$$\sigma_d = \sqrt{\frac{\Sigma(d - \bar{d})^2}{n - 1}}, d = x_i - y_i$$

11.5 **In an exercise a standard is set at 5 unit as acceptable and 10 samples are taken : 9.0, 7.8, 9.7, 8.3, 10.3, 8.0, 10.0, 7.5, 10.5 and 11.5. The concern is whether the population exceeds the 5-unit standard at 0.01 significance level.**

11.6 **In a survey of productivity of wheat on 24 agricultural plots, 12 plots were selected as treated with fertiliser and the other 12 selected were untreated. Following is the summary data:**

	Mean	Std. deviation
For treated plots	5.2 kg	3.6 kg
For untreated plots	5.0 kg	3.9 kg

Find out the significance of the difference in productivity between the treated and untreated plots.

12

Parametric Statistics In Geo-science Systems III (Analysis of Variance)

12.1 COMPARISON OF MORE THAN TWO MEANS AND ANALYSIS OF VARIANCE

We have already done in the last two chapters some tests of comparing the means of two samples to decide if there are significant differences between them. When such a comparison is made, in many cases we do not want to be limited to two sets of data or two conditions. In this chapter we introduce the statistical technique which can help us in enlarging testing of problem to include multiple sets of data (that is three or more groups of specimens for sampling) for their statistical significance. This technique known as 'Analysis of Variance' (abbreviated as ANOVA) is basically an arithmetical procedure about the sources of variability in data by a purposeful partitioning of the total sum of squares of deviations from the mean into components associated with defined sources of variation. Thus if we wish to find out whether a variable x_1 is a function of variables x_2 and x_3 we can determine how x_1 varies for changes in x_2 or for changes in x_3 by traditional experiments of two sample confidence intervals and they are discussed earlier in which number of comparisons of each mean with every other would be three viz, x_1 and x_2, x_1 and x_3 and x_2 and x_3.

However the problem arises when the number of groups is larger say 5 or more. For example, if we have to compare means of ten groups ($k = 10$) the number of pairwise comparisons which would be required under the z or the t test would be $45 = \left(\frac{n(n-1)}{2}, n \text{ is } 10\right)$. In this large number of comparisons, the probability of making a type I error, (by rejecting the true hypothesis that two population means are equal), also becomes large. In a test with a 10 percent significance level four or five of the comparisons would on the average lead to rejecting the hypothesis, through chance alone, even though all were true. The second difficulty is that there may be too few observations in each statistical sample to yield good estimates of the population variance σ. Hence some method of testing difference between the means of all the k groups at the same time would prove very valuable. The analysis of variance and the corresponding test of significance based upon "F distribution" permit us to do this simultaneously.

Another important consideration which limits the z or the t test and warrants the use of the more sophisticated F test of significance, is the case of situations in which two or more experimental variables or one experimental and one or more central variables are simultaneously operating and not only comparison of means within each variable is required but also the joint operation or interaction of two or more varable is of interest.

Analysis of variance in its task of comparing the variances of k number of samples ($k \geq 2$) try to determine whether the samples pertain to the analysis of designed experiments. An "experiment design" is a procedure for gathering data systematically according to a formal plan carefully conceived and executed in the field for the purpose of evaluating different kinds of variability. Such designed experiments are common in agriculture, in geological mapping of a particular rock outcrop in k number of areas to select a number of samples, n for each k outcrop and then to measure a particular attribute as a key indicator for finding whether all the k rock outcrops can be assigned to a single formation. A more common example of designed experiment may be a herd of 16 cows are randomly partitioned into four different groups, each group is fed differently according to some predetermined criteria and the milk of the 16 cows are periodically compared to find out which of the four groups provides the highest yield.

There are many methods of *designed experiments* which are not discussed here. The analysis of variance is used first of all in comparing the variances of two samples to determine whether they are significantly different from one another. Secondly, it is used in the conventional statistical sense to describe tests which compare the means of several groups ($k \geq 2$) and/or the effect of two or more sources of variation on a given phenomenon. In this chapter, two tests of the second type are discussed after the "variance test" is introduced, namely "one-way" and "two-way" analysis of variance, the latter being a generalization of the former since in the latter more than one factor at a time is studied in one experiment.

12.2 *F* DISTRIBUTION

The sampling distribution of the analysis of variance is known as "F distribution", originally conceived by the noted British statistician Sir Ronald A. Fisher in 1923 and later on proposed by Professors Snecdor of USA and P.C. Mahlanobis of India, calling it F distribution after Fisher. When F test was first reported in 1923 its early applications were in the field of agriculture. Since then the F test found many fields of application in many areas of experimentation where we need to compare alternative treatments or materials or methods according to some predetermined criteria.

The analysis of variance as the name indicates, deals with variance rather than with standard deviations. and standard errors. Hence the *F* distribution is a family of theoretical frequency distributions for the ratio of two variances, calculated from independent random sample means drawn from two populations labelled 1 and 2 whose observations are normally distributed and whose variances are equal. Conceptually, this distribution is obtained as follows.

Variation is inherent in nature. The total variation in any set of numerical data is due to a number of causes which may be grouped into (i) assignable causes and (ii) chance causes; the variation for the former can be detected and measured whereas the latter cannot be traced separately. The ANOVA consists in the estimation of the amount of variation due to independent factors and then comparing these estimates with these two causes. From the first population, with variance σ_1^2 all possible samples of a given size n_1 are drawn and, for each sample the sample variance s_1^2 with (n_1-1) *d.f.* is calculated. Likewise, from the second population also with the same variance σ^2 all possible samples of a given size n_2 are drawn and for each sample the sample variance s_2^2 with (n_2-1) *d.f.* is calculated. Then, the *F* test is conducted by a ratio in which every sample variance s_1^2 is divided by every possible sample variance s_2^2 so that we have the *F* statistic:

$$F = \frac{s_1^2}{s_2^2} \qquad \text{... (12.1)}$$

The frequency distribution of all possible *F* ratios is the *F* distribution. Thus the *F distribution* is a theoretical frequency distribution for the ratio of two variances, calculated from independent random samples drawn from two populations whose observations are normally distributed and whose variances are equal. It seems reasonable that the sample variances will range more from trial to trial if the number of observations used in their calculation is small. Therefore the shape of the *F* distribution would be expected to change with changes in sample size. This returns us to the idea of *d.f.* and in this situation the *F* distribution is dependent upon two values of ν associated with each variance in the ratio. Also, it should be apparent that the *F* distribution cannot be negative because it is the ratio of the two positive numbers. If the samples are large the average of *F* ratio should be close to 1.0.

Appendix Table VII gives the *F* distribution values in various combinations of the number of *d.f.* upto infinity for 0.05 and for 0.01 significance levels. In Appendix Table VII, each column corresponds to a different number of *d.f.* in the numerator, shown as (ν_1) and each row corresponds to a different number of *d.f.* in the denominatior (ν_2). It can be noted that the table entries decrease downward in each column, because the average variance of the denominator s_2^2 decreases. Likewise, except for very small *d.f.* in the denominator, the table entries decrease toward the right with increasing *d.f.* in the numerator. For infinitely large number of *d.f.* in both numerator and denominator (∞ , ∞) the value of the *F* ratio is 1.0, recorded in the lowermost right hand corner entry, because both s_1^2 and s_2^2 become equal to the comman variance σ^2 for infinitely large samples. Except for this entry, which is the same for both the sub-tables of 0.05 and 0.01 significance levels, each entry gets larger in the higher significance level than in the lower. Now, suppose we wish to find out the tabulated value of $F_{0.05}$ (5,10). First, we look for the column headed 5 at the top of the *F* distribution tables and then we go down this column to read 3.33 at the row 10. Again, following the same rules, $F_{0.01}$ (5,10) is 5.64. It should be mentioned that this $F_{0.01}$ distribution table is asymmetric so that, for example, $F_{0.01}$ (3,5) = 12.06 but $F_{0.01}$ (5,3) = 28.24. The concept of the critical region in the *F* test is like that of the *t* test since in both the tests, the critical region is in the right-hand tail. Accordingly if the calculated value of *F* is found to be higher than the tabulated value of *F* at a significance level, it is concluded that there is a significant difference within the sample.

It can be seen that compared with the Student's *t* distribution values (Appendix Table VI) for $v_1 = 1$, the *F* values is the square of the corresponding *t* value (in both cases, assuming a two-tailed test). For example, for 4 *d.f.* (i.e. with *d.f.* = 4), the *t* value is 2.78 at 0.05 significance level but the corresponding *F* value is 7.71 which is square of the *t* value. Therefore, from this perspective, the *t* test can be viewed as the variance test in which there is always one *d.f.* associated with the numerator. Thus, from the Appendix tables VI and VII, we have the following examples:

Probability level	Value of *t* statistic	Value of *F* statistic
0.05	2.57 for *d.f.* = 5	$6.61 = (2.57)^2$ for $v_1 = 1.\ v_2 = 5$
0.05	2.13 for *d.f.* = 15	$4.54 = (2.13)^2$ for $v_1 = 1.\ v_2 = 15$
0.05	1.96 for *d.f.* = ∞	$3.84 = (1.96)^2$ for $v_1 = 1.\ v_2 = \infty$

So notionally we can write : $t_\alpha(n_2) = \sqrt{F_\alpha\ (1.n_2)}$ (12.2)

that is, if a variable *t* follows a *t distribution* with particular *d.f.*, the variable t^2 follows an *F* distribution with 1 as the *d.f.* in the numerator as well as the same particular number of *d.f.* in the denominator. Finally the fact that a *t* distribution with *d.f.* equal to infinity is the normal distribution. This shows that the entry in the *F* statistical table with 1 and infinity *d.f.* is the square of the corresponding entry in the table for normal distribution.

12.3 DIFFERENCE OF VARIANCE RATIO TEST

In any test of analysis of variance, the parameter under consideration is thus not the mean but the variance. The question asked is : could the sample variances have been drawn from the same population?

As mentioned earlier, for application of this variance ratio test, it requires four assumptions – (i) that the populations from which the samples are drawn are normally distributed (it is not required if the sample sizes are large), (ii) that the variances for the populations from which the sample have been drawn that is, the populations have a common variance, σ^2; this assumption is needed in order to combine or pool the variances within the groups into single 'within group' source of variation, (iii) samples are randomly selected and (iv) they are measured on interval scale. Once these requirements are satisfied, the significance of the 'difference of variance ratio test' can be applied to (a) two independent random samples as well as to (b) *k* number of independent random samples. In the '*k*' number of samples, individual sample or group sizes should not be much larger than another, otherwise there can be a biased interpretation of importance of the independent variable. We will be now discussing these two tests below separately.

12.3.1 Difference of Variance Ratio Test For Two Samples

Many times, we take two samples and we want to test the significance of the difference between the two variances for finding whether the two sample variances are from the same population. In this we compare the spreads of the two samples. The null hypothesis is that there is no significant difference between these two populations and both of the populations are equally variable. The test statistic is the ratio of the two sample variances, s_1^2 / s_2^2 is distributed as F statistic with $v_1 = (n_1-1)$ and $v_2 = (n_2-1)$ *d.f.* Though this is a very simple test, care should be taken that the larger variance is always placed in the numerator because F values are taken for values larger than 1.0 only. Thus guaranteeing that F values have to be larger than 1.0, we rewrite the Eq. (12.1) for F statistic as

$$F = \frac{\text{Larger variance}}{\text{Smaller variance}} = \frac{s_1^2}{s_2^2} \quad \text{... (12.3)}$$

and while consulting the F distribution value given in Appendix Table VII, we concern ourselves with one limit – the upper limit in the F distribution. A critical region is defined as before for a selected level of significance, and if the computed F value falls in this region, the null hypothesis is rejected. We give solution to a specific problem of comparison of two variances below. We are now testing the hypothesis:

$$H_0 : \sigma_1^2 = \sigma_2^2$$

against $$H_1 : \sigma_1^2 \neq \sigma_2^2$$

The null hypothesis states that the parent populations of the two samples have equal variances, the alternative hypothesis states that they do not.

Example 12.1 In studying the influence of lithology on steepness of slopes 100 samples are taken on each of two areas: one in a region A of shale and the other in a region B of interbedded sandstones and shales. If the variances of these two sample data for region A and region B are found to be 4.42° and 5.78°, test the hypothesis at 0.05 significance level that the difference in slope variability on the two rock types of the regions is statistically significant.

Placing the larger variance in the numerator of Eq. (12.3), we find $F = 5.78°/4.42° = 1.71$. This value is much more than the value of $F(99,99)$ at 0.05 significance level which is approximately $F \geq 1.39$. Hence since the computed statistic falls in this region, the null hypothesis is rejected and the conclusion is that the difference in slope variability on the two rock types is statistically significant.

12.3.2 Difference of Variance Ratio Test for *k* Independent Samples – Variance Decomposition and the Logic of Analysis of Variance

Upto this point we have only considered techniques for comparing two samples, yet many problems involve groups of observations. Hence we can have the extension of test of the significance of difference between two variances to the case of *k* number of independent samples, ($k > 2$ samples) which involves one of the more powerful forms of statistical analysis, the analysis of variance. In this analysis of variance of *k* number of independent samples, the total variance observed in the experimental data is partitioned into different components, each component is assignable to a known source, cause or factor. We may assess the relative magnitude of variation resulting from different sources and ascertain whether a particular part of variation was greater than expectation under null hypothesis. However, it must be remembered that analysis of variance divides the total sum of squares, $\Sigma(x_{ij} - \bar{x})^2$ into two additive parts, (i) a *within groups variation* which relates to the deviations between observations and their respective sample means and (ii) a *between groups variation* associated with the differences between the sample means and the total mean for all observations. Once these two additive parts are calculated, they are converted into "variance estimates" by *d.f.* Thus the main difference between *variance* and *variance estimate* is that the former is obtained by dividing the sum of squares by '*n*', the size of the samples while the latter is obtained by dividing the sum of squares by *d.f.*

The 'logic of analysis of variance' is the *partition of the total variance*. It is based on a very important assumption regarding how the individual *j*th observation is determined from the *i*th population. In the analysis of variance context, it is assumed that each individual observation represents the 'additive' effects of 3 factors which are

(1) a factor common or constant to all observations is due to their membership in a common population. For example, in a specific problem of assigning a single environmental background to the three maritime eco-systems: the open ocean, continental shelf and tidal estuaries, it was decided that the capacity of an eco-system to support life may be measured by its annual rate of production of organic matter and accordingly for these three eco-systems 10, 8 and 6 samples of net production of organic matter (g/m²/year) are taken (Table 12.1).

In Table 12.1 the problem then is to estimate the population mean μ_i in each of the three marine environments and the common population variance σ^2 and then to devise a null hypothesis '*H*' so that for each μ_i

$$\mu_1 = \mu_2 = \mu_3$$

and μ_1 'which is the first factor,' represents the mean of the population from which the observations are taken.

Table 12.1 : Observations of Net Production of Organic Matter (g/m²/year) in 3 Marine Environments

Open Ocean	Continental Shelf	Tidal Estuaries
108	279	1694
24	234	1278
138	504	624
158	348	529
165	198	3222
267	364	1503
147	602	
55	343	
149		
389		

(2) The second factor is unique to each individual observation. This factor is assumed to vary between all scores and may be thought of as error effect i.e. the effects of random sampling error, measurement error etc. The effect of this 'error factor' on any score is represented by e_{ij} where the *ij* subscript denotes the, '*i*th observation on the *j*th population.'

Now if any given *j*th observation from the *i*th population is represented as x_{ij} then we can write any given score x_{ij} in an additive form as

$$x_{ij} = \mu_i + e_{ij} \quad \text{... (12.4)}$$

where μ_i is the factor one representing the mean of the population from which the observation is taken and e_{ij} is the error factor in x_{ij} (e.g. the fourth observation from the third population, the tidal estuaries in the above data is $x_{34} = 529$). The assumption is made for all the random fluctuations e_{ij} that their mean is 0 and that their variance is equal to the common population variance σ^2 or $\mu_e = 0$ and $\sigma_e^2 = \sigma^2$

Now the expression for an observation, x_{ij} may be rewritten

$$x_{ij} = \mu + (\mu_i - \mu) + e_{ij}\,. \quad \text{... (12.5)}$$

where μ is the mean of the three population means μ_i values. If the μ_i values are all equal the terms in parentheses $(\mu_i - \mu)$ are all zero. If they are not equal, one measure of their difference is the variance of the population means, σ_μ^2 which is

$$\sigma_\mu^2 = \frac{(\mu_1 - \mu)^2 + (\mu_2 - \mu)^2 + (\mu_3 - \mu)^2}{2} \quad \text{... (12.6)}$$

where the number 2 in the denominator is equal to the number of means minus one i.e. k–1.

The hypothesis H_0 that

$\mu_1 = \mu_2 = \mu_3$

may be rewritten

$\sigma_\mu^2 = 0$

for, if the variance of population means is equal to zero the means must also be equal to each other. Now if σ_μ^2 is equal to zero, the average value of the variance of the three sample means in $\sigma^2/3$. If $\sigma_\mu^2 > 0$, because the three population means are unequal the average value of the variance of the three sample means must be larger than $\sigma^2/3$. In fact, this average value is equal to $\sigma^2/3 + \sigma_\mu^2$.

(3) The factors μ_i and e_{ij} are conceived as parameters whose values are not known directly. However, they may be estimated from sample data so we may substitute sample estimates for the various factors. For example, we can substitute the sample total mean $\bar{x}$ for the common population mean, u. We can substitute $(x_j - \bar{x})$ for the effect due to membership in category j. Similarly we can substitute $(x_{ij} - \bar{x}_j)$ for the effect due to the error factor, e_{ij}.

Thus in the analysis of variance of k number of independent samples, the total variance in a set of data is decomposed into the independent estimates of the population variance:

(a) the proportion of the variance estimate between the samples or groups known as *between group variance estimate* – this is the between group variation as estimated by how different the means of the various samples (groups) are from another, and

(b) the proportion of the variance estimate within the samples (groups) known as the *within group variance estimate* – this is the within group variation estimate as based on how different each of the scores in a given sample (group) is from the other scores in the same group.

Generally, the variance between the samples (groups) is greater than that within the samples (groups). If there is a considerable difference between these variances, there must be a specific cause for it.

To explain the procedure, let us assume k independent samples or, groups with sizes n_1, n_2, ..., n_k (*i.e.* the number of observations in the samples are $n_1, n_2, ..., n_k$). Let the mean of these samples or group be $\bar{x}_1, \bar{x}_2, ..., \bar{x}_k$ and the total mean of the samples or groups $\bar{x}_T$. The frequency distributions of these samples or groups will each have a normal character. Any one observation, say the ith observation of the jth sample or group, denoted by x_{ij} can be located with reference to the total mean of the samples or groups $\bar{x}_T$ and the mean for the jth sample or group $\bar{x}_j$. Thus x_{ij} can be written as $x_{ij} - \bar{x}_T$ and $x_{ij} - \bar{x}_j$. The mean for its sample or groups can similarly be located relative to the total mean $\bar{x}_T$, so that $\bar{x}_j - \bar{x}_T$ and the location of x_{ij} relative to $\bar{x}_T$ are then given by

$$x_{ij} - \bar{x}_T = (x_{ij} - \bar{x}_j) + (\bar{x}_j - \bar{x}_T) \quad ... (12.7)$$

Figure 12.1 illustrates the decomposition of an observation for sample x_i with respect to total variance. Squaring the two sides in Eq. 12.7, we get

$$(x_{ij} - \bar{x}_T)^2 = (x_{ij} - \bar{x}_j)^2 + (\bar{x}_j - \bar{x}_T)^2 + 2(x_{ij} - \bar{x}_j)(\bar{x}_j - \bar{x}_T) \quad ... (12.8)$$

If Eq. (12.8) is summed up for all observations in all samples, we get the total variation. Thus, we have

$$\Sigma_j^k \Sigma_i^n (x_{ij} - \bar{x}_T)^2 = \Sigma_j^k \Sigma_i^n (x_{ij} - \bar{x}_j)^2 + \Sigma_j^k \Sigma_i^n (\bar{x}_j - \bar{x}_T)^2 + \Sigma_j^k \Sigma_i^n 2(x_{ij} - \bar{x}_j)(\bar{x}_j - \bar{x}_T) \quad ... (12.9)$$

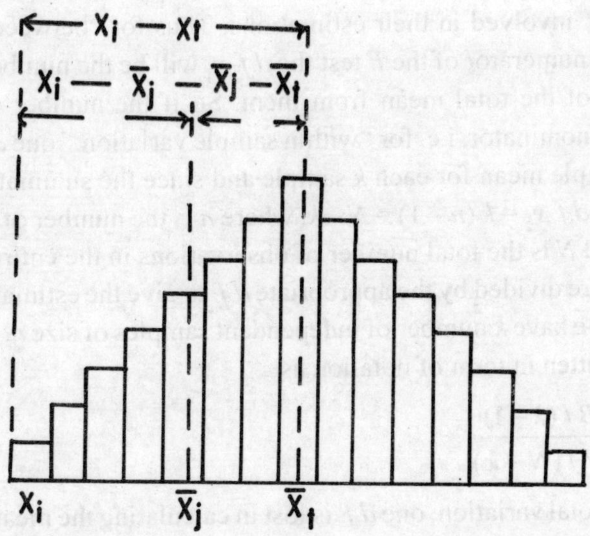

Fig. 12.1 : Decomposition of an Observation to Total Variance

Since $\sum_j^k \sum_i^n (x_{ij} - \bar{x}_j) = 0$ because the positive and negative deviations around the mean of the sample or group j cancel each other, the above expression for the total variation can be simplified as

$$\Sigma\Sigma(x_{ij} - \bar{x}_T)^2 = \Sigma\Sigma(x_{ij} - \bar{x}_j)^2 + \Sigma n_j(\bar{x}_j - x_T)^2 \quad \text{... (12.10)}$$

The total variation is thus the sum of the squared deviations of each observation from its sample or group mean and the squared deviations of the sample or group means from the total mean that is, total variance is the variation of all measurements about their total mean. So the total variation in Eq. (12.10) is written as

Total variation = Within groups variation + Between groups variation ... (12.11)

Hence we can write "within groups variation" as

$$\Sigma\Sigma(x_{ij} - \bar{x}_j)^2 = VW \quad \text{... (12.12)}$$

Similarly, the formula for the "between groups variation" as

$$\Sigma n_j(\bar{x}_j - \bar{x}_T)^2 = VB \quad \text{... (12.13)}$$

Note: when all the n_j categories are equal for the sample or groups

$$VB = n\Sigma(\bar{x}_j - \bar{x}_T)^2 \quad \text{... (12.14)}$$

Now, to compute an F test, the above variation components require to be standardized according to the *d.f.* involved in their estimations. Thus for "between sample variation," which will be in the numerator of the F test the *d.f.* v_1 will be the number of groups less one for the calculation of the total mean from them. So if the number of groups is k, then $v_1 = k - 1$. For the denominator, i.e. for "within sample variation," one *d.f.* will be lost in the calculation of a sample mean for each k sample and since the summation extends over all the samples, we get *d.f.* $v_2 = k(n-1) = N - k$, where n is the number of observations within a sample groups and N is the total number of observations in the entire data. Thus the two variance estimates are divided by the appropriate *d.f.* to give the estimates of the population variance. Hence if we have k number of independent samples of size n, then the formula for the F test can be written in term of notation as

$$F = \frac{VB/(k-1)}{VW/(N-k)} \qquad \text{... (12.15)}$$

Finally, for the total variation, one *d.f.* is lost in calculating the mean. Therefore, *d.f.* for the total variation is equal to $N - 1$.

All of the parameters needed to find this difference of variance ratio test for k independent samples have now been introduced and we can test the data given in the Table 12.1 which is reproduced below along with the computations.

Table 12.2 : Test of the Difference of Variance Ratio for 3 marine environments

Number of samples	Observations of net production of organic matter (g/m²/year)		
	Open ocean (1)	Continental shelf (2)	Tidal estuaries (3)
1	108	279	1694
2	24	234	1278
3	138	504	624
4	158	348	529
5	165	198	3222
6	267	364	1503
7	147	602	
8	55	343	
9	149		
10	389		
Total Σx_i	1600	2872	8850
No. of observations. n	10	8	6
Mean $\bar{x}_i$	160	359	1475

Total mean $\bar{x}_T = 555.08$

Now, the Total variation = Total sum of squares

$$= \sum_{n}^{10}(x_1 - 555.08)^2 + \sum_{n}^{8}(x_2 - 555.08)^2 + \sum_{n}^{6}(x_3 - 555.08)^2$$

$= 1657800.1 + 436000.93 + 9836196.8 = 11929998$ with *d.f.* for total variation $= N - 1 = 24 - 1 = 23$

Again variation within samples $= \sum_{n}^{10}(x_i - 160)^2 + \sum_{n}^{8}(x_i - 359)^2$

$$+ \sum_{n}^{6}(x_i - 1475)^2 = 96918 + 128422 + 4758680 = 4984020$$

and *d.f.* for variation within samples = (10 – 1)+(8 – 1)+(6 – 1) = 21. Finally, variation between samples

$$= \sum_{n}^{10}(\bar{x}_1 - 555.08)^2 + \sum_{n}^{8}(\bar{x}_2 - 555.08)^2 = \sum_{n}^{6}(\bar{x}_3 - 555.08)^2$$

$$= 10\,(160 - 555.08)^2 + 8\,(359 - 555.08)^2 + 6(1475 - 555.08)^2$$

$$= 1560882.1 + 307578.93 + 5077516.8 = 6945977.8$$

with *d.f.* for variation between samples $= k - 1 = 3 - 1 = 2$

$$So, F = \frac{\text{Between sample variance}}{\text{Within sample variance}} = \frac{6945977.8 / 2}{4984020 / 21}$$

$$= \frac{3472988.9}{237334.29} = 14.63$$

Now the critical region for F with $v_1 = 2$ and $v_2 = 21$ at $\alpha = 0.05$ is 3.47. Since the calculated F value is 14.63 which is higher than the critical value so we reject the null hypothesis and conclude that the net production of organic matter of the three marine environments are significantly different.

Example 12.2 In a traffic survey, the number of commuters using two suburban stations of a city in the morning as well as in the evening were recorded and the following are the mean and variance of the sample means:

Station A

Morning period: Mean 1200 and variance 600

Evening period : Mean 700 and variance 750

Station B

Morning period : Mean 900 and variance 950

Evening period : Mean 1300 and variance 650

Test whether the difference in the number of commuters are due to the difference in the locations of the stations. The number of commuters recorded in the survey was for 30 days in each sample.

The above test is regarding that of homogeneity in the means. For testing this hypothesis, we take the morning and evening periods of stations A and B as 4 periods and test whether there is any difference between these periods.

The total number of observations, $N = 30 \times 4 = 120$.

The grand mean of these periods $\bar{x}_T$ is

$$\bar{x}_T = \frac{1200 + 700 + 900 + 1300}{4} = 1025$$

Since the means for each period given in the data is already based on 30 observations, from Eq. (12.14) variations between the periods, *VB* is computed as

$$\begin{aligned} VB &= 30\,[(1200-1025)^2 + (700-1025)^2 + (900-025)^2 + (1300-1025)^2] \\ &= 30\,(30625 + 105625 + 15625 + 75625) \\ &= 6825000 \end{aligned}$$

Finally from Eq. (12.12) the variations within the periods, *VW* is given by

$$VW = \sum_{j=1}^{4} \sum_{i=1}^{30} (x_{ij} - \bar{x}_j)^2$$

But since the data provides us the variance, $\sum_{i=1}^{30} (x_{ij} - \bar{x}_j)^2$ for $j = 1, 2, 3$ and 4, the Eq. (12.12) will be obtained by multiplying 30 with the variance of the *j*th period. Thus

$$VW = 30 \times 600 + 30 \times 750 + 30 \times 950 + 30 \times 650 = 88500$$

Now to compute the *F* test, the above variation components are required to be standardised according to the *d.f.* involved in this estimation. For *VB*, the *d.f.* will be the number of sample groups less one i.e. 3 and for *VW*, the *d.f.* is the total number of *d.f.* in the entire data minus the *d.f.* for *VB* i.e. 119 – 3 equal to 116.

The above calculations and the *F* statistic are given below in tabular from.

Çomponents	Variations	*d.f.*	Estimates of variance	*F* ratio	*F* limit at 5%
Between periods	6825000	3	2275000	2981.6	2.68
Within periods	88500	116	763		

Since the calculated value far exceeds the critical value at $F_{0.05}$ (3,116), one can conclude that there are significant differences between the sample means.

12.4 RANDOMIZATION AND ONE-WAY ANOVA

Now that the *F* distribution which is essential for testing the sample variance ratio for $k \geq 2$ independent samples has been introduced, the one-way ANOVA test can be presented. In any designed experiment in agriculture, it is necessary for an objective comparison between experimental treatments, condition of the random allocation of treatments to various plots is required to be fulfilled. This is 'randomization' and when this is applied by replication to a set of randomly arranged plots, the desire on part of the researcher is to average out as far as possible the effects of environmental differences so as to give the various treatments equal scope to show their merit because a lower environmental (or, local) error means that a smaller real difference between treatments can be detected to be significant. In the present example the designed experiment involves the testing of plot yield variation in a single factor, difference in yield in blocks. So we can term the analysis of variance calculations as the one-way analysis of variance (or, 'one-way ANOVA') since only one factor or variable of classification is considered for discrimination that is, scores on a dependent variable x_{ij} are grouped by sub-sample categories and then it is observed that the reason this one factor may be important is that several possible levels or types of treatments can occur within that factor. Hence the four categories or treatments, each with a sample size of three given in Table 12.3 is an illustration of the one-way analysis of variance or ANOVA.

Now testing the one-way ANOVA in Table 12.3, we have to prove that the overall classification does not indicate significant difference from one block to another at a particular confidence level. This one-way ANOVA can be computed by the variance ratio test discussed earlier. But the computation for within sample and between sample variations seem to be tedious and hence in solving the above one-way ANOVA we like to deduce an alternative computational method, known as the *deviation score method*. In this method, the total sum of squares of the plot yields from the general mean, SST, based on $d.f. = t.r - 1$ (with block as t and number of plots as r) can be partitioned into (one) the sum of squares of the plot based on $n_j - 1$ *d.f.* due to block (statistically taken as treatment) differences and (two) sum of plot yields, within blocks based on $(n_j - 1)\,(n_i - 1) = N - k$ *d.f.* due to inherent variation within the similarly treated plots in different blocks. The first type of variation is due to assignable causes which can be detected and controlled by human endeavour and the second type of variation is due to chance causes which are inherent and not controllable.

Now we give below the computation by the deviation score method.

Computational Procedures

Step 1 : The categories are listed in four columns here, add up each column and then add the column sums to give the total sum $= \sum_{j}^{k}\sum_{i}^{n} x_{ij}$ where x_{ij} is a value in the *i* row and *j* column.

Step 2 : Square the total sum and divide it by *N*, the total number of samples to get the mean of the square of the total sum. This is assumed as correction factor, *C.F.*

$$\frac{\left(\sum_{j}^{k}\sum_{i}^{n} x_{ij}\right)^2}{N} = C.F. \quad \text{... (12.16)}$$

Step 3 : Square all the values in the columns, sum each column and finally sum all the column squares. From this subtract the mean of the square of the total sum. This is the "sum of total variance" *SST**, given by

$$SST = \sum_{j}^{k}\sum_{i}^{n}(x_{ij} - \bar{x}_{ij})^2 = \left(\sum_{j}^{k}\sum_{i}^{n} x_{ij}^2\right) - C.F. \quad \text{... (12.17)}$$

where $\bar{x}_{ij}$ is the total mean

Step 4 : Next we find the between group variation. It is measured by sum of the squared differences between the sample mean of each group $\bar{x}_i$ and the total mean $\bar{x}_{ij}$, weighted by the sampled size n_i in each group. The between group variation which is usually called the "Sums of Squares of variance", *SSB* is computed as:

---

* From Eq. (3.5) we know

Variance $\sigma^2 = (\Sigma x_i^2 / N - \bar{x}^2)$

The sum of the sequences is given by

Variance $\times N = \Sigma x_i^2 - N\bar{x}^2 = \Sigma x_i^2 - N \cdot \frac{\Sigma x_i}{N} \cdot \frac{\Sigma x_i}{N} = \Sigma x_i^2 - \frac{(\Sigma x_i)^2}{N}$

$$SSB = \sum_j^k \sum_i^n (\bar{x}_j - \bar{x}_{ij})^2 = \sum_j^k \frac{\left(\sum_i^n x_{ij}\right)^2}{n_j} - C.F. \qquad \text{... (12.18)}$$

Step 5 : Since the sum of squares of total variance, *SST* is equal to "between sum of squares" and "within sum of squares" [as given in Eq. (12.11)], the latter can be calculated by subtracting *SSB* from *SST*

$$SSW = SST - SSB \qquad \text{... (12.19)}$$

or, from Eq. (12.12) as

$$SSW = \sum_j^k \sum_i^n (x_{ij} - \bar{x}_j)^2$$

$$= \sum_j^k \sum_i^n \left(\sum x_j^2 - \frac{(\sum x_j)^2}{n_j} \right) = \sum_j^k \sum_i^n x_{ij}^2 - \sum_j^k \frac{\left(\sum_i^n x_{ij}\right)^2}{n_j} \qquad \text{... (12.20)}$$

Note that the first term in the Eq. (12.20) is the first term of Eq. (12.17) for *SST* and the last term is the first term in the Eq. (12.18) for *SSB*, therefore in Eq. (12.19) for writing *SSW* as equal to *SST–SSB* is justified.

Finally, following is the model of one-way analysis of variance using deviation score method :

			Categories or columns for the variable					Total by
			1	2	3	.....	k	rows
Obser-	a	Replic-	x_{a1}	x_{a2}	x_{a3}	....	x_{ak}	...
vations	b	ates	x_{b1}	x_{b2}	x_{b3}	....	x_{bk}	...
	⋮	⋮	⋮	⋮	⋮	⋮	⋮	⋮
	i		x_{n1}	x_{n2}	x_{n3}	.....	x_{nk}	....
Column total			...	...	...		...	Grand total $\sum_i^n \sum_j^k x_{ij}$

Degrees of Freedom and Critical Region

Degrees of Freedom is the same as mentioned in logic of analysis of variance ratio test given earlier.

Under the H_0 that three or more category means are equal that is $\mu_i = \mu_2 = \mu_3 \ldots = \mu_k$, we would expect the category means that are computed from sample data should be approximately the same. In reality, if there are differences, for H_0 to be true, the between category sum of squares should be relatively small reflecting only sampling error. The within category sum of squares, on other hand is sensitive only to sampling error, no matter whether the H_0 is true or false but the between category sum of squares is sensitive to differences in means as well as sampling error. So when H_0 is false the sample category means differ widely than the relatively small differences, expected from sampling error alone so that the F test ratio becomes large.

Computing the Test Statistic

The steps for computation outlined above is worked out now for the one-way ANOVA example given below :

Treating the yield of a crop as an index of quality rating of agricultural regions, three randomized plots were taken in four blocks I, II, III and IV each and the following data were collected :

Table 12.3 : Quality Rating of Blocks

I	II	III	IV
99	70	90	99
96	65	80	96
95	60	45	85

Test the above data in the one-way ANOVA method and find out the significance whether the quality rating values differ from block to block.

Steps 1 and 2 :

Category 1	Category 2	Category 3	Category 4
$\Sigma x_1 = 290$	$\Sigma x_2 = 195$	$\Sigma x_3 = 215$	$\Sigma x_4 = 280$

The correction factor, $C.F. = \dfrac{\left(\sum_j^k \sum_i^n x_{ij}\right)^2}{N} = \dfrac{(980)^2}{12} = 80033.3$

Step 3:

$\sum x_1^2 = 28042,\ \sum x_2^2 = 12725,\ \sum x_3^2 = 16525\ \sum x_4^2 = 26242$

Now, sum of squares of total variance, *SST*

$$= \sum_j^k \sum_i^n x_{ij}^2 - C.F. = 83534 - 80033.3 = 3500.7 \approx 3501$$

Step 4:

$$SSB = \sum_j^k \left(\frac{\left(\sum_i^n x_{ij}\right)^2}{n_j}\right) - C.F. = \left(\frac{290^2}{3} + \frac{195^2}{3} + \frac{215^2}{3} + \frac{280^2}{3}\right) - 80033.3$$

Step 5 :

$SSW = SST - SSB = 3500.7 - 2216.6 = 1284.1$

or

$$SSW = \sum_i^n \left[\sum_j x_j^2 - \frac{\left(\sum_j x_j\right)^2}{n_j}\right] = \left(28042 - \frac{84100}{3}\right) +$$

$$\left(12725 - \frac{38025}{3}\right) + \left(16525 - \frac{46225}{3}\right) + \left(26242 - \frac{78400}{3}\right)$$

$$= 8.67 + 50.00 + 1116.67 + 108.67 = 1284.1 \approx 1284$$

Finally the computation of the F statistic for this one-way analysis of variance is given in Table 12.4.

Table 12.4 : Testing in One-way Analysis of Variance

Sources of variation	Sum of Squares	Degrees of freedom	Estimates of Variance of sum of square	F Statistic
SST, Sum of Squares of Total	3501	$N - 1 = 11$		
SSB, Sum of Squares of Between	2217	$k - 1 = v_1 = 3$	$SSB/v_1 = s_B^2 = 739.0$	$S_B^2 / S_W^2 = 4.60$
SSW, Sum of Squares of Within	1284	$N - k = v_2 = 8$	$SSW/v_2 = S_W^2 = 160.5$	

Decision

The decision rule is to reject the null hypotheses of no difference between the categories of blocks if at the α level of significance

$$F = \frac{\text{Between group variance estimate}, S_B^2}{\text{Within group variance estimate}, S_W^2} \quad \text{... (12.21)}$$

that is, the critical value of the F distribution with $k-1 = v_1$ and $N-k = v_2$ $d.f.$ at a particular significance level is larger than the computed F value.

Returning to the data of Table 12.4, since the F statistic is 4.60 which is less than the critical value of F at 0.01 level of significance, equal to 7.6 so at the higher level, the quality rating values of the blocks are not significant. However, it should be noted that if a significance level of 0.05 is selected, the critical value is 4.07 which is less than the calculated value of 4.60. Hence at the lower level the quality rating values differ significantly from block to block.

In the above example all the samples are of the same size but the samples can be different as in the example on variance ratio test for three marine environments in Table 12.2. In this case, the analysis is carried out in the same way as before except that the difference in the size of the samples should be noted when the sums of squares between sample and within sample variances are calculated. The one-way analysis of variance is appropriate if we are able to select replicates randomly within the samples and make measurements in random order so that we can test the hypotheses that the populations are identical.

Example 12.3 Following is the data on wheat yields (in units) for 3 soil types:

Sandy Soil	Clay Soil	Loamy Soil
18	21	21
20	23	17
16	14	24
21	24	23
22	21	25
18	20	22
19	19	23
17	21	26
21	23	23

Test the significance if any between the wheat yields on three different soil types.

In this example the designed experiment involves testing the plot yield variation in a single factor difference in wheat yield in three different soil types and computations for one-way ANOVA leading to F statistic is carried out in the following table using deviation score method:

Table 12.5 : One-Way ANOVA

Compoment	Source of Variation	Sum of Squares	Degrees of freedom	Estimate of variance	F ratio observed	F ratio for $\alpha = 0.05$
SST	Sum of Squares of Total	214.07	26			
SSB	Sum of Squares of Between	58.59	2	29.30	4.49	3.40
SSW	Sum of Squares of Within	156.78	24	6.53		

Since the critical value of $F_{(2,24)}$ at 0.05 significance level is 3.40 so we conclude that there is a significant difference in wheat yields among the three soil types.

12.5 RANDOMIZED BLOCKS AND TWO-WAY ANOVA : SINGLE OBSERVATION

An analysis of variance as mentioned earlier basically pertains to the analysis of designed experiments in which data are collected systematically according to a formal plan that is carefully conceived and executed. Such designed experiments are common in agricultural researches of yield of new varieties, seed treatment, cultivation practises etc. The purpose of designed experiments is to evaluate different kinds of variability, for which the field on which the trial (experiment) is to be carried out is divided into as many 'blocks' of the same size and shape as there are to be 'replications' (repetition of the treatments under experiment) and each of the blocks into as many plots of the same size and shape as there are treatments. If there are '*t*' treatment and '*r*' replications there will be '*tr*' plots altogether. The treatments are allocated randomly to the '*t*' plots in the '*r*' blocks. The plot yields obtained from the '*tr*' plots furnish the data for the comparison of treatments by using the analysis of variation technique.

Randomized block is thus a statistically designed scientific investigation through which the experimental material is first of all arranged in homogeneous groups named *blocks* to which experimental treatments are applied randomly. The homogenized blocks make the within block variability small, the among blocks variability large and the randomized block variability to be removed from the statistical analysis. In simplest case randomized block procedure for reducing unexplained variability may be for two treatments to which we have applied *t test* in the earlier chapter. But generally the randomized block is for more than two treatments.

The general principle of the randomized block design may best be explained with the aid of an illustration. Suppose that the two improved strains of a cereal numbered 1 and 2 developed by an agricultural university were put under a yield trial in randomized blocks against the existing strain which formed strain number 3. There were four replications. The plot sizes were uniform, 3.50m × 1.80m.

Figure 12.2 gives the plan of experiment, the strains sown and yield obtained from each experimental plot in 10 gm units. In this Fig. 12.2 summation proceeds down the strains to find sample totals and means and also across samples to find the totals and means for the strains.

Strains	Block I	Strains	Block II
3	95	1	70
2	96	2	65
1	99	3	60
2	80	3	85
3	45	1	99
1	90	2	96
	Block III		Block IV

Fig. 12.2 : Plan of Randomized Block Trial with Strains of a Cereal.

The randomized block design above yield 12 (=3 × 4) observations. For the purpose of analysis, it is necessary now to tabulate the yield figures according to strains and block in the following manner:

Table 12.6 Plot yields in a Randomized Block Design

	Samples				
	Quality Rating of Blocks				
(Factor I)	(Factor II)				
Strain	I	II	III	IV	Strain total
1	99	70	90	99	= 358
2	96	65	80	96	= 337
3	95	60	45	85	= 285
$N = 12$, Block Total Σx_j	290	195	215	280	$\Sigma\Sigma x_{ij} = 980$

We find we look at the same data as before (i.e. Table 12.3) except that the three readings were not taken under the same identical conditions. The three inputs we add this time are the two independent varieties of improved strains and one existing strain. Thus we can assume that there can be another classification in addition to variation due to the agricultural regions as *Factor II* and variation between the strains (of the cereal) as *Factor I*. These two factors do not have independent or, what is technically termed, *additive effects* but interact with each other or are independent in their effects. The only effective approach for answering these questions lie in investigating the effect of the two factors together by comparing in one and the same experiment all the possible combinations of the level of both the factors.. This approach is known as the *factorial approach* of experimentation. The present example is known as two-way ANOVA because of two sources of independent variations, agricultural inputs as factor I and agricultural relationships as factor II as their relationship to a dependent variable, the *quality ratings*. In this factorial design the first factor has three levels and the second factor has four levels and this example can be known as 3 × 4 factorial design. Fig. 12.3 illustrates the pattern of summation for both one-and two-way ANOVAs.

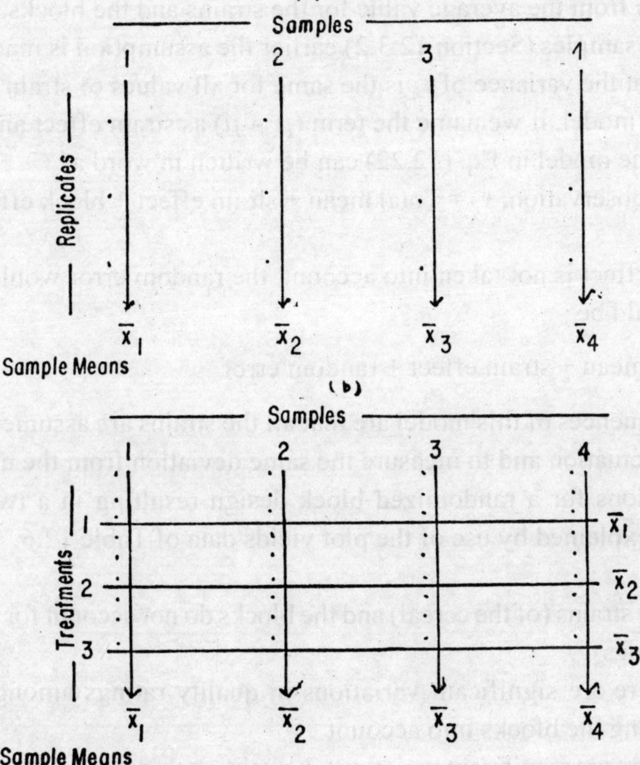

Fig. 12.3 : Pattern of Summation in ANOVA

(a) Summation proceeds down replicates to find sample means, $\bar{x}_i$: One-way ANOVA (b) Summation proceeds down treatments to find sample means, $\bar{x}_i$ and also across samples to find treatment means, $\bar{x}_j$: Two-way ANOVA

The last column of Table 12.6 gives strain totals while the last row gives the block totals. The grand total of all the 12 plots is obtained from both the strain total as well as the block total and it is found to be 980 (the double procedure serving as a check). Now the plot yields (as observations) are arranged to be as homogeneous as possible within the blocks, although the different blocks may be very heterogeneous. These each of the three strains (including the two improved ones) as *treatment* were randomly applied within each block. As the within-block variability is small, the among-blocks variability may be large, and the randomized block design allows the effect of the among-block variability to be removed from the statistical analysis.

Before solutions to this randomized block design are discussed, a mathematical model is given here. The model for a plot yields observation x_{ij} is

$$x_{ij} = \mu + (\mu_s - \mu) + (\mu_b - \mu) + e_{ij} \qquad \text{... (12.22)}$$

where μ the mean of all observations, μ_s is the mean of the observation of the strains, μ_b is the mean of observations of the blocks and e_{ij} is the random fluctuation due to the deviation of an observation from the average value for the strains and the blocks. As in the variance ratio model for k samples (Section 12.3.2) earlier the assumption is made that the mean of e_{ij} is zero and that the variance of e_{ij} is the same for all values of strain and block.

In this linear model, if we name the term $(\mu_s - \mu)$ as strain effect and the term $(\mu_b - \mu)$ as block effect, the model in Eq. (12.22) can be written in word as :

A plot yield observation, x_{ij} = Total mean + strain effect + block effect + random error ... (12.23)

If the block effect is not taken into account, the random error would be larger and the linear model would be

x_{ij} = general mean + strain effect + random error ... (12.24)

Other consequences of this model are that all the strains are assumed to have about the same random fluctuation and to measure the same deviation from the mean of the blocks.

The calculations for a randomized block design resulting in a two-way analysis of variance is now explained by use of the plot yields data of Table 12.6.

H_0 = The strains (of the cereal) and the blocks do not account for variation in quality ratings

H_1 = There are significant variations in quality ratings among the three strains, taking the blocks into account

H_2 = There are significant variations in quality ratings among the four blocks, taking the three strains into account.

Significance level : 0.05 or α = 5 percent and two-way analysis of variance test is applied.

Computing test statistic

Step 1 : The categories are listed in four columns; each column and column sums are added to give the total sum = $\Sigma\Sigma x_{ij}$, the rows are also added to give the sum of each row.

Step 2 : We calculate the mean of the squares of total sum which is known as the correction factor *C.F.* following the same method as in one-way analysis [Eq. (12.16)].

Step 3 : The sum of squares of total variance, *SST* is also calculated in the same way as in Eq. (12.17) in one-way analysis.

Step 4 : The sum of squares of between variance, *SSB* is also calculated as in Eq. (12.18) in one-way analysis. But since the categories in this two-way analysis is taken as factor II, *SSB* is known here as the sum of squares for factor II, SS_2 .

Step 5 : Since in the present analysis we do not have repeated values or replications such as in the earlier example of one-way analysis, we cannot compute the 'sum of squares (of variance) within samples,' *SSW* as in the one-way analysis.

Thus before calculating anything else we now require to calculate the two sum of squares, the sum of squares for factor I, SS_1 and the sum of squares for interaction, *SSI*.

The sum of squares for factor I, SS_1 is computed as

$$SS_1 = \sum_{i}^{n} \frac{(\text{Sum of each row})^2}{\text{Total number of samples in each row}} - \text{Mean of the squares of total sum}$$

$$= \sum_{i=1}^{n}\left(\frac{(\Sigma x_1)^2}{n_1} + \frac{(\Sigma x_2)^2}{n_2} + \frac{(\Sigma x_3)^2}{n_3}\right) - \text{correction factor} \quad \text{... (12.25)}$$

Step 6: Now, the sum of squares for interaction, *SSI* is a part of variability of the levels in the two factors and is obtained by subtracting the sum of SS_1 and SS_2 from *SST*, i.e.

$$SSI = SST - (SS_1 + SS_2) \quad \text{... (12.26)}$$

This sum of squares for interaction, *SSI* is the new second type of variation. This sum of squares is better known as "residual" or *error variation* since it is deduced by subtracting SS_1 and SS_2 from *SST*, as mentioned earlier. Table 12.7 below illustrates the above two-way analysis of variance without replications by its sum of squares and the computation leading to the *F* statistic itself.

Table 12.7 : Two-way Analysis of Variance : Single Observation

Components	Sources of variation	Sum of squares	Degrees of freedom	Estimate of variance of sum of squares (or, Mean squares)	*F* ratio	5% of *F* limit
SST	Total sum	3500.7	$N-1=11$			
SS_2	Factor II (Between columns)	2216.6	$n_j-1=3$	738.9	$\frac{738.9}{96.2}=$ 7.68*	at (2.6) = 4.76
SS_1	Factor I (Between) rows	706.5	$n_i-1=2$	353.3	$\frac{353.3}{96.2}=$ 3.67	at (2.6) = 5.14
SSI	Residual or error	577.6	$(n_i-1).(n_j-1)=6$	96.2		

* Significant

To determine if these two-way classifications are significant, the *F* values are computed by estimating the variances as usual in one-way analysis but in two-way analysis the sum of squares of total variance, *SST* is paritioned into three components instead of two components in one-way analysis. These are : (i) The sum of squares of variance between the columns of factor II, SS_2 based on (n_j-1) *d.f.*, (2) the sum of squares of variance between the rows of factor I based on (n_i-1) *d.f.* and (3) the residual or error sum of squares *SSI* based on $(n_i-1)(n_j-1)$ [equal to $N-k$] *d.f.* due to variation between similarly treated plots in different factor II columns representing the uncontrolled variation affecting the treatments. Hence, in the Table 12.7 after obtaining the estimate of variance for each of these three sum of squares, mean squares are obtained for the factors I and II by dividing their respective estimates by the estimate of the error sum of squares. These mean of the square values for both factors I and II are calculated and compared with their corresponding critical values at 0.05 significance level in Table 12.7 to conclude that there are significant variation in quality ratings among the four blocks because the variance ratio for factor II is significant at five percent for the blocks.

This above two-way analysis of variance of rating of quality agricultural blocks in a randomzied plot yields trial with strains of a cereal which can be conducted as two simple one-way experiment providing information about the strains and the blocks. Thus more experiments would be needed to achieve the same degree of accuracy as that obtained by a two-way analysis. Hence two-way analysis is more economical than one-way analysis.

Secondly two-way analysis enables us to study the interaction of the factors. Some varieties of strains of the cereal (as agricultural inputs) in the above example may increase the quality rating of some blocks but decrease it with others. Third, the conclusions reached have broader applications. This is due to the fact that the behaviour of each factor is studied with varying combinations of other factors. Thus the results in two-way analysis or factorial design are more useful than those obtained by holding all other factors constant in one-way analysis. Below we give one more example of the two-way ANOVA. Finally we should say that in two-way ANOVA, we introduce the concept of comparing between samples.

Example 12.4 In the following a random sample of five properties were assessed between three regions and the numerical scores are:

No. of Properties	Region 1	Region 2	Region 3
1	78	82	79
2	102	102	99
3	68	74	70
4	83	88	86
5	95	99	92

Assess the significance of difference between the regions.

This is a two-way ANOVA in which it is necessary to tabulate the numerical scores according to properties and the regions which are done in the tabulation given below :

No. of Properties (Factor I)	Number of regions (Factor II)			Σx_i	$\bar{x}_i$
1	78	82	79	239	79.67
2	102	102	99	303	101.00
3	68	74	70	212	70.67
4	83	88	86	257	85.67
5	95	99	92	286	95.33
Σx_j	426	445	426	$\Sigma x_{ij} = 1297, N = 15$	
$\bar{x}_j$	85.2	89.0	85.2	$\bar{x}_{ij} = 86.47$	

We assume : $H_0 : \mu_1 = \mu_2 = \mu_3$

H_1 : not all the properties are equal

H_2 : not all the regions are equal

If j denotes the columns, i the rows and N the total number of observations the compilations for the sum of squares of variance are as follows:

$$SST = \sum_{j}^{k}\sum_{i}^{n}(x_{ij} - \bar{x}_{ij})^2 = \sum_{j}^{k}\sum_{i}^{n} x_{ij}^2 - \frac{\left(\sum_{j}^{k}\sum^{n} x_{ij}\right)^2}{N}$$

where the second term is known as the correction factor

$$=113977.00 - 112147.27 = 1829.73$$

Now, $SS_2 = \sum_{j}^{k}\sum_{i}^{n}(\bar{x}_j - \bar{x}_{ij})^2 = \frac{\sum_{j}^{k}\left(\sum_{i}^{n} x_{ij}\right)^2}{n_j} - C.I$

$$= \left(\frac{426^2}{5} + \frac{445^2}{5} + \frac{426^2}{5}\right) - 112147.27 = 112195.40 - 112147.27 = 48.13$$

$$SS_1 = \sum_{i=1}^{n}(\bar{x}_i - \bar{x}_{ij})^2 = \left(\frac{239^2}{3} + \frac{303^2}{3} + \frac{212^2}{3} + \frac{257^2}{3} + \frac{286^2}{3}\right) - 112147.27$$

$$= 113906.33 - 112147.27 = 1759.06$$

Finally, $SSI = SST - SS_2 - SS_1 = 1829.73 - 48.13 - 1759.06 = 22.54$

We tabulate all these values in the following table to calculate the F statistic.

Component	Source of variation	Sum of squares	Degrees of freedom	Estimates for sum of squares	F ratio	5% of limit
SST	Total sum	1829.73	14	130.70		
SS_2	Between Regions	48.13	2	24.07	8.54* (2,8)	4.46
SS_1	Between Properties	1759.06	4	439.77	155.95* (4,8)	3.84
SSI	Interaction between Properties and Regions	22.54	$(k-1)(i-1) = 8$	2.82		

The two asterisks indicate that there are significant differences in interaction between the properties among the regions and hence we can conclude that the classification into 3 regions is significant rejecting the null hypotheses of $\mu_1 = \mu_2 = \mu_3$.

Example 12.5 In Example 11.7, we compared the annual rainfall at two places A and B for 24 years. If these two places A and B are not considered to be independent samples since there is season-wise correspondence between the annual rainfall values, can we decide that the differences (in annual rainfall values) are significant due to the places, to the seasons or to all other indistinguishable causes?

The total variation in the data can be expressed as

$\sum_{i}^{24}\sum_{j}^{2}(x_{ij}-\bar{x}_T)^2$ which can be again put as

$$\sum_{i}^{24}\sum_{j}^{2}(x_{ij}-x_i-x_j+\bar{x}_T)+(x_i-\bar{x}_T)+(x_j-\bar{x}_T)^2$$

which is obviously equal to

$$\sum_{j}^{24}\sum_{j}^{2}(x_{ij}-x_i-x_j+\bar{x}_T)+\sum_{i}^{24}\sum_{j}^{2}[(x_i-\bar{x}_T)^2+(x_j-\bar{x}_T)^2]$$

since the other three terms of the expansion will be equal to zero each.

Hence we can rewrite the above expansion as follows:

$$\sum_{i}^{24}\sum_{j}^{2}(x_{ij}-\bar{x}_T)^2=\sum_{j}^{24}\sum_{j}^{2}(x_j-x_i-x_j+\bar{x}_T)^2+2\sum_{i}^{24}(x_i-\bar{x}_T)^2+24\sum_{j}^{2}(x_j-\bar{x}_T)^2$$

Now the sum of squares of total variation is partitioned into three components which for the above equation can be written as:

Total variation = Variation due to (interaction + season + place)

Now if these four variations are expressed by deviation score method we have:

Total variation, $\sum_{i}^{24}\sum_{j}^{2}(x_{ij}-\bar{x}_T)^2=\sum_{i}^{24}\sum_{j}^{2}x_{ij}^2-48\bar{x}_T^2$

where $48\bar{x}_T^2$ is the correction factor ($C.F.$)

Similarly, the variation due to places is taken as factor II and it is $2\sum_{i}^{24}(x_i-\bar{x}_T)^2$ which can be written as :

$$2\sum_{i}^{24}x_i^2-48\bar{x}_T^2, i.e. 2\sum_{i}^{24}x_i^2-C.F.$$

and the variation due to seasons is the factor I and it is $24\sum_{j}^{2}(x_j-\bar{x}_T)^2$ which can be written as :

$$24\sum_{j}^{2}x_j^2-48\bar{x}_T^2, \text{ i.e. } 24\sum_{j}^{2}x_j^2-C.F.$$

The sum of squares due to interaction is usually calculated by subtracting the sum of squares due to place and due to season from sum of squares of the total variance.

The mean squares are obtained for all these sum of squares by dividing each sum of square by its corresponding *d.f.* For example, the mean squares for the total variation is obtained by dividing the sum of squares for the total variation with its *d.f.* = 47; all these different sum of squares and their mean squares leading to the F statistic itself is now presented in the two-way analysis of variance table for single observation below.

Components	Source of variation	Sum of squares	Degrees of freedom	Estimates of variance of sum of squares	Mean squares	(F ratio)	5% of F limit
SST	Total sum	5590.21	47	118.94			
SS_2	Factor II (Between columns)	16.37	1	16.37	$\frac{16.37}{3.38}$ = 4.84	(1,23)	4.28
SS_1	Factor I (Between seasons)	5496.10	23	238.96	$\frac{238.96}{3.38}$ = 70.70*	(23,23)	< 2.05
SSI	Regional or Error Interaction between places & seasons)	77.74	23	3.38			

* Significant

From the above tabulation we find the two places are significantly different just at five percent since at 0.01 significance level the critical F value is 7.88 and hence more than the tabulated value.

12.6 THE FACTORIAL CONCEPT AND TWO-WAY ANOVA WITH *n* OBSERVATIONS PER CELL

In the two-way analysis of variance earlier we have considered two factors which do have interactive or dependent effect upon each other. But since we took only one observation per cell, this does not enable us to obtain an estimate of or make a test for the interaction effect between 2 factors A and B, (say) under study. For this we need more than one observations for some or all of the cells. Thus the only effective approach for answering these questions of interactions or independence together is by comparing in one and the same experiment all the possible combinations of the levels of both the factors. This approach is known as the *factorial concept* of experimentation. We illustrate this concept by extending the scope of the example on quality rating of the four blocks. As in the earlier example, we take the three different strains of a cereal as factor I whose yields are studied in response to application of four types of fertiliser. This latter group is taken as the factor II and we thus have a factorial design of 3 × 4 or 12 treatment combinations. Now each of these 12 treatment combinations are in randomised block design with two replications. If the factor I and factor II are designated as x and y respectively, the 12 treatments which we can have for each replication are x_1y_1, x_1y_2, x_1y_3, x_1y_4, x_2y_1, x_2y_2, x_2y_3, x_2y_4, x_3y_1, x_3y_2, x_3y_3 and x_3y_4.

The Table 12.8 gives the plan of experiment, strains sown and their yields in 10 gm units which are obtained from each experimental plot of replication of two blocks.

Table 12.8 : Layout and yields of Cereal in Varieties of Strains-cum-Applications of Fertilisers

Block I		Block II	
Plots	Yields	Plots	Yields
x_1y_1	99	x_3y_1	75
x_1y_2	70	x_1y_3	75
x_2y_3	80	x_2y_1	86
x_2y_2	65	x_2y_2	70
x_1y_4	99	x_3y_3	40
x_3y_1	95	x_1y_2	65
x_3y_3	45	x_3y_4	70
x_3y_2	60	x_1y_1	80
x_2y_4	96	x_2y_3	78
x_1y_3	90	x_1y_4	86
x_3y_4	85	x_2y_4	80
x_2y_1	65	x_3y_2	40

In the above, yields are in 10 units, plot sizes are uniform, 3.50 m × 1.80m.

Now if we arrange the above data on 12-treatment combination with two replications, we get the data for 'two-way analysis of variance with n observations per cell', n as equal to 2 here. The data are given in Table 12.9.

Table 12.9 : Quality Rating of Yield of a Cereal by Varieties of Strains-cum-Fertiliser, Each Treatment Rated Twice

Strains of the cereal (Factor I)	Types of fertilisers (Factor II)				
	y_1	y_2	y_3	y_4	
x_1	99	70	90	99	
	80	65	75	86	$x_{1\cdot}$
x_2	96	65	80	96	
	86	70	78	80	$x_{2\cdot}$
x_3	95	60	45	85	
	75	40	40	70	$x_{3\cdot}$
Total	$y_{1\cdot}$	$y_{2\cdot}$	$y_{3\cdot}$	$y_{4\cdot}$	Grand Total

Now the information what we want to seek from the above data is to answer three explicit questions whether the data indicate significant differences: (1) among the three strains on one hand, (2) among the types of fertilisers applied on the other and (3) whether there is any interactions between these two factors, the strains and the fertilisers.

These points can be brought out by a simple extension of the earlier two-way analysis of variance by further subdividing the sum of squares of total variance, *SST* and its *d.f.* into components corresponding to sum of squares (1) between subclass effects or errors, *SSBE* and (2) within subclass effects or errors, *SSWE*. For repeated observations of all the categories, we can obtain a separate independent measure of inherent (or smallest) variations called between subclass effects or errors *SSBE*. It is a measure of interrelationships or interdependence among the two different classification given by

$$SSBE = \sum_{j}^{k} \sum_{i}^{n} \frac{(\sum x_{ij})^2}{r} - \sum_{j}^{k} \sum_{i}^{n} \frac{(x_{ij})^2}{N} \quad \text{... (12.27)}$$

where r = number of repeated samples in each category, k and n = number of columns and rows respectively and N = total number of trials/observations equal to $r \times k \times n$.

Now, we calculate the sum of squares for types of fertilisers (factor II), SS_2 and sum of squares for strains (factor I) SS_1 as before in the earlier two-way analysis of the variance test (Eqs. 12.18 and 12.25). The sum of squares for interaction, *SSI* is calculated differently here by subtracting the total for SS_1 and SS_2 from *SSBE* instead from *SST* (in Eq. 12.26) Hence, it is

$$SSI = SSBE - (SS_1 + SS_2) \quad \text{... (12.28)}$$

This interaction term, *SSI* is tested for significance at the outset by comparing it to the sum of squares, for the *within subclass effects* or *errors*, *SSWE* where

$$SSWE = SST - SSBE \qquad \text{... (12.29)}$$

Table 12.10 gives the summary table for testing the interaction.

If the interaction is not significant, the small amount of variation associated with interaction probably results from sampling error. It is therefore conventionally added into the sum of squares for the within subclass errors, *SSWE* to become "within sum errors", *SSW* as

$$SSW = SSWE + SSI = (SS_1 + SS_2) \qquad \text{... (12.30)}$$

Table 12.10 : Testing for Interaction Effects In Two-way ANOVA

Components	Sources of variation	Sums of squares	Degrees of freedom	Estimates of variance	*F* ratio	5% of *F* limit
SSI	Interaction		$(n-1) \times (k-1) = v_1$	$\frac{SSI}{v_1} = S_1^2$	$\frac{S_1^2}{S_{ES}^2}$	
SSWE	Effect/error within sub-class		$N - n_i - k_j = v_2$	$\frac{SSWE}{v_2} = S_{ES}^2$		

Finally, one proceeds to test the factors I and II for their effects by dividing their estimated values with the estimated within sub-class errors, *SSW* to calculate the corresponding *F* ratios. In case, the interaction is significant, the *F* ratios for the factors are obtained by dividing each by estimated within sub-class errors, *SSWE*. Now we set the assumptions and compute the test statistic as follows.

Assumptions

1. $n_i = 3$, $k_j = 4$, $r = 2$ and $N = 24$
2. Since this two-way classification has more than one observation per cell, there are three sets of hypotheses to be tested.

   For row means:

   $H_0 =$ Effects of the factor I (strains of the cereal) do not account for significant variations in quality ratings

$H_1 =$ Effects... do account for.... in quality ratings.

For column means:

$H'_0 =$ Effects of agricultural regions do not account for significant variations in quality ratings.

$H_2 =$ Effects....... do account for ...... in quality ratings.

For interactions between row and columns:

$H''_0 =$ Interaction effect in not significant

$H_3 =$ Interaction effect is significant

Computing the test statistic

The additivity assumption will be considered first by testing for interaction. Hence, the sums of the two replications and of their squares for each cell as well as for their row and column totals are given in Table 12.11.

Table 12.11 : Sums and Squares for Each Cell and Their Totals

Strains of the cereal	Types of fertilizers				Row totals
	y_1	y_2	y_3	y_4	$\Sigma x_{1i}, \Sigma x_{1i}^2$
x_1	179	135	165	185	664
	16201	9125	13725	17197	56248
x_2	182	135	158	176	651
	16612	9125	12484	15616	53837
x_3	170	100	85	155	510
	14650	5200	3625	12125	35600
Column totals	531	370	408	516	1825
$\Sigma y_{1j}, \Sigma y_{1j}^2$	47463	23450	29834	44938	145685

Beginning with the sum of squares of total variance according to Eq.(12.17), we have

$$SST = 145685 - \frac{(1825)^2}{24} = 145685 - 138776$$

$$= 6909 \text{ (with } d.f. = N - 1 = 23)$$

The sum of squares for rows (i.e. for factor I) is with a divisor of 8, since each row total is a total of 8 plots; hence as in Eq. (12.25), we have

$$SS_1 = \frac{664^2 + 651^2 + 510^2}{8} - 138776 = 140599.6 - 138776.0$$

$= 1823.6$ (with *d.f.* = No. of strains in Factor I – 1 = 3 –1 = 2)

Similarly the sum of squares for columns (i.e. for factor II) is with a divisor of 6, since each column total is a total of 6 plots; hence as in Eq. (12.18), we have

$$SS_2 = \frac{531^2 + 370^2 + 408^2 + 516^2}{6} - 138776 = 141930.2 - 138776$$

$= 3154.2$ (with *d.f.* = No. of fertilizers in Factor II – 1 = 4 – 1 = 3)

Now, from Eq. (12.27), we can have sum of squares for between sub-classes errors (or effects) *SSBE* with a divisor of 2, since each cell total is a total of 2 plots, hence we have

$$SSBE = \frac{179^2 + 135^2 + 165^2 + \ldots + 100^2 + 85^2 + 155^2}{2} - 138776$$

$= 144577.5 - 138776.0$

$= 5801.5$ (with *d.f.* = No. of treatments i.e. group or cells – 1 = 12–1 = 11)

The sum of the squares for interaction, *SSI* is calculated according to the Eq. (12.28) as

$SSI = SSBE - (SS_1 + SS_2) = 5801.5 - (1823.6 + 3154.2)$

$= 823.7$ (with *d.f.* equal to $v_1 = (n_i - 1)(k_i - 1) = 6$

This interaction term, *SSI* is now tested for significance errors, *SSWE*, where

$SSWE = SST - SSBE$ (as in Eq. 12.29)

$= 6909 - 5801.5 = 1107.5$

(with *d.f.* $= v_2 = N - n.k = 24 - 12 = 12$)

Note : *SSWE* is also equal to sum of squares of deviation of each replication in each cell. Thus

$SSWE = [(99 - 89.5)^2 + (80 - 89.5)^2] + [(70 - 67.5)^2 + (65 - 67.5)^2] +$

$\ldots [(85 - 77.5)^2 + (70 - 77.5)^2] = 1107.5$

At this stage, the test for interaction effect is done by finding the estimated interaction and within sum errors which finally gives us the F ratio equal to 1.49. This value compared to the F statistic at $\alpha = 0.05$ significant level is 3.00. Hence the interaction is not significant indicating that the data do not reveal definite evidence of differential response of varieties of strains (factor I) to the types of fertilizers (factor II). This test of significance of the interaction is recorded in the ANOVA Table as shown in Table 12.12.

Since in the present example, there is no significant interaction effect, we compute the sum of squares for within sub-class errors, *SSW* as in Eq. 12.30. We have

$SSW = SSWE + SSI$ (or, $SST - SS_1 - SS_2$)
= 1107.5 + 823.7 (or, 6909 – 1823.6 – 3154.2)
= 1931.2 (with sum of *d.f.* for *SSI* and *SSWE* = 6 + 12 = 18).

Finally, we proceed to test the significance of the two sets of factors for their effects and the interactions between these two factors. We tabulate the above sums of squares, estimate their variances and compute the *F* ratios to compare with their corresponding critical values at a significance level in the following "nested" ANOVA table since it combines the three one-way ANOVA. In Table 12.12 the sum of squares for the total and its total *d.f.* have been partitioned into 4 pieces: SS_1, SS_2, *SSWE* and *SSI*.

Table 12.12 : Two-way Nested ANOVA Table (with *n* observations per cell)

Components	Source of variation	Sums of squares	Degrees of freedom	Estimates of variances	F ratio	5% F value
SST	Total sum	6909	23			
SS_1	Factor I (between rows)	1823.6	2	911.8	8.498*	F (2.18) = 3.55
SS_2	Factor II (between columns)	3154.2	3	1051.4	9.799*	F (3, 18) = 3.16
SSBE	Between subclass errors/ effect	5801.5	11			
SSWE	Within subclass effects effects	1107.5	12	92.2	1.490 Interaction not significant	F(6, 12) = 3.00
SSI	Inter-action	823.7	6	137.2		
SSW	Within sum of errors/ effects	1931.2	18	107.3		

*Significant

F ratios and Tests of significance

The first null hypothesis H_0 to be tested is whether the three varieties of strains of the cereal has come from populations with equal means. In this case the *F* ratio is SS_1/SSW

= 8.498 but since the critical value for *d.f.* (2,18) is 3.55, we reject H_0 and conclude that the strains are significant, they have not come from population with equal means and so they have a different effect on the quality rating scores.

The second null hypothesis H_0' is that the factor II, the type of fertilizers do belong to the same population. To test this, the F ratio is SS_2/SSW = 9.799. In this case also the computed value is more than the critical value so that the factor II inputs have a different effect on the quality rating scores.

Regarding the third null hypothesis H_0'' that whether the interaction effect is zero, the F test is $SSI/SSWE$ = 1.49 which is less than the corresponding critical value of F at 0.05. Hence we retain H_0'' and we say that the interaction is insignificant. This unambiguous information could not have been obtained without making the critical analysis of the results in the way indicated above.

Example 12.6 The following is a hypothetical data on factorial concept with two factor-I and three factor-II variables. The number of replications is five so that in the data there are 30 values. The entry in the data is

Factor-I	Entries in factor -II : Value of y in		
	1	2	3
x_1	10,11,12,13,14	12,13,14,16,15	14,15,16,17,18
x_2	12,13,14,15,16	14,15,16,17,18	30,31,32,33,34

Furnish an ANOVA table and test the interaction effect.

We treat the data as the data with N = 5 observations per cell. We compute the F statistics and enter them in a two-way ANOVA table associated with replications. This table is given below.

The three F ratios computed in the above ANOVA table and their comparison with the corresponding five per cent critical F value lead to the conclusion that the interaction effect is significant but neither of the factor I and II set of variables is significant at five per cent level. Our conclusion is though there are unidentified variables, in the interaction effect, the factor I and II variables are not likely to produce the same difference in values, of y in the total population.

Sources of variance	Sums of squares	*d.f.*	Extimates of variance	*F* ratio	5% values
SST	1406.65	29			
SS_1	333.34	1	333.34	$\frac{333.34}{163.33} = 2.04$	(1.2) = 18.51
SS_2	686.65	2	343.33	$\frac{343.33}{163.33} = 2.10$	(2.2) = (19.00)
SSI	326.66	2	163.33	$\frac{163.33}{2.50} = 65.33$*	(2.24) = 3.40
SSWE	60.00	24	2.50		

LIST OF FORMULAE

1. $F \text{ Statistic} = \frac{\text{Larger variance}}{\text{Smaller variance}} = \frac{s_1^2}{s_2^2}$ at $d.f$ *with* $v_1 = n_1 - 1$ and $v_2 = n_2 - 1$

2. Setting up of ANOVA Table for one-way analysis (with *k* samples and total number of items *N*) : by variance estimate method:

Source of Variation	Sum of Squares	*d.f.*	Estimate of Variance	*F* ratio
Total Variation, VT	$\sum_j^k \sum_i^n (x_{ij} - \bar{x}_T)^2$	$N - 1$	$VT/N - 1$	
Within group Variation, VW	$\sum_j^k \sum_i^n (x_{ij} - \bar{x}_j)^2$	$N - k$	$VW/N - k$	$F = \frac{VB/(k-1)}{VW/(N-k)}$
Between group variation, VB	$\sum_j^k n_j (\bar{x}_j - \bar{x}_T)^2$	$k - 1$	$VB/K - 1$	

3. Setting up of ANOVA Table for one-way analysis (with *k* samples and total number of items *N*) : by coding method for computation—

Source of Variation	Sum of Squares
Sum of squares of total variance, *SST*	$\sum_{i}^{k}\sum_{j}^{n} x_{ij}^2 - \frac{(T)^2}{N}, j = 1, 2, 3, \ldots$ and $N = K.n$
Sum of squares of within variance, *SSW*	$\sum_{j}^{k}\sum_{i}^{n} x_{ij}^2 - \sum_{j}^{k} \frac{(T_j)^2}{n_j}$
Sum of squares of between variance, *SSB*	$\sum_{j}^{k} \frac{(T)^2}{n_j} - \frac{(T)^2}{N}$

where T = Grand Total, i.e. $\sum_{j}^{k}\sum_{i}^{n} x_{ij}, T_j$ is the column total = $\sum_{j=1}^{n} x_{ij}$ and n_j is the number of terms/trials in the jth column, note $\frac{(T)^2}{N}$ is generally known as 'correction factor' *C.F.*

4. Setting up of ANOVA Table for two-way analysis (without repeated items for factors)

Source of variation	Sum of squares	*d.f.*	Estimate of	*F* ratio
Total variation, *SST*	as earlier			
Sum of squares between columns i.e. factor II, SS_2	as earlier	$n_j - 1$	$\frac{SS_2}{n_j - 1}$	$\frac{\text{Estimated } SS_2}{\text{Estimated } SSI}$
Sum of squares between rows i.e. factor I, SS_1	$\sum_{i}^{n} \frac{(T_i)^2}{n_i} - \frac{(T)^2}{N}$	$n_i - 1$	$\frac{SS_1}{n_i - 1}$	$\frac{\text{Estimated } SS_1}{\text{Estimated } SSI}$
Sum of squares of residual or error variation, *SSI*	$SST - (SS_1 + SS_2)$	$(k - 1)(n - 1)$	$\frac{SSI}{(k-1)(n-1)}$	

5. Setting up of ANOVA Table for two-way analysis (with repeated items for factors)

Total variation, *SST* as in *SST* of one-way ANOVA above

Sum of squares between columns, i.e. factor II, SS_2 as in SS_2 of two-way ANOVA above

Sum of squares between rows, i.e. factor I, SS_1 as in SS_1 of two-way ANOVA above

Sum of Squares between sub-class errors (or effects) *SSBE*

$$\sum_{j}^{k}\sum_{i}^{n}\frac{(\sum x_{ij})^2}{r}-\frac{(T)^2}{N}$$

Sum of squares within sub-class errors (or, effects)

$SSWE = SST - SSBE$ at $(N - n_i - k_j) = v_2$; $SSWE/v_2 = S_{ES}^2$

Sum of squares for interaction, $SSI = SSBE - (SS_1 + SS_2)$

at $(x_i - 1)(k_j - 1) = v_1$ for $SSI/v_1 = S_1^2$

At this stage, it is required to do the testing for significance of interaction effect. In case $\frac{S_1^2}{S_{ES}^2}$ at *d.f.* (v_1, v_2) is significant, the *F* ratio for SS_2, SS_1 and *SSI* is computed by dividing each by estimated *SSWE*, S_{ES}^2.

But if *SSI* is not significant, the above *F* ratios for SS_1, SS_2 and *SSI* are computed by dividing the corresponding estimated values with the estimated values of sum of squares within sub-class errors, *SSW* which is equal to the sum of *SSI* and *SSWE* divided by the sum of *d.f.* for *SSI* and *SSWE*.

EXERCISES

12.1. Two sets of scores regarding the yield of a crop in kg. per unit plot in an area are as follows:

Plots: Set one : 39, 45, 49, 53, 57, 60, 61, 65, 71, 78, 82, 60

Plots: Set two : 24, 28, 31, 37, 53, 59, 64, 70, 75, 75, 83, 90, 91

Are the two sets of yield of the crop per unit plot more variable?

12.2 The following are the scores for three groups : A, B and C.

Group A	Group B	Group C
63	43	53
47	57	61
67	59	41
81	46	43

Apply a homogeneity test of variance.

12.3 A place can be reached along four different routes and the following are the number of minutes taken in reaching the place on five different occasions on each route:

Route 1 : 15, 17, 16, 18, 19

Route 2 : 17, 20, 24, 20, 24

Route 3 : 14, 16, 16, 15, 18

Route 4 : 11, 15, 14, 14, 20

Test the hypothesis that there is no difference in the true average time it takes to reach the place along four different routes.

12.4 Crop yields in kg per plot taken in random from four different regions, namely low plains, deltas, mountain and irrigated deserts give the following data:

Low plain	Delta	Mountains	Irrigated deserts
13	8	3	4
6	2	5	5
5	10	10	5
6	9	6	6
10	11	6	2

Can we reject the null hypothesis that the four sample means do not differ among themselves?

12.5 Following is the data on 3 regions regarding an event :

Region 1 : 18, 20, 16, 21, 22, 18, 19, 17 and 21

Region 2 : 21, 23, 14, 24, 21, 20, 19, 21 and 23

Region 3 : 21, 17, 24, 23, 25, 22, 23, 26 and 23

Test the regions are not different.

12.6 In the following a random sample of 5 properties are taken as being assessed by 3 appraisers.

Property No.	Appraiser 1	Appraiser 2	Appraiser 3
1	78	82	79
2	102	102	99
3	68	74	70
4	83	88	86
5	95	99	92

Test whether the appraiser designing into 3 categories is appropriate.

12.7 The industrial production (in unit of Rs. 10,000) per employee by five regions and four industry groups in a country in a particular year is as follows:

Region	Industry groups			
	Group A	Group B	Group C	Group D
A	11	5	10	9
	7	7	8	18
	10	3	12	20
B	12	7	14	16
	9	9	15	18
	13	7	15	14
C	12	7	14	16
	10	10	15	18
	12	7	15	14
D	12	6	12	13
	10	9	10	18
	9	7	18	24
E	13	8	14	15
	10	8	11	14
	12	7	15	18

Test the hypothesis that there are significant variations either amongst the five regions or amongst the four groups of industries or the interaction effects amongst the regions and industries are significant.

12.8 In the table below are given the yields of six varieties of a crop (in some standard unit) in a 4 replicate experiment:

	Treatments					
	1	2	3	4	5	6
A	19	15	16	14	13	14
B	11	14	13	15	17	13
C	15	17	16	20	18	22
D	17	19	13	16	17	18

Analyse the data to test whether the treatments and/or the blocks provide the homogeneity in yields.

13

Non-parametric Statistics in Geo-science Systems I

(Chi-square Test)

13.1 NON-PARAMETRIC STATISTICS

In the preceding chapters we considered a variety of hypothesis-testing situations in terms of three tests mainly, the z test, t test and F test. These tests require quantitative measurement of their data like rainfall, discharge etc. and they come under the so called "classical" or "parametric statistics", the latter requires basically three conditions together: (i) the measurements of the observations are available on interval or ratio scale, (ii) the background populations of the observations (which form the parameter) should be distributed normally so that the standard deviations of the observations are used predictively and (iii) the observations be independent of each other.

But very often the researchers in agriculture or in geography or in social science, in particular may like to deal with data of the qualitative type in which the observations are classified into several mutually exclusive classes or groups (based on nominal or ordinal scaling) and the frequencies in these classes are given. The principles underlying the statistical treatment of such data are essentially similar to those dealt with the parametric statistics, but the assumptions of normality underlying the use of the parametric statistics cannot be met and consequently the problems encountered in connection with the qualitative data are somewhat different from what we have discussed in parametric statistics earlier. Thus in handling the qualitative data we may require a study of such features as "randomness, trend, independence, symmetry or goodness or fit" rather than the significance of testing of hypotheses about particular population parameters in parametric statistics.

For such purposes, non-parametric methods of hypothesis testing have been levied and from the present chapter onwards we set out the methods of dealing with such problems.

When classical or parametric methods of hypothesis testing are not applicable, appropriate non-parametric procedures can be selected. Non-parametric statistical methods are broadly defined as either : (1) those which are distribution-free statistics i.e. they do not have the restrictive assumption of requirement that the data should be drawn from a population which is normally distributed, or, (2) those which are not concerned with the characteristics of population (known as "parameters") or, (3) those for which the data are of (insufficient strength) being not measured on interval or ratio scale to warrant meaningful arithmetic operations.

The use of non-parametric statistical methods offers numerous advantages. A detailed listing of the advantage and disadvantages of non-parametric tests is given below.

1. *Scope of application*: Non-parametric tests are based on fewer and less number of stringent assumptions (which are more easily met) than do the classical parametric tests. Hence they enjoy wide appiicability to much larger class of population and yield a more general broad-based set of conclusions.
2. *Simplicity of derivation*: Parametric tests require a level of competence in mathematics on behalf of the person working whereas most non-parametric methods are generally non-mathematical. Hence non-parametric methods are generally easy to apply and quick to compute when the sample sizes are small or moderate. Sometimes they are as simple as just counting, ranking, addition or subtraction.
3. *Scale of measurement required*: Non-parametric methods may be used on all types of data – it usually requires the data at least on rank form (ordinal scaling) though sometimes qualitative data (on nominal scaling) can also be used. Again the data that have been measured precisely on interval or ratio scaling but have been failed to comply with the assumptions of a parametric test may be converted into nominal or ordinal measurements for its analysis by using a non-parametric test.
4. *Susceptibility to violations of assumptions* : As assumptions are fewer with non-parametric test, they are less susceptible to violations. The effects of violations are equally applicable to both parametric and non-parametric statistics, but the effect is less with distribution-free statistics.
5. *Economy and speed of computations* : Non-parametric methods may be more economical than the parametric ones, since the researcher requires to collect data in non-parametric methods which are more grossly measured (that is, qualitative data or data in rank form), thereby making the survey more economical (in terms of money, time and labour spent) and quick to compute the results. Thus the non-parametrical tests are often used for pilot studies and in situations which desire quick answers.
6. *Statistical efficiency and sample size* : When samples are small, say 10 or so, the distribution-free statistics are easier, quicker but slightly less efficient as compared to their classical equivalents. At small sample sizes, the violation of the parametric

assumptions are most devastating, and hence in these cases the non-parametric tests are the only alternative. But with an increase in the sample size, non-parametric tests are more time-consuming and tedious and hence become much less efficient statistical tests in comparison to parametric tests.

It should be now apparent that non-parametric methods may be advantageously employed in a variety of situations. On the other hand, non-parametric methods suffer from some disadvantages: (1) they waste information, (2) they tend to be less too conservative i.e. they tend to lead to acceptance of the null hypothesis more often than they should, and (3) they are less efficient than their parametric counterparts. The *measure of efficiency* is the relative number of observations needed to achieve equally accurate results by a given method with 100 per cent efficiency. Thus the mean of samples drawn at random from a normal probability distribution is the most efficient estimate of the mean of that distribution and can be said to have 100 per cent efficiency. By comparison, the efficiency of the median of a sample is about 64%. Thus a method which is only 50% efficient requires twice as many observations to achieve an accuracy comparable to a fully efficient method.

13.1.1 Classification of Non-parametric Tests

Now that we have discussed the advantage and disadvantages of non-parametric statistics it is apparent that such procedures do not provide a panacea for all problems of statistical data analysis. Nevertheless, they comprise an important and useful body of statistical techniques. In this chapter we shall describe the chi-square test, the oldest and the most widely used of the non-parametric tests to numerous problems of frequency data. In the next six chapters we discuss another six non-parametric tests. Yet there is no limit on the number that can further be proposed. Moreover, these tests do not on the whole duplicate one another, rather each of them is specially discussed to a different research context. Two major criteria help in choosing a particular non-parametric test: (1) the scale of measure of the data, and (2) the type of research in question in terms of the number of samples. If the research is on one sample observation and is concerned with a population distribution regarding its relevance, the research is a *one-sample test.* On the other hand when we are concerned with two samples, it is a *two-sample test*. Similarly we can have *three-* (*or multi-*) *sample test*. A classification of non-parametric tests is given in Table 13.1.

13.1.2 Standard Procedure

The use of non-parametric tests thus enables us to be independent of using a series of assumptions about the population values. Though there are many non-parametric tests, for

Table 13.1 : A Classification of Non-Parametric Tests

Level of measurement in data	Questions about differences		Questions about strength and relationships		Questions about form and relationships	
	One-sample test	Two-sample test	Three-sample test	Two samples	More than two samples	
Nominal scale	χ^2 test	χ^2 test	χ^2 test	φ coefficient	φ coefficient	–
Ordinal scale: Weak ordering	Kolmogorov-Smirnov test	Kolmogorov-Smirnov test Mann-Whitney U test	Kruskal-Wallis test	Spearman's rank correlation coefficient test		
Strong ordering		Kendall's rank correlation test Wilcoxon rank test				

drawing probabilistic inferences from a set of statistical tables, we require a standard procedure. This standardised procedure involves six basic steps in a statistical test which are as follows:

1. We state our *null hypothesis*, H_0 which is typically a *null* statement asserting no significant differences between populations. Against H_0, research hypothesis (or *alternate hypothesis)* written symbolically as H_1 is stated which asserts significant differences between populations. H_1 can be one-sided or two-sided, depending upon the situation.
2. Secondly we specify our *level of significance* in statistics – this is a precise statement of the degree of difference, α we accept as evidence for rejecting H_0 and hence accepting H_1. This level of significance has been explained in the context of parametric statistics in Chapter 10 and hence to understand the level of significance we have to go back to probability, which is frequently mentioned on a scale from 0 to 1. Thus, the 95 per cent confidence is often called the 0.05 level of significance, meaning that the particular result has only a 5 per cent chance of occurrence through

random variation. Similarly, the 99 per cent confidence level is referred to as the 0.01 level of significance, i.e. the probability that H_0 will be correct is only one per cent.

3. Now, we choose a proper statistical test. The test selected will involve the computation of a statistic from our sample data. The computed one is simply a measure of the difference between samples.
4. After counting the test statistic, critical values are picked up in the corresponding statistical table (given in the Appendix Tables in the book along with the tables of the normal, t and F critical values) at the significance level $\alpha = 0.05$ and $\alpha = 0.01$ under H_0 conditions.
5. Finally, the statistical test will consist of deciding whether one should accept or reject the null hypothesis, H_0. This is done by comparing the probability that H_0 is incorrect after the significance level chosen. If the probability is lower than our significance level we conclude that we cannot reject H_0 i.e. there is no real difference between the samples, as given in the sets of data.

13.2 CHI-SQUARE TEST

Frequency distribution which record the number of observations into certain categories is a common way of summarising data. The results from many statistical investigations take the form of the number of observations (nominal scale) or members of the sample taken fall into various classes (ordinal scale). In these cases we may wish to test whether the frequency in any particular class is in agreement with the totals expected in the class upon a given hypothesis as, for instance, in comparing an observed frequency distribution with a theoretical one like the normal probability distribution. This is the 'goodness of fit' between observation and theory. This is made possible by calculating the value of quantity called Chi-square (χ^2) which is dependent on the observations between the observed and the expected class totals. The distribution of this quantity is known and so the probability of observing a given value of under a particular set of conditions can be calculated like in the case of z, t and F tests. Thus chi-square test is a simple technique which works by comparing the actual frequencies observed with a regular frequency expected.

So like the binomial test, the chi-square test is a method for comparing "counted data" in which individual observations (on nominal scale) are assigned to categories (i.e. are differentiated on a nominal scale) and the number of frequency in each category are counted. In 1900, Karl Pearson proposed this test to establish the accordance/discrepancy between theory and observed facts.

13.2.1 Calculations of Chi-square Test

Let O_j stands for the observed total number of items or frequencies in a class and E_j the corresponding expected frequencies in the particular class. Then the chi-square statistic is given by

$$\chi^2 = \frac{(O_1 - E_1)^2}{E_1} + \frac{(O_2 - E_2)^2}{E_2} + \frac{(O_3 - E_3)^2}{E_3} + ... + \frac{(O_k - E_k)^2}{E_k}$$

$$\text{or } \chi^2 = \sum_{j=1}^{k} \frac{(O_j - E_j)^2}{E_j} \quad ... (13.1)$$

where the symbol χ is the greek letter "chi" (pronounced as "Kie") and χ^2 is read as chi-square, the symbol $\sum_{j=1}^{k}$ stands for the summation over all classes. This χ^2 distribution is a sample-based statistic and it is similar to distribution of all possible samples of size n from normal population having a mean μ and standard deviation σ, in which we standardise each observation within a sample to the standard normal form:

$$z = \frac{x - \mu}{\sigma}$$

and their squaring and summing yield

$$\sum z^2 = \sum_{i=1}^{n} \left(\frac{x_i - \mu}{\sigma} \right)^2 \quad ... (13.1a)$$

so that the value of $\sum z^2$ form a χ^2 distribution.

It is thus apparent from Eq. (13.1) that the quantity χ^2 is derived by : (i) obtaining the difference between an observed frequency and expected frequency, (2) squaring the difference and (3) dividing the squared difference by expected frequencies. It can be seen that for a complete agreement with the expected frequencies, $\sum O_j = \sum E_j$ = Total frequency, N, the value of the χ^2 score will be zero; but chance deviations are bound to occur and we snall obtain positive values of the score due to sampling fluctuations. On the other hand, larger values of chi-square occur where actual and expected frequencies differ considerably. Therefore more the sample results deviate from what would be expected if independence of the samples are true, the larger will be the value of chi-square and *vice versa*. Also whether or not this chi-square value is greater than the values one would expect by chance is decided by reference to the table of the distribution of chi-square in sampling. This table is due to Karl Pearson and it has been tabulated in Appendix Table VIII. This table shows the value of χ^2 that will be exceeded in sampling with given probabilities. This value depends on the number of classes of the observed distribution that can be filled up arbitrarily. The latter is also called the number of degrees of freedom (*d.f.*) on which the value of chi-square is based. Appendix Table VIII accordingly gives, corresponding to each value of the number of *d.f.* from 1 to 150 the value of chi-square that will be exceeded with various probabilities, p at different levels of significance. Like the F distribution, chi-square is not centered around zero, but it is entirely positive which is due to deviations of expected from observed frequency in each category in chi-square formula are squared. Consequently, chi-square tests are always one-tailed and the region of rejection is on the right. For greater values of *d.f.* more than 30, the significance of chi-square can be tested by calculating the quantity

$$\sqrt{2\chi^2} - \sqrt{2(d.f.)-1} \qquad \text{... (13.2)}$$

and treating it as a normal deviate and judging the chi-square value as significant on the five per cent level whenever the calculated normal deviate exceeds 1.55.

A couple of points ought to be noted about the sampling distribution of χ^2. This distribution is similar to the *t* distribution having defined by the one parameter, the *d.f.* As with the *t* distribution, the *d.f.* are in each application and as the *d.f.* increases, the distribution curve in χ^2 becomes more symmetrical like that in *t* distribution. Like *t* distribution, the χ^2 distribution is a sample-based distribution and hence it is not a single distribution but a family of distribution curves, the exact shape of each of which depends upon a *d.f.* For the number of *d.f.*, the same rule applies in both the distributions except that in χ^2 distribution the number in the sample in *t* distribution is replaced by the number of the frequency classes in χ^2 distribution. Thus if *k* is the number of frequency classes *d.f.* will be equal to $k - 1$. Any value of chi-square which is based on the difference between the sample and the population estimated is so large that normally one supposes that it is not likely to have occurred by chance. As a result, the critical region conventionally lies in the upper tail of the sampling distribution. Hence any chi-square value based on the estimated values will have to be less than the corresponding upper limit given in the chi-square table so that the null hypothesis is accepted. On the other hand, if any value of chi-square comes out to be greater than the values given in the chi-square table, (i.e. calculated $\chi^2 \geq \chi^2_\alpha$) we will be in position to reject the null hypothesis. The following properties of the χ^2 distribution are important:

(a) For *d.f.* greater than 30, the significance of χ^2 can be treated as a standardised normal deviate i.e. $z = \sqrt{2\chi^2} - \sqrt{2(d.f.)-1}$ given in Eq. (13.2) earlier.

(b) If χ^2_1 and χ^2_2 are independent random variables having chi-square distribution with v_1 and v_2 *d.f.*, then we can express χ^2 as an additive property, $= \chi^2_1 + \chi^2_2$ distributed as chi-square with $d.f. = n_1 + n_2$.

(c) It can be shown that the statistic

$$\chi^2 = \sum_{j=1}^{k} \frac{(f_j - F_j)^2}{F_j} \qquad \text{... (13.3)}$$

where f_j is the observed frequency in the jth of k classes and F_j is an expected (or theoretical) frequency for the same class, is distributed approximately with $d.f. = (k-1)$ for one-sample test where k is the number of classes. This simply means that given the groups or categories for a frequency table, once the frequencies for all but one category are known, the final category frequency automatically follows. When the parameters of a theoretical distribution are derived from the same empirical evidence as observed frequencies, further *d.f.* are lost so that it becomes equal to $k - 1 - m$, where m is the number of parameters estimated. Thus the *d.f.* vary from one chi-square test to another. This is also discussed later.

13.2.2 Requirements of Chi-square Test

The chi-square test is a powerful tool for analysing the univariate count data as well as the bivariate relationships, especially when the observations are independent through which the coefficient of correlation cannot be worked out. For the application of this chi-square test, the following four requirements should be satisfied in the data:

1. The data must be in the form of ordinal and it should be based on the frequencies or counts obtained in each number of categories, as such it should not be based on relative frequencies (proportion), percentages or rates per thousand etc. The gross inapplicability of the chi-square test for the latter purpose would be obvious when it is noted that by a change in the scale of measurement, for example, from frequencies to relative frequencies, the chi-square value calculated according to the Eq. (13.1) will be changed, altering very often the verdict about significance. Statistics appropriate to tests with measurement data such as z, t or F, on the other hand, remain unaffected by change of scale.
2. There must be at least two mutually exclusive categories into which the observations are placed and they must be independent and random.
3. The total numbers observed in the chi-square data must exceed 20 and preferably greater than 40.
4. The expected frequency in anyone category for a 2×2 problem (2 row by 2 columns) must not be less than 5 normally. However, if there are five or more categories, then not more than 20% of the expected frequencies may be less than five and there should be no category with an expected frequency less than one. It should be noted that this rule is true only for expected frequencies and not for the frequencies actually obtained in the data which can be of any size.

For a 2×2 problem, in case the frequency expected in any class is 5 or less, it is generally possible to pool the frequencies in the adjacent classes so as to obtain the expectation of pooled frequencies greater than 5. The pooled frequencies will have to be treated as belong to a single class and the *d.f.* on which chi-square is based would consequently be reduced. This is a disadvantage and hence pooling of frequencies in the adjacent classes should not therefore be carried out indiscriminately unless when it is essential and justifiable.

Finally, the assumed hypothesis must be carefully defined and clearly understood so that the result can be correctly interpreted. It should be noted also that the question of *tailedness* of the alternative hypothesis H_2 does not arise in the context of chi-square tests, and because of the manner of its execution the direction of departure is immaterial.

13.2.3 Advantages of Chi-square Test

Being based on some sort of frequency groupings, the chi-square test is a powerful statistical tool widely applicable for experimental data in agricultural and for geographical analyses, because both the experimental and geographical data can be arranged in nominal and ordinal classes.

The statistical significance of differences between sample data is assessed in terms of sample mean values and standard errors of the sample means in Chapters 10 and 11 earlier. However, the chi-square test enables us to compare the frequency distributions rather than the mean values. Moreover, they always assume that the body of data are markedly skewed and the chi-square test effects a comparison in terms of frequency distribution. This test has another advantage over the *t* test that the whole range of the data is considered and not merely the mean and standard deviation.

13.2.4 Uses of Chi-square Test

The chi-square test is generally used in the single variable conditions. The single variable test is essentially a test of independence of the distribution. In one-sample test it can be the test for the existence of a relationship between one set of variables. Another common use of the chi-square test is in testing independence of simultaneous classification of a body of data in two different ways which are known as *contingency*. If there are *r* rows and *c* columns the table is said to be an $r \times c$ table. The simplest contingency table is a 2×2 known as 'fourfold table'. The chi-square test is also primarily used for testing how well the observed frequencies fits a theoretical distribution. This test of chi-square is of wide applicability to numerous problems of significance in frequency data like the chi-square test of *goodness-of-fit*. These different uses of chi-square tests are separately discussed below.

13.3 ONE-SAMPLE TEST

This is a single sample test. It is an extension of the binomial or 1×2 table to the more $1 \times n$ table. Often, it is assumed that the population of a variable/ phenomenon is uniformly distributed on an even basis in terms of an equal number. The χ^2 value in this case tests whether the observed frequencies of a given phenomenon differ significantly from the frequencies which can be expected according to an assumption or any *a priori* hypothesis, that is the hypothesis formulated prior to the experiment. The assumed hypothesis is required to be defined carefully so that it can be tested by the chi-square test, for example, it can be on a basis of 50:50 or any *a priori* basis. The one-sample test is illustrated by Example 13.1.

Example 13.1. A sample of 100 farms was listed from an area of diversified relief and their frequencies in the groups of physical character of terrain were as follows:

S. No.	Terrain sites	No. of farms O	Types of terrain as % to total E
1.	Plain	10	10
2.	Rolling plain	50	35
3.	Plateau	1	10
4.	Steep Slopes	14	25
5.	Hills	25	20

The above table also includes the percentage area occupied by the above terrain with respect to the total area. Can we assume that the farms are preferentially located with respect to the terrain sites?

In this case it is reasonable to assume that the observed distribution can be expected to be in conformity of the proportions of different types of terrain, that is we can set up a null hypothesis that there is no significant difference between the observed frequencies of farm sites and the expected frequencies of farm sites in context of the five different type of terrains forming a proportion with respect to the total area. Thus we proceed to calculate the chi-square value as follows :

	Group 1	Group 2	Group 3	Group 4	Group 5	Total
Observed frequency, O	10	50	1	14	25	100
Expected frequency, E	10	35	10	25	20	100
$O-E$	0	15	9	11	5	
$\frac{(O-E)^2}{E} =$	0	6.42	8.1	4.84	1.25	20.61

$$\text{So } \Sigma \frac{(O-E)^2}{E} = 20.61$$

Note : The sum of the differences of observed and expected frequencies in chi-square test is always zero, i.e. $\Sigma(O-E) = \Sigma O - \Sigma E = N - N =$ zero. This is an important check on the computation of chi-square.

The χ^2 is based on $k - 1 = 5 - 1 = 4$ *d.f.* here and the critical value of χ^2 at this 4 *d.f* and at 0.001 significance level is 18.47 which is less than the calculated value of chi-square. Hence the null hypothesis would be rejected and the "inverse" of the null hypothesis is accepted, i.e. there is some preferential location of farm sites in relation to the various terrains.

13.4 TEST OF INDEPENDENCE OF 2 × 2 CONTINGENCY TABLE

Example 13.1 above in which observed frequencies occupy a single column is called a *one-way classification* table. Since, the number of columns is *k*, this is also called a $1 \times k$ table. The generalisation to be considered now is the situation where an individual observation is characterised in two ways, each of which is a binomial type of classification, resulting in a 2×2 table (also known as fourfold table) at simplest. When a group of individual is classified in two ways the results of the classification can be set out as in Table 13.2.

Table 13.2 : Contingency Table

Class	x_1	x_2	x_3 etc
y_1	n_{11}	n_{21}	n_{31}
y_2	n_{12}	n_{22}	n_{32}
y_3	n_{13}	n_{23}	n_{33}
etc.			

Such a table giving the simultaneous classification of a body of data in two different ways (that is, in two qualitative variables) is called *contingency* or *cross-classification table*. If there are *r* rows and *c* columns, the contingency table is said to be an $r \times c$ table. At simplest it is a 2×2 fourfold table. For example, we may wish to determine whether literacy is related to the sex of the population. For this purpose, we take samples of individuals and classify them in four ways: male sex and literacy, male sex and illiteracy, female sex and literacy, and female sex and illiteracy. Thus we have two nominal factors of classification of sex – male and female resulting in the division of sample observations into different ways – literate and non-literate. Observations which we have in this two-way table is a count and frequency data. This two-way table is known as a *bivariate frequency distribution* because it is the result of cross-classifying two nominal variables – sex and literacy which are contingent on each other. Thus there are four possible combined categories or cells : male-literate, male non-literate, female-literate and female non-literate. When the frequencies in the contingent table are summed by row and columns, one obtains univariate frequency distribution.

Table 13.3 gives the classification of 180 persons surveyed according to sex and literacy. We can state the null and research hypothesis in an alternative form. The statement of no difference between two groups (sex and literacy) here also means that the two variables are independent, i.e. there is no relationship between the variables is the null hypothesis and the chi-square test is called a *test of independence*. If the null hypothesis is rejected, we accept the research hypothesis, H_1 that there is a relationship between the variables.

Table 13.3 : Classifications of persons by sex and literacy

	Literacy		
Sex	Literate	Non-literate	Total
Male	96	24	120
Female	24	36	60
Total	120	60	180

In this contingency table where two attributes of surveyed persons : sex and literacy are cross-classified with the purpose of studying their relationships, a value of chi-square can be calculated by following the approach given below.

In the above 2×2 contingency table the categories of one attribute are used as column headings and the categories of the other as row headings. The cell frequencies thus indicate the number of cases classified into each of the joint categories of two variables. Now if this 2×2 contingency table is extended into a $r \times c$ contingency table, we have r classes of one attribute and c classes of another attribute giving rc classes or cells in the contingency table.

In this 2×2 or $r \times c$ contingency table, the expected frequency requires little calculation since no *a priori* expected frequencies can be involved in this case like that in the one-sample (or, variable) test. Instead, the expected frequencies are found by multiplying the sum of the each row by the sum of the corresponding column in which the observed frequency occurs (i.e. the row total times the column total) and dividing this by the sum of all the observed frequencies. Thus

$$E = \frac{\sum r . \sum c}{N} \qquad \text{... (13.4)}$$

where $\sum r$ = total number in the row, $\sum c$ = total number in the column and N = total sample size.

Hence in a $r \times c$ contingency table if the row totals are $\Sigma r_1, \Sigma r_2, \Sigma r_3, ...$, and column totals are $\Sigma c_1, \Sigma c_2, \Sigma c_3, ...$ and the total sample size (or, the total of frequencies) is N, then for a class corresponding to the fourth row and the second column, the expected frequencies will be $(\Sigma r_4 \,.\, \Sigma c_2)/N$. Thus the theoretical or expected frequency of 80 for the cell "male-literate" can be arrived by taking the total number in the appropriate row (120), multiplying it by the total number in the appropriate column (120) and dividing by the overall number of person surveyed (180). The computation of the theoretical frequency for each cell in the above 2×2 sex-literacy contingent table is illustrated as follows:

$$E_{11} = \frac{120 \times 120}{180} = 80 \qquad E_{12} = \frac{60 \times 120}{180} = 40$$

$$E_{21} = \frac{120 \times 60}{180} = 40 \qquad E_{22} = \frac{60 \times 60}{180} = 20$$

This chi-square method of analysis uses the squared difference between the observed frequency distribution O_{ij} and the theoretical frequency E_{ij} in each cell. If there is no difference in the population proportions of the two groups (sex and literacy), then the squared difference between the observed and theoretical frequencies should be small. If the proportions of the two groups are sufficiently different, then the squared difference between the observed and theoretical frequencies should be large. Of course, what is a large difference is relative; the same difference would mean more in a cell with only a few observations than in a cell where there are many observations. Therefore, an adjustment is made for the size of the cell. The squared difference is divided by the expected frequency for each cell. Hence the test statistic is $(O_{ij} - E_{ij})^2/E_{ij}$ summed over all the cells of the contingency table. That is

$$\chi^2 = \sum_{j}^{r}\sum_{j}^{k} \frac{(O_{ij} - E_{ij})}{E_{ij}} \quad \text{where } E_{ij} = \frac{\Sigma r_i \,.\, \Sigma c_j}{N} \qquad \text{... (13.5)}$$

where O_{ij} is the observed frequency in a cell and E_{ij} is the theoretical or expected frequency in a cell.

Thus this equation for the test statistic is similar to that one-sample expression [Eq. (13.3)] but includes a double summation as the additions take place over rows and columns.

This χ^2 statistic approximately follows a chi-square distribution, with the degrees of freedom for the general case in a $r \times c$ contingency table this will be

$$d.f. = (r-1)(c-1) \qquad \text{... (13.6)*}$$

Now, the null hypothesis of equality between the observed and the theoretical frequencies will be rejected when the computed value of the test statistic is greater than the critical value of the chi-square with the approximate number of *d.f.* In the above 2×2 contingent table of sex and literacy, we have

$$\chi^2 = \frac{(96-80)^2}{80} + \frac{(20-40)^2}{40} + \frac{(24-40)^2}{40} + \frac{(36-20)^2}{20}$$
$$= 3.20 + 6.40 + 6.40 + 12.80 = 28.8$$

Now for a *d.f.* = (2 – 1) (2 – 1) = 1 we have a critical value of 6.63 at 0.01 level of significance which is lower than the computed value.

Hence we reject the null hypothesis to conclude that there is a definite link up between the variables, sex and literacy.

However, it is not necessary to calculate the expected frequencies for the calculation of the χ^2 from the 2×2 (or, fourfold) table. The chi-square can be directly calculated from the observed frequencies with the help of the following formula. If *a, b, c* and *d* denote the frequencies of observations at the nominal scale in the four cells of a 2×2 table and the observations are counted in terms of presence or absence in a collection of sampling units, *SU*s, we can have the

$$\chi^2 = \frac{(ad-bc)^2\,(a+b+c+d)}{(a+b)(c+d)(a+c)(b+d)} \quad \text{i.e.} = \frac{N\,(ad-bc)^2}{mnrs} \qquad \text{... (13.7)}$$

This formula involves, apart from the factor $(ad - bc)^2$, that is, square of difference of the cross products, the four marginal totals and the grand total. Making use of this Formula (Eq. 13.7), in the 2×2 data in sex and literacy of Table 13.3, we have

---

*To see why *d.f.* in the contingency table is equal to the number of rows minus 1 time the number of columns minus 1, consider how we calculate the *rc*, expected frequencies in the table. The expected frequency for each cell is the row total times the column total divided by the overall total. How many of these values are independent ?

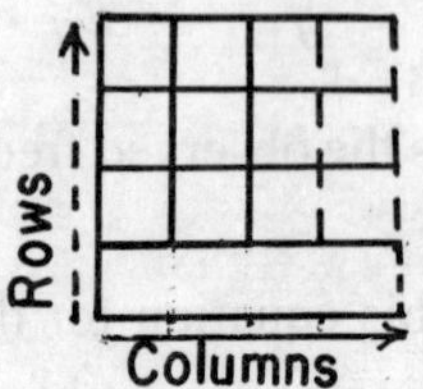

In the diagram here, the number of these values which will be taken as *d.f.* can be apparent easily. Starting with the overall total *rc,* we need $r-1$ additional row totals (since the final one can be calculated independently by subtraction) and $c-1$ additional column totals (again the final one can be found by subtraction). We have therefore used up $1 + (r-1) + (c-1)$ independent pieces of information in calculating the expected frequencies. This leaves : the number of cells – [1 + number of row and column parameters estimated] $= rc - 1 - [(r-1)+(c-1)] = rc - 1 - r + 1 - c + 1 = r(c-1) - 1(c-1) = (r-1)(c-1)$. These cells cover a shaded rectangle of $(r-1)$ rows and $(c-1)$ columns in the above diagram.

$$\chi^2 = \frac{(3456-576)^2\,(180)}{(120)(60)(120)(60)}$$

which on simplification gives the same value as before namely 28.8. Now since the critical χ^2 value for 1 *d.f.* at 5 percent probability level is 3.84 for which this calculated χ^2 value is higher so we reject the null hypothesis and conclude once again that sex and literacy are related or associated.

Note that there are two types of relation/association : (1) positive – if observed $a > E(a)$ and (2) Negative – if observed $a < E(a)$. In the sex-literacy example above, a is 96 but $E(a)$ is $rm/N = 120 \times 120/180 = 80$ that is, observed frequency is more than the expected so that we can conclude that sex and literacy are positively related.

13.4.1 2 × 2 Contingency Table and Yate's Correction

The critical values of χ^2 are derived from a probability distribution which is a continuous smooth curve. But with frequency data the distribution of the statistic χ^2 in $\Sigma \frac{(O-E)^2}{E}$ distribution is essentially discrete. This approximation of the discreate distribution to a theoretical continuous χ^2 distribution is fairly satisfactory when the *d.f.* are more than one and the expected frequencies in the various classes are more than 5. But when the contingency table is small 2 × 2 giving *d.f.* equal to one and the expected cell frequencies are small, 5 or less than 5, this approximation leads to rounding errors so that the use of χ^2 with *d.f.* = 1 leads to overestimation of significance. This is corrected very simply by Yate's Correction

$$\chi^2 = \sum_i^r \sum_j^c \frac{\left(\left|O_{ij} - E_{ij}\right| - 0.5\right)^2}{E_{ij}} \qquad \text{... (13.8)}$$

For the direct method of χ^2 computation to a 2 × 2 contingency table, *Yate's Correction for continuity* would be

$$\chi^2 = \frac{\left(\left|ad - bc\right| - N/2\right)^2 .(N)}{(a+b)(c+d)(a+c)(b+d)} \qquad \text{... (13.9)}$$

where N is the sum of the total size.

For example, the Yate's correction when applied to above 2 × 2 sex-literacy data, we get a corrected value of χ^2 equal to 27.03.

Example 13.2 Two batches of 24 guinea-pigs each, one inoculated and the other not inoculated, were exposed to the infection of a disease. The following were the results

	Dead	Survived	Total
Inoculated	4	20	24
Not inoculated	16	8	24

Can the inoculation be regarded as effective?

We apply the direct method of computing the chi-square, and let a, b, c and d denote the frequencies in the data

$$\chi^2 = \frac{(ad-bc)^2\,(a+b+c+d)}{(a+b)(c+d)(a+c)(b+d)}$$

$$= \frac{(32-320)^2}{(24)(24)(20)(28)}(48) = \frac{3981312}{322560} = 12.34$$

Now the Yate's correction for continuity will give us a value for χ^2. Following the Eq. (13.9) we have the corrected χ^2

$$\chi^2 = \frac{(|32-320|-48/2)^2\,(48)}{(24)(24)(20)(28)}$$

$$= \frac{(264)^2\,(48)}{322560} = \frac{3345408}{322560} = 10.37$$

Thus the value of χ^2 without correction is higher than the value with correction making the former significant at the highest level 0.001 which is not so with the value of χ^2 corrected for continuity. Thus for a 2×2 contingency table, the continuity correction factor has a big effect on the value of χ^2, so the computation of chi-square without taking into account the Yate's correction for continuity, leads to an exaggeration in significance level.

Example 13.3. Following is the presence-absence data for 2 species from 5 quadrats

Species A	Species B	
	Present	Absent
Present	2	2
Absent	1	0

Test whether the occurrence of the species A and B in the 5 quadrats is independent.

For testing the above hypothesis we compute the χ^2 statistic.
$\chi^2 = 5\,|\,(2)(0)\ -(2)\,(1)\,|^2/(4)\,(1)\,(3)\,(2) = 0.83$ and with Yate's correction
$\chi^2 = 5[\,|\,(2)(0) - (2)\,(1)\,| - (5/2)]^2/(4)\,(1)\,(3)\,(2) = 0.0521$

Note that the Yate's correction has a big effect on the χ^2 statistic. But since the critical value of χ^2 at 1 *d.f.* is 3.84, we do not reject the null hypothesis to say that occurrence of species A is not associated with the occurrence of species B.

13.4.2 2 × 2 Contingency Tables and the Rules for Comparing sets of Nominal Data

When the frequencies in a 2 × 2 table are small, even Yate's correction for continuity does not lead to satisfactory tests and it is necessary to calculate the level of probability using R.A Fisher's Exact Probability test.

In *Fisher's Exact Probability test* for comparing the observations in the two groups: groups 1 and 2, we have the null hypothesis H_0 that the frequency distribution of observations arose by chance so that there is no significant difference between the group 1 and group 2. Now the probability, p of our partioning of the four frequencies (a, b, c and d) arising by chance is given by

$$p = \frac{(a+b)!(c+d)!(a+c)!(b+d)!}{n!a!b!c!d!} \qquad \text{... (13.10)}$$

where n is the sum of the total size.

We illustrate this Fisher's test in an example given below.

Example 13.4 Following is the record of the presence or absence of a fossil in 12 samples each of two different sedimentary environments: (1) off-shore facies and (2) near-shore facies

Sedimentary environments	Present (+)	Absent (–)	Total
1. Off-shore facies	2	10	12
2. Near-shore facies	8	4	12
Total	10	14	24

The research workers guess that the fossil is found preferentially in the near-shore facies. Does the above data upheld this guess?

We erect a null hypothesis H_0 that there is no difference between the two environments saying that above frequencies in the data arose by chance. From Eq. (13.10) of Fisher's Exact Probability test, we have

$$p = \frac{12!12!10!14!}{24!2!10!8!4!} = \frac{12!12!14!}{24!2!8!4!}$$

$$= 0.01666$$

i.e. the probability of observed frequencies arising by chance is only 1.67 in 100. So we should reject the null hypothesis at 0.05 significance level to conclude that there is a significant difference between the two sedimentary environments.

We can calculate further values of p with deviations more extreme than the above in the following two tables :

Groups	+	–	Total
I	0	12	12
II	10	2	12
Total	10	14	24

Table (i)

Groups	+	–	Total
I	1	11	12
II	9	3	12
Total	10	14	24

Table (ii)

In the above two tables, with the marginal (row or column) totals remaining the same, we have given different partioning of frequencies. In the contingency table (i), the frequency call for the group I includes a zero also. For all these tables we can ask how likely the partioning of frequencies occur by chance? This probability must include the whole "extreme tail" of probability values, that is, to represent our observations plus any more extreme situation by Fisher's Exact Probability test. Really we are calculating what fraction of all possible partionings that can occur is represented by our sample. So we compute:

for the table (i) probability of occurrence is

$$p = \frac{10!14!12!12!}{24!} \cdot \frac{1}{0!10!12!2!} = 0.00034$$

Similarly, for the table (ii) the probability of occurrence is

$$p = \frac{10!14!12!12!}{24!} \cdot \frac{1}{1!9!11!3!} = 0.00135$$

and we conclude that the result is significant.

Thus the Fisher's Exact Probability test is useful for treating small sets of 2×2 data, but it becomes very tedious to compute when the smallest cell frequencies depart far from zero. In addition, the factorials of quite small numbers like 20! becomes very large (2.432902E 18) requiring the facilities of scientific calculator. Fortunately, we are having alternatives regarding what to use for a 2×2 contingency test? It can be said that

(1) If $n < 20$, use Fisher's Exact Probability test.
(2) If n is less than 40 but more than 20 and if all frequencies are more than 5, use χ^2 test, otherwise use Fisher's test.
(3) If $n > 40$, use the χ^2 test using Yate's continuity correction.

Finally, it can be said that if the co-occurrence of groups (or, species) is not independent, the groups (or species) are determined to be either positively or negatively associated. But this association (+ve or – ve) with regard to their joint occurrence in *SU*s, may have a strong negative covariation (i.e. when one species abundance increases, the other's decreases), hence it is important to make a sharp distinction between species association and species covariation. Unfortunately these two terms– "association" and "correlation" – are too often inappropriately interchanged. The term "correlation" responds for measuring the relative intensity of covariation in pairs *X* and *Y*.

13.4.3 Test of Independence of $r \times c$ Contingency Table

The test of independence in a $r \times c$ contingency table is carried out on the same lines in a 2×2 table as illustrated in Example 13.5 below.

Example 13.5 Rainfall and runoff data measured 200 times in a river basin are cross-tabulated as given below:

Runoff	Rainfall (in cm)				
(in cm)	4.0 – 5.0	5.0 – 6.0	6.0 – 7.0	7.0 – 8.0	8.0 – 9.0
1.0 – 2.0	27	9	0	0	0
2.0 – 3.0	19	31	12	5	0
3.0 – 4.0	7	14	19	14	6
4.0 – 5.0	0	6	5	13	7
5.0 – 6.0	0	0	1	4	1

Test the significance of the null hypothesis that rainfall and runoff data as two variables are independent.

We find that in the above rainfall and runoff data, n is 200 and it is 5×5 contingency table and there are seven cells which do not have any frequency value. We use the chi-square formula Eq. (13.5) for $r \times c$ contingency table, that is,

$$\chi^2 = \sum_{i}^{r}\sum_{j}^{k}\frac{(O_{ij}-E_{ij})^2}{E_{ij}} \quad \text{where } E_{ij} = \frac{\sum r_i . \sum c_j}{N}$$

From the above data, we get the row totals and the column totals and the grand total, N. The expected frequencies for each cell positions are also calculated and they are put in brackets alongwith the observed ones.

Runoff	Rainfall 4.0–5.0	5.0–6.0	6.0–7.0	7.0–8.0	8.0–9.0	Row Total
1–2	27(9.54)	9 (10.80)	0 (6.66)	0 (6.48)	0 (2.52)	36
2–3	19(17.76)	31 (20.18)	12 (12.40)	5 (12.06)	0 (4.69)	67
3–4	7(15.90)	14(18.00)	19(11.10)	14(10.80)	6(4.20)	60
4–5	0(8.21)	6(9.30)	5(5.74)	13(5.50)	7(2.17)	31
5–6	0(1.59)	0 (1.80)	1 (1.11)	4 (1.08)	1 (0.42)	6
Column Totals	53	60	37	36	14	200

For these 5×5 cells we compute the χ^2 for each cell. These χ^2 values are

31.96	0.30	6.66	6.48	2.52
0.09	5.91	0.01	4.13	4.69
4.98	0.89	5.62	0.95	0.77
8.21	1.17	0.10	9.87	10.75
1.59	1.80	0.01	7.89	0.64

The total of these above values give us the computed χ^2 for the data which totals 117.99. Now for $d.f. = (5-1)(5-1) = 16$ since at $\alpha = 0.05$ significance level the critical value of χ^2 is 26.30 so we reject H_0 and conclude that rainfall and runoff are dependent variables.

For classifying a group of individuals or species (which are more than 2) simultaneously for testing of their independence or association, variance ratio (*VR*) model was proposed by D. Schluter in 1984., This *VR* index of independence or association is derived from presence-absence data as given in following table of presence (= 1) or absence (= 0) of m number of species (i = 1,2,3, ..., m) in n number of quadrats (j = 1, 2, 3, ..., n):

Groups/species	Quadrats or (SUs) (1)	(2)	(3)	(n)	Total
(1)	1	0	1.....	0	n_1
(2)	1	0	1.....	1	n_2
(3)	0	1	0.....	0	n_3
.	.	.	.	.	
.	.	.	.	.	
.	.	.	.	.	
(m)	0	0	1.....	1	nm
	T_1	T_2	T_3	T_n	

To test the null hypothesis that there is no association among the groups/species i.e. they are independent, we compute, the total sample variance for the occurrences of the m groups (or species) in the sample as

$$\sigma_T^2 = \sum_{i=1}^{m} p_i\,(1-p_i) \qquad \text{... (13.11)}$$

where $p_i = \dfrac{n_i}{N}$ with n_i = groups/species total row-wise and N = Total number of sampling units as in the above table

Next, we estimate the variance in total groups/species number as

$$s_T^2 = \frac{1}{N}\sum_{j=1}^{N} (T_j - t)^2 \qquad \text{... (13.12)}$$

where t is the mean number of species or groups per quadrat (or, SU)

Now, the variance ratio,

$$VR = \frac{s_T^2}{\sigma_T^2} \qquad \text{... (13.13)}$$

This variance ratio serves as an index of overall association among the groups/species. The expected value under the null hypothesis of independence is 1. $VR > 1$ suggests that group species exhibit a positive association, $VR < 1$ indicates a negative net association.

Schluter in 1984 provides a statistic, W to test for significant departures from the expected value of no association, where W approximates a χ^2 distribution. This statistic, W tests whether deviation of VR from 1 are significant. For example, if the species are not associated, then there is 90 per cent probability that W lies between limits given by the chi-square distribution:

$$\chi^2_{.05,\, N} < W < \chi^2_{.95,\, N} \quad \text{where } W = (N)\,(VR) \qquad \text{... (13.14)}$$

Example 13.6 Following is an ecological data matrix of the presence or absence of 3 species in 5 quadrats.

Species	Presence-absence					Totals (n_i)
	1	2	3	4	5	
(1)	1	1	1	1	0	= 4
(2)	0	1	1	1	1	= 4
(3)	1	0	1	0	1	= 3
Totals = T_j	2	2	3	2	2	= 11

We now compute the total sample variance for the occurrence of species in the samples

$\sigma_T^2 = (4/5)\,(1 - 4/5) + (4/5)\,(1 - 4/5) + (3/5)\,(1 - 3/5) = 0.56$

We compute the mean number of species for the 5 quadrats

$= [2 + 2 + 3 + 2 + 2]\,/5 = 2.2$

Now, $s_T^2 = (1/5)\,[(2 - 2.2)^2 + (2 - 2.2)^2 + (3 - 2.2)^2 + (2 - 2.2)^2 + (2 - 2.2)^2] = 0.16$

Thus, the variance ratio is, $VR = \dfrac{0.16}{0.56} = 0.285$ which suggests a net negative association among the species. To test this deviation from Eq. (13.14), we compute:

$W = (5)(0.2.85) = 1.43$

Under the hypothesis of no association, there is a 90 per cent probability that W should be between the limits.

$1.14 < W < 11.07$

Thus, we accept the hypothesis of no association.

13.5. GOODNESS-OF-FIT TEST

Goodness-of-fit test finds out how well the observed frequency distribution fits a theoretical frequency distribution as obtained by probability laws. For example, for fitting a normal distribution to a set of observed frequencies, we had no basis for comparing the "goodness-of-fit" other than inspection. In this context, the chi-square test and Kolmogorov-Smirnov test can be used to provide with a more precise answer to the question of whether the observed frequencies and those to be expected on the basis of a normal differ significantly or not. Both the tests are appropriate to a null hypothesis that the observed frequencies come from a distribution function which is compatible with a specified theoretical distribution (like normal distribution mentioned earlier) function i.e. the observed distribution is the same as the specified distribution. The acceptability of the null hypothesis is judged on the similarity of the observed values and the values expected for the specified theoretical distribution. The basic difference between the chi-square and the K-S test is that the former compares the differences between the two frequency distributions and the latter compares the differences between the two cumulative frequency distributions.

The distribution of the variable constitutes a sample and goodness-of-fit test decides whether the samples could have been drawn from a probability distribution. The type of scores involved in the data decides the use of the chi-square or the K-S test. When the data in the distribution is arranged in the nominal order, the chi-square is used. On other hand, where the data and the classes are rankable, the K-S test is applicable.

For the goodness-of-fit test of a data set with the χ^2 statistic we are restricted to probability distributions only. We can use this test against a data set for any specific curve, such as log-normal, Poisson, exponential, negative binomial or any arbitrary distribution. The test procedure remains the same, although the *d.f.* must be adjusted to account for the number of parameters estimated.

13.5.1. Procedure for Goodness-of-Fit Test

The following procedure can be adopted for goodness-of-fit test by chi-square.

(1) The expected frequencies are calculated from a knowledge of the hypothesised parent population. These expected frequencies need not be equal categorywise but they must sum to the sample size (N).

(2) For each category, the chi-square value is calculated and they are summed to the test statistic in Eq. (13.3).

(3) The *d.f.* in evaluating the critical value of chi-square depends upon the question is posed. If the only restriction is that the total of expected frequencies remains the same, then the *d.f.* is number of classes minus one. If further restriction is placed that the mean and standard deviation must remain the same which is assumed in a normal distribution, then the *d.f.* is equal to the number of classes minus three. Thus, *d.f.* for chi-square distribution is *d.f.* = (No. of classes, k over which statistic is calculated) –1– (No. of parameters estimated m, over which χ^2 statistic is calculated), that is

$$d.f. = k - 1 - m \qquad \text{... (13.15)}$$

(4) If the value of chi-square is larger than it's critical value for the *d.f.* at a significance level, the differences between these two sets of frequencies are not statistically significant.

We illustrate this application of chi-square in goodness-of-fit test by an example below.

Example 13.7 In a record of the number of times a river floods in rainly season over a period of 140 years, following data were compiled:

No. of years	No. of floods
19	0
39	1
42	2
23	3
11	4
6	5

Can we have the hypothesis that the observed frequencies are distributed as the Poisson distribution with the same mean and the total frequency?

Expected frequencies are calculated by the probability function of the Poisson distribution and these are as follows:

If the individual flood occurrences are assumed *x* and the number of years in which the floods occurs as f_x then the total number of floods is $\sum_{x=0}^{5} (x.f_x) = (0)(19) + (1)(39) + (2)(42) + (3)(23) + (4)(11) + (5)(6) = 266$; So *z*, the average number of floods per year = 266/140 = 1.9. Now we calculate the expected frequencies of flood occurence by Poisson distribution as follows :

r	$p(r)$	$p(r) \times 140$
0	$p(0) = e^{-1.90} = 0.1496$	20.94 years
1	$p(1) = \frac{1.90}{1} p(0) = 0.2842$	39.79 years
2	$p(2) = \frac{1.90}{2} p(1) = 0.2700$	37.80 years
3	$p(3) = \frac{1.90}{3} p(2) = 0.1710$	23.94 years
4	$p(4) = \frac{1.90}{4} p(3) = 0.0812$	11.37 years
5	$p(5) = \frac{1.90}{5} p(4) = 0.0309$	4.32 years

Using these expected frequencies, the value of chi-square is calculated as given below. The *d.f.* for the goodness of fit test is $k - 1 - m$ (Eq. 13.15) where k = the number of classes in the distribution and m = the number of estimates required form the original data to calculate the expected frequencies. In the present case, we lose one *d.f.* by making the estimated frequencies and another by specifying the mean of the Poisson distribution used. There are six classes in the data, so allowing for the two restraints, there are $k - 2 = 6 - 2 = 4$ *d.f.*

No. of floods	No. of years Observed	Expected	$(O-E)$	$(O-E)^2$	$(O-E)^2/E$
0	19	20.94	– 1.94	3.76	0.18
1	39	39.79	– 0.79	0.62	0.02
2	42	37.80	+ 4.20	17.64	0.47
3	23	23.94	– 0.94	0.88	0.04
4	11	11.37	– 0.37	0.14	0.01
5	6	4.32	– 1.68	2.82	0.65
				Total χ^2	= 1.37

Now at the 5 per cent significance level we find that at *d.f.* = 4, chi-square value must exceed 9.49 before we reject the null hypothesis. But since the calculated chi-square value is 1.37, much less than 9.49, we can conclude that there is no reason to reject the null hypothesis and that the Poisson distribution can be used adequately to describe the observed distribution. Poisson distribution is also proved if mean of the distribution becomes equal to variance. Now, variance following Eq. (3.9) is

$$s^2 = \frac{\Sigma f_x . x^2 - N\bar{x}^2}{N-1} = [(39)(1)^2 + (42)(2)^2 + (23)(3)^2 + (11)(4)^2 + (6)(5)^2 - (140)$$

$$(1.90)^2] \div 139 = 1.687 \text{ which is close to mean.}$$

13.6 CORRELATION COEFFICIENT FOR NOMINAL DATA CONTINGENCY

We have seen above that nominal data can be arranged in a contingency table with minimum two groups, 2 × 2 and their chi-square as a test of association can be worked out. But these chi-square tests do not, of course, determine the strength of relationship. Now if we provide a simple extension of the chi-square we can convert the chi-square to a correlation coefficient called the *Contingency coefficient* or "phi (ϕ) coefficient". This coefficient determines the degree of association/correlation between the two variables which are at the nominal level of measurements and are generally dichotomous in form. Coefficient is used when dichotomous classes of two variables are ordered so that they may be assigned 'scores' of 0 and 1 – the larger the value, stronger is the relationship between the variables. In Table 13.3 we dichotomise the persons surveyed in the data into two categories: sex and literacy. The *phi coefficient* is a special form of Pearson product moment correlation coefficient, *r*. This coefficient tells us about the strength of a relationship between the two variables, whereas the chi-square test shows whether there is any significant association between the variables.

For a 2 × 2 contingency table, $\phi = \sqrt{\frac{\chi^2}{N}}$ ... (13.16)

This phi has a value of "zero" when there is "no relationship" but it does not reach 1.0. For contingency table larger than 2 × 2 the phi coefficient is,

$$\phi = \sqrt{\frac{\chi^2}{N(L-1)}} \qquad \text{... (13.16a)}$$

where N is the total number of samples, $(L-1)$ is the product of $(r-1)$ and $(c-1)$, r and c are the row and column numbers.

However, when the rows and columns of a contingency table is larger than 2 × 2, the phi coefficient is not applied because ϕ can attain a value more than 1.0 which is undesirable. In this context for larger contingency tables of ordered categories, statistics such as Kendal's Tau (τ) or Kruskal's Gamma (γ) tests are applied.

For the 2×2 sex-literacy data in Table 13.3, we computed $\chi^2 = 28.8$. If we work out the phi cofficient between sex and literacy for this 2 × 2 contingency table, we have the phi coefficient (ϕ)

$$\varphi = \sqrt{\frac{\chi^2}{N}} \doteq \sqrt{\frac{28.8}{180}} = 0.4$$

Thus the phi coefficient of 0.40 indicates that the correlation between sex and literacy is a weak one but the χ^2 test concludes that there is a definite association between sex and literacy.

Example 13.8 A data set is having marginal frequencies 100, 100, 164 and 40. Find out the contingency coefficient of the data.

For computation of the phi coefficient we treat the data as a single sample test for chi-square. We take the mean of the frequency totals of the data set as $\frac{100+100+164+40}{4} = \frac{404}{4}$ = 101. Hence the χ^2 value is equal to

$$\frac{(100-101)^2}{101}+\frac{(100-101)^2}{101}+\frac{(164-101)^2}{101}+\frac{(40-101)^2}{101}=76.16$$

Hence, the phi coefficient, $\varphi=\sqrt{\frac{76.16}{404}}=0.43.$

13.6.1 Cross-Association

When we have two sets of nominal data, like that in stratigraphy, we cannot have the measures of similarity that presume the data are continuously distributed. The problem can be circumvented by substituting a measure known as "cross-association." For example, the data consist of a series of states such as lithologic type i.e. sandstone, shale, limestone, coal and siltstone encountered in a stratigraphic sequence. These states are mutually exclusive and cannot be ranked in a meaningful way. With cross-association, two such sequences are moved past one another, and the degree of correspondence between the overlapped segments is calculated. Thus we have two sequences as chain 1 and chain 2 and there are m possible categories into which observations can be classed. If we denote the number of observations in the kth state of chain 1 as x_k the total length of chain 1 is $n_1 = \sum_{k=1}^{m} x_{1k}$. In chain 2, there are also m categories, each with x_{2k} observations, giving a total length of chain 2 is $n_2 = \sum_{k=1}^{m} x_{2k}$. Now we find the probability, p that a given number of observations from the two sequences will match, as

$$p=\frac{\text{Sum of the products of each chain of } m \text{ categories}}{\text{Product of chain lengths}}$$

$$=\frac{\sum_{k=1}^{m} x_{1k}\ x_{2k}}{n_1\ n_2} \quad \text{... (13.17)}$$

So, the probability of mismatch is $(1-p)$. Now p times the number of comparisons will give us the expected number of matches, E and similarly the expected number of mismatches, E'. Using these quantities of O, E and O', E' for matches and mismatches, in a χ^2 test, we have

$$\chi^2 = \frac{(O-E)^2}{E}+\frac{(O'-E')^2}{E'} \text{ with } d.f = 1 \quad \text{... (13.17a)}$$

Example 13.9 In a stratigraphy sequence, we code sandstone = 1, shale = 2, limestone = 3, coal = 4, siltstone = 5, we have 2 sections as:

Chain (1) 5 1 2 1 2 5 2 3 2 4 2 5

Chain (2) 3 2 1 5 2 5 1 2 1 2 3 2

From which we get:

Category	Chain 1	Chain 2	$x_{1k} \cdot x_{2k}$
1. Sandstone	2	3	6
2. Shale	5	5	25
3. Limestone	1	2	2
4. Coal	1	0	0
5. Siltstone	3	2	6
Total:	$n_1 = 12$	$n_2 = 12$	$\sum_{i=1}^{m} x_{1k} \cdot x_{2k} = 39$

From this table, we compute, $p = \dfrac{39}{12 \times 12} = 0.27$ so, $1 - p = 0.73$. Now in the two chains, the observed sequence is 2, for which E is $12 \times 0.27 = 3.2$, so $O' = 12 - 2 = 10$ and $E' = 12–3.2 = 8.8$

Hence $\chi^2 = \dfrac{(2-3.2)^2}{3.2} + \dfrac{(10-8.8)^2}{8.8} = 0.61$ and we conclude that the stratigraphic chains are of random sequences.

13.7 BISERIAL CORRELATION

The phi coefficient deals with correlation between two dichotomous variables like persons categorised by two nominal scales of sex and literacy in Table 13.3. Like the phi coefficient, very often we come across data in which one variable is measured on a continuous scale while the other is a dichotomous variable on a nominal scale so that it can only take the values of 1 or 0. Any relationship between a set of numerical values on one hand and two nominal categories on the other is known as *point-biserial correlation coefficient*, r_b. For example, it is decided to test whether the population size of villages is a factor for the villages to have a village post office or not. 10 villages are sampled by population and the information are collected about the presence or absence of post office in the village. Following are the data where population is represented by a continuous variable and the variable X represent the presence ($X = 1$) or absence ($X = 0$) of post office.

Village	Population (Y)	Post Office (X)
1.	200	1
2.	200	1
3.	250	0
4.	250	0
5.	250	1
6.	300	0
7.	300	1
8.	320	1
9.	310	1
10.	750	0

Thus the continuous variable (Y) is dichotomised into two subgroups depending on the village with a post office ($X = 1$) and those without ($X = 0$). The point biserial coefficient r_b is then given by the equation

$$r_b = \frac{\bar{Y}_1 - \bar{Y}_0}{s_Y} \cdot \sqrt{\frac{n_1 . n_0}{N(N-1)}} \qquad \text{...(13.18)}$$

where N = total number of observations, n_0 = No. of observations with an X value of 0, and n_1 = No. of observations with an X value of 1, $\bar{Y}_0$ = mean of Y values with $X = 0$, $\bar{Y}_1$ = mean of Y values with $X = 1$, $\bar{Y}$ = mean of all the Y values and s_y is the standard deviation of all the data, N equal to

$$\sqrt{\frac{\Sigma Y^2 - N(\bar{Y})^2}{N-1}}$$

Now we categories the data for the necessary computation

Y	Y_0	Y_1	Y^2
200		200	40000
200		200	40000
250	250		62500
250	250		62500
250		250	62500
300	300		90000
300		300	90000
320		320	102400
310		310	96100
750	750		562500
Total 3130	1550	1580	1208500
Mean 313	387.5	263.3	

$$\overline{Y} = 313, \qquad N = 10$$

$$\overline{Y}_0 = 387.5, \qquad n_0 = 4$$

$$\overline{Y}_1 = 263.3, \qquad n_1 = 6$$

$$(\overline{Y})^2 = 97969$$

$$\text{So, } r_b = \frac{263.3 - 387.5}{s_Y} \cdot \sqrt{\frac{24}{10(10-1)}} \quad \text{where } s_Y = \sqrt{\frac{1208500 - 979690}{9}} = 159.45$$

$$\text{Hence, } r_b = \frac{-124.2}{159.45} \cdot \sqrt{\frac{24}{90}} = -0.40$$

Thus the point biserial coefficient indicates a negative relationship. But we can apply a *t test* for significance of the r_b value by the use of the following formula

$$t_{N-2} = r_b \cdot \sqrt{\frac{N-2}{1-r_b^2}} \qquad \text{... (13.19)}$$

where N–2 is the *d.f.* which is equal to 8 here. This t value is 1.23 and the critical value of t at *d.f.* = 8 and 0.05 significance level is 2.31. Hence we accept the null hypothesis that population size of the villages is not a factor for the villages to have a post office in a village and look for the other factors that may condition the survival of village post offices.

The nature of the point biserial statistic, relating as it does a continuous variable with nominal data, means that certain assumptions do underlay its application. Thus the values of the continuous variables should be normally distributed. Again the two sub-samples of the dichotomous variable should not be vastly different in terms of the number of observations in each sub-set. More equal these sub-groups are, the more accurate the test will be.

This particular test is well used in analysis of field observations where a particular continuous variable is dichotomised in two nominal categories for which observations are taken. We illustrate the application of this test in another example below.

Example 13.10. A test was conducted whether aspect of slope could be a factor in slope steepness in areas of homogeneous rock. A random sample is taken on 24 locations for each of which the aspect is observed and the mean angle of the principal slope angle is measured. The following is the data :-

Aspect	N	N	S	S	N	S	N	S	N	N	S	N	S	S	N	N	S	S	N	S	S	N	S	N
Slope Angle	7	11	9	5	17	8	13	13	21	9	10	9	12	7	14	9	4	6	18	9	12	22	4	14

Slope	Aspect (North)	Aspect (South)	$(\text{Slope})^2$
7	7		49
11	11		121
9		9	81
5		5	25
17	17		289
8		8	64
13	13		169
13		13	169
21	21		441
9	9		81
10		10	100
9	9		81
12		12	144
7		7	49
14	14		196
9	9		81
4		4	16
6		6	36
18	18		324
9		9	81
12		12	144
22	22		484
4		4	16
14	14		196
Total 263	164	99	3437

So, $r_b = \frac{\overline{Y}_n - \overline{Y}_s}{s_Y} \cdot \sqrt{\frac{n_n . n_s}{N(N-1)}}$ where $\overline{Y}_n = 13.66, \overline{Y}_s = 8.25, n_n = 12, n_s = 12$ and $N = 24$

Now $s_Y = \sqrt{\frac{\Sigma Y^2 - N\overline{Y}^2}{N-1}} = \sqrt{\frac{3437 - (24)\,(10.96)^2}{23}} = 4.91$

Hence, $r_b = \frac{(13.66 - 8.25)}{4.91} \cdot \sqrt{\frac{144}{552}} = 0.56$

Now the significance of this value of r_b is tested by

$$t = 0.56\sqrt{\frac{24-2}{1-(0.56)^2}} = 3.17$$

The critical value of t at $\alpha = 0.05$ significance level for $d.f. = 22$ is 2.07, so we conclude that at 0.05 significance level slope angle is related to aspect of the slope but since at $\alpha = 0.001$, the computed value is lower than the critical value of 3.79 so we cannot be very decisive about the rejection of the null hypothesis.

13.8 CONCLUSION

The contingency test shows that the chi-square analysis can accommodate the *multinominal distribution*, a many-variable development of the binomial distribution. However, the chi-square test is so sensitive to the very small cell frequencies, this test may not be found applicable in cases.

This operation restriction on the use of chi-square tests with small samples can be overcome partially by combining some of the classes for which expected frequencies work out less than 5 in 20% of the cells. But since some of the information in the data is lost because of regrouping the classes it should be avoided and the Kolmogorov-Smirnov test can be used instead of the chi-square test. This K-S test can accommodate very small samples because cell frequencies are summed cumulatively and directly taken into account in specifying the critical value of the test. This test we discuss now in the next chapter.

Finally, it should be mentioned that as the analysis of variance allows us to compare three simultaneously or more samples variances based on the interval or ratio data, the chi-square test by the contingency table provides the same facility with three or more sample observations based on nominal data.

LIST OF FORMULAE

Test Statistic χ^2

For one-sample test $\chi^2 = \sum \frac{(O-E)^2}{E}$ with $d.f. = k - 1$

For two or more than two-sample test

$$\chi^2 = \sum_i^r \sum_j^k \frac{(O_{ij} - E_{ij})^2}{E_{ij}} \quad \text{where } E_{ij} = \frac{\sum r_{ij} . \sum c_{ij}}{N} \quad \text{and } d.f. = (r-1)(c-1)$$

The critical value of χ^2 beyond $d.f. = 30$ is z equal to

$$\sqrt{2\chi^2} - \sqrt{2(d.f.)-1}$$

Chi-square test for independence of 2×2 *contingency table with* $d.f. = 1$

(a) by direct method, $\chi^2 = \frac{(ad-bc)^2 (a+b+c+d)}{(a+b)(c+d)(a+c)(b+d)}$

(b) by Yate's correction, $\chi^2 = \sum_i^r \sum_j^c \frac{(|O_{ij} - E_{ij}| - 0.5)^2}{E_{ij}}$

(c) by Fisher's Exact Probability test

$$p = \frac{(a+b)!(c+d)!(a+c)!(b+d)!}{n!a!b!c!d!}$$

Chi-Square test for independence of $r \times c$ contingency table with d.f. = N
Same as for two or more than two-sample test
For Goodness-of-Fit test

$$\chi^2 = \sum \frac{(O-E)^2}{E} \quad \text{as in one-sample test but } d.f. = k - 1 - m$$

Contingency or *Phi coefficient of correlation*

(a) for 2×2 table, $\varphi = \sqrt{\frac{\chi^2}{N}}$

(b) for $r \times c$ table, $\varphi = \sqrt{\dfrac{\chi^2}{N(L-1)}}$

Point Biserial Correlation Coefficient

$$r_b = \frac{\bar{Y}_1 - \bar{Y}_0}{s_Y} \cdot \sqrt{\frac{n_1 . n_0}{N(N-1)}}$$

Test of significance of r_b is $t_{N-2} = r_b \sqrt{\dfrac{N-2}{1-r_b^2}}$

Variance ratio Test of presence – absence data

$$VR = \frac{s_T^2}{\sigma_T^2} \text{ where } s_T^2 = \frac{1}{N} \sum_{j=1}^{N} (T_j - t)$$

and $\sigma_T^2 = \sum_{i=1}^{m} p_i (1 - p_i)$ with $p_i = \dfrac{n_i}{N}$, N being the total number of quadrats $\chi^2_{.05, N} < W$

$< \chi^2_{95, N}$ where $W = (N)(VR)$

EXERCISES

13.1 From a 1 : 50,000 geomorphological map, equal areas were selected consisting of sandstone, marl, shale, schist and limestone. Within each area all farms were visited and any farm having its 75% income from milk was classified as dairy farm. Following is the data of the dairy farms rock typewise.

Rock type	No. of dairy farms counted
Sandstone	8
Marl	10
Shale	6
Schist	3
Limestone	3

Can it be concluded that the reliance on milk production as a main element of the farm income is affected by the rock type?

13.2 In five drainage basins stream-ordering was done according to Strahler and following were the number of initial tributaries:

Drainage basins	No. of initial tributaries
Basin 1	32
Basin 2	62
Basin 3	49
Basin 4	51
Basin 5	23

Test the extent to which the observed differences in the number of initial tributaries are due to chance.

13.3 On the basis of the hypothesis that a good rainfall in July brings a good rainfall in August and September and *vice versa*, the monthwise rainfall data of July, August and September were collected for a place during 1900–1950. An attempt is made to qualify the combined August and September rainfall data against the July rainfall data in three categories of observations: wet, normal and dry. Following are the data:

	August and September		
July	Wet	Normal	Dry
Wet	8	11	2
Normal	4	4	8
Dry	5	2	7

Can we carry out a test for association and interpret the result? Determine the contingency coefficient and comment.

13.4 In a fieldwork slope angle measurements were done using Abney Level randomly for a number of sample sites on sandstone and shale rocks and the data were grouped as follows:

Sl. No.	Slope angle (in degrees)	No. of slope measurements on	
		Sandstone	Shale
1.	0 – 5	21	9
2.	5 – 10	18	11
3.	10 – 15	9	15
4.	15 – 20	8	13
5.	over 20	1	5

Test the hypothesis that the difference of slope angles are the result of chance variations and not due to the difference between the nature of rocks. At what level of significance is your result?

13.5 Following is a 2 × 2 data of soil type against parent material:

Soil type	Parent Materials	
	Alluvium	Non-alluvium
Glayed soil	15	5
Brown soil	5	30

Determine whether soil types and parent materials are related to each other. Apply the Yate's correction and Fisher's exact probability test.

13.6 A survey of all the farms in three separate valleys revealed that in some of the valleys more farmers tended to concentrate on livestock than in others. Before reasons were sought to explain the pattern it was decided to test the possibility that the variations observed were due to chance. The farms were roughly of the same size. The data collected were as follows:

Valley	A	B	C	D	E
Area	16	25	39	20	17
No. of farms with 75% income from livestock	9	7	12	3	10

Calculate the probability that this distribution is due to chance variations.

13.7 In a mountainous surrounding, a random sampling of abandoned farms was done in relation to the aspect of sites of the farm from the point of insolation and in relation to the height. The following data were collected:

Altitude in meters	Aspect and frequency of sampled abandoned farms			
	North	South	East	West
100 – 150	13	0	14	1
151 – 200	21	2	17	9
201 – 300	20	9	21	10
301 – 400	30	13	29	15

(i) Test that the frequency of the abandoned farms are irrespective of the altitude aspect and due to chance variations.

(ii) In this example, the influence of the aspect was examined through the whole circle of 360° and we choose the four cardinal points north, south, east and west. Would you expect any change from the answer in (i) if you take only the north and south aspects?

13.8 To find whether the bedrock character of the terrain has any association with settlement, the following data are taken about the villages, by their number, bedrock-wise and area-wise.

S.No.	Types of bedrocks	Observed frequency of villages	% of Total
1.	Grit	6	25
2.	Limestone	20	32
3.	Shale	16	43

Examine from the above table whether the number of villages on the three types of rock are so different that the observed variations being due to chance is acceptably low?

13.9 The distribution of tropical cyclones in a coastal area over a period of 70 years is as follows:

No. of cyclones in a year	Frequency
0	1
1	6
2	10
3	16
4	19
5	5
6	8
7	3
8	1
9	1

Test whether the above data can be fitted to a Poisson distribution.

13.10 The yield per acre in a flood plain for 40 sample farms were taken to determine whether or not the yield is normally distributed. The data are as follows:
8.46, 4.90, 9.85, 6.62, 6.21, 8.28, 7.06, 10.27, 8.06, 7.31, 8.16, 6.99, 7.90, 6.19, 6.60, 8.04, 9.48, 8.90, 9.37, 7.99, 6.66, 6.82, 8.20, 6.88, 9.02, 6.50, 7.52, 9.39, 9.04, 8.66, 8.77, 6.98, 8.08, 8.29, 7.62, 9.02, 7.15, 6.64, 9.12, and 8.87.
The above yield data are in 1,000s of kg. Test whether the data can be fitted to a normal distribution.

13.11 The rainfall data of a place were compared with its mean rainfall. In this comparison the 250 days' data were used and the amounts by which days' rainfall differ from the normal is classified as follows:

Deviations (cm)	No. of days
– 0.4 to – 0.2	7
– 0.2 to 0	11
0 to 0.2	28
0.2 to 0.4	49
0.4 to 0.6	78
0.6 to 0.8	41
0.8 to 1.0	20
1.0 to 1.2	11
1.2 to 1.4	5

Can we conclude with the hypothesis that the observed frequency of deviation of rainfall from mean is distributed adequately by the normal distribution with the same mean and standard deviation?

13.12 In a yield survey of paddy, 886 plots were sampled for testing whether the type of manuring is independent of the supply of irrigation or not. The data were cross-tabulated in the following way :

Manure	Irrigated	Unirrigated
No manure	123	413
Farmyard manure	81	223
Oilcakes	8	6
Other manure	14	18

14

Non-parametric Statistics in Geo-science Systems II (Kolmogorov-Smirnov Test)

14.1 THE K-S TEST

There are a number of non-parametric tests for testing the significance of the difference between the two independent samples. Not all of them test the same null hypothesis. Some are specific for testing if the two populations which are represented by two samples differ in the central tendency while others are better suited to test if the two populations differ in any respect, such as central tendency etc. We shall present here the *Kolmogorov-Smirnov test* ("K-S test") which is one of the most powerful non-parametric tests for differences between two cumulative frequency distributions of the observed and estimated ones. Like chi-square, this test deals with data in the form of frequencies, but from the classification of non-parametric test given in the earlier chapter, it can be seen that this test requires ordinal data i.e. data on ordered classes whereas chi-square test requires nominal data. Secondly, this test cannot be used to compare more than two samples in a single test. To counter these two limitations, this test has two advantages over the chi-square: (1) by ordering of the classes from smallest to the largest or *vice versa*, this test can use more information and so this test is preferred in many situations, again (2) this test can also accommodate very small samples and unlike chi-square test, in this test there are no limitations on frequencies required in categories, because cell frequencies are summed to total ($= n$) and they directly take into account in specifying the critical values. Thus where the chi-square test's sensitivity to very small cell frequencies make itself unsuitable when expected frequencies work out at less than 5 in 20 per cent of the cells, this K-S test can be applied.

14.1.1 Computing the *D* Statistic

As indicated in the chi-square test, there is often the need to test a sample against some predetermined population for which the distribution characteristics are known. Thus, quite frequently we make use of a variable which is continuous in nature, but the data is grouped into a relatively small number of ordered classes, when there are four or more such ordered groups, the K-S test is particularly suitable.

The principle behind this test is very simple. The statistic used in this test is the maximum number of the differences between the cumulative frequencies (or proportions) of the two distributions. This is done by first ranking the observations from the smallest to the largest for the two distributions, summing them together and expressing the result in fractions as a proportion of the respective totals, so that the cumulative distributions in both the sets range from 0 to 1.0. The maximum difference between two such distributions in terms of the absolute value of the difference is taken ignoring the signs. Thus the test statistic D is

$$D = \text{Max} \mid O_i - E_i \mid \qquad \ldots (14.1)$$

where O_i is an observed cumulative distribution function and E_i is an expected cumulative distribution function.

Obviously, one would expect same deviation between an observed (empirical) and an expected distribution as a result of sampling variations. Hence the rejection of the null hypothesis depends upon the deviations being large. Therefore if the maximum difference D is larger than would be expected as compared to the critical value of D given in the Appendix Table IX at a significance level, it indicates that the gap between the distributions has become so large that the null hypothesis is to be rejected. Like the chi-square test, this statistic can be used both to compare an empirical frequency distribution with a theoretical frequency distribution (one-sample test) or to compare two observed cumulative distributions with one another (two-sample test). Again this test can be directional or non-directional in two-sample test only.

14.1.2 Requirements for use of K-S test

The requirements of the K-S test are:

1. The data must be in the form of ordinal scale or at a higher level of measurement.
2. The observations must be independent events and there must be at least two mutually exclusive categories into which the observations are placed.
3. The population from which a sample is drawn should have a continuous distribution — however, this assumption can be relaxed without much effect.

We now discuss this K-S test with for its one- and two-sample cases.

14.2 ONE-SAMPLE TEST

In the one-sample test, as mentioned earlier, comparison is done between the cumulative proportions of the empirical frequency distribution with that of a theoretical frequency distribution. The rationale of this one-sample test is that if the null hypothesis is true, the cumulative proportions of the empirical distribution in corresponding classes would not differ much from the theoretical distribution. It is only when the maximum difference in the two cumulative proportions exceeds a critical value there is ground for rejecting the null hypothesis. This one-sample test is non-directional, that is, it can be performed in two-tailed test. We now present this one-sample test with the help of one example.

Example 14.1 The rainfall in one particular month is compared with the frequency of the daily rainfall amounts between given limits over a long term period. The data is as follows:

Rainfall in a day (in cm)	Ranked categories			
	0.00	0.01–0.04	0.05 – 0.20	> 0.20
Group I : Percent of days (total)	50	25	20	5
Group II : No of days (for one particular month)	12	6	9	3

It is clearly apparent that there is some difference in the month's rainfall distribution from the overall distribution, but estimate whether this difference is statistically significant?

Assumptions

Level of measurement	:	Percentage of days ordered into four groups of rainfall between given limits
Model	:	Independent random samples
Hypothesis	:	The particular month in question does not differ radically from the overall period as regards the frequency of daily rainfall amounts (= H_o)

Sampling Distribution

The K-S test is applied here. The data belongs to one-sample test.

Significance Level and the Critical Region

0.01 level of significance and a two-tailed test are selected.

Computing the Test Statistic

To apply the one-sample K-S test, we require to change both sets of the above values (total and one particular month) into proportions:

Rainfall in a day (in cm)	0.00	0.00 – 0.04	0.05 – 0.20	> 0.20
Expected overall proportions	0.50	0.25	0.20	0.05
Observed proportions for the month	0.40	0.20	0.30	0.10

Following this first step, the second step is to convert both these into cumulative proportions and then list the differences for each rainfall category as given below :

Cumulative Proportions

Expected overall	0.50	0.75	0.95	1.00
Observed for the month	0.40	0.60	0.90	1.00
Difference $D = \mid O_i - E_i \mid$	0.10	0.15	0.05	0.00

Therefore, the largest value on the difference:

$$D = \text{Max} \mid O_i - E_i \mid = 0.15$$

From Appendix Table IX it can be seen that the computed D value for a sample size of 30 days (= n) does not exceed the critical value of D at any significance level. Hence the difference is not found to be statistically significant.

14.3 TWO-SAMPLE TEST

This test is similar to the one-sample test, except that attention is focussed on the maximum difference between the two observed cumulative distributions. This test is used to determine whether the samples are drawn from the same or different proportions and this test is sensitive to the population difference with respect to the location, dispersion or skewness. Unlike the one-sample test, both one- and two-tailed tests can be performed in the two-sample case.

The number of observations, n in the two samples, sample I and sample II have to be equal until the sample sizes are large, $n > 35$. The two-tailed test is concerned with the maximum absolute difference between the two cumulative distribution functions (irrespective of sign) as in the one-sample case. The test criterion also ramains the same for the two-sample case. But the one-tailed test, in contrast, is based on maximum difference in a specified direction between two cumulative distributions, as before i.e.

$$D = \text{Max}\left|O_{n_1} - O_{n_2}\right| \qquad \text{... (14.2)}$$

where n_1 and n_2 are the sample categories.

The test criterion for the one-tailed test in this two sample test makes use of the chi-square distribution. So, for a one-tailed test, use is made of the calculated D by the above method and we evaluate χ^2 given by

$$\chi^2 = 4D^2 \frac{n_1 n_2}{n_1 + n_2} \qquad \text{... (14.3)}$$

where the *d.f.* associated with χ^2 are always 2 in this particular approximation formula. If this calculated χ^2 is less than the χ^2 value with $d.f. = 2$, the null hypothesis is retained. H_0 is rejected when $\chi^2 \geq \chi^2_\alpha$. We now explain the two-sample case with reference to both the one- and two-tailed tests in the Example 14.2 below.

Example 14.2 Following is the annual rainfall at 2 places A and B near each other in the same rainfall tract for 24 seasons. Rainfall is in centimetres.

Station at A : 39.59, 19.93, 23.91, 29.38, 43.09, 25.34, 49.35, 39.62, 42.90, 53.35, 57.66, 37.05, 34.14, 38.01, 52.40, 32.20, 47.81, 33.98, 39.46, 37.78, 63.24, 39.04, 60.51 and 38.08

Station at B : 39.48, 17.81, 24.47, 24.32, 41.18, 23.41, 45.13, 42.83, 46.94, 51.51, 57.50, 34.35, 34.29, 38.65, 50.32, 29.94, 45.25, 34.13, 40.68, 35.54, 57.24, 42.05, 55.35 and 37.45

Consider the data of annual rainfall distribution at these two places as independent. Suppose that either for reason the recording cannot be treated as very accurate or that the assumption of normality of distribution is considered doubtful and we wish to take recourse to the two-sample case of the K-S test which does not require the assumption of normality of population distributions of the data.

To apply the K-S test to the annual rainfall distribution data at two places A and B, we adopt a procedure below:

(a) We choose a suitable class interval of 13 cm for the data and prepare frequency distributions for both the samples, A and B, using the same interval. The frequency values are divided by $n = 24$ for the frequency proportions and these are shown in columns (2) and (3) of Table 14.1.

Table 14.1 : Frequency distribution

Class interval (in cm)	Cumulative frequency distribution		
	for A = (O_{n_1})	for B = (O_{n_2})	$\left\|O_{n_1} - O_{n_2}\right\|$
(1)	(2)	(3)	(4)
< 20	1/24	1/24	
20–25	2/24	4/24	2/24
25–30	4/24	5/24	1/24
30–35	7/24	8/24	1/24
35–40	15/24	12/24	3/24
40–45	17/24	16/24	1/24
45–50	19/24	19/24	
50–55	21/24	21/24	
55–60	22/24	24/24	2/24
60–65	24/24	24/24	

(b) These frequency proportions are cumulated and then we calculate the differences of the cumulative frequency proportions in corresponding classes, $(O_{n_1} - O_{n_2})$. These are given in column (4).

(c) We find the maximum difference, D which is $3/24 = 0.125$. For critical values of D in the K-S we look up the Appendix Table IX and for the one-tailed test, the D statistic is approximately distributed as χ^2 with $d.f. = 2$ as given in Eq 14.3. For the one-tailed test, if H_0 is not accepted, the research hypothesis may be either:

H_1: There is a significant difference in annual rainfall distribution at the two places, A and B.

H_2: The station A receives significantly larger amount of rainfall annually than the station B (i.e. the direction of the difference is specifically involved in an one-tailed test)

Now on applying the D value (i.e. $D = 0.125$) in Eq. (14.3), we find the computed value of χ^2 as 0.75 which is less than the critical value of χ^2 at 0.05 significance level (equal to 5.99 at 2 *d.f.*). Therefore the null hypothesis is accepted and it is unlikely that any significant difference exists between these two stations A and B. This is also supported from the Appendix Table IX since the computed value 0.125 of the D statistic is less than the critical value at 0.05 significance.

14.4 GOODNESS-OF-FIT TEST

Like chi-square, goodness-of-fit test finds an important use in K-S test. Unlike the chi-square, in K-S test even a data as small as 5 can be fitted to a probability distribution and in K-S test, this test of fitness can be done both for one- and two-sample cases. In Example 14.3 we illustrate the principle of application of the *D* statistic.

Example 14.3 Sixteen objects are located within the circumference of a circle. These objects occur along the unit radius from the centre of the circle and the unit distances are as follows

0.19, 0.06, 0.02, 0.02, 0.03, 0.01, 0.01, 0.04, 0.08, 0.01, 0.02, 0.02, 0.04, 0.08, 0.08 and 0.13

Can we infer that the above distribution of distances is a representative of uniform distribution?

This is a one-sample case and the null hypothesis in this case is that distribution of 16 (= *n*) objects from the centre of the circle is an uniform distribution, i.e. $H_0 : F_x(x) = F_u(x)$ for all 16 (= *x*) objects.

Now for along one unit distance from the centre of the circle to its circumference, if the 16 objects are spaced uniformly we can have the distribution of the 16 points at every 1/16 = 0.0625 unit distance. Hence we give the expected uniform distribution cumulated against the observed cumulated distribution in the table below :

Expected uniform spacing	Observed spacing	Expected cumulative spacing	Observed cumulative spacing	|Difference|
(1)	(2)	(3)	(4)	(5)
0.0625	0.19	0.0625	0.19	0.1275
	0.06	0.1250	0.25	0.1250
	0.02	0.1875	0.27	0.0825
	0.02	0.2500	0.29	0.0400
	0.03	0.3125	0.32	0.0075
	0.01	0.3750	0.33	0.0450
	0.01	0.4375	0.34	0.0975
	0.04	0.5000	0.38	0.1200
	0.08	0.5625	0.46	0.1025
	0.01	0.6250	0.47	0.1550
	0.02	0.6875	0.49	0.1975
	0.02	0.7500	0.51	0.2400
	0.04	0.8125	0.55	0.2625
	0.08	0.8750	0.63	0.2450
	0.08	0.9375	0.71	0.2275
	0.13	1.0000	0.84	0.1600

Hence, the Difference (maximum) = $D = 0.2625$. Now since for $n = 16$, the critical value of D statistic at $\alpha = 0.05$ is 0.328, so we accept the null hypothesis and conclude that the actual distribution function of distances is not different from the uniform distribution. In Fig. 14.1a, we plot the location of the 16 objects within the unit radius of a circle. The Fig. 14.1b shows the graphical illustration of the location of these 16 points superimposed with the straight line of the points at uniform spacing of 0.0625.

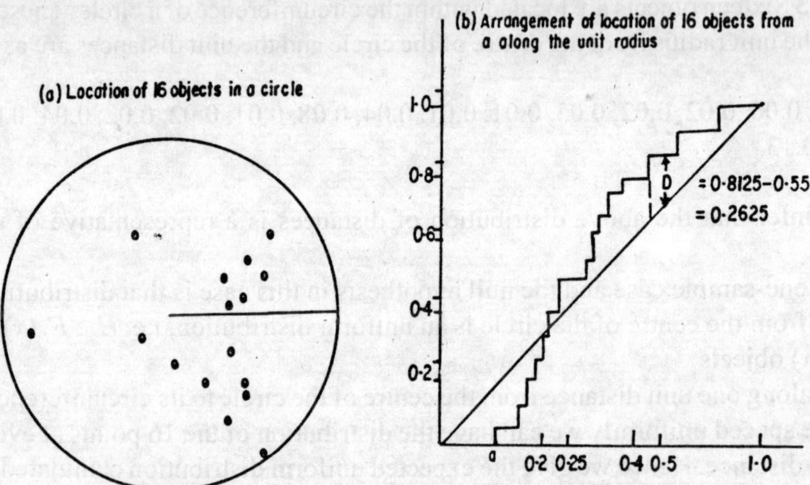

Fig. 14.1 : Graphical Illustration for a K–S One-Sample Example

14.5 K-S TEST FOR LARGE SIZE SAMPLES

When the sample sizes are large (more than 35 items), the method of operating the K-S test remains the same, but the critical probability limits of D are derived rather differently. Also, with samples of over 35 items for the two-sample case there is no longer the need for both samples to be of the same size. If necessary, the sample sizes will have to be rendered equal by omitting at random the excess number of observations from the larger sample, losing some information in the process. Smirnov has indicated that for a two-tailed test, in order to reject the hypotheses at different significance levels, we need the value of D to be larger than as shown in Table 14.2.

Table 14.2 : Critical values of D in K-S Test
(for Large samples : Two-tailed test)

Level of significance	Critical values of D	
	For Sample of unequal size	For the same sample size
0.20	$1.07\sqrt{\frac{n_1+n_2}{n_1 n_2}}$	$\frac{1.07}{\sqrt{n}}$
0.15	$1.14\sqrt{\frac{n_1+n_2}{n_1 n_2}}$	$\frac{1.14}{\sqrt{n}}$
0.10	$1.22\sqrt{\frac{n_1+n_2}{n_1 n_2}}$	$\frac{1.22}{\sqrt{n}}$
0.05	$1.36\sqrt{\frac{n_1+n_2}{n_1 n_2}}$	$\frac{1.36}{\sqrt{n}}$
0.01	$1.63\sqrt{\frac{n_1+n_2}{n_1 n_2}}$	$\frac{1.63}{\sqrt{n}}$
0.005	$1.73\sqrt{\frac{n_1+n_2}{n_1 n_2}}$	$\frac{1.73}{\sqrt{n}}$
0.001	$1.95\sqrt{\frac{n_1+n_2}{n_1 n_2}}$	$\frac{1.95}{\sqrt{n}}$

Example 14.4 A class teacher gives the same test to a class of 40 students and to another class of 50 students. The scores so obtained have been shown in the two frequency distributions given below:

Score	19	18	17	16	15	14	13	12	11	10
Frequencies in class A	3	5	6	10	8	3	2	1	0	2
Frequencies in class B	0	0	0	5	4	8	10	13	10	0

We take the frequencies accumulated and divide the cumulated frequency by the respective size of the students in the classes.

Score	19	18	17	16	15	14	13	12	11	10
Class A	0.075	0.200	0.350	0.600	0.800	0.875	0.925	0.950	0.950	1.00
Class B	0.000	0.000	0.000	0.100	0.180	0.340	0.540	0.800	1.00	1.00

Maximum Difference, $D = 0.620$

Now following Table 14.2 we calculate the critical value of D at different significance levels for this example of unequal samples, $n_1 = 40$ and $n_2 = 50$

Levels of significance	0.20	0.15	0.10	0.05	0.01	0.005	0.001
Critical Values of D	0.226	0.242	0.259	0.289	0.345	0.367	0.414

Now since the computed value 0.620 for D statistic is larger than any critical value above, so we conclude that the scores obtained by the two classes are significantly different.

LIST OF FORMULAE

Test statistic D : for one-sample test, $D = \text{Max} \mid O_i - E_i \mid$

for two-sample test, $D = \text{Max} \left|(O_{n_1} - O_{n_2})\right|$

For one-tailed test of two-sample case:

$\chi^2 = 4D^2 \dfrac{n_1 n_2}{n_1 + n_2}$ for H_o to be rejected $\chi^2 \geq \chi^2_\alpha$ at $d.f. = 2$

EXERCISES

14.1 In an attempt to quantify the settlement pattern, a grid overlay was placed over a settlement map and with some amalgamation of adjacent and partly-occupied grids, at the edge of the map, the map was found covered with 158 grids. The frequency of grids with 0, 1, 2, 3, 4, 5, 6 and 7+ towns were counted. Since the present test is concerned with whether the settlements are randomly distributed, the expected frequencies were calculated according to Poisson distribution. The observed and expected frequencies are given below.

No. of towns in a quadrat	Observed No. of quadrats	Expected No. of quadrats ($= p \times 158$)
0	85	59.61
1	36	58.10
2	16	28.31
3	9	9.20
4	6	2.24
5	2	0.44
6	3	0.06
7+	1	0.03

Apply the K-S test to find the null hypothesis, i.e. the settlements are randomly distributed.

14.2 The location theory of industries implies that industries involving weightloss in converting raw materials into a final product are attracted towards raw materials sources. In an analysis, the industries are put into two groups: (1) located wholly or partly at raw material sources, and (2) not located at raw materials sources. The groups are then classified according to their material index (an index of weightloss during manufacturing) using eleven categories ranked from A with considerable weightloss during manufacturing through to K where an increase in weight takes place. The resulting data consisted of 65 industries are given below:

Industry	Material Index A	B	C	D	E	F	G	H	I	J	K	Total
1. Located wholly/ partly at raw materials	2	4	1	3	4	4	4	3	3	3	0	= 31
2. Not located at raw materials	3	3	3	2	1	6	2	2	1	3	8	= 34

Test the hypothesis that industries located wholly or partly at raw material locations tend to involve greater weightloss and have higher materials indices than those not located at raw material locations.

14.3 A random sample is taken sequentially of sample size $n = 5$ for a random variable, x_i with the mean and standard deviation specified as $\mu = 49.3$ and $\sigma = 16.133$, test the hypothesis that the distribution of 5 values of x_i like 27.9, 37.6, 44.4, 55.2 and 65.9 does not differ from a normal distribution with the two parameters as specified.

15

Non-parametric Statistics in Geo-science Systems III (Mann-Whitney Test)

15.1 MANN-WHITNEY TEST

The *Mann-Whitney* test is one of the most versatile of the non-parametric tests which is used as an alternative to the parametric two-sample t test for finding the significance of differences in the means of the two independently drawn samples. To use the two-sample t test, however, it is necessary to make a set of stringent assumptions such as that the two independent samples be randomly drawn from normal populations having equal variances and that the data be measured on interval scale. However it is frequently the case in field surveys that only ordinal type of data can be obtained. Since the classical t test could not be used in such situations, a very simple but powerful non-parametric t test like the Mann-Whitney test may be used. In both the parametric t test and the non-parametric Mann-Whitney test, the test statistics are computed from the samples and they are compared with the sampling distribution of all possible sample outcomes. However in this Mann-Whitney test, no assumption need be made concerning the distribution of the data regarding its normality. This test can be employed when measurement is on ordinal scale, or when interval measurement have been placed in rank order. In addition, this test is effective with small, medium and large size samples equally.

15.1.1 Computing the U Statistic

To perform the Mann-Whitney test all the score in the two groups of samples of size n_1 and n_2 $(n_1 \neq n_2)$ are combined in one composite distribution in a decreasing order of size i.e. ranked from highest (considering algebraic signs when negative measures are present). Thus the ranks are assigned in such a manner that rank 1 is given to the smallest of the $n = n_1 + n_2$ combined observations, rank 2 is given to the second smallest, and so on until rank n is given to the largest. If in case several values are tied, then we would assign each of the tied observation the mean of the ranks which they jointly occupy, e.g if 6th, 7th and 8th values are identical, we would assign each the rank (6 + 7 + 8)/3 = 7

For convenience we must establish that whenever two sample sizes are unequal, $n_1 \neq n_2$ we will let n_1 represent the "smaller" sized sample and n_2 the "larger" sized sample. Mann-Whitney statistic U_1 is merely the sum of the ranks assigned to the n_1 observation in the smaller group. The statistic U_1 is thus computed corresponding to the smaller sample group (with the sum of ranks ΣR_1 and with sample n_1) according to

$$U_1 = n_1 n_2 + \frac{n_1(n_1+1)}{2} - \Sigma R_1 \qquad \text{... (15.1)}$$

Having done this, the U_2 statistic is computed for the larger sample group (with the sum of ranks ΣR_2 and with samples n_2)

$$U_2 = n_1 n_2 + \frac{n_2(n_2+1)}{2} - \Sigma R_2 \qquad \text{... (15.2)}$$

Thus on every occasion this test is used, there will be two values obtainable for the test statistic, U. The lower of two values, U_1 and U_2 has to be selected as the *U statistic* when consulting the statistical table for the critical values of U. If the higher value of U is calculated, it will be found that this value does not appear in the significance table. If this happens, the higher value U_2 can easily be transformed to the lower value, U_1 (= the U statistic) as follows:

$$U_1 = n_1 n_2 - U_2 \qquad \text{... (15.3)}$$

Equation (15.3) is also a quick check to clarify the smaller of the two U values.

The probability of sampling distributions of U will be obtained from Appendix Table X with reference to n_1 and n_2. For small-sized sample, n_1 or $n_2 \leq 8$, the probabilities read off in Tables A to F of Appendix Table X must be less than the rejection level if U is to be regarded as significant. For medium-sized samples, $9 \leq n_2 \leq 20$, if the calculated value of U is equal or less than the critical value at a significance level, null hypothesis is rejected and *viçe versa*.

For the small- and medium-sized values, in the U test for ranking, the rank of 1, as mentioned earlier, is usually given to the data with the lowest value (if the highest value is given to rank of 1 and U_1 and U_2 are calculated, they will be reversed).

15.1.2 *U* Transformation for Samples exceeding 20

The *U* test is described above for the cases where each of the sample group contained 20 or less than 20 observation, but if either sample group (n_1 or n_2) exceeds 20, the sampling distribution of *U* rapidly approaches a normal distributions. The following large sample approximation formula may be used for testing the null hypothesis when sample sizes are outside the range of Appendix Table X for Mann-Whitney *U* statistic

$$z = \frac{U - \bar{x}_u}{\sigma_u} \qquad \text{... (15.4)}$$

where *U* is smaller of the computed value U_1 and U_2, $\bar{x}_u$ is the mean value of *U* equal to

$$\frac{n_1 \, n_2}{2} \qquad \text{... (15.5)}$$

and σ_u is the standard error equal to

$$\sigma_u = \sqrt{\frac{n_1 \, n_2 \, (n_1 + n_2 + 1)}{12}} \qquad \text{... (15.6)}$$

Thus when either n_1 or n_2 sample size exceeds 20, it is possible to make use of the normal distribution for calculating a *z* score and then determining the significance of *U* from the normal distribution table as given in Appendix Table I(a) or I(b). Thus to find the significance of *U* for n_1 or $n_2 \geq 20$, we are able to calculate a *z* score using

$$z = \frac{U - \dfrac{n_1 \, n_2}{2}}{\sqrt{\dfrac{n_1 \, n_2 \, (n_1 + n_2 + 1)}{12}}} \qquad \text{... (15.7)}$$

The *U* value in this *z* score can be either the lower or higher value of the *U* score. Numerically, the result will be the same. The difference is that if the lower value of *U* is taken, the sign of the *z* score will be negative. This is obvious because *z* relates to the symmetrical normal curve and a positive score simply indicates the area above the mean and *vice versa*. Based on the level of significance selected, the null hypothesis may be rejected if the computed *z* value falls in the appropriate region of rejection, depending on whether a two-tailed or a one-tailed test is used .

15.2 CONCLUSION

Finally, it can be said that the Mann-Whitney *U* test is a powerful test and an excellent substitute for the *Student's t test*. Also, if ties occur between observations in the sample groups, the tied ranks do not greatly affect the outcome of the U test unless the same value occurs in the long run. We conclude this chapter by giving three worked examples, one each for the small-, medium- and large-sized samples.

Example 15.1 (for small-sized samples) The values in one-sample set are 15, 14, 23, 16, 20 and 23 and the values in the other sample set are 24, 19, 21, 13, 17, 19, 22 and 18. Test the hypothesis H_0 at the 0.05 level of significance.

To perform the test of the hypothesis we do the Mann-Whitney U test. The combined ranking of the two sample sets is done below.

Subset 1	Ranks	Subset 2	Ranks
15	3	24	14
14	2	19	7.5
23	12.5	21	10
16	4	13	1
20	9	17	5
23	12.5	19	7.5
		22	11
	$\Sigma R_1 = 43.0$	18	6
	$n_1 = 6$		
			$\Sigma R_2 = 62.0$
			$n_2 = 8$

We now obtain $U_1 = 26$ and $U_2 = 22$ and so the test statistic U is 22. Now since n_1 and $n_2 \leq 10$, so we consult the Table F of Appendix X where the probability read off for the $U = 22$ and $n_1 = 6$ is 0.426 which is more than the significance level 0.05, hence the null hypothesis that the two sample sets are not significantly different, cannot be rejected.

Example 15.2 (for medium-sized samples) Two set of samples, X and Y are given below regarding soil moisture content in two areas of a forest:

X	14	12	13	10	7	6	4	–	–
Y	18	16	15	14	19	7	8	6	3

Test whether the samples are taken from a common population.

Assumptions

Level of measurement	:	Data is in an ordinal scale
Model	:	Independent random samples
Null Hypothesis	:	Samples have been drawn from population having the same common distribution

Sampling Distribution

The Mann-Whitney U test is applied here.

Significance Level and Critical Region

0.05 level of significance is adopted.

Computing the Test Statistic U

All the soil moisture values in the two subsets X and Y are combined together and they are ranked as follows

										Total ranks
X	11.5	9	10	8	5.5	3.5	2	–	–	$\Sigma R_1 = 49.5$
Y	15	14	13	11.5	16	5.5	7	3.5	1	$\Sigma R_2 = 86.5$

n_1 and n_2 are equal to 7 and 9 respectively.

$$\text{So, } U_1 = n_1 n_2 + \frac{n_1(n_1+1)}{2} - \sum R_1$$

$$= 63 + 28 - 49.5 = 41.5$$

$$\text{and } U_2 = n_1 n_2 + \frac{n_2(n_2+1)}{2} - \sum R_2$$

$$= 63 + 45 - 86.5 = 21.5 \text{ (= } U \text{ statistic)}$$

Decision

From Table H of Appendix Table X, it can be seen that for $n_1 = 7$ and $n_2 = 9$ at the significance level $\alpha = 0.05$, the critical value of U is 12 which is less than the calculated value of U. Therefore, the null hypothesis is retained.

Example 15.3 (for large-sized samples) Following are the sets of two samples, A and B, of size =12 each

Sample A: 53, 38, 69, 57, 46, 39, 73, 48, 73, 74, 60 and 78
Sample B: 44, 40, 61, 52, 32, 44, 70, 41, 67, 72, 53 and 72

Treat the samples as large size samples and test the hypothesis at 10% that they come from populations with the same mean.

As before we do the combined ranking of the sample set (with $n_1 = n_2 = 12$) as follows

Sample A: 11.5, 2, 17, 13, 8, 3, 21.5, 9, 21.5, 23, 14 and 24
Sample B: 6.5, 4, 15, 10, 1, 6.5, 18, 5, 16, 19.5, 11.5 and 19.5

The total of above ranking for sample A is $\Sigma R_1 = 167.5$ and $\Sigma R_2 = 132.5$, Now $U_1 = 54.5$ and $U_2 = 99.5$, so the computed U statistic is 54.5.

If we treat the samples as of medium size from the Table H of the Appendix Table X for $n_1 = n_2 = 12$, the critical value is 37 against which the computed value of 54.5 for U is found larger. Hence we can conclude that the null hypothesis of the two sample sets sharing a common population is to be retained.

Now if we treat the samples large, the sampling distribution of U will approximate a normal distribution. Keeping this in view, we calculate the mean and standard deviation taking the null hypothesis that the two samples come from identical populations as under

$$\bar{x}_u = \frac{n_1\, n_2}{2} = 72 \text{ and } \sigma_u = \sqrt{\frac{n_1\, n_2\, (n_1 + n_2 + 1)}{12}} = 17.32$$

As the alternative hypothesis is that the mean of two populations are not equal, a two-tailed test is appropriate. Accordingly the limits of the acceptance region at 0.10 level of significance can be worked as under :

Since z value for 0.45 of the area under the normal curve is 1.64, we have following limits of acceptance region :

upper limit = $\bar{x}_u + 1.64\, \sigma_u = 72 + 1.64\, (17.32) = 100.40$

lower limit = $\bar{x}_u - 1.64\, \sigma_u = 43.60$

Since the observed U statistic is 54.5 which is in the acceptance region, we accept the null hypothesis at 10% level. The other value of U is 89.5 and the conclusion remains the same.

LIST OF FORMULAE

U computed is the lower of U_1 and U_2 where $U_1 = n_1 n_2 + \dfrac{n_1 (n_1 + 1)}{2} - \Sigma R_1$ and

$$U_2 = n_1\, n_2 + \frac{n_2\, (n_2 + 1)}{2} - \Sigma R_2$$

In case U_1 is unknown and U_2 known, $U_1 = n_1 n_2 - U_2$. For large sample size, U distribution becomes normal distribution with

$$z = \frac{U - \bar{x}_u}{\sigma_u}$$

where $\bar{x}_u = \frac{n_1 n_2}{2}$, and $\sigma_u = \sqrt{\frac{n_1 n_2 (n_1 + n_2 + 1)}{12}}$

EXERCISES

15.1 Stone roundness coefficients of samples from a stream confluence with its tributary are in two sample sets:

n_1 = 51, 32, 23, 6, 13, 13, 19, 47, 27, 20, 7, 12, 13, 19, 39, 27, 24, 9, 11, and 17

n_2 = 58, 10, 52, 47, 23, 27, 38, 30, 57, 16, 47, 45, 25, 28, 33, 30, 56, 19, 47, 44, 27, 29 and 30

Test the hypothesis that two set of samples indicate that pebbles forming the bank of the main river is supplied not from its stream but from higher up the main river channel.

15.2 In a study of farm income, 19 farms were taken from the subsets: 9 from the chalk, 10 from the shale.

X (on chalk): 791, 918, 501, 403, 556, 715, 90, 326, 756

Y (on shale): 648, 76, 36, 128, 20, 139, 72, 118, 104 and 116

Test whether the distribution of farm income is the same irrespective of the locational character of them.

16

Non-parametric Statistics in Geo-science Systems IV (Wilcoxon-Ranked Pair Test)

16.1 WILCOXON-RANKED PAIR TEST

The *Wilcoxon-Ranked Pair Test* is used to test the significance of the difference between matched paired values. It may be chosen over its respective parametric counterpart – the one-sample *t* test or the paired-difference *t* test – when it is possible to obtain data measured at a higher level than an ordinal scale but does not believe that the assumptions of the parametric tests are sufficiently met. When the assumptions of the *t* test are violated, the Wilcoxon procedure is likely to be more powerful in detecting the existence of significant differences than its corresponding *t* test. Moreover, even under conditions appropriate to the *t* test, this test has proved to be almost as powerful.

For considering samples as *paired*, data considering the same place but taken at different times are regarded as pairs. For example, if the yield of crop in two areas are taken from each of the same areas but in different years and it is then desired to find out whether better seeds, better agricultural techniques or some other factors had resulted in significant difference in the yield, we would be dealing with paired values. Similarly, the employment structure in a city can be regarded as paired if a comparison is made between two census dates. An example of paired values where the time component is not involved can be the comparison of a series of slope angles matched on opposite side of a valley in a homogeneous material. Again, on the time scale, the changing angles of slope on beach profiles from season to season is another example where the samples can be paired for the application of this test.

16.1.1 Computing the *T* Statistic

This test is based upon the differences between matched paired values both *A* and *B* and while matching paired values for the individualities, it makes use of the sign and magnitude of the rank of differences between the matched paired values, $A - B$. Now the difference $(A - B)$ may be taken for analysis to test the hypothesis that there is no difference between the scores of A and B and we perform the Wilcoxon-Ranked Pair test in the following six-step procedure:

1. For each item in a sample of n items we obtain a "difference score D" (to be described in sections 16.2. and 16.3. for the one-sample and paired-sample tests respectively afterwards).
2. We then neglect the + and – signs and obtain a set of absolute differences $|D|$.
3. We omit from further analysis any absolute difference score of zero (i.e. $D = 0$) and take "N" be the number of absolute non-zero differences, $N \leq n$.
4. We then assign ranks R from 1 to N to each of the $|D|$ such that the smallest absolute difference gets ranks 1 and the largest gets rank N. In case two or more $|D|$ are equal, they are each assigned the "average rank" of the ranks they otherwise would have individually been assigned had ties in the data not occurred.
5. We now reassign the symbol + or – to each of the non-zero absolute difference scores $|D|$, depending on whether D was originally positive on negative. Next the positive as well as negative matched-pair differences are summed.
6. The smaller of these matched-pairs difference is termed the statistic T. Thus $T = \sum R^+$ provided $\sum R^+ < \sum R^-$ or *vice versa*. ...(16.1)

The Wilcoxon-Ranked Pair test can be used with small samples ($N = 6$ to $N = 25$), N is the number of non-zero absolute difference in the samples for which distribution tables are given in Appendix Table XI. This table can be used for obtaining the critical values of the test statistic T for both one- and two-tailed tests at various levels of significance.

For samples of $N \leq 100$, the Appendix Table XI may be used for obtaining the critical values of the test statistic T for both one- and two- tailed tests at various levels of significance. For a two-tailed test and for a particular level of significance, if the observed value of T is equal to or less than the critical value, the null hypothesis may be accepted. For a one-tailed test in the negative direction the decision rule is to reject the null hypothesis if the observed value of T is less than or equal to the critical value.

16.1.2 Transformation of *T* statistic for Samples Exceeding 25

When N, the number of pairs of samples exceeding 25, Appendix Table XI need not be used to establish the significance for samples of $N > 25$, since the test statistic T is approximately normally distributed. However when this happens, the T statistic can be found in the way described above, but it has to be converted to a z score using the formula for sample mean $\bar{x}_T$ and standard deviation of the sample σ_T as given by

$$z = \frac{T - \bar{x}_T}{\sigma_T} \qquad \text{... (16.2)}$$

where T = computed T statistic, the smaller of ΣR^+ and ΣR^-, $\bar{x}_T$ and σ_T the mean and standard deviation of the T statistic respectively.

Now since the sum of the first N integers $(1, 2, \ldots, N)$ is given by $N(N + 1) / 2$, the Wilcoxon-statistic T may range from a minimum of 0 (when the sum of the ranks for positive differences ΣR^+ and the sum of the ranks for negative differences ΣR^- are not different) to a maximum of $\frac{N(N+1)}{2}$ (when all the difference scores D are either all positive or all negative). If the null hypothesis is true, we would expect the test statistic T to take on a value close to its mean

$$\bar{x}_T = \frac{N(N+1)}{4} \qquad \text{... (16.3)}$$

while if the null hypothesis is false, we would expect the observed value of the test statistic to be close to one of it extremes, σ_T which is

$$\sigma_T = \sqrt{\frac{N((N+1)(2N+1)}{24}} \qquad \text{... (16.4)}$$

Thus using the above expressions [Eqs. (16.3) and (16.4)] in the formula for

$$z = \frac{\text{Difference of } T \text{ from sampling distribution}}{\text{Standard deviation of sampling distribution}}$$

$$z = \frac{T - N(N+1)/4}{\sqrt{N(N+1)(2N+1)/24}} \qquad \text{... (16.5)}$$

where N = the number of pairs of observation with a non-zero difference and T = the test statistic as smaller of the sum of matched pair differences, positive or negative.

Now based on the level of significance selected, the null hypothesis may be rejected if the computed z score falls in the appropriate region of rejection depending on whether a two-tailed or one-tailed test is used. Accordingly a result is taken as significant at $\alpha = 0.05$ if z in the Eq. (16.5) is more than 1.96.

The only difference between the one-sample and the paired-sample test is described below. Their only difference lies in how the ranks are computed. After that step is done, the computation procedures for the test statistic T are identical.

16.2 ONE-SAMPLE TEST

If one is interested in testing a hypothesis regarding population median M_e based on data from a single set of sample, the Wilcoxon-Ranked Pair test may be used. The test of the null hypothesis may be one-tailed or two-tailed.

The assumptions necessary for performing the one-sample test are :

1. The observed data $(x_1, x_2, \ldots x_N)$ constitute a random sample of N independent values from a population with unknown median.
2. The underlying random phenomenon of interest be continuous.
3. The observed data be measured at a higher level than the ordinal scale.
4. The underlying population be approximately symmetrical but not necessarily be normal .

We now explain this one-sample test with the help of Example 16.1 below.

Example 16.1. The level of water pollution in a river is measured by the biochemical oxygen demand (BOD) method. The data is obtained for 21 sampling points along downstream of a river which is as follows:

0.6, 0.8, 3.1, 3.6, 6.6, 5.4, 2.2, 1.8, 1.4, 6.7, 4.6, 3.8, 2.9, 3.0, 2.4, 2.2, 2.4, 2.2, 1.7, 1.5 and 1.5.

Is there any evidence that the median value of BOD will exceed 2.2 ?

The following null and research hypotheses are established:

$$H_0 : \text{Median} \leq 2.2$$
$$H_1 : \text{Median} > 2.2$$

and the test is one-tailed.

To perform the one-sample test the first step is to obtain a set of difference scores D between each of the value of BOD and the hypothesized median M_e – that is, $D = x - M_e$

The remaining steps of the six-step procedure are done as follows:

Difference scores D is placed in column 2, noting the sign. All the zero differences are ignored, now the values of difference scores are placed in column 3 irrespective of the sign. The rank of each positive differences R^- $(x > M_e)$ are now placed in column 4 and the rank of each negative differences R^- $(x < M_e)$ in column 5. The smaller of the two sums ΣR^+ and ΣR^- is termed as the T statistic.

Stated value of BOD, x	$D = x - 2.2$	$\|D\|$	Ranks of R^+ (A > B)	R^- ($A < B$)
1	2	3	4	5
0.6	– 1.6	1.6		13.5
0.8	– 1.4	1.4		11.5
3.1	0.9	0.9	10.0	
3.6	1.4	1.4	11.5	
6.6	4.4	4.4	17.0	
5.4	3.2	3.2	16.0	
2.2	zero			
1.8	– 0.4	0.4		3.0
1.4	– 0.8	0.8		8.5
6.7	4.5	4.5	18.0	
4.6	2.4	2.4	15.0	
3.8	1.6	1.6	13.5	
2.9	0.7	0.7	6.0	
3.0	0.8	0.8	8.5	
2.4	0.2	0.2	1.5	
2.2	zero			
2.4	0.2	0.2	1.5	
2.2	zero			
1.7	– 0.5	0.5		4.0
1.5	– 0.7	0.7		6.0
1.5	– 0.7	0.7		6.0
		$\Sigma R^+ = 118.5$	$\Sigma R^- = 52.5$	$T = 52.5$

Appendix Table XI gives, the critical value of T. If the observed value of T (which is 52.5) is more than the critical value of T, the null hypothesis can be true. At α level selected at 0.05, the critical value of T statistic for two-tailed test of $N = 18$ is 40 which is less than the computed T value of 52.5. Hence the null hypothesis is accepted and we can say that the true median of the BOD values may not exceed 2.2.

It is interesting that the large-sample approximation formula Eq. (16.5) for the test statistic $T (= 52.5)$ yields excellent results in the present example, for $N = 18$, the number of absolute non-zero differences, D

$$z = \frac{T - N(N+1)/4}{\sqrt{N(N+1)(2N+1)/24}}$$

$$= \frac{52.5 - 85.5}{\sqrt{527.25}} = \frac{-33.0}{22.96} = -1.44$$

which is less than the critical value of 1.96, hence the null hypothesis would be accepted.

16.3 PAIRED-SAMPLE TEST

In research it is frequently of interest to examine differences between two related groups. For example, in a geomorphological analysis of an area, a comparison can be done about the role of the aspect of site in slope angles, for which repeated measurements on slopes facing certain directions like northwest site and southwest site can be obtained in the data below

NW slope (in °) : 10, 7, 9, 10, 15, 12, 8 and 9
SW slope (in °) : 11, 7, 13, 8, 12, 15, 10 and 15

We thus have situations, involving matched (paired items) on the same attribute are obtained. For this type of situations the Wilcoxon-Ranked Pair sample test can be used when the *t* test is not appropriate and we can test the hypothesis that slope angles are not significantly different due to difference in aspects.

We apply the six-step procedure to this two-sample test and get a table as follows:

Check points	NW Slope (A)	SW Slope (B)	Difference (D)	$\lvert D \rvert$	R^+ ($A > B$)	R^- ($A < B$)
1	10	11	–1	1		1.0
	7	7	0	Discarded from sample list because $A = B$		
2	9	13	– 4	4		6.0
3	10	8	2	2	2.5	
4	15	12	3	3	4.5	
5	12	15	– 3	3		4.5
6	8	10	– 2	2		2.5
7	9	15	– 6	6		7.0
					$\Sigma R^+ = 7.0$	$\Sigma R^- = 21.0$

Hence the T statistic is 7.0. Now from Appendix Table XI, the critical value for $N = 7$ at 0.05 significance is 2. Since $T > 2$, the null hypothesis is accepted that there is no significant difference in slope angle due to change of aspect of the site.

Now we apply this paired-sample test to a large-sized sample, $N > 25$ below.

Example 16.2. A sample survey was taken in 31 selected plots where two brands of fertilizer (A and B) were compared regarding the yields of a crop per hectare. Each plot was divided into one-hectare subplots with brand A fertilizer assigned to one plot and brand B fertilizer to the other. The data, the yield in kg. per hectare for the individual plots are listed in the following table.

Selected Plots	Yield in kg. per hectare for subplots	
	Fertilizer A	Fertilizer B
1	2	3
1	658	675
2	631	656
3	582	578
4	600	600
5	670	687
6	685	701
7	740	747
8	736	751
9	603	590
10	594	588
11	611	590
12	625	650
13	675	709
14	656	679
15	640	652
16	670	677
17	732	728
18	693	718
19	683	660
20	668	677
21	654	666
22	660	662
23	672	697
24	648	663
25	631	638
26	638	622
27	598	625
28	644	697
29	617	675
30	590	646
31	560	620

Test the significance of the difference in the two brands of fertilizers A and B in terms of the difference in yield capacity.

As in the previous example, the test statistic T is calculated as follows:

Selected Plots	A – B	| A – B |	Ranks of	
			R^+ (A > B)	R^- (A < B)
1	2	3	4	5
1	– 17	17		16.5
2	– 25	25		22.5
3	4	4	2.5	
4	Dropped from sample list because A = B			
5	– 17	17		16.5
6	– 16	16		14.5
7	– 7	7		6.0
8	– 15	15		12.5
9	13	13	11.0	
10	6	6	4.0	
11	21	21	18.0	
12	–25	25		22.5
13	–34	34		26.0
14	–23	23		19.5
15	–12	12		9.5
16	–7	7		6.0
17	4	4	2.5	
18	–25	25		22.5
19	23	23	19.5	
20	–9	9		8.0
21	–12	12		9.5
22	–2	2		1.0
23	–25	25		22.5
24	–15	15		12.5
25	–7	7		6.0
26	16	16	14.5	
27	–27	27		25.0
28	–53	53		27.0
29	–58	58		29.0
30	–56	56		28.0
31	–60	60		30.0
Total :			Σ72.0	Σ393.0

Therefore, the test statistic is ΣR^+ equals to 72.0 and it can be seen that the number of pairs in the sample is reduced by one from the original 31 to 30 because one pair of value is same ($D = 0$) and hence dropped. Now substituting this value of T in Eq. 16.5, we get a z score for $N = 30$.

$$z = \frac{T - N(N+1)/4}{\sqrt{N(N+1)(2N+1)/24}} = \frac{72.0 - 232.5}{48.62} = -4.78$$

It is therefore possible to reject H_0 at the 0.05 level of significance since the calculated z (= 4.78) is larger than the critical z (which at the 0.05 significance level is 1.96). Thus we can conclude that there was a significant difference in the two brands of fertilizer in terms of difference in the yield capacity.

Note : This can also be verified from the Appendix Table XI where the critical value of T is 137 at 0.05 significance level for which T computed is 72.

LIST OF FORMULAE

T statistic for sample sizes $N \leq 25$ is the minimum of the sum of rank scores, ΣR^+ and ΣR^- for N = number of pairs of observations with a non-zero difference, D

T transformation: for large samples, $N \geq 25$

$$z = \frac{T - N(N+1)/4}{\sqrt{N(N+1)(2N+1)/24}}$$

EXERCISES

16.1 Two different modes of agricultural development packages A and B were tried on 16 sets of plots. Each of the plots is asked to score the result of the two packages. The scores are given in rows (2) and (3) in the data below :

Plot No.:	1	2	3	4	5	6	7	8	9	10	11	12	13	14	15	16
Score for package A:	5	8	6	8	8	7	8	4	4	8	6	7	6	7	7	8
Score for package B:	5	6	7	3	6	5	4	7	8	6	6	4	6	8	5	6

Test the hypothesis that there is no significant difference in the two packages A and B.

16.2 Examine the following data of yield of a group in 12 plots ranked in one bad year and in one good year. Ranks are given on a six-point scale.

Plot→ Year↓	1	2	3	4	5	6	7	8	9	10	11	12
Bad	2	6	5	4	4	1	6	3	3	3	4	6
Good	3	6	2	4	1	2	1	6	1	5	5	3

Examine the proposition that there is no rank difference between the plots in terms of good or bad years.

16.3 The level of water pollution in a river is measured in two ways: (one) by the biochemical oxygen demand (BOD) and (other) by ammoniacal nitrogen (AN). These values are observed for sampling point along downstream of a river at 2 periods in an interval of 3 years as follows.

BOD.	1st period :	4. 4. 9. 15. 10. 12. 9. 7. 7. 14. 20. 13. 13. 13. 10. 11. 9. 11. 9. 8 and 8
	2nd period :	5. 5. 12. 25. 11. 13. 9. 9. 9. 15. 15. 14. 15. 16. 13. 11. 12. 13. 11. 11 and 13
AN	1st period :	0.6. 0.8. 3.1. 3.6. 6.6. 5.4. 2.2. 1.8. 1.4. 6.7. 4.6. 3.8. 2.9. 3.0. 2.4. 2.2. 2.4. 2.2. 1.7. 1.5 and 1.5
	2nd period :	1.1. 0.7. 5.6. 11.1. 9.7. 6.6. 2.9. 2.4. 1.6. 7.7. 7.3. 5.6. 4.8. 3.6. 2.7. 2.4. 2.9. 2.5. 2.3. 1.8 and 1.4

Test the significance in difference in level of water pollution by the two methods.

17

Non-parametric Statistics in Geo-science Systems V

(Kruskal-Wallis Test)

17.1 KRUSKAL-WALLIS TEST

The direct generalization of the two-sample Wilcoxon Ranked Pair test to its extension of the many sàmple situation is known as the "H test" developed in 1952 by W. H. Kruskal and W. A. Wallis. Named after these two inventors, the 'Kruskal-Wallis test' is an appropriate test whenever we have more than two independent samples and an ordinal scale of measurement and we have to assess the contribution made by each of them in the total variability of data. In fact, this test is an adaptation of the analysis of variance F test with a single criterion of classification for use with ranked data. Thus this test enjoys the same power properties relative to the one-way classification in the analysis of variance, the latter has been considered earlier in Chapter 12. As with its two-sample counterpart, the Mann-Whitney test, this method is useful not only for intrinsically ordinal data, but also for the interval and ratio scale data that have been reduced to ranks to avoid problems of normality or small sample size.

17.1.1 Computing the H Statistic

To employ the H test, the measurements need only be ordinal from k number of sample groups, and the common population distributions need only be continuous–their common shapes are irrelevant. On the other hand, to utilise the parametric test, the level of measurement must be more sophisticated and we must assume that k number of samples are coming from underlying normal population having equal variances, that is, common shapes.

The steps in applying this test is basically a very simple one. To perform this test we must first rank together all the observations under the samples from the smallest to the largest, irrespective of the groups of samples such that rank 1 is given to the smallest of the combined observations and rank n_k to the largest of the combined observations (where $N = n_1 + n_2 + \ldots + n_k$). The sums of the ranks are then computed for each of the nominal-scale categories of the variables within each of the k number of independent samples and they are designated as $R_1, R_2, \ldots, R_k$. If any values are tied, they are assigned the average of the ranks. These R-values are then included within the equation and a test statistic H is computed in order to measure the degree to which the various sums of ranks differ from what could be expected under the null hypothesis. The H statistic is computed from the formula

$$H = \left[\frac{12}{N(N+1)} \sum_{j=1}^{k} \frac{R_j^2}{n_j} \right] - 3(N+1) \qquad \ldots (17.1)$$

where N is the total number of observations over the combined samples i.e. $N = n_1 + n_2 + \ldots + n_k$

n_j is the number of observations in the jth sample, i.e. $j = 1, 2, \ldots, k$.

R_j^2 is the square of the some of the ranks assigned to the jth sample.

Part of the formula (Eq. 17.1) may look familiar, $\sum_{j=1}^{k} (R_j^2 / n_j)$ is very much like part of the formula for computing the between groups sum of squares, *SSB* in the analysis of variance. In fact, if we were to compute the between sum of squares using the ranks instead of the raw data and then multiply by the constant $12/N(N+1)$ to subtract $3(N+1)$ from the product, we would get "*H*". *H* test is, therefore, a measure of discrepancy among the mean scores of the groups as in ANOVA.

A "few samples" for the *H* statistic, in Appendix Table XII is defined as a situation in which there are not more than three samples (i.e. $k = 3$) and not more than five values (i.e. $n_j = 5$) in each sample group.

If there are more than five values ($n_j > 5$) in each sample group and/or if the number of samples group are more than three ($k > 3$), the sampling distribution of H is approximated by χ^2 distribution with *d.f.* equal to $(k - 1)$ i.e. one less than the number of sample groups. We use the χ^2 distribution table given in the Appendix Table VIII by replacing the H statistic in Eq. (17.1) with χ^2 as in below

$$\chi^2 = \left[\frac{12}{N+(N+1)} \sum_{j=1}^{k} \frac{R_j^2}{n_j} \right] - 3(N+1) \qquad \ldots (17.2)$$

with $d.f. = k - 1$

The general null hypothesis H_0 is that all the k populations of measurements on the same attribute are identical with the same mean and each of the different arrangements of the ranking among the k samples is equally likely. On other hand, the alternative hypothesis

H_1 is that they are not all identical having different distribution characteristics and at least one of the populations of measurement leads to larger values than at least one of the other populations. Since the value of $\sum_{j=1}^{k} \frac{R_j^2}{n_j}$ for k samples, as stated earlier, is a measure of the variability of the sample group's rank totals, when the populations have different means, we expect $\sum_{j=1}^{k} \frac{R_j^2}{n_j}$ to be large. And in principle, greater inter-group differences provide large H statistic, which, if greater than the critical value, cause H_0 to be rejected. Thus for $k > 3$ and/or $n_j > 5$, when χ^2 distribution approximates H distribution, if the computed χ^2 value in Eq. (17.2) exceeds the critical χ^2 value at $d.f. = k - 1$, the null hypothesis H_0 is rejected.

This Kruskal-Wallis test is having one limitation that this test does not isolate the pairs that have different mean nor by accepting the null hypothesis does it exclude the possibility that for one pair of samples the null hypothesis could be rejected by the Wilcoxon-Ranked Pair test at the same significance level.

17.1.2 Correction for Tie

Equation (17.1) or (17.2) for the H test is for samples without tie as it was assumed that the underlying random phenomenon of interest is on a continuous scale. But in many practical cases some tied values do occur. So long as there are few tied values, the H test requires no correction. But when a large number of tied values occur in the samples (e.g. more than 25% of size of the samples, N), the H value becomes slightly smaller and the test becomes less rigorous. The correction required in this case is simply to divide H [i.e. Eq. (17. 1)] by

$$1-\frac{\Sigma(t^3-t)}{(N^3-N)}$$ so that

$$H_C = H / \left[1-\frac{\Sigma(t^3-t)}{N^3-N}\right] \qquad \text{... (17.3)}$$

where t is the number of observations tied for a given rank in each sample group and H_c is the correct H statistic.

We give below two examples on H test, one example without any tied rank in the samples and the other with ties. In the Example 17.1 below the number of sample groups are $k = 3$ and sample sizes are $n_1 = 5$, $n_2 = 4$ and $n_3 = 3$.

Example 17.1. In a geomorphological study of slope angles of screes in a glaciated area, three bedrock areas, namely, limestone, schist and granite were selected and the following slope angles were measured in these bedrock areas.

	Limestone	Schist	Granite
Slope angles	17	21	30
	20	18	26
	19	25	22
	24	28	
	23		

Test the hypothesis that the angle at which the scree slope develops is related to the type of the bedrock from which it is formed, rather than to the process of shattering common to all.

Assumptions

Level of measurement : Nominal and ordinal scales in continuous measurement
Model : Independent random sampling from continuous population
Critical region : Two-tailed test of significance level of $p = 0.05$

Computation of Test Statistic

The sampling distribution will follow the Kruskal-Wallis test. The three samples are ranked and the sum of the ranks are calculated as follows.

Procedure for Computation

1. We rank all samples irrespective of the group, i.e. from 1 to N
2. After all the ranks are assigned, we then obtain the sum of the ranks for each sample group : $R_1, R_2, \ldots, R_k$
3. Next we calculate $\sum_{j=1}^{k} \frac{R_j^2}{n_j} = \frac{R_1^2}{n_1} + \ldots + \frac{R_k^2}{n_k}$

We observe in the given data, there is no tied rank. The computed data are tabulated below:

	Lime-stone	Rank	Schist	Rank	Granite	Rank
Slope angles	17	1	21	5	30	12
	20	4	18	2	26	10
	19	3	25	9	22	6
	24	8	28	11		
	23	7				
Sum of Ranks $= (\Sigma R_j)$		23		27		28

4. Now we compute the H statistic, since there is no tie, H computed need not be corrected.

Therefore, $$H = \frac{12}{N(N+1)} \sum_{j=1}^{k} \frac{R_j^2}{n_j} - 3(N+1)$$

$$= \frac{12}{12(12+1)}\left(\frac{23^2}{5} + \frac{27^2}{4} + \frac{28^2}{3}\right) - 3(12+1)$$

$$= 3.26$$

5. Since there are only three groups and not more than five scores in each group, we compare the H statistic (calculated) with the critical value of H given in Appendix Table XII. H must be larger than the critical value given for significance.

Decision

Referring to Appendix Table XII we see that with the three sample sizes, $n_1 = 5$, $n_2 = 4$ and $n_3 = 3$ the critical value at 0.05 level is 5.6308 which is more than the calculated value of H (= 3.26). So it is not possible to reject the null hypothesis and we conclude that the screes developed from different bedrocks do not form different slope angles.

Example 17.2 Samples of beach sand were collected from half way between the mean high and low water shorelines of six beaches. The samples were sieved to grain size and they were weighed as given in the data tabulated below :

Sand	Samples from six beaches by weight (in g)					
Grain-size by ϕ-units	Sample A	Sample B	Sample C	Sample D	Sample E	Sample F
– 2 to – 1	105	282	125	307	60	291
– 1 to 0	123	221	. . .	85	105	324
0 to 1	. . .	248	. . .	201	45	112
1 to 2	145	252	55	184	355	108
2 to 3	78	148	121	355	. . .	165
3 to 4	160	131	77	175	. . .	380
4 to 5	82	273	67	. . .	55	. . .

Test the hypothesis that the samples are drawn from the same population.

We do the combined ranking as follows :

Sample A	Sample B	Sample C	Sample D	Sample E	Sample F
10.5	29.0	16.0	31.0	4.0	30.0
15.0	25.0	...	9.0	10.5	32.0
...	26.0	...	24.0	1.0	13.0
18.0	27.0	2.5	23.0	33.5	12.0
7.0	19.0	14.0	33.5	...	21.0
20.0	17.0	6.0	22.0	...	35.0
8.0	28.0	5.0	...	2.5	...
$\Sigma R_1 =$ 78.5	$\Sigma R_2 =$ 171.0	$\Sigma R_3 =$ 43.5	$\Sigma R_4 =$ 142.5	$\Sigma R_5 =$ 51.5	$\Sigma R_6 =$ 143.0

Now we assume that variations in the amount of particles of different sizes are due to chance and that the samples are drawn from one population, the level of significance selected is 0.01. Since it is a large size sample, k being more than 3 and the number of samples in the categories are more than 5, so we do the χ^2 approximation to the H distribution and we compute

$$\chi^2 = \left[\frac{12}{N(N+1)} \sum_{j=1}^{k} \frac{R_j^2}{n_j}\right] - 3(N+1)$$

$$= \frac{12}{35(35+1)} \cdot \left[\frac{78.5^2}{6} + \frac{171.0^2}{7} + \frac{43.5^2}{5} + \frac{142.5^2}{6} + \frac{51.5^2}{5} + \frac{143.0^2}{6}\right] - 108.0$$

$$= 122.91 - 108.0 = 14.91$$

Now for $d.f. = k - 1 = 5$ for which the χ^2 table shows that at 0.01 significance level it is 15.08, so we cannot reject the null hypothesis.

But since the critical value of χ^2 is marginally higher than the computed χ^2 value and since there are 3 pairs of tied ranks, we can have the correction for tie

$$1 - \frac{\Sigma(t^3 - t)}{N^3 - N} = 1 - \frac{(2^3-2)+(2^3-2)+(2^3-2)}{(35^3-35)}$$

$$= 1 - \frac{18}{42840} = 0.99958$$

and so H corrected for ties is : $H_c = \frac{14.91}{0.99958} = 14.92$

LIST OF FORMULAE

H statistic : for small samples, $k = 3$ and $n_j > 5$

$$H = \left[\frac{12}{N(N+1)} \sum_{j=1}^{k} \frac{R_j^2}{n_j} \right] - 3(N+1)$$

H statistic : for large samples, $k > 3$ and $n_j > 5$

$$\chi^2 = \left[\frac{12}{N(N+1)} \sum_{j=1}^{k} \frac{R_j^2}{n_j} \right] - 3(N+1)$$

Adjustment Factor : If 25% of the ranked observations are ties, divide H by

$$1 - \left[\sum_{i=1}^{t} (t_i^3 - t_i) / (N^3 - N) \right]$$

EXERCISES

17.1 The organic content of soil at 15 cm was determined for 8 sample sites in each of three woodland types:(1) the first location was high forest of mixed woodland undisturbed for over 100 years, this was used as control, (2) the second location class was oak forest planted as replacement in a section of the high forest 100 years ago, (3) the third location class was pine forest also planted in cutover high forest 100 years ago. The data are as follows :

(a) High forest	(b) Oak forest	(c) Pine forest
7.1	9.5	14.0
6.3	10.3	6.0
7.8	19.1	7.5
13.4	12.0	6.4
9.6	10.2	10.7
14.2	13.6	8.1
12.0	8.6	5.2
11.7	12.2	8.8

Test the hypothesis that the measurement of populations were identical.

17.2 In an area under study it is known that four different tribal groups exist and it is suspected that the size of the village unit may vary with the tribe. The sample is stratified proportionally to the estimated average number of people in villages occupied by these four different tribes, with the results given below:

Tribe A	Tribe B	Tribe C	Tribe D
150	150	300	350
550	500	550	800
250	400	400	300
150	350	250	350
	250	350	700
	450	550	550
			450
			500
Average Population per village =			
275	350	400	500

As can be seen from this above table, not only does the average population per village vary with the tribe, but also the size of the sample differs from one tribe to the other. Test whether the tribal groupings affect the average population per village.

17.3 Following is a data on wheat yield in quintal kg./hectare in three soil types: sandy, clay and loamy soil.

Sandy soil : 18, 20, 16, 21, 22, 18, 19, 17 and 21
Clay soil : 21, 23, 14, 24, 21, 20, 19, 21 and 23
Loamy soil : 21, 17, 24, 23, 25, 22, 23, 26 and 23

Test the hypothesis that wheat yields are different.

18

Non-parametric Statistics In Geo-science Systems VI (Spearman's Rank Correlation)

18.1 SPEARMAN'S RANK CORRELATION

The statistical methods described so far in the previous chapters have been all concerned with the study of a single variable–its probability distribution, estimation and testing of hypothesis, However, central to the study of any geo-science systems, is the evaluation of the relationships which exist between spatially distributed phenomena. For instance, there can be two apparently related sets of numerical information in which one item from each set forms a pair and each of the pairs refer to a particular place or a particular event in time. like irrigation (in terms of irrigated area) and yield of crops in an area can be related to each other and can form a pair of data sets which is known as "bivariate distribution".

There are many occasions where data sets are only available in the form of ranks (ordinal) and actual values are not known. There are also occasions when the ranked data is more convenient to use rather than any observed interval or ratio data.

For such data there are simple "correlation techniques" that can measure the degree of association between the two variables. Earlier in Chapter 13, we have dealt with the correlation coefficients (the phi coefficient and point biserial coefficient) for nominal data arranged in 2 × 2 contingency tables. In this chapter and in Chapter 19 we deal with two non-parametric coefficients of correlation – the Spearman's rank correlation (ρ) and Kendall's rank correlation (τ). These two non-parametric correlation techniques are used in cases where the data is on the ordinal scale (as mentioned above) and as well as where data is available on the interval or ratio scale but deviate from normality and transformations also do not prove to be effective.

18.1.1 Computing the ρ Statistic

The *Spearman's rank correlation* test is thus a test for correlation using ranked data in the ordinal scale of measurement. Since this test use the rank order and not the actual values for determining the association between the two sets of values it is called a "rank correlation." Named after the British psychologist Charles Edward Spearman, this test developed in 1904, is the earliest test among the various statistical methods based on ranks. This test is widely used because it is simple to use and quick to apply since the data is to be ranked even when the data is collected on interval scale.

Let us suppose that we obtain a sample of n subjects, each measured on two variables, X and Y. To investigate the degree of possible association between X and Y, we must focus on how the values of X "relate" to the values of Y. A "direct" or "positive correlation" is said to exist if subjects possessing high values of X also possess high values of Y while subjects having low values of X also contain low values of Y. At the other extreme, an "indirect" or "negative correlation" is said to exist if subjects possessing high values of X also contain low values of Y, and *vice versa.*

To study the amount or degree of correlation between the two variables X and Y, we may compute the Spearman's coefficient of rank correlation, ρ (pronounced as "rho"), provided that in our data we have attained at least an ordinal level of measurement for each variable.

The Spearman's coefficient of rank correlation may be obtained using the following five steps:

Step 1. If the data is not already ranked, let the variate, X and Y are placed in rank order – rank order 1 is usually allotted to the smallest value of each variable, X and Y, rank 2 to the next, and so on..., the rank of n to the largest. If two or more values of the variable X or Y are tied, they are each assigned the average of the rank positions they otherwise would have been assigned individually had ties not occurred.

Step 2. For each of the n values of the variables X and Y, a "set of rank difference scores" is obtained, that is,

$$d = R_X - R_Y$$

Step 3. Since the sum of the difference in rank of paired values is zero, that is, $\sum d = 0$, we obtain $\sum d^2$ that is, the sum of each of the squared rank difference scores.

Step 4. Finally, the Spearman's coefficient of correlation, ρ is obtained :

$$\rho = 1 - \frac{6\sum d^2}{n^3 - n} \qquad \text{... (18.1)}$$

The rank correlation coefficient is a relative measure which varies from –1 through 0 to +1. Where the rankings are identical, a correlation coefficient of + 1.0 is obtained. Where the rankings of one variate are in reverse order to those of the second, a correlation coefficient of – 1.0 is obtained. For no correlation, the value of the second term in Eq. (18.1) will be exactly equal to unity, resulting in a zero value for ρ. Thus the coefficient is symmetric about zero in the intervals of +1 and –1.

18.1.2 Correction for Tie

When two or more of the values of a variable (*X* or *Y*) are equal or "tied," they are given the average of the rank values that would have been assigned if no ties recurred. For example 19, 17, 17, 5, 3, 3, 3 and 2 will be ranked as : 1, 2.5, 2.5, 4, 6, 6, 6 and 8 respectively. In this example, the two 17 have ranks 2 and 3 in the sequence. Therefore, the average of 2 and 3 is 2.5 which is assigned to them. This procedure of giving the same rank is satisfactory so long as rank ties are infrequent. If they are common, this rank correlation coefficient becomes less efficient and the Eq. (18.1) requires to be modified by adding a factor

$$\sum \frac{t(t^2-1)}{12} \text{ to } \sum d^2$$

where t is the number of times a particular value is repeated. Thus in a set of data if two values are repeated 2 times in *X*-series and 3 times in *Y*-series, the correction to $\sum d^2$ will be

$$\left[\frac{2(2^2-1)}{12} + \frac{3(3^2-1)}{12}\right] = \frac{1}{2} + 2 = 2.5$$

So ρ will be calculated as

$$\rho \text{ (corrected)} = 1 - \frac{6\sum d^2 + \sum t(t^2-1)/12}{n^3 - n} \qquad \text{... (18.2)}$$

and in the present case it will be:

$$\rho \text{ (corrected)} = 1 - \frac{6\sum d^2 + 2.5}{n^3 - n}$$

18.2 TESTING RANK CORRELATION COEFFICIENT

For testing the significance of a computed Spearman's correlation coefficient, ρ (rho), we turn to Appendix Table XIII which gives the critical value of ρ for a number of pairs of scores (n) of 4 to 100 at the four different levels of significance. To evaluate whether or not a significant relationship between X and Y exists, we may test the null hypothesis of independence against the alternative that there is a relationship, positive or negative between X and Y. This is a two-tailed test. On the other hand, if a one-tailed test is needed we may have the one-sided alternative of increasing ranks of X with increasing ranks of Y (i.e. a positive correlation between X and Y, $\rho > 0$) or the one-sided alternative of increasing ranks of X with decreasing ranks of Y in which case there will be a negative correlation.

When the sample size is not very small, equal or greater than 10, the significance of the Spearman's rank correlation coefficient can be tested by the following large sample approximation formula:

$$z = \rho\sqrt{n-1} \qquad \text{... (18.3)}$$

and, based on the level of significance selected, the null hypothesis may be rejected if the computed z value falls in the appropriate region of rejection, depending on whether a two-tailed or one-tailed test (for negative or positive correlation) is employed.* We now illustrate the use of the Spearman's rank correlation procedure with the help of the following Example 18.1.

Example 18.1. Cereal yields for two districts of an area are given below for a ten-year period.

Districts	Cereal yields over the years (in 100 Kg/ha)									
	1	2	3	4	5	6	7	8	9	10
X	25	26	34	25	24	28	27	29	28	29
Y	21	17	35	21	19	22	26	22	26	26

Correlate the yields of cereal of the two districts by Spearman's method.

Spearman's rank correlation coefficient table for the above data is worked out below:

Districts	Rank order by Spearman's rank correlation method									
	1	2	3	4	5	6	7	8	9	10
X	2.5	4	10	2.5	1	6.5	5	8.5	6.5	8.5
Y	3.5	1	10	3.5	2	5.5	8	5.5	8.0	8.0
Difference (= d) in rank	−1	3	0	−1	−1	1	−3	+3	−1.5	0.5
d^2	1	9	0	1	1	1	9	9	2.25	0.25
									Total d^2 = 33.50	

* Note for the significance test of ρ, the Appendix Table XIII can also be used for sample size $n = 4$ to $n = 100$.

Now using Eq. (18.1) we have $\rho = 1 - \frac{6\sum d^2}{n^3 - n} = 1 - \frac{6(33.50)}{10^3 - 10}$

$$= 1 - 0.20303 = 0.79697$$

To test the null hypothesis for the sample size $n = 10$ here, we may either use the Appendix Table XIII or the large sample approximation formula, Eq. (18.3). Thus from the Appendix Table XIII we find that the above value of $\rho = 0.79697$ is significant at 0.05 significance level but at the higher 0.01 significance level, the critical value of ρ is 0.794 which is still marginally smaller than the calculated value 0.79697. Hence both at the 0.05 as well as at 0.01 significance levels, we reject the null hypothesis. This character of correlation coefficient will also be apparent from the Eq. (18.3) for which

$$z = \rho\sqrt{n-1}$$

$$= (0.79697)(\sqrt{10-1}) = 2.39091$$

We find that at 0.05 significance level the one-tailed test has a critical z value of 1.645 which is much lower than the above computed z value of 2.39. But at 0.01 significance level the critical z value is 2.326 which is marginally smaller than 2.39. Hence again we reject the null hypothesis by the large-size sample approximation formula also and conclude that there is evidence of a positive correlation in yields of the cereal of the two districts X and Y.

18.3 USE OF THE ρ STATISTIC

The use of Spearman's rank correlation coefficient ρ is equivalent to computing the *Pearsonian product moment correlation coefficient* (to be discussed in Chapter 20) if the ranks in the "rank-of-order" data in the former is treated as scores in the latter and there are no tied ranks in the data. The Spearman's method is a short cut for determining the relationship in a *bivariate data* and it is quick to apply also.

This rank correlation thus measures the degree to which the two series of data or measurements are related to any continuously increasing or decreasing relationship. This is illustrated in Fig. 18.1 where all the relationships including the curvilinear one have $\rho = 1$ although only the linear relationship would have value equal to 1 in the product moment correlation coefficient. Thus one should be cautious while using the rank data (rather than actual values) for correlation analysis.

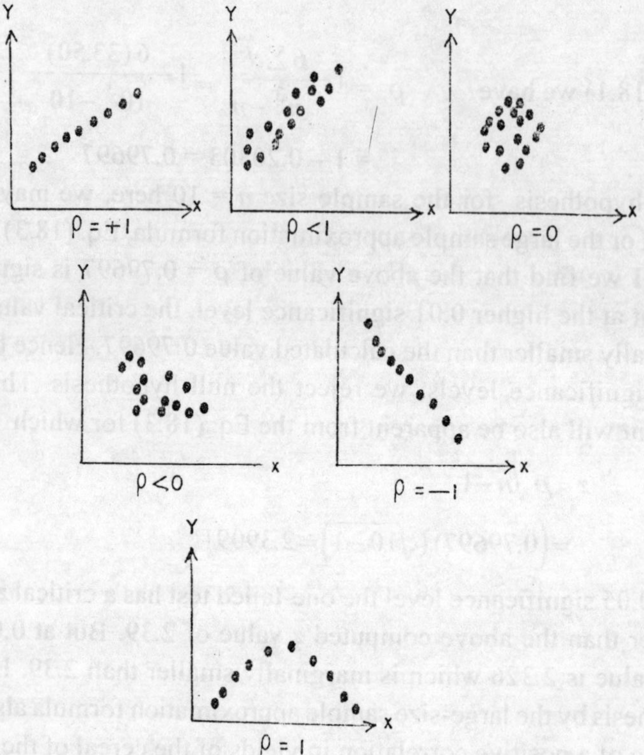

Fig. 18.1 : Scatter Diagram for Various Degrees of Rank Correlation

18.4 LIMITATION OF SPEARMAN'S RANK CORRELATION

We have mentioned earlier that the Spearman's rank correlation coefficient is sensitive to the rank ties in the X and Y variables in the data. If the ties are common, this correlation coefficient becomes less efficient because the tied values cause $\sum d^2$ to be too slow. For moderate number of tied values in the data, the correlation to $\sum d^2$ by the Eq. (18.2) makes the discrepancy to be slight. But in case of many ties in the data, this rank correlation index becomes unsuitable. For example, if all the units in one set have equal rank (which would be 5.5 with 10 units), the Spearman's rank correlation index would be always 0.5 irrespective of the ranking on the second variable.

Another limitation of this method is that it does not use the data to the full : the data is used on the relative rank of the values and hence it is not a measure of the real numerical relationship such as that of the product moment correlation coefficient. Nevertheless, it is still the easiest and the quickest method to compute for sets of data which are not very big. And it becomes particularly important on cognitive-perception and behavioural studies in general for which ordinal data is the common scale of measurement.

LIST OF FORMULAE

Spearman's rho, $\rho = 1 - \dfrac{6\sum d^2}{n^3 - n}$

Correction for tied rank of variables X and/or Y

Add $\sum \dfrac{t(t^2 - 1)}{12}$ to $\sum d^2$ in equation for rho (= ρ)

Significance test for rho, for sample size $n \geq 10$

$z = \rho\sqrt{n-1}$

EXERCISES

18.1 10 places were ranked by three environmentalists in terms of environmental pollution by ranks of one for the least polluted area increasing to 10 for the area polluted to the maximum.

Ranks of ↓	Places									
	A	B	C	D	E	F	G	H	I	J
X	1	6	5	10	3	2	4	9	7	8
Y	3	5	8	4	7	10	2	1	6	9
Z	6	4	9	8	1	2	3	10	5	7

Using rank correlation method, discuss which pair of environmentalists has the nearest approval to common lineage in perception of environmental pollution of the areas.

18.2 Rainfall and runoff data (both measured in cm) were tabulated for 16 hydrological stations of a river basin. The data is as follows.

Rainfall : 46.4, 63.0, 48.8, 60.1, 50.6, 57.5, 55.5, 57.0, 60.8, 48.3, 59.0, 41.7, 66.7, 56.4, 58.3 and 55.7

Runoff : 31.9, 46.8, 34.2, 47.5, 35.2, 40.5, 41.3, 43.5, 44.8, 38.5, 39.1, 26.5, 46.5, 43.4, 40.9 and 41.3

Test the significance of correlation between rainfall and runoff by Spearman's method.

18.3 For 14 medium sized countries (including 7 countries from western Europe with large populations and 7 other countries randomly selected), correlation is sought between percentages of population growth and per capita GNP growth rate. The data was collected for two periods: (one) during the 1960s and the (another) during the 1970s.

Country	Population growth rate		Per capita GNP growth	
	(during 1960s)	(during 1970s)	(during 1960s)	(during 1970s)
1	2	3	4	5
Brazil	3.0	2.9	1.6	6.0
Nigeria	2.4	2.5	– 0.3	4.4
West Germany	1.0	0.1	3.4	2.4
Great Britain	0.7	0.1	2.0	1.9
Italy	0.8	0.7	4.0	2.0
France	1.1	0.6	3.7	3.1
Mexico	3.5	3.3	3.4	1.3
Spain	0.9	1.1	6.5	3.1
U.A.R	2.5	2.2	1.6	6.3
Myanmar	2.1	2.2	1.6	1.7
Yugoslavia	1.1	0.9	4.2	5.0
Afghanistan	2.0	2.2	–0.3	2.7
Algeria	2.3	3.2	–3.5	2.6
Netherlands	1.3	0.8	3.0	2.3

Find out the significance of correlation between the two variables for the two different periods by Spearman's methods and comment.

19

Non-parametric Statistics in Geo-science Systems VII (Kendall's Rank Correlation)

19.1 KENDALL'S RANK CORRELATION

The "Kendall's rank correlation coefficient". is a distribution-free-statistic which is simple to calculate and can be used with very small samples, provided measurement has been achieved at least on the ordinal scale. *Kendall's rank correlation coefficient* is denoted by τ (pronounced as "tau"). Kendall's "tau" like Spearman's correlation coefficient ρ (rho) varies between –1.0 and + 1.0. The assumptions for the Kendall's tau test are the same as the Spearman's rho test. Here also we have a two-dimensional random variable (X, Y) with measurements made on them but Kendall's tau has a somewhat different operation which is given below.

19.1.1 Computing the *S* Statistic

The calculation of Kendall's rank correlation coefficient, τ is quite simple. It involves first the computation of the "*S* statistic" by looking at all possible pairs of cases and noting whether the ranks are in order or not. In this test the rank of 1 is given to the highest value in the set (which is unlike in the non-parametric test discussed so far like in Mann-Whitney, Wilcoxon-ranked pair and in Spearman's rank correlation tests in which the ranking is from lowest to the highest). For example, we have the following state or ranks with respect to the following two variables:

X:	2	1	3	5	4	6
Y:	1	2	4	5	3	6

Now the ranks of one of the variables, say X are put in descending order i.e. from 1 to the largest rank along with the appropriate pair of Y. Thus

X:	1	2	3	4	5	6
Y:	2	1	4	3	5	6

Let us inspect each of the ranks of Y beginning at the left, in this case with 2, and record the number of ranks to the right that are greater than 2, allotting each time +1. These are the ranks 4, 3, 5, 6 which contribute +4. Similarly, we record the number of ranks to the right which are smaller than 2, allotting each time –1. In this case, there is only one rank 1, which contributes –1. Thus, when a given pair is ordered in the same way, it is known as a "concordant pair" (n_c). When this is ordered oppositely, it is known as a "discordant pair" (n_d). This process is repeated for each of the ranks of Y in the data set.

Now, the concordant and disconcordant pairs for each ranks of Y variable are totalled. The sum of the concordant and disconcordant pairs, is known to be equal to the *S statistic*. Therefore

$$S = \sum n_c + \sum n_d \qquad \text{... (19.1)}$$

The total number of possible combinations of any two values of the variables X and Y in the total set can be paired as

$$(n-1) + (n-2) + (n-3) \ldots + 2 + 1, \text{ for } n \text{ number of pairs} = \frac{n(n-1)}{2} \qquad \text{... (19.2)}$$

[*Note*, the sum of the n number of pairs of the variables X and Y in a set of data is : $[(1 + 2 + 3 + \ldots + n) = n(n-1)/2]$

Now, if we divide Eq. (19.1) by the Eq. (19.2), we get Kendall's coefficient tau as

$$\tau = \frac{\sum n_c + \sum n_d}{n(n-1)/2} \qquad \text{... (19.3)}$$

This above expression for Kendall's tau is appropriate when there are no ties on X and Y variables. Values of Kendall's tau lie along a scale from – 1.0 to + 1.0 and analogous to the Spearman's rank correlation coefficient (in Chapter 18) and the Pearson's product moment correlation coefficient (in Chapter 20) but the actual values differ. Now, coming back to our example, we can compute the n_c and n_d for all ranks of Y as follows:

Ranks of Y_i values	No. of higher ranks to the right (n_c)	No. of lower ranks to the right (n_d)
2	+4	–1
1	+4	0
4	+2	–1
3	+2	0
5	+1	0
	$\sum n_c = 13$	$\sum n_d = -2$

Therefore, the Kendall's rank correlation coefficient τ will be

$$\tau = \frac{13-2}{(6)(6-1)/2} = 0.73$$

Thus the two variables *X* and *Y* seem to be related. When two variables are completely unrelated, the positive and negative contributions to *S* will exactly cancel each other (i.e. $\Sigma n_c = \Sigma n_d$) and "tau" will be zero.

It may be seen that for small samples (like $n = 6$ above) the calculation of the Kendall's correlation coefficient, τ itself is sufficient to determine the significance of association between the two variables although the test of significance for the sample size of 5 may be necessary for comparison with other variables.

The critical values of *S* given in Appendix Table XIV show the probabilities associated with the values of the *S* statistic for $n = 4$ and $n = 10$. It can be seen from this Appendix Table that the larger the value of *S*, the smaller is the probability that in a sample of 6 in the above example, the association of the variables *X* and *Y* is due to chance or random variations. Thus for $n = 6$ here, the probability that *S* will be equal to or greater than 11 is 0.028 or less than 3 percent. Hence we conclude at 0.05 significance level all the six pairs of samples are not drawn at random.

19.1.2 *S* Transformation for Large Samples

For large samples ($n > 10$), the sampling distribution of tau approximates to the normal distribution and the probability can be established by calculating a *z* value corresponding to *S* or τ from either of the expressions as given below:

$$z = \frac{S}{\sqrt{(2n+5)(n^2-n)/18}} \quad \text{... (19.4)}$$

or

$$z = \frac{\tau}{\sqrt{2(2n+5)/9n(n-1)}} \quad \text{... (19.5)}$$

19.1.3 Correction for Tie

The effect of ties upon the value of the Kendall's rank correlation coefficient is to make the values of *S* and τ slightly smaller and less significant. It has been mentioned earlier that in the case of many ties, the Spearman's rank correlation test becomes less rigorous. In this case, a safer method is to use the Kendall's rank correlation after applying a simple correction. For each group of tied ranks, the correction can be found by $\frac{1}{2}\Sigma t(t-1)$ where t is the number of times a particular value is repeated in the group. The correction factor (c_x or c_y) for each variable (X or Y) is the sum of correction values of each group of tied ranks and can be expressed as

$$c_x \text{ or } c_y = \frac{1}{2}\Sigma t(t-1) \qquad \text{... (19.6)}$$

The value of τ corrected for ties is found in form of a slightly modified formula of Eq. (19.3) as

$$\tau_{corr} = \frac{S}{\sqrt{\frac{1}{2}n(n-1)-c_x}\,.\sqrt{\frac{1}{2}n(n-1)-c_y}} \qquad \text{... (19.7)}$$

where c_x and c_y represent the correction for ties of each variable.

Now we apply the tie correction to the Example 18.1 on Spearman's rank correlation below and calculate the Kendall's tau on it.

From the Example 18.1 on cereal yields of the two districts we arrange the district-wise data in rank order, giving rank 1 to the highest value for each group. The ranked values are given within brackets besides the yields are as follows:

Districts	Cereal yields over the years in 100Kgs/ha (Rank order by Kendall's)									
X	25 (8.5)	26 (7.0)	34 (1.0)	25 (8.5)	24 (10.0)	28 (4.5)	27 (6.0)	29 (2.5)	28 (4.5)	29 (2.5)
Y	21 (7.5)	17 (10.0)	35 (1.0)	21 (7.5)	19 (9.0)	22 (5.5)	26 (3.0)	22 (5.5)	26 (3.0)	26 (3.0)

Now the ranks of district X are put in descending order, i.e. from 1 to the largest value (34 here) alongwith the appropriate pair of Y. Thus

X	1.0	2.5	2.5	4.5	4.5	6.0	7.0	8.5	8.5	10.0
Y	1.0	3.0	5.5	5.5	3.0	3.0	10.0	7.5	7.5	9.0

Therefore, $S = (9+6+4+4+4+4+0+1+1)-(0+0+2+2+0+0+3+0+0)$ $= 33 - 7 = 26$

Before calculating τ, we require to apply "tie correction." For the district X, we note that 3 values are repeated twice so, the correction for X will be

$$c_x = \frac{1}{2}[2(2-1)+2(2-1)+2(2-1)] = 3.$$

Again for the district Y, 2 values are repeated 2 times and 1 value 3 times, so,

$$c_y = \frac{1}{2}[2(2-1)+2(2-1)+3(3-1)] = 5.$$

Hence, the corrected Kendall's rank correlation will be :

$$\tau_{corr} = \frac{S}{\sqrt{\frac{1}{2}n(n-1)-c_x}\sqrt{\frac{1}{2}n(n-1)-c_y}}$$

$$= \frac{S}{\sqrt{\frac{1}{2}.10(10-1)-3}\sqrt{\frac{1}{2}.10(10-1)-5}} = \frac{26}{\sqrt{42}.\sqrt{40}}$$

$$= 0.63$$

For testing significance of this $\tau = 0.63$, we can treat the sample size $n = 10$ as large size sample and if a normal approximation is done with the help of Eq. (19.5), z value computed will be 2.53, which is more than 1.645, the critical value of z at 0.05 significance and also more than 2.326 at 0.01 significance level. Hence we conclude in the same way as we did in the Spearman's test about the significance of the correlation but τ value corrected is a better estimate of the association between yields of cereals of the two districts than the Spearman's rho value.

Example 19.1 **Growth rate of population and per capita growth rate in GNP are given for 14 medium-sized countries below :**

Sl. No.	Medium-Sized countries	Growth rate of Population	Per capita Growth rate in GNP
1.	Brazil	3.0	1.6
2.	Nigeria	2.4	–0.3
3.	W. Germany	1.0	3.4
4.	U. K.	0.7	2.0
5.	Italy	0.8	4.0
6.	France	1.1	3.7
7.	Mexico	3.5	3.4
8.	Spain	0.9	6.5
9.	U.A.R.	2.5	1.6
10.	Myanmar	2.1	1.6
11.	Yugoslavia	1.1	4.2
12.	Netherlands	1.3	3.0
13.	Algeria	2.3	–3.5
14.	Afghanistan	2.0	–0.3

Test the significance of the relationship between population and per capita GNP growth rates of the countries by Kendall's rank correlation method.

We index the population growth rate as X variable and the per capita growth rate in GNP as Y variable. Assigning the rank 1 to the maximum value of the variables we can substitute values of the X and Y by the ranks as

	Countries													
	1	2	3	4	5	6	7	8	9	10	11	12	13	14
X	2	4	11	14	13	9.5	1	12	3	6	9.5	8	5	7
Y	10	12.5	5.5	8	3	4	5.5	1	10	10	2	7	14	12.5

In above ranking it can be seen that one value has occurred two times in X variables; for Y variable the number of times the ties have occurred are three : two values repeated two times and one value has repeated three times.

The test statistic S can now be calculated if the ranks of X are put in a sequence with appropriate Y. So ranks of X and Y are rewritten, for convenience as follows:

X	1	2	3	4	5	6	7	8	9.5	9.5	11	12	13	14
Y	5.5	10	10	12.5	14	10	12.5	7	2	4	5.5	1	3	8

Therefore, $S = (8 + 3 + 3 + 1 + 0 + 1 + 0 + 1 + 4 + 2 + 1 + 2 + 1) - (4 + 7 + 7 + 8 + 9 + 7 + 7 + 5 + 1 + 2 + 2 + 0 + 0) = 27 - 59 = -32$

Now applying correction to X and Y values for ties, we get $c_x = 1$ and $c_y = 5$. Substituting $S = -32$, $n = 14$ and the correction factors $c_x = 1$ and $c_y = 5$ in Eq. (19.7) we get

$$\tau_{corr} = \frac{S}{\sqrt{\frac{1}{2}n(n-1) - c_x} \cdot \sqrt{\frac{1}{2}n(n-1) - c_y}}$$

$$\frac{-32}{\sqrt{(91-1)} \cdot \sqrt{(91-5)}} = -\frac{32}{87.97727}$$

$$= -0.36373$$

Since the sample size n is more than 10, we can substitute $\tau_{corr} = -0.36373$ in z transformation, that is,

$$z = \frac{\tau}{\sqrt{2(2n+5)/9n(n-1)}} = \frac{-0.36373}{\sqrt{66/1638}} = -\frac{0.36373}{\sqrt{0.04029}}$$

$$= -1.812$$

Referring to the probabilities in Appendix Table 1 (b) with $z = -1.812$, the probability that $\tau = -0.36373$ has occurred through chance is 0.036 or less than four percent. We may therefore have $\tau = -0.36373$ at the 0.05 percent level to reject the null hypothesis and we accept that negative relationship exists between population growth rate and per capita GNP growth rate of the countries.

19.2 USES OF KENDALL'S TAU

The correlation by tau is one of the rank order correlations. Thus this correlation implies that an increase in the score on one variable will often be accompanied by an increase in the score on the other variable but no inference can be made about the size of the increase. Moreover, like in Spearman's, the function relating the two variables is not necessarily linear.

In the above Example 19.1, if the Spearman's rank correlation coefficient is applied, "ρ" becomes equal to -0.555 and from Appendix XIII we can see that H_0 should be rejected and there is a significant negative correlation between the growth of population and per captia growth rate in GNP. Therefore, the significance of both Spearman's rho and Kendall's tau will be the same but in a comparison between them it can be seen that τ becomes more rigorous. Hence, for small-sized samples with $n < 20$, Kendall's tau should be used in preference to Spearman's ρ. Beyond sample size $n = 20$, the ease of calculation makes ρ more suitable.

LIST OF FORMULAE

Kendall's rank correlation coefficient, τ (= tau), for small-size samples, $4 < n < 10$,

$$\tau = \frac{S}{\frac{1}{2}n.(n-1)}$$

where $S = \Sigma n_c + \Sigma n_d$

Kendall's rank correlation coefficient, τ (tau), approximated to normal transformation for which probability is

$$z = \frac{S}{\sqrt{(2n+5)(n^2-n)/18}}$$

or

$$= \frac{\tau}{\sqrt{2(2n+5)/9n(n-1)}}$$

Adjustment factor : Correction for variables X and Y, c_x or $c_y = \frac{1}{2}\Sigma t(t-1)$ to be used for corrected tau as

$$\tau_{corr} = \frac{S}{\sqrt{1/2.n.(n-1) - c_x} . \sqrt{1/2.n(n-1) - c_y}}$$

EXERCISE

19.1 Test the significance of the associations between rainfall and runoff data given in the Exercise 18.2 of the earlier Chapter 18.

20

Regression Analysis in Geo-science Systems I (Product Moment Correlation)

20.1 PRODUCT MOMENT CORRELATION

The *Regression Analysis* which includes *correlation* is one of the oldest statistical tools used in geo-sciences particularly in hydrology. This chapter deals with correlation and presents discussion necessary for understanding of the *simple linear correlation coefficient,* its computation for both the listed and the grouped data in *bivariate distribution*, testing its significance and its uses.

It is to be noted that in the earlier chapters correlation analysis have been done but they all basically aim to measure the degree to which the two variables taken on the ordinal scale are related. But the correlation analysis attempted here is essentially included in parametric statistics and hence the bivariate data which are taken in the present chapter are required to be on interval/ratio scale rather than on ordinal scale earlier.

Again, earlier in non-parametric statistics, correlation between two sets of data was measured in terms of strength of their association and their significance by the test statistic of Spearman's rho *(ρ)* and Kendall's tau (τ). In many problems, however, there is the need to compare sets of data in terms of the extent to which a change in one set is or is not reflected by a change in the other. In such a case, the degree of association of the bivariate data sets is known as correlation and it is denoted by *correlation coefficient, r* which is a single number measuring the relation between the two sets of data and it reflects the degree to which changes in one direction or the other (+ or –) and magnitude in one set of data are associated with comparable changes in the other set. An index of this sort is provided when the data sets are on interval/ordinal scale and it is termed as *Pearsonian product moment correlation coefficient.* This product moment correlation itself considers the original data points themselves but as we know from the earlier chapters, Spearman's rho and Kendall's tau considered the overall characteristic of rank of the data sets and not the individual data points. Hence the product moment correlation coefficient is the most powerful of all the tests of correlation (e.g. Spearman's rho and Kendall's tau) and there is no constraint in the formula which precludes its application.

Before the product moment correlation coefficient, denoted by r is calculated, the simplest method of seeing the relationship is to plot the two variables against each other on a graph. The graph so obtained is known as *scatter diagram* (Fig. 20.1). The smaller the dispersion of the points on the scatter diagram, more closely are the variables related and the correlation

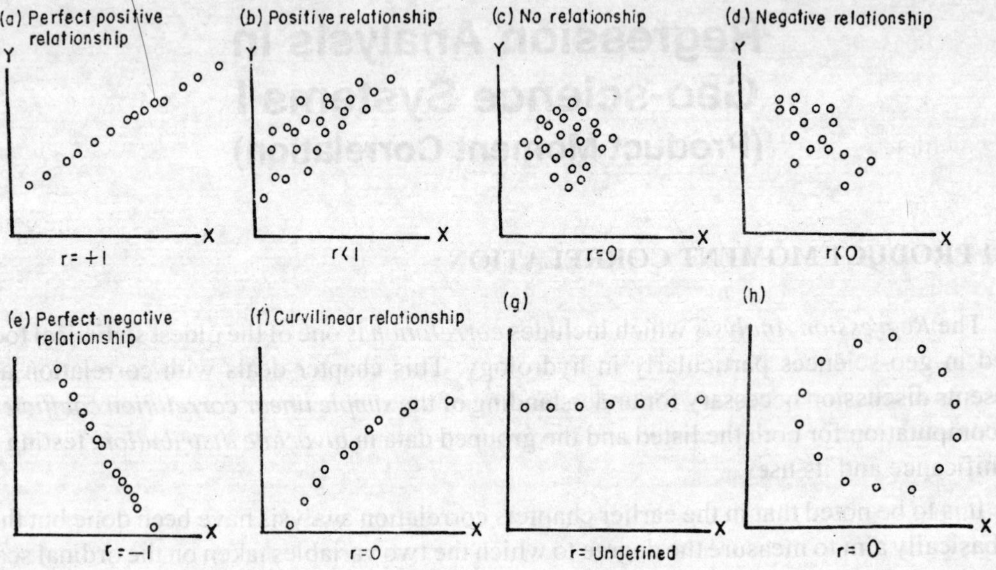

Fig. 20.1 : Scatter Diagram for Various Degrees of Product Moment Correlation

(a) Perfect positive relationship
(b) Positive relationship
(c) No relationship : random
(d) Negative relationship
(e) Perfect negative relationship
(f) Curvilinear relationship
(g) Relationship undefined
(h) Circular relationship

will be greater. Thus when increasing Y is associated with increasing X, we say that the two variables are directly or positively correlated with a correlation coefficient of $r = +1.0$. But since correlation coefficient r is having a direction also, so when an increasing X results in a decrease in Y, variables are said to be inversely or negatively related with a correlation coefficient of $r = -1$. In this latter case, the cluster of the points would slope in the opposite direction.

In Fig. 20.1a, the spread of the points are upward and perfectly around a linear line and we are supposed to get a positive correlation coefficient, $r = +1.0$. The spread of the points are around a linear line and point downward in Fig. 20.1e which indicates a negative correlation coefficient, $r = -1.0$. In Fig. 20.1b the points can be said to lie in a "narrow belt" indicating a positive relationship and the same narrow belt of points but in a "reverse direction" indicates a negative relationship in Fig. 20.1d. A crude scattering of points, i.e. a random distribution of *X* and *Y* values in scatter in Fig. 20.1c is indicative of a low correlation coefficient and *r* approaches 0 that is a zero correlation. When an increase in *X* results in a decreasing increase in *Y*, the scattering of points would indicate a curvilinear trend (Fig. 20.1f). An interesting extreme is shown in Fig. 20.1g in which one variable *Y* does not change, it is "invariant." If we calculate the correlation it will encounter division by zero, so the correlation is undefined in this situation. In Fig. 20.1h, the *X* and *Y* lie on a circle, so the relationships between them can be expressed as $Y = \sqrt{r^2 - x^2}$ where *r* is the coefficient for radius of a circle, the centre of which is assumed to lie at the origin. But since this circular relationship between *X* and *Y* is not linear, and correlation coefficient (which itself is an expression of the linear relationship between two variables), so like in Fig. 20.1f, the correlation coefficient if computed, it will be zero.

The points in Fig. 20.1b and Fig. 20.1d are said to lie in a "narrow belt", but the definition of "narrow" is a matter of subjective interpretation and individual judgement.

To provide an objective measurement of the narrowness of the belt of concurrent points for values (*X*, *Y*) of the two variables in a scatter diagram, one requires to determine the strength of the numerical relationship between the two variables, i.e. the coefficient of correlation, *r*. This coefficient of correlation *r* is stated in abstract units divorced from the units of measurement of the variables and this abstract scale for measuring the degree of correlation runs from $-1 \le r \le +1$, the sign indicates the nature of correlation, direct (+) or indirect (–) and the number, the amount of correlation.

20.2 SIMPLE LINEAR CORRELATION COEFFICIENT

The first step in calculating the degree of correlation in terms of the simple linear product moment correlation coefficient is to find the covariance i.e. the amount by which the variates (*X* and *Y*) differ from the mean of the variables in the numerator. Thus a number of points may contribute significantly to the variances of both *X* and *Y*, while the other points may contribute little. This introduces the *concept of covariance* which is the mean of the sum of the products of contributions of each point to the individual variance. This covariance itself is a *second order moment.* We also require to know the standard deviation of *X* and *Y* values in the denominator for standardizing the variables so that the two linearly dependent variables *X* and *Y* has unit variance. Therefore, the simple linear product moment correlation coefficient is computed as the ratio of the covariance in *X* and *Y* to the product of their standard deviation as follows:

$$r_{XY} = \frac{\text{Covariance of } X \text{ and } Y}{\text{Standard Deviation of } X \text{ and } Y} \qquad \text{... (20.1)}$$

which in terms of moment measures can be written as

$$r_{XY} = \frac{M_{XY}}{\sqrt{M_{XX} \cdot M_{YY}}} \qquad \text{... (20.1a)}$$

$$= \frac{\frac{1}{n}.\Sigma(X-\bar{X})(Y-\bar{Y})}{\sqrt{\frac{\Sigma(X-\bar{X})^2}{n}}.\sqrt{\frac{\Sigma(Y-\bar{Y})^2}{n}}} \qquad \text{... (20.2)}$$

which can be rearranged to give the raw score formula for r_{XY} as

$$r_{XY} = \frac{\Sigma XY - \frac{\Sigma X . \Sigma Y}{n}}{\sqrt{\left[\Sigma X^2 - \frac{(\Sigma X)^2}{n}\right]\left[\Sigma Y^2 - \frac{(\Sigma Y)^2}{n}\right]}} \qquad \text{... (20.3)*}$$

where $\Sigma XY - \frac{\Sigma X . \Sigma Y}{n}$ is the covariance of X and Y defined as the sum of the cross products of the variables minus the product of sum of the pairs divided by the number of pairs. Hence if X_i has a positive departure from its means and Y_i a negative departure (or *vice versa)* then the contribution to covariance will be negative. The covariance thus measures the "joint size" of X and Y (relative to their means) unlike the variation. Again, the covariance

---

$$* \; \Sigma(X-\bar{X})(Y-\bar{Y}) = \Sigma(XY - X\bar{Y} - \bar{X}Y + \bar{X}\bar{Y})$$

$$= \Sigma XY - \bar{Y}\Sigma X - \bar{X}\Sigma Y + n.\bar{X}\bar{Y}$$

$$= \Sigma XY - \Sigma X . \frac{\Sigma Y}{n} - \frac{\Sigma X}{n} . \Sigma Y + n.\frac{\Sigma X}{n}\frac{\Sigma Y}{n}$$

$$= \Sigma XY - 2\frac{(\Sigma X)(\Sigma Y)}{n} + \frac{(\Sigma X(\Sigma Y)}{n}$$

$$= \Sigma XY - \frac{\Sigma X . \Sigma Y}{n}$$

which is again equal to $\Sigma XY - n.\bar{X}\,\bar{Y}$.

Similarly according to the footnote for the expansion of Eq. (3.3) in Chapter 3. $\frac{\Sigma(X-\bar{X})^2}{n}$ is equal to $\frac{\Sigma X^2}{n} - \bar{X}^2$ or $\frac{n\Sigma X^2 - (\Sigma X)^2}{n^2}$

must equal but cannot exceed the product of standard deviation of its variables, so correlation coefficient, r_{XY} ranges from + 1 to –1.

Finally, we have

$$r_{XY} = \frac{n\Sigma XY - (\Sigma X)(\Sigma Y)}{\sqrt{\left[n\Sigma X^2 - (\Sigma X)^2\right]\left[n\Sigma Y^2 - (\Sigma Y)^2\right]}} \quad \text{... (20.4)}$$

It can be seen that identical results are obtained by calculating the mean of the product of the z score values of X and Y. This is written as follows:

$$r_{XY} = \frac{\Sigma z_X \cdot z_Y}{n} \quad \text{... (20.5)}$$

where z_X and z_Y are standardized form of variables X and Y, defined for example $z_X = \dfrac{X - \overline{X}}{\sigma_x}$ so that the Eq. (20.5) becomes

$$r_{XY} = \frac{\Sigma(X - \overline{X})(Y - \overline{Y})}{n.\sigma_X \cdot \sigma_Y} \quad \text{... (20.6)}$$

Note in Eq. (20.1) to Eq. (20.6), n stands for the number of pairs of the observations. In these equations the raw score values of r_{XY} namely ΣXY, ΣX, ΣY, $(\Sigma X)^2$, $(\Sigma Y)^2$, ΣX^2, ΣY^2 look formidable, but they are in fact the easiest to use in practice with the help of a desk calculator.

Again, as it has been told earlier that the sign of r_{XY} (+ or –) is determined by the sign of the numerator, i.e. by covariance. For perfect positive or negative correlation, the covariance should equal the product of the standard deviation. If the sum of the cross products is zero, the variables X and Y are linearly independent and the correlation coefficient is zero. However, it does not follow that the variables are independent if the sum of the cross products is zero. If the points (X, Y) fall symmetrically around and all along a circle (as in Fig. 20.1h) r_{XY} is zero, or the linear dependence is zero. However, in this case there is a high correlation for the function which has a circle as the trend line (Fig. 20.1h).

The calculation of Pearson's product moment linear correlation coefficient is given in Example 20.1 by both the methods of raw score values and the product of the z score values. It can be seen that the values of the correlation coefficient by both the methods are identical.

Example 20.1 Calculate the Pearsonian correlation coefficient between the following 10 values of rainfall and runoff (in cm) by the raw score values and by the product of the z score values:

Rainfall (X): 6.5; 6.3; 4.8; 4.4; 6.0; 2.2; 6.4; 7.3; 5.2 and 4.8

Runoff (Y): 3.8; 3.6; 2.8; 2.6; 3.7; 1.2; 2.2; 3.0; 3.1 and 3.0

To find the correlation coefficient between these two sets of values, it is necessary to create columns of squares and cross products which is done below for both the methods.

Sl.No.	Raw Score Method					z Score Method		
	X	Y	XY	X^2	Y^2	z_X	z_Y	$z_X.z_Y$
1.	6.5	3.8	24.70	42.25	14.44	+ 0.828	+ 1.216	+ 1.0092
2.	6.3	3.6	22.68	39.69	12.96	+0.679	+0.946	+0.6424
3.	4.8	2.8	13.44	23.04	07.84	–0.440	–0.135	+0.0595
4.	4.4	2.6	11.44	19.36	06.76	–0.739	–0.405	+0.2995
5.	6.0	3.7	22.20	36.00	13.69	+0.455	+ 1.081	+ 0.4921
6.	2.2	1.2	02.64	04.84	01.44	–2.381	–2.297	+ 5.4690
7.	6.4	2.2	14.08	40.96	04.84	+0.754	–0.946	+0.7129
8.	7.3	3.0	21.90	53.29	09.00	+1.425	+0.135	+0.1926
9.	5.2	3.1	16.12	26.04	09.61	–0.142	+0.270	+0.0383
10.	4.8	3.0	14.40	23.04	09.00	–0.440	+0.135	–0.0595
Total	53.9	29.0	163.60	308.51	89.58			

Raw Score Method

Mean, $\bar{X} = 5.39$; Mean, $\bar{Y} = 2.90$

$(\Sigma X)^2 = 2905.2$ and $(\Sigma Y)^2 = 841$ and $n = 10$

$$\text{So, } r = \frac{n\sum XY - (\sum X)(\sum Y)}{\sqrt{[n\sum X^2 - (\sum X)^2][n\sum Y^2 - (\sum Y)^2]}}$$

$$= \frac{1636.0 - 1563.1}{\sqrt{(3085.1 - 2905.2)(895.8 - 841)}} = \frac{72.9}{99.29} = 0.7342$$

z Score Method

We know $z_X = \dfrac{X - \bar{X}}{\sigma_X}$ *and* $z_Y = \dfrac{Y - \bar{Y}}{\sigma_Y}$

Now from above: $\Sigma z_X \cdot z_Y = 7.3535$

where from raw score method, Mean $\bar{X} = 5.39$; Mean $\bar{Y} = 2.90$

standard deviation, $\sigma_X = 1.34$, standard deviation, $\sigma_Y = 0.74$

$$\text{So, } r = \frac{\Sigma z_X . z_Y}{n} = \frac{7.3535}{10} = 0.7353$$

[*Note* the r value in these two methods are nearly identical. The exact value is not attained here because of the approximations done while taking the products or in doing the divisions.]

20.3 COMPUTATION OF CORRELATION COEFFICIENTS FROM GROUPED DATA

If the number of values in the sample pair is large and if a desk calculator is not available, the computation of the product moment correlation coefficients can become extremely tedious. In such cases it may be more convenient to make use of grouping the data in terms of frequencies and for the paired data we use a sort of "double bivariate frequency distributions table". To illustrate this *grouped data* and the *double bivariate frequency distribution table,* we go by the data given in the Example 20.2 afterwards for the two variables, arable acreage *(X)* and yield *(Y)* of farms. We start illustrating this bivariate (grouped) data by a *correlation table* which is essentially a "double entry" table (Table 20.1). The preparation of this correlation table is essentially by the data in Example 20.2.

In Table 20.1 we have a set of class intervals, displayed in the topline corresponding to the scores in arable acreage and designated as the *X*-variable. In the left hand columns a similar set of intervals is placed for yield *(Y)* of the farms in 1,000 kg. Note the magnitudes of the *X*-variable increase from right to left and those for the *Y*-variable increase from bottom to the top: both these arrangements correspond to the arrangement (in a reverse manner of course) in a scatter diagram.

The main body of Table 20.1 consists of a grid of which each cell corresponds to a class interval on the *X*-scale and another on the *Y*-scale. The figure, if any, within each cell indicates the number of cases having *X* and *Y* values in the corresponding classes. We have set a class interval value designated by Y' at the right of the mid-class column Y_0 and similarly a set designated by X' below the mid-class X_0 row. These X' and Y' are the deviations of the mid-class values of *X* and *Y* from their respective assumed mean in terms of their class intervals.

In order to complete the calculation of mean and standard deviation of the correlation coefficient we have in addition the columns f_x, $f_x X'$, $f_x X'^2$, $f_c X'Y'$ at the bottom and f_y, $f_y Y'$, $f_y Y'^2$, $f_c X'Y'$. Note both f_x and f_y show the total number of items falling in each *X* and *Y* interval respectively. Hence we have the frequency distribution of both the variables. Note these figures both in the f_x row and f_y column add up to the total number of cases in the data (Table 20.1).

For computing the correlation coefficient from this *correlation table* we have the computations which are straightforward extension of the procedures introduced in calculating the covariance of *X* and *Y* and their respective standard deviations. Thus, we write the

covariance of *X* and $Y = \left[\Sigma XY - \dfrac{\Sigma X . \Sigma Y}{n}\right]$ as equal to:

$$C_x C_y \left[\frac{\Sigma f_c X'Y'}{n} - \left(\frac{\Sigma f_x . X'}{n}\right)\left(\frac{\Sigma f_y . Y'}{n}\right)\right] \qquad \text{... (20.7)}$$

Similarly, standard deviation of X.

$$= C_x \sqrt{\frac{\Sigma f_x . X'^2}{n} - \left(\frac{\Sigma f_x . X'}{n}\right)^2} \qquad \text{... (20.8)}$$

X, Arable acreage

Y, Yield (in 1000 Kg.)

Class Interval Y	Mid-Class Y_0	Class Interval X	139-120	119-100	99-80	79-60	59-40	39-20	19-0				
		Mid-Class x_0	129·5	109·5	89·5	69·5	49·5	29·5	09·5				
		$x' \rightarrow$ / $Y' \downarrow$	2	1	0	−1	−2	−3	−4	fy	fy·Y′	fy·Y′²	$f_c x'Y'$
119 − 100	109·5	3	/ = 2 ⌐2							2	6	18	12
99 − 80	89·5	2	/ = 1 ⌐4							1	2	4	4
79 − 60	69·5	1		/ = 1 ⌐1	/// = 3 ⌐0	/ = 1 ⌐−1				5	5	5	0
59 − 40	49·5	0		/ = 1 ⌐0	/ = 1 ⌐0	//// = 4 ⌐0	// = 2 ⌐0			8	0	0	0
39 − 20	29·5	−1				//// = 4 ⌐4		////̸ = 5 ⌐5		9	−9	9	19
19 − 0	09·5	−2					/ = 1 ⌐4	//// = 4 ⌐24	////̸ ////̸ = 10 ⌐80	15	−30	60	108
		fx	3	2	4	9	3	9	10	40	−26	96	143
		fx.x′	6	2	0	−9	−6	−27	−40	−74			
		fx.x′²	12	2	0	9	12	81	160	276			
		$f_c x'Y'$	16	1	0	3	4	39	80	143			

Table 20.1 : Correlation Table

and standard deviation of Y

$$= C_y \sqrt{\frac{\Sigma f_y . Y'^2}{n} - \left(\frac{\Sigma f_y . Y'}{n}\right)^2} \qquad \text{... (20.9)}$$

where C_x and C_y are the intervals in the X and Y class groups, X' and Y' are the deviation of X_0, Y_0 from their assumed mean. $\overline{X}, \overline{Y}$ in terms of their class intervals and $\Sigma f_c . X'Y'$ is the grand total of the frequencies of columns and rows.

Now once we compute the above three values by following the definition of the correlation coefficient as given in Eq. (20.1), we can find out the r_{XY} for the grouped data. In this method certain inaccuracies will be introduced because of grouping the data but as it will be found out from the Example 20.2 below, the inaccuracies will be slight.

Example 20.2 Prepare a correlation table for the following data between the arable acreage and the corresponding yield of 40 farms.

Arable acreage X	Yield (in 1,000 kg) Y	Arable acreage X	Yield (in 1,000 kg) Y
125	118	135	90
80	73	86	68
4	3	8	6
91	53	80	73
39	35	58	45
4	0	71	24
6	4	25	27
7	0	20	12
35	33	19	14
100	50	100	71
132	119	39	35
30	27	24	16
65	27	75	20
67	53	70	61
5	1	8	4
7	6	21	13
30	14	54	40
61	51	65	44
72	50	42	6
14	7	60	36

Calculate the correlation coefficient between the arable acreage and yield of the farms.

The data for the arable acreage and yield is first listed in ascending order as follows:

X	Y	X	Y	X	Y	X	Y	X	Y
4	0	14	7	35	33	65	27	80	73
4	3	19	14	39	35	65	44	86	68
5	1	20	12	39	35	67	53	91	53
6	4	21	13	42	6	70	61	100	50
7	0	24	16	54	40	71	24	100	71
7	6	25	27	58	45	72	50	125	118
8	4	30	14	60	36	75	20	132	119
8	6	30	27	61	51	80	73	135	90

Now, the arable acreage and yield data are assumed as X and Y respectively and placed in classes row-wise and column-wise in the bivariate frequency table (Table 20.1). Thus on the left in Table 20.1, we group the class interval Y, mid-value Y_0 and the Y' score all on the Y variable. Similarly as mentioned earlier at the top we give the X, X_0 and X' values for the X variable. For these two variables, in the example here seven columns and six rows form this table of $7 \times 6 = 42$ gridded cells. Each of these rows and columns represent an interval of 20 units. The cell frequencies of pairs of occurrences f_c are displayed in the cells by "tally marks" and similarly the corresponding total, e.g. the frequency of occurrences of farms with 79-60 arable acreage and 59-40 thousand kg of yield is "N" = 5 from the above listed data. The mid-values X_0 and Y_0 are converted into X' and Y' respectively with the help of the following formulae

$$X' = \frac{X_0 - \bar{X}_0}{C_X} \text{ and } Y' = \frac{Y_0 - \bar{Y}}{C_Y}$$

where $\bar{X}_0$ and $\bar{Y}_0$ are the assumed means.

The X' and Y' values are placed in the correlation table row-wise and column-wise respectively. The total of the class frequencies f_x are listed along the bottom of the column of the cells and similarly the total of the class frequencies f_y are listed along the right-hand edge of rows of the cells. Since each pair of scores from the listed data above is shown in the correlation table by a tally mark, so for finding the cell in which to place the first farm in the listed data with a score of 4 on the X-scale and 0 on the Y-scale, we run up the left hand intervals till we come to the Y class interval 19–0. Now we move to the right till we come to the X class interval 19–0 and we place a tally mark (/) in the first Y-cell from the bottom and the last X-cell from the left. In this cell we made a tally for every other farm which has a farm average up to 19 acres and a yield up to 19,000 kg and we got 10 tallies in this cell. In a similar manner by running our eyes through the pairs of X, Y values, we make a tally for each pair of scores. When we have used the entire data set, we enter in the cells the numbers totalling all the 40 pairs of data in the example and totalling of the tallies also give these 40 pairs. Finally, as mentioned earlier we total the class frequencies, f_x and f_y. The successive values of $f_x X'$, $f_x X'^2$, $f_y Y'$ and $f_y Y'^2$ are then calculated and inserted around the correlation table.

The sum of the above three sets of entries are quite familiar (i.e. variable $f_x, f_x X'$ and $f_x X'^2$ are used for mean and standard deviation of the X variable and similarly $f_y, f_y Y', f_y Y'^2$ for the same with respect to Y variable. But column 4 and row 4 are new. For 79 - 60 arable acres (corresponding to $X' = -1$) and yield of 59 - 40 thousand kg (corresponding to $Y' = 0$), the frequency f_c is 4 and $f_c X' Y'$ equals to zero. These $f_c X'Y'$'s are displayed in the lower right hand corner of the corresponding cell within the correlation table. The column 4 and row 4 are simply the sum of the products of the entries of $f_c X'Y'$ in the lower right hand corner of the cells of X' and Y'. The total sum of rows in row 4 (at the most bottom edge of the correlation table) should tally exactly with the total sum of columns in column 4 (at the right most column of the correlation table): in the present example they are all equal to 143. Another check is that the sum of rows in row 1 should tally exactly with the sum of columns in column 1.

Thus, for this grouped data with 7 classes for X and 6 classes for Y have

$$r_{XY} = \frac{20 \times 20\left[\frac{143}{40} - \left(\frac{-74}{40}\right)\left(\frac{-26}{40}\right)\right]}{20.\sqrt{\frac{276}{40} - \left(\frac{-74}{40}\right)^2}\,.20.\sqrt{\frac{96}{40} - \left(\frac{-26}{40}\right)^2}}$$

$$= \frac{400(3.575 - 1.2025)}{(37.2961)(28.1247)}$$

$$= \frac{949}{1048.9428} = 0.9047$$

Note: Taking into account individual values of the variables, the correlation by Eq. (20.4) will be found equal to 0.9155, thus grouping of the paired data do not increase the correlation coefficient further.

If the listed data of our above example are actually plotted in a scatter diagram by running our eyes through the pairs of the set of X and Y variates, we get the correlation table as the exact counterpart of the scatter diagram.

20.4 CORRELATION ANALYSIS

Earlier we have diagrammatically represented various degrees of linear correlation between X and Y variates by eight possible patterns of scatter diagrams (Fig. 20.1a to h). Again when scores of X and Y are found to lie in a narrow belt to be sure of the exact measure of the numerical relationship, we have computed the correlation as a ratio between the covariance of the variates and the product of the standard deviation of the variates. Now in order to analyse this measure of correlation, let us have the scatter diagram again but on a new set of axes $X' = X - \bar{X}$ and $Y' = Y - \bar{Y}$ (instead of the original X and Y axes in the earlier one) with the new origin at $\overline{X}, \overline{Y}$ (Fig.20.2). Then the points on the scatter diagram will be seen scattered over the four quadrants of the $\overline{X}, \overline{Y}$ plane. For pair of points X, Y lying in quadrants I and III, $\Sigma X'Y'$ will be positive, since in quadrant I, X' and Y' of a point are both positive while for points in quadrant III, they are both negative. Again, for points in quadrants II and IV, X' and Y' of a point are of opposite sign (X negative but Y positive in quadrant II and *vice versa* in quadrant IV). Thus for these two quadrants $\Sigma X'Y'$ will be negative. The net result of this is that if the points in quadrants I and III, are more numerous or have generally larger products $X'Y'$ then $\Sigma X'Y'$ will be positive, whereas $\Sigma X'Y'$ will be negative if the same is true for points in quadrants II and IV. Finally, in the case of zero correlation, the points (X', Y') will be equally distributed over the four quadrants and then $\Sigma X'Y'$ will become zero.

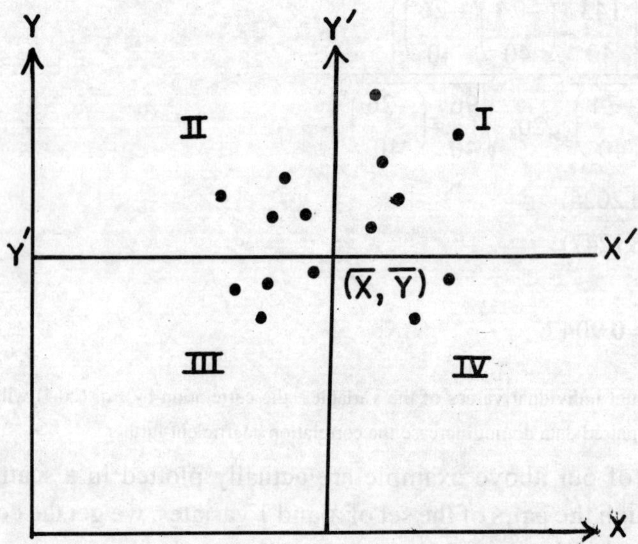

Fig. 20.2 : The Four Quadrants of Scatter Diagram with respect to X' and Y' axes

Thus it appears that $\Sigma X'Y' = \Sigma\ (X - \overline{X})\ (Y - \overline{Y})$ may be taken as a measure of the correlation between X and Y. But this is not quite so firstly because if we change n, the number of pairs of X, Y values, $\Sigma X'Y'$ is changed and so it is desirable that we divide $\Sigma X'Y'$ by n and the result is the covariance of X and Y [vide the numerator in Eq. (20.2)]. The second drawback is that this covariance as well as $\sum X'Y'$ depends heavily on the units of measurement of X and Y. For example, if all the values of X are multiplied by two, the covariance will increase two times, though the degree of relationship between X and Y will remain unchanged. This makes it necessary to introduce factors to take account of this stated undesirable property. This is done by dividing the covariance by the standard deviations of both X and Y variables and thus finally we arrive at the Eq. (20.1) stated earlier.

20.5 COEFFICIENT OF DETERMINATION

If the correlation coefficient is squared we get a value known as the *coefficient of determination* denoted by r^2. When r^2 is multiplied by 100, it gives the percentage of the variance in Y which is associated with the variance in X and *vice versa.* Thus in Example 20.2, the coefficient of correlation is 0.9047 then the coefficient of determination is 0.8184 and this would indicate 81.84% of the variation in the X variable can be determined from the Y variable and *vice versa.* Therefore, the coefficient of determination is the proportion of the variation of the one variable that is explained by the other variable to the original variation. Thus, it is the ratio of between the explained variation (i.e. the original variation 100% minus the residual variation) to the original variation:

$$r^2 = \frac{\text{Explained Variation}}{\text{Original Variation}}$$

$$= \frac{\text{Original variation} - \text{Residual variation}}{\text{Original variation}} \quad \ldots (20.10)$$

The details of this formula and the coefficient of determination is discussed later in Chapter 21.

20.6 PROPERTIES OF THE PEARSONIAN CORRELATION COEFFICIENT

1. It is clear from the definition of the formula for linear correlation coefficient of X and Y (Eq. 20.1) that it is a pure number with value of r ranging from –1.0 through zero to + 1.0. Thus r is independent of the units of measurement of X and Y.
2. The correlation coefficient of X and Y is symmetric in X and Y. i.e. $r_{xy} = r_{yx}$.
3. The Pearsonian product moment correlation is the most powerful of all the tests of correlation. It is said that Spearman's rank correlation is only 91 per cent as efficient as Pearsonian r. This means that if in a sample of 100 cases from a bivariate normal population, the Pearsonian r is significant, it will require a sample of 110 cases of the same data to achieve the same level of significance for the rank order coefficient.

20.7 CORRELATION SIGNIFICANCE TEST

In all cases of correlation, there is always the probability that the coefficient obtained could have recorded by chance i.e. its significance is in suspect because of the probability of a chance occurrence. Therefore, the correlation coefficient must be tested to see whether or not a chance occurrence of this magnitude is likely as a result of the set of data analysed. When a sample is of any size with not less than 10 paired values, the *significance of the product moment correlation* can be tested by the use of Student's t distribution, using Eq.

(20.11) as follows:

$$t_{n-2} = r.\sqrt{\frac{n-2}{1-r^2}} \quad \text{... (20.11)}$$

where n = number of pairs of data studied and where the degrees of freedom *d.f.* are always $n-2$. For convenience, the value of r is always taken as positive.

Now if the computed values of t do not exceed the critical values of t at the given number of degrees of freedom, null hypothesis can be postulated i.e. H_0 is $r = 0$ and that the two variables are not linearly related in the population. On the other hand, if a Pearsonian r is statistically significant, the significance denotes some degree of linear relationship in terms of significantly "high" or "strong." It should be noted in Eq. (20.11) that the larger the sample size, n the smaller the absolute value of the correlation coefficient needed for the statistical significance. For example, for $n = 12$ (i.e. $d.f. = 10$), a correlation of 0.5760 or larger is needed for significance at 95% confidence level, while for $n = 102$ ($d.f = 100$), a correlation of 0.1950 or larger is needed for .05 significance level. This implies that the importance of obtaining statistical significance can easily be exaggerated. Thus with n = 102, a correlation of 0.1950 is a very weak correlation though it is statistically significant. Therefore, we need to have not only a statistical significance but also a high absolute value of correlation itself.

20.8 REQUIREMENTS FOR USE OF PEARSONIAN CORRELATION COEFFICIENT

In order to employ correctly the Pearsonian correlation coefficient as a measure of association between X and Y variables, the following criteria should be satisfied in the data:

1. A straight line relationship between X and Y with the value of r ranging from –1.0 through zero to +1.0.
2. X and Y variables must be measured at the interval or ratio scale.
3. X and Y variables must have been taken at random, otherwise a test of significance cannot be applied.
4. Testing the significance of the Pearsonian r requires that both X and Y variables should be normally distributed. However, when the sample is large, $n > 30$, this requirement becomes of minor importance.
5. Correlation coefficient does not make any distinction between variables X and Y as dependent or independent. It is only a measure of association of the two variables.
6. Correlation measure r is a linear correlation and a value of r near zero indicates that there is no linear correlation between the two variables. But this does not mean they are unrelated for there may be a high non-linear correlation between them. Hence it is advisable that one should draw the scatter diagram to see whether the X and Y variables are linearly related or not before using r as a correlation coefficient

between the two variables.

7. Correlation coefficient *r* of the sample available is only an estimate of the correlation coefficient for the universe.

20.9 CORRELATION MATRIX

When we compute data on a number of variables say *m* (at a point of time) for a number of regions say *n* ($m > n$), we tabulate the rectangular data into a matrix X, its rows *m* represent the attributes (*i*) of each region for the variables (*j*) in the *n* columns.

Now given this data matrix (row, $m \times$ column, *n*), if we like to determine the linear combinations of the original *n* variables of the data matrix, X then the Pearsonian correlation coefficient may be computed from the raw score of the data matrix as

$$r_{ij} = \frac{\sum_{k=1}^{n} (X_{ki} - \bar{X}_i)(X_{kj} - \bar{X}_j)}{\sqrt{\sum_{k=1}^{n} .(X_{ki} - \bar{X}_i)^2} \sqrt{\sum_{k=1}^{n} (X_{ki} - \bar{X}_j)^2}} \quad \text{... (20.12)}$$

where the notation $\sum_{k=1}^{n}$ refers to summation over all the entities, $\bar{X}_i$ and $\bar{X}_j$ are the mean values of the variables X_i, X_j.

Now if we define z_{ki} and z_{kj} as the standard deviate of the pair of variables (taken) then following Eq. (20.5) we can substitute the above raw-score formula as

$$r_{ij} = \frac{1}{n} \sum_{k=1}^{n} z_{ki} z_{kj} \quad \text{... (20.13)}$$

So far, we have considered only 2 variables. In the *correlation matrix,* we compute correlation coefficients between every possible pair of variables (*ij*) and they are arranged into a square symmetrical matrix, $R_{n \times n}$ known as correlation matrix. This matrix contains all the information regarding the pairwise linear relationships between the variables.

For two variables, we know that we can perceive the degree of linear association between the variables by plotting them on a two-dimensional scatter diagram. If 3-variables are considered, the 3rd variable is constructed as an axis at perpendiculars to the other two, and the entries will form some sort of 3-dimensional swarm of points. Beyond the 3-dimension, since no further presentation like this sort is possible so a row of the data matrix may then be considered as a vector that gives the coordinates of an entity (i.e. the region here) in *n*-dimensional space. We will consider about the vectors and the angles between the vector as correlation coefficient in more detail in Chapter 22 on spatial correlation.

LIST OF FORMULAE

Product Moment Correlation Coefficient of Bivariate data

For listed data,

$$r_{XY} = \frac{M_{XY}}{\sqrt{M_{XX} \cdot M_{YY}}}$$

or

$$= \frac{\sum XY - (\sum X . \sum Y)/n}{\sqrt{\left[\sum X^2 - \frac{(\sum X)^2}{n}\right]\left[\sum Y^2 - \frac{(\sum Y)^2}{n}\right]}}$$

or

$$= \frac{\sum z_X . \sum z_Y}{n}$$

For grouped data,

Co-variance of X and $Y = C_x\, C_y \left[\frac{\sum f_c . X'Y'}{n} - \left(\frac{\sum f_x . X'}{n}\right)\left(\frac{\sum f_y . Y'}{n}\right)\right]$

Standard Deviation of $X = C_x \sqrt{\frac{\sum f_x . X'^2}{n} - \left(\frac{\sum f_x . X'}{n}\right)^2}$

Standard Deviation of $Y = C_y \sqrt{\frac{\sum f_y . Y'^2}{n} - \left(\frac{\sum f_y . Y'}{n}\right)^2}$

Coefficient of Determination, $r^2 = \frac{\text{Explained Variation}}{\text{Original Variation}}$

Correlation Significance test, $t_{n-2} = r . \sqrt{\frac{n-2}{1-r^2}}$

EXERCISES

20.1 The product moment correlation coefficient measured between 50 pairs of measurements is found to be $r = 0.25$. Comment on the significance of the result. The Spearman's rank correlation coefficient for the same data is $r_\rho = 0.82$. Comment on the significance of the result and the reason for its wide divergence in the value of r.

20.2 By plotting the points and observing the scatter, what guesses would you make for the value of the correlation coefficient in each of the following three cases?

(a)		(b)		(c)	
X	*Y*	*X*	*Y*	*X*	*Y*
5	20	48	41	16	9
11	31	2	21	11	21
13	37	67	68	8	25
22	50	99	4	4	34
16	46	71	67	17	5
8	22	66	92	4	33
20	40	43	83	12	16
25	45	27	85	11	18
7	32	39	12	3	31
17	27	36	31	8	23

Calculate the value of *r* in the three cases.

20.3 Calculate the correlation coefficient for the following bivariate frequency table:

	Y			
X	20 - 30	30 - 40	40 - 50	50 - 60
10-15	10	15	8	4
15-20	8	10	4	2
20-25	10	5	5	3
25-30	5	10	2	4

20.4. The following is the volume of the river runoff (in 100 cumecs) measured against the unit depth of rainfall (in mm) in the catchment area at a discharge site.

Rainfall	Runoff	Rainfall	Runoff
710	5.4	540	6.0
540	5.7	714	5.3
670	7.0	720	5.1
760	9.8	900	10.0
496	10.0	780	9.5
760	7.1	745	12.5
975	14.0	710	12.5
805	8.0	660	13.5
785	11.5	775	10.5

Calculate the product moment correlation coefficient by the raw score and z score method.

20.5 Following is the bivariate frequency table obtained from a summary of height of a green plant and the dry weight of the group regarding 350 experimental samples:

Weight of dry crop of the plant, W (in gms.)	Height of the green plant, H (in cm)							
	111.5	127.5	143.5	159.5	175.5	191.5	207.5	Total
1.175	12	25	15	1				53
2.775	1	4	33	59	29	3		129
4.375	1		4	28	35	14	2	84
5.975				2	20	18	1	41
7.575				1	1	14	5	21
9.175					4	8	2	14
10.775						3	2	5
12.375							3	3

Plot the above bivariate frequencies data, draw the isolines of equal bivariate frequencies and interpret the isolines in perspective of the coefficient of correlation you compute.

[**Hint:** By geometrical interpretation, the more the bivariate-frequency isolines are concentrated along a straight line, the greater is the correlation coefficient. The correlation coefficient will be unity for the case where all isolines coincide with the trend line. In the present example, the r computed will be 0.7554 and hence a fair concentration of the isolines can be expected. *Note* when the isolines are in circle forms, r is zero and there is no significant linear correlation.]

20.6 Find the coefficient of correlation from the following paired data by assumed mean:

X	50	3	47	52	55	49	48	51	54	46
Y	95	90	97	92	93	94	96	95	92	96

[**Hint:** Let assumed mean of X is 50 and of $Y = 94$, then transform X value into $X - 50 = X'$ and Y to $Y - 94 = Y'$.

Now use the conventional formula [Eq. (20.3)] to calculate the correlation coefficient.

20.7 A computer while calculating correlation coefficient between 2 variables X and Y from 25 pair of scores obtained the following results:

$\Sigma X = 125$, $\Sigma X^2 = 650$, $\Sigma Y = 100$, $\Sigma Y^2 = 460$, $\Sigma XY = 508$

It was found afterwards at the time of the checking that it had copied down two pairs for (X, Y) as (6, 14) and (8, 6) instead of the actual values (8, 12) and (6, 8). Find out the correct correlation coefficient.

20.8 Test the relationship of basin area (sq. km) and total stream length (km) of the basin:

Stream length :	12.0	16.4	15.8	21.0	17.5	26.3	23.0	25.4	28.6	30.2
Basin area :	6.3	7.3	9.5	10.5	11.4	11.7	12.5	13.6	14.5	16.3

20.9 Two places are having different locations: one is in a valley and the other on the mountains. If these two places are indexed as A and B and we collect the summer precipitation for both of them separately for a period of 30 years, the data will be as below:-

(Figures are in cm)

Station A: Summer rainfall

20.73, 36.73, 28.40, 20.19, 20.70, 32.97, 14.10, 35.74, 19.43, 24.69, 28.09, 28.12, 30.07, 25.48, 20.93, 17.68, 28.42, 21.36, 19.20, 18.30, 28.42, 34.82, 12.19, 19.66, 13.16, 20.45, 31.32, 21.51, 16.10 and 36.91.

Station B: Summer rainfall

23.98, 40.94, 23.52, 30.15, 21.62, 43.54, 19.43, 36.86, 31.65, 30.71, 33.48, 34.39, 25.68, 31.32, 25.73, 27.61, 42.75, 29.74, 24.18, 22.63, 33.63, 34.52, 13.21, 27.69, 16.74, 18.19, 26.57, 22.17, 23.39 and 40.41.

Test the hypothesis that the computed correlation coefficient is not significantly different from zero.

21

Regression Analysis in Geo-science Systems II (Linear Regression)

21.1 LINEAR REGRESSION

In the previous chapter, we have discussed the interdependence (or, correlation) of the two variables X and Y and presented a measure of this interdependence (that is, a degree of association), called the Correlation Coefficient. We also mention about a significance test of it to show whether the degree of association which is estimated is likely to be more than a matter of chance. But what neither the correlation coefficient nor its significance test can tell is the way in which the two set of variates are related. Thus neither (1) they can be used to predict one set of variates from a knowledge of the other, nor (2) they can signal any anomalies in the relationship between individual pairs, nor (3) they can estimate an unknown value of the other.

If we wish to pursue the above mentioned types of functional relationship between the variables (two or more) we must turn to regression analysis and the study of residuals. *Regression* in the present chapter is defined as a functional relationship between one independent variable X and the other a dependent variable Y, so that for a set of data for X variable, there will be always a concurrent distribution of Y variable which are related to each other by a correlation measure. It should, however, be stressed at the outset that both are of limited value unless the two variables are significantly correlated.

Rarely are the two sets of data in any geo-science systems so perfectly correlated that their individual values when plotted on a scatter diagram all fall on a trend (or, straight) line. For example, the sons of all tall fathers and of short fathers may not be as tall or as short as their tall or short fathers in all cases, i.e. they all "regress" towards their mean. Therefore, to give the closest approximation to the relationship between the set of data which have already fallen more or less along a trend line, a *line of best fit* is required to be inserted on a scatter diagram. This line which best fits a scatter of points is known as the *regression line.* It may be merely a summary expression of the relationship between the two variables or actually a trend line estimated by mathematical equation and fitted to all values of Y versus X. Note when correlation is positive, the regression line rises from left to right, but in the case of negative correlation it rises from right to left instead. Below we now discuss the geometrical and the mathematical ways of fitting the regression line.

21.2 METHODS OF CONSTRUCTING REGRESSION LINES

A pure functional relationship between variables assumes that all points would follow a curve, without spread (or, a regression line). In such context, drawing the regression line is at its easiest. But as the spread of points increase in the scatter diagram, drawing the regression (or, the "best fit") line by graphical methods become difficult. In that case, for the line to be really a best fit, a more precise technique is required. Four techniques will now be outlined first two, of which are graphical and very simple while the last two are more sophisticated and they require mathematical computation.

1. Regression lines for ranked data
2. Regression lines by the method of semi-average
3. Regression lines by the least square method
4. Regression lines by the correlation coefficient method

21.2.1 Regression Lines for Ranked Data

When paired data are ranked and plotted as points on a graph, we get a scatter diagram with the *X*- and *Y*-scales based on rank. A straight edge is positioned on the plotted points till it appears to summarise the linear pattern of the scatter diagram and the regression line is then drawn. In a scatter diagram, the "best fit" regression line of the set of paired data $[X, Y]$ is a perfect function $Y = f(X)$. It is simply a diagonal, running from the origin of the scatter diagram if the data are positively correlated and between the extremities of the two rectangular axes if the data are negatively correlated. The regression line drawn for the ranked data, however, is of limited value, but it does focus attention upon the differences between the data sets: the further a point is from the diagonal the greater will be the difference.

21.2.2 Regression Lines by the Method of Semi-average

When the paired data are on an interval scale, a method of reducing, though not entirely eliminating the element of guess work in drawing the regression line is by the *method of semi-average.* The calculation is simple and the result is more reliable than that achieved by relying solely on "eyeballing". In this method, drawing the regression line involves the plotting of three points in the scatter diagram: the mean, and the lower and upper semi-averages. The mean is simply the points whose coordinates are equal to the means of the paired variates, *X* and *Y*. The lower semi-average has coordinates which are determined as: the mean of all values of X below the mean $\overline{X}$ and the mean of all the Y below the mean $\overline{Y}$. Similarly, the upper semi-average has the mean of all the values of X above the mean $\overline{X}$ and the mean of all the values of Y above the mean $\overline{Y}$ as the coordinates.

Now it becomes much easier to draw by eye a "best fit" line through the three points which are themselves nearly in a straight line than through a large number of points in the data with a much less linear scatter than in the earlier technique. The regression line thus established helps in generating informations to map areal or locational differences for many situations, provided paired data are involved e.g. the relationship between annual temperatures and latitudinal positions or between the crop yield and rainfall of places.

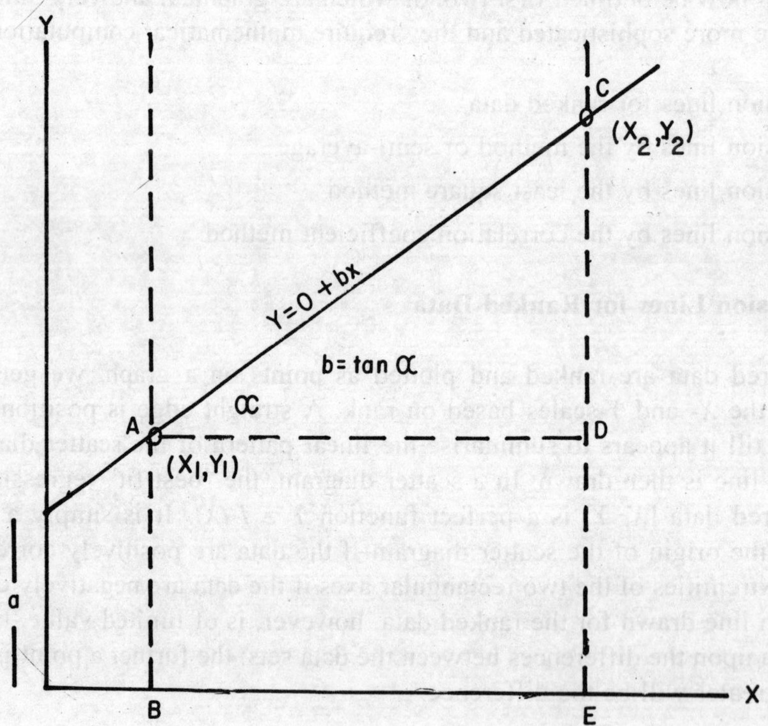

Fig. 21.1 : Constants *a* and *b* in a Regression Line

21.2.3 Regression Lines by the Least Square Method

For a highly scattered data as in Fig. 20.1b or d of Chapter 20, we require an objective for fitting the data to get the "best fit" to the data. Now if we ask "what is a best fit?" the answer surely is "a fit that makes the total deviations from the mean equal to zero" i.e $\Sigma(Y - \overline{Y}) = 0$. But this criterion is bad since the fit may be intuitively a good one as in Fig. 20.1a or bad one as in Fig. 20.1b. The only way to overcome this problem is to minimise the sum of squares of the deviations from the mean a minimum i.e. $\Sigma (Y - \overline{Y})^2 = \Delta Y^2$ = a minimum. This is the famous "Least Square" criterion, which requires that a trend (or regression) line be chosen to fit a set of n scores for two numerical variables, X and Y on the scatter diagram so that Y may be partially predictable from X. In this way as told before that the trend line serves two mathematical characteristics : (1) that the algebraic sum of deviations of values around such a line is zero and (2) that the sum of the squares of the deviations of a variable Y for given value of the other variable X is at a minimum (see also Chapter 3). It is because of these two properties that this method is known as the *method of least square.*

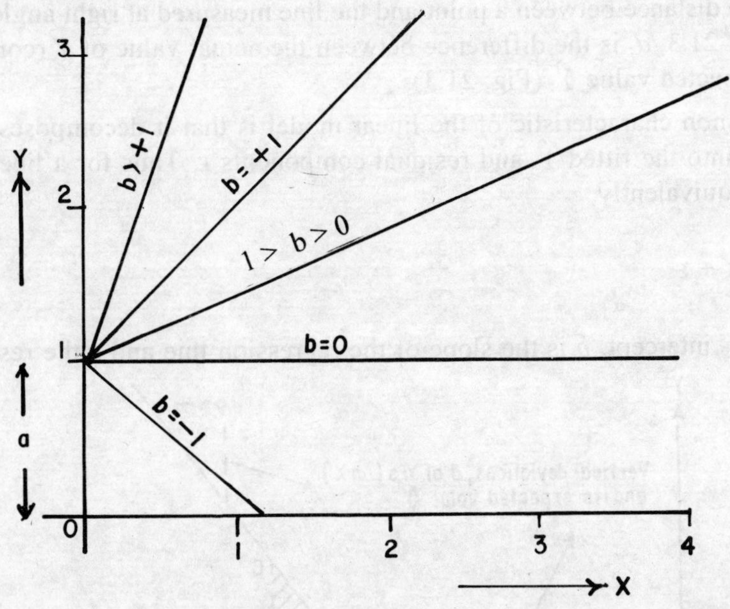

Fig. 21.2 : Various Linear Functions, (all with the same "a" but with different "b" values)

Any straight line (i.e. linear) graph drawn on the X and Y axes, can be represented by a linear function of $Y = f(X)$ in the form of

$$Y = a + bX \qquad ...(21.1)$$

where a and b are the two constants with a determining the intercept of the straight line on the Y-axis and b is its slope, (tan α) indicating the rate at which Y change with a unit change in X (Fig. 21.1). When a is equal to zero, the line $Y = a + bX$ passes through the "origin" of the graph. Again, when the slope or gradient of the line is 45° so that tan 45° = 1 and hence b becomes equal to one. Thus, with $b = 1$, the line will have an upward slope of 45° to the X-axis, so that a unit increase in X will be matched by a unit increase in Y plus the constant a and the gradient is 1-in-1 (Fig. 21.2). We can now think of various linear functions in the Fig. 21.2 all with the same intercept value a, but with different b values. Thus with b greater than 1, Y increase more rapidly than X increases, so the line will be at an angle in excess of 45°. With b less than 1, Y will increase less rapidly than X increases so the line will slope at a lesser angle than 45°. The b coefficient will be negative in those cases where Y decreases per unit increase in X and the line slopes down. Note when b is equal to zero, the line is horizontal (Fig. 21.2).

We determine the two unknown constants a and b, in the above Eq. (21.1) from a set of n observed values of X and Y. It is evident that there will be many possible choices of the values of (a, b). We choose that pair (a, b) which will make Eq. (21.1) the most suitable. In the set of n scores of X and Y we observe Y along with X and let $\hat{Y}$ (read Y hat) denote the predicted value of Y. So $(\hat{Y} - Y)$ be the *error of prediction*, ε or *residual*. A residual is the distance between a point and the line measured at right angles to one of the axes. In Fig. 21.3, d_i is the difference between the actual value of Y (corresponding to X) and its expected value $\hat{Y}$ (Fig. 21.3).

Now a common characteristic of the linear model is that it decomposes dependent variable values into the fitted $\hat{Y}$ and residual components ε. Thus for a linear function $Y = a + bX$ or equivalently,

$$\hat{Y} = a + b\,X \pm \varepsilon \qquad \text{...(21.2)}$$

where a is the Y- intercept, b is the slope of the regression line and ε the residual.

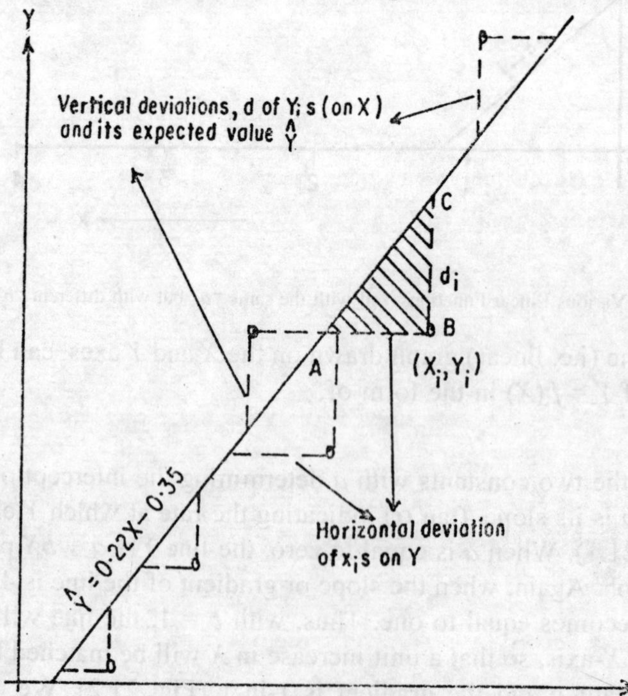

Fig. 21.3 : Summation of Deviations for Regression Line (Minimize areas ABC shaded)

Now the regression line for which these errors are a minimum may be supposed to be the most desirable. In this context although we need to combine residuals for all n scores of X and Y but a simple sum of $\sum_{i=1}^{n} \varepsilon_i$ will not do. Since positive residuals can offset negative ones, a line that fits quite poorly can have a sum of residuals close to zero. Indeed it can be easily shown that, regardless of its orientation, any line passing through the mean of the variables (the point $\bar{X}$, $\bar{Y}$) has $\Sigma\, \varepsilon_i = 0$. Such a line satisfies the

$$\bar{Y} = a + b\bar{X} \qquad \text{...(21.3)}$$

Subtracting this Eq. (21.3) from Eq. 21.2, we have

$$\hat{Y} - \bar{Y} = b\,(X - \bar{X}) \pm \varepsilon$$

or

$$\pm\,\varepsilon \;= (\hat{Y} - \bar{Y}) - b\,(X - \bar{X}) \qquad \text{...(21.4)}$$

Now, summing over the set of n scores produces

$$\Sigma\,\varepsilon = \Sigma\,(\hat{Y} - \bar{Y}) - b\Sigma\,(X - \bar{X}) = 0 - b\,(0) = 0 \qquad \text{...(21.5)}$$

A straight forward solution to get $\Sigma\,\varepsilon$ equal to zero in our problem is to transform the residuals so that they are all positive, which may be accomplished either by taking absolute value $\Sigma|\varepsilon_i|$ or by squaring $\Sigma\varepsilon_i^2$.

The former though attractive is relatively unwieldy algebraically, the latter on other hand is tractable algebraically provided we go by the principle of least square stated earlier. The least square principle is essentially a criterion for minimization. In this principle the regression line is that line about which the total sum of squares of the deviations between the regression line and each point, i.e. $\Sigma\,(Y - \hat{Y})^2 = \Sigma\,\varepsilon^2$ is at a minimum. For this reason, the regression line (as mentioned earlier) is often called the *line of best fit* to data in statistics.

Let us regard the sum of squared residuals, $\Sigma\,\varepsilon^2$ as a function of coefficient a and b, since each choice of values for these coefficients determines a value for the sum of squares:

$$s(a,\, b) = \Sigma\,\varepsilon^2 = (Y - a - bX)^2 \qquad \text{...(21.6)}$$

Now this sum of squares function in Eq. (21.6) is a quadratic function and thus it has either a minimum or maximum value of a and b. For our minimization criterion, we minimize the above sum of the squares function. Hence on differentiating $s(a, b)$ with respect to the regression coefficients a and b, we get

$$\frac{ds(a,b)}{da} = \Sigma\,(-1)\,(2)\,(Y - a - bX) \qquad \text{...(21.7)}$$

and

$$\frac{ds(a,b)}{db} = \Sigma\,(-X)\,(2)\,(Y - a - bX) \qquad \text{...(21.8)}$$

Since in *maxima* and *minima* of calculus a stationary point of function cannot be a maximum, we set these two partial derivatives with respect to a and b equal to zero and we get further:

$$\frac{ds}{da} = -2\Sigma Y + 2na + 2b\Sigma X = 0 \qquad ...(21.9)$$

$$\frac{ds}{db} = -2\Sigma XY + 2a\Sigma X + 2b\Sigma X^2 = 0 \qquad ...(21.10)$$

Finally on dividing by 2 and setting the negative terms on r.h.s., we have

$$\Sigma Y = na + b\Sigma X \qquad ...(21.11)^*$$

and $\Sigma XY = a\Sigma X + b\Sigma X^2$ ...(21.12)**

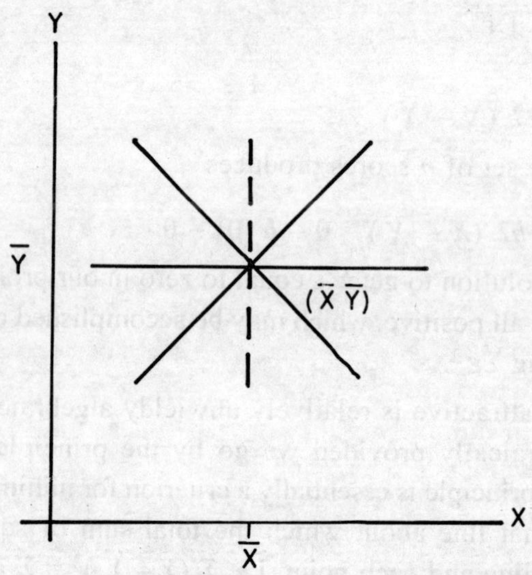

Fig. 21.4 : No Determinate Estimate of Y (because of no spread of X)

*We can also have the two linear equations for solving the regression coefficients, a and b if we write $Y_i = a + bX_i$ as

$$Y_1 = a + bX_1$$
$$Y_2 = a + bX_2$$
$$\cdots$$
$$Y_n = a + bX_n$$

On summing all these n pairs of values, we get $\Sigma Y = na + b\Sigma X$ as in Eq. 21.11.

Now if we multiply $Y = a + bX_i$ by the coefficient of b, i.e. by X_i as

$$X_1Y_1 = aX_1 + bX_1^2$$
$$X_2Y_2 = aX_2 + bX_2^2$$
$$\cdots$$
$$X_nY_n = aX_n + bX_n^2$$

**On summing all these n pairs of values, we get $\Sigma XY = a\Sigma X + b\Sigma X^2$ which is the second linear equation in Eq. 21.12.

Thus, mathematically it is proved that the required values of a and b must satisfy the set of two linear equations in Eq. (21.9) and Eq. (21.10). Now, to compute the values of a and b, we can have the statistical method. Accordingly, let us translate X scores into deviations from its mean, that is, we define a new variable $x = X - \bar{X}$. Now since $\Sigma x = \Sigma(X - \bar{X}) = 0$, so $\Sigma Y = na + b\Sigma X$ gets reduced to $\Sigma Y = na + b\Sigma x = na + 0$ so $a = \frac{\Sigma Y}{n} = \bar{Y}$

Again, $\Sigma XY = a\Sigma X + b\Sigma X^2$ is reduced to $\Sigma xY = b\Sigma x^2$ so $b = \frac{\Sigma xY}{\Sigma x^2}$. Thus, with x values measured as deviations of $X - \bar{X}$, the least square values of a and b are:

$$a = \bar{Y} \qquad \text{...(21.13)}$$

and, $$b = \frac{\Sigma xY}{\Sigma x^2} \qquad \text{...(21.14)}$$

Now, since a is the average value of Y, so we can see that our fitted regression line must pass through the bivariate point $(\bar{X}, \bar{Y})$, which may be interpreted as the centre of gravity of the set of n scores for X and Y variables.

There can be a set of data in which values of X may show little or no variation against changing values of Y. As a limiting case, we can treat all the values of X's concentrated on one single value $\bar{X}$. In this case since all $X = \bar{X}$ then all $x = (X - \bar{X})$ will be zero and the sum of squares (in Eq. 21.6) does not depend on b at all and it follows that any b will do equally in minimising the sum of squares so that b in Eq. (21.14) is not defined at all as all x values are zero and Σx^2 in the denominator is zero. In this case geometrically there can be any number of differently sloped lines passing through $\bar{X}, \bar{Y}$ which fit Y variable equally well and the best prediction becomes $\bar{Y}$ for the same value. Figure 21.4 shows this unreliable estimate because of no spread (or variation) in X value.

We can compute the coefficients a and b by treating the two linear equations in Eqs. (21.11) and (21.12) as simultaneous equations, we can have the value of b first as

$$b^* = \frac{n\Sigma XY - (\Sigma X)(\Sigma Y)}{n\Sigma X^2 - (\Sigma X)^2}$$

Now, substituting the above value for b in $\Sigma Y \approx na + b\Sigma X$ [of Eq. (21.11)] we have: $a = 1/n. (\Sigma Y - b\,\Sigma X)$.

---

* If we divide $b = \frac{n\Sigma XY - (\Sigma X)(\Sigma Y)}{n\Sigma X^2 - (\Sigma X)^2}$ by n^2, we get b in the form of:

$$b = \frac{\text{Covariance of }(X, Y)}{\text{Variance}(X)}$$

Again, since $\Sigma Y/n = \overline{Y}$ and $(\Sigma X)^2 = n^2.(\overline{X})^2$ we can further simplify the above two equations for a and b to write

$$a = \overline{Y} - b\overline{X} \qquad \text{...(21.15)}$$

and $$b = \frac{\Sigma XY - n.(\overline{X})(\overline{Y})}{\Sigma X^2 - n.(\overline{X})^2} \qquad \text{...(21.16)}$$

For any set of paired data, there are two least square regression (or best fit) lines which minimize the sum of squares of residuals for the X- and Y-variables respectively. If the values of Y are to be predicted from given values of X, the line which minimizes the Y-residuals is used, if values of X are to be predicted from given values of Y and we need the regression line of X-on-Y which is

$$\hat{X} = a'Y + b' \qquad \text{...(21.17)}$$

(instead of $Y = aX + b$ in the earlier case). Note in this regression equation for X-on-Y we use the same expressions for a' and b' as in the earlier two sets of equations, but with the X's and Y's reversed. Thus, except in the case of perfect correlation, $r = 1.0$, the regression lines, Y-on-X, X-on-Y are different lines. Note Fig. 21.5a shows the regression line of Y on X as obtained by minimising the sum of squares of the vertical distances of the points from the regression line $\hat{Y} = aX + b$ itself and Fig. 21.5b shows the horizontal distances of points on the scatter diagram from the regression line $\hat{X} = a'Y + b'$.

21.2.4 Regression Lines by Correlation Coefficient Method

In Eq. (21.16), we have b coefficient for regression of Y-on-X which can be written as

$$b_{yx} = \frac{\text{Covariance}(X,Y)}{\text{Variance}(X)}\frac{\text{Std. dev. } Y}{\text{Std. dev. } Y}$$

$$= \frac{\text{Covariance}(X,Y)}{(\text{Std. dev. } X)(\text{Std. dev. } Y)}\frac{\text{Std. dev. } Y}{\text{Std. dev. } X}$$

$$= \text{Correlation} \cdot \frac{\text{Std. dev. } Y}{\text{Std. dev. } X} = r \cdot \frac{s_y}{s_x} \qquad \text{...(21.18)}$$

Thus the slope of the regression line of Y on X can be explicitly written as $b_{yx} = r. \dfrac{s_y}{s_x}$. Now substituting this value of b_{yx} [Eq. (21.18)] and of the other coefficient a [Eq. (21.15)] in Eq. (21.1) we have :

$$\hat{Y} - \overline{Y} = r \cdot \frac{s_y}{s_x}(X - \overline{X}) \qquad \text{...(21.19)}$$

where $(\hat{Y} - \overline{Y})$ are the residuals which are perpendicular to the X-axis (Fig. 21.5a), r is the product moment correlation, $\overline{X}$ and $\overline{Y}$ are the means and s_x and s_y are the standard deviations of the given values of X and Y respectively. Thus, the unknown value Y differs from the mean of its set of data $\overline{Y}$ by the same amount as the known value X differs from its average $\overline{X}$, modified by the ratio of the two standard deviations and the correlation coefficient. Therefore, Eq. (21.19) not only requires the correlation coefficient but also the mean and standard deviation values for the two sets of data. But these values have been calculated for the correlation coefficient itself. Now, by rearranging Eq. (21.19) we have the regression line of *Y*-on-*X* rewritten as

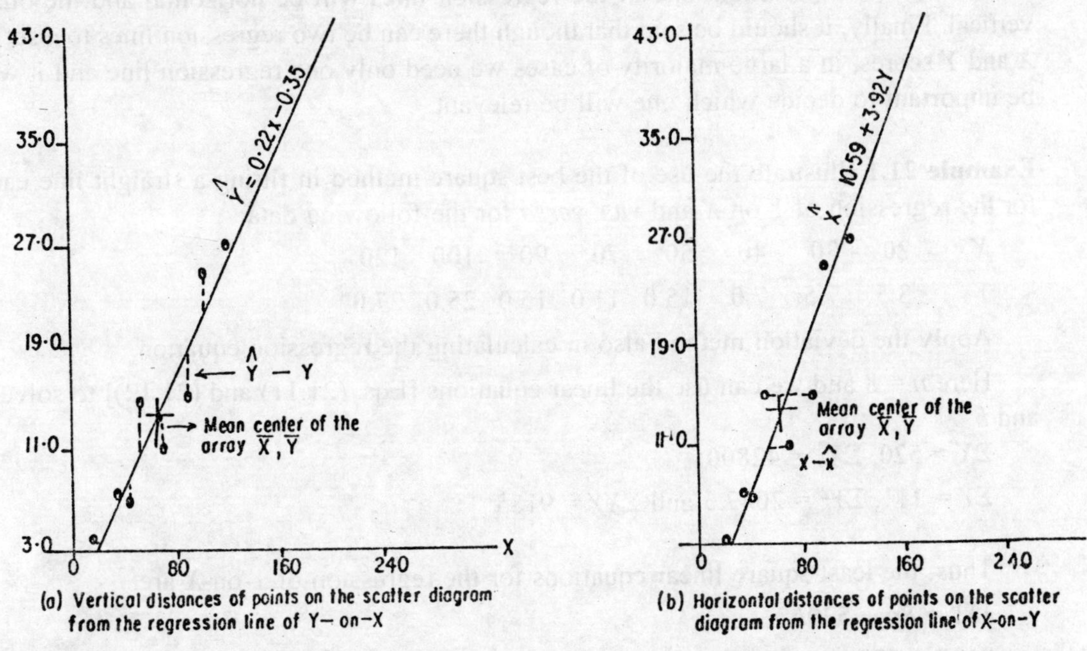

Fig. 21.5 : Regression Lines

$$\hat{Y} = \overline{Y} + r.\frac{s_y}{s_x}(X - \overline{X}) \qquad ...(21.20)^*$$

Similarly, for the regression equation of *X* on *Y* we have $b' = b_{xy} = r.\ s_x/s_y$ and $a' = \overline{X} - b\overline{Y}$ which on substituting in Eq. (21.17), we get:

$$\hat{X} - \overline{X} = r.\frac{s_x}{s_y}\ (Y - \overline{Y}) \qquad ...(21.21)$$

This again can be rewritten as

$$\hat{X} = \overline{X} + r.\frac{s_x}{s_y}\ (Y - \overline{Y}) \qquad ...(21.22)$$

---

* Since $\hat{Y} = \overline{Y} + r.\dfrac{s_y}{s_x}(X - \overline{X})$ so if we sum all the values of *X* from 1 to *n* we get $\Sigma\hat{Y} = n\overline{Y} + r.\dfrac{s_y}{s_x}\Sigma(X - \overline{X})$. Now $\Sigma(X - \overline{X}) = 0$. so on dividing both sides by *n*, we get $\overline{\hat{Y}} = \overline{Y}$ i.e. the mean of the observed values, $\overline{Y}$ is equal to the mean of the corresponding predicted values $\hat{Y}$.

These two regression lines in Eqs (21.20) and (21.22) as has been stated earlier will always intersect at the point representing the mean of the two sets of data $\overline{X}$ and $\overline{Y}$. Figure 21.5 illustrates this where the data in Example 21.1 is worked out. Apart from the intersection of the two regression lines at ($\overline{X}$, $\overline{Y}$), another point to be seen is that the angle between these two regression lines is very small because the correlation coefficient worked out in Example 21.2 shows that $r = 0.9265$. If the two regression lines are drawn then in case of no relation, one of the regression lines will be horizontal and the other vertical. Finally, it should be said that though there can be two regression lines to a set of X and Y scores, in a large majority of cases we need only one regression line and it will be important to decide which one will be relevant.

Example 21.1 Illustrate the use of the best square method in fitting a straight line each for the regression of Y on X and *vice versa* for the following data:

X	20	30	40	50	70	90	100	120
Y	3.5	7.5	7.0	15.0	11.0	15.0	25.0	27.0

Apply the deviation method also in calculating the regression equation.

Here $n = 8$ and we can use the linear equations [Eqs. (21.11) and (21.12)] to solve a and b.

$\Sigma X = 520$, $\Sigma X^2 = 42800$

$\Sigma Y = 111$, $\Sigma Y^2 = 2042.5$ and $\Sigma XY = 9185$

Thus, the least square linear equations for the regression of Y-on-X are:

$$111 = 8a + 520b$$

$$9185 = 520a + 42800b$$

Treating these two as simultaneous equations, we solve them algerbraically that is, we multiply the first equation by 65 and subtracting the second from the first, we eliminate a to calculate b as shown below:

$$111 = 8a + 520b \; [\times 65]$$

$$9185 = 520a + 42800b$$

or

$$7215 = 520a + 33800b$$

$$9185 = 520a + 42800b$$

on subtraction, $-1970 = -9000b$ or $b = 0.2189$

Now we substitute this value of b in the first equation $111 = 8a + 520b$ that is, $111 = 8a + 113.8222$ or $a = -0.3528$ and the equation of the straight line which provides the best fit in the sense of the least square for Y-on-X is:

$$\hat{Y} = 0.2189X - 0.3528$$

Similarly, for solving the equation for regression of X-on-Y by the least square method, we use the model $\hat{X} = a'Y + b'$ in Eq. (21.17) where we use the above expression for a and b [i.e. Eqs. (21.11) and (21.12)] but with the X's and Y's reversed. Thus we have

$\Sigma Y = 520,\ \Sigma Y^2 = 42800$

$\Sigma X = 111,\ \Sigma X^2 = 2042.5$ and $\Sigma XY = 9185$

and we have least square linear equations as follows:

$520 \;\; = 8a + 111b$

$9185 = 111a + 2042.5b$

Again solving these two equations algebraically, we have $a = 10.5909$ and $b = 3.9214$. Thus the equation of the straight line which provides the best fit in the sense of least square for X-on-Y is

$\hat{X} = 10.5909 + 3.9214\ Y$

Now for the deviation method, we create a new x scores for the given data of the X variable. Thus the new bivariate data will be:

x	–45	–35	–25	–15	5	25	35	55
Y	3.5	7.5	7.0	15.0	11.0	15.0	25.0	27.0

since in the above, each x_i is the difference of the corresponding X_i value and the mean of the X_i values, that is, $x_i = X_i - 65$.

With respect to this new set of data for x and Y variables, we can compute a by Eq. (21.13) and b by Eq. (21.14) so that

$$a = \bar{Y} = \frac{111}{8} = 13.8750$$

and $$b = \frac{\Sigma xY}{\Sigma x^2} = \frac{1970}{9000} = 0.2189$$

And we obtain the trend equation $\hat{Y} = 13.8750 + 0.2189x$ with respect to our transformed variable x and the variable Y. But since $x = X - 65$ so we substitute this in this trend equation $\hat{Y} = 13.8750 + 0.2189\ (X - 65)$ to get the final trend equation $\hat{Y} = 0.2189X - 0.3528$.

Again for the trend equation for X-on-Y we reverse the data in terms of X for given Y and Y for given X and solving them finally we get

$\hat{X} = 10.5909 + 3.9214\ Y$

(*Note* this last equation tallies with its counterpart totally but the earlier one for Y-on-X differs slightly).

Example 21.2. For the data X and Y sets in the Example 21.1 above, calculate the two regression lines, one for Y-on-X and the other for X-on-Y, by the correlation coefficient method.

No.	X	Y	XY	X^2	Y^2
1.	20	3.5	70	400	12.25
2.	30	7.5	225	900	56.25
3.	40	7.0	280	1600	49.00
4.	50	15.0	750	2500	225.00
5.	70	11.0	770	4900	121.00
6.	90	15.0	1350	8100	225.00
7.	100	25.0	2500	10000	625.00
8.	120	27.0	3240	14400	729.00
Total =	520	111.0	9185	42800	2042.50

For convenience, we will use Eq. (20.3) for determining the correlation coefficient.

$$r_{XY} = \frac{\Sigma XY - \dfrac{\Sigma X.\Sigma Y}{n}}{\sqrt{\left[\Sigma X^2 - \dfrac{(\Sigma X)^2}{n}\right]\left[\Sigma Y^2 - \dfrac{(\Sigma Y)^2}{n}\right]}}$$

$$= \frac{9185-(520)(111)/8}{\sqrt{(42800-33800)(2042.50-1540.125)}}$$

$$= \frac{9185-7215}{\sqrt{(9000)(502.375)}} = \frac{1970}{2126.3525}$$

$$= 0.9265$$

Again from above $s_x = 94.8683$, $\overline{X} = 65$, $s_y = 22.4137$ and $\overline{Y} = 13.875$. Therefore, from Eq. (21.20), we have regression for Y-on-X as:

$$\hat{Y} = \overline{Y} + r.\frac{s_y}{s_x}(X - \overline{X}) = 13.875 + (0.9265)\left(\frac{22.4137}{94.8683}\right)(X-65)$$

$$= 13.875 + 0.2189\,(X - 65)$$

$$= 13.875 + 0.2189X - 14.2282$$

$$= 0.2189\,X - 0.3532$$

Similarly for regression of X-on-Y we can have Eq. (21.22) to compute $\hat{X} = 10.5909 + 3.9214Y$.

Example 21.3 Fit the linear equation $\hat{Y} = aX + b$ as regression of Y-on-X to the 8 pair of bivariate scores (X,Y) in a least square frame and calculate the constant of the equation by partial differentiation.

We know by least square principle, the sum of squared residuals $\Sigma\varepsilon^2$ which equals the sum of squared deviation between the predicted and the observed Y values can be regarded as a function of coefficients a and b. Accordingly, from Eq. (21.6), we can write:

$$s(a, b) = \Sigma\varepsilon^2 = \Sigma(Y - \hat{Y})^2 = \Sigma(Y - a - bX)^2$$

Hence for the 8 pairs of score (X, Y) given, we compute the squares of deviations for each pair and then sum them. Following are the calculations:

X	Y	Predicted Y or $\hat{Y}$		Deviation	Square of Deviation
20	3.5	$20a + b$	$20a+b - 3.5$	$400a^2 - 140a + 40ab$	$+ b^2 - 7b + 12.25$
30	7.5	$30a + b$	$30a+b - 7.5$	$900a^2 - 450a + 60ab$	$+ b^2 - 15b + 56.25$
40	7.0	$40a + b$	$40a+b - 7.0$	$1600a^2 - 560a + 80ab$	$+ b^2 - 14b + 49$
50	15.0	$50a + b$	$50a+b - 15.0$	$2500a^2 - 1500a + 100ab$	$+ b^2 - 30b + 225$
70	11.0	$70a + b$	$70a+b - 11.0$	$4900a^2 - 1540a + 140ab$	$+ b^2 - 22b + 121$
90	15.0	$90a + b$	$90a + b - 15.0$	$8100a^2 - 2700a + 180ab$	$+ b^2 - 30b + 225$
100	25.0	$100a + b$	$100a + b - 25.0$	$10000a^2 - 5000a + 200ab$	$+ b^2 - 50b + 625$
120	27.0	$120a + b$	$120a + b - 27.0$	$14400a^2 - 6480a + 240ab$	$+ b^2 - 54b + 729$
				$\Sigma = 42800a^2 - 18370a + 1040ab$	$+ 8b^2 - 222b + 2042.5$

The above summation equation can be written in the form of a quadratic, in terms of either a or b. Starting with terms of a, we write : $\Sigma = 42800a^2 + (1040b - 18370)a + (8b^2 - 222b + 2042.5)$. Again for terms of b : $\Sigma = 8b^2 + (1040a - 222)b + (2042.5 - 18370a + 42800a^2)$.

Now, if we wish to make (each of) the summations minimum, to satisfy the principle of least squares, we set the above summation as S and assume $S = px^2 + qx + q$. Now to set the first differential $ds/da = 0$ and thus setting the first summation equation (in terms of a) to zero, we differentiate:

$$\frac{ds}{da} = 2px + q = 0 \text{ which on rewriting, } x = -\frac{q}{2p}$$

Now by analogy with the 1st summation equation

$$-\frac{q}{2p} = \frac{-(1040b - 18370)}{85600} = a, \; p \text{ being } 42800$$

or, $85600a + 1040b = 18370$ (i)

This is the partial differential $\frac{\sigma_\Sigma}{\sigma a}$ of the sums of squares of the deviations holding b constant. Similarly using the second summation equation and following the procedure just outlined, we get

$$16b + 1040a = 222 \quad \text{(ii)}$$

which is the partial differential $\frac{\sigma_\Sigma}{\sigma b}$ of the sums of squares of the deviations holding a constant now. Thus we have a pair of equations:

$$1040b + 85600a = 18370 \text{ and } 16b + 1040a = 222$$

from which we can solve $a = 0.2189$ and $b = -0.3528$ and so the trend equation $\hat{Y} = 0.2189X - 0.3528$.

21.3 STANDARD ERRORS AND CONFIDENCE LIMITS

As already stated, unless two variables correlate perfectly (so that their regression on each other, Y-on-X and X-on-Y coincide), it is not possible to interpolate or predict the variable Y (for an X-value given) with complete accuracy and reliability. It is possible, however, to estimate with a given level say 68.26 per cent or 95 per cent, the range within which a value predicted will lie very much in the same way the confidence limits are calculated for estimates of population parameters. This is done by using what is called the *standard error of estimates*—this is a measure of variation of points about the regression line. In fact, the standard error of estimate is an extension of the standard error of the mean discussed in Chapter 10.

It is that portion of the variance in Y not accounted for by X with its standard deviations. This is done in two ways as follows: (1) The standard error estimate, $s_{\hat{y}}$ of Y-on-X is the residual variance not accounted from the regression of Y on X. This is given by

$$s_{\hat{y}} = s_y . \sqrt{1 - r^2} \quad \text{...(21.23)*}$$

where s_y is the standard deviation for $Y = \sqrt{\Sigma(Y - \bar{Y})^2 / n}$.

* We know from Eq. (21.2) the residual component in regression of a set of paired data is ε for which the residual variance can be written as:

$$\text{var}(\varepsilon) = \frac{1}{n}\Sigma\varepsilon^2 = \frac{1}{n}\Sigma(Y - \hat{Y})^2$$

Now from Eq. (21.20), $\text{var}(\varepsilon) = \frac{1}{n}\Sigma[(Y - \bar{Y}) - r.\frac{s_y}{s_x}(X - \bar{X})]^2$

$$= \frac{1}{n}\Sigma(Y - \bar{Y})^2 - 2r\cdot\frac{s_y}{s_x}\cdot\frac{1}{n}\Sigma(X - \bar{X})(Y - \bar{Y}) + r^2\cdot\frac{s_y^2}{s_x^2}\cdot\frac{1}{n}\Sigma(X - \bar{X})^2$$

$$= s_y^2 - 2r\cdot\frac{s_y}{s_x}\cdot(r.s_x.s_y) + r^2\frac{s_y^2}{s_x^2}\cdot s_x^2$$

$$= s_y^2 - 2r^2.s_y^2 + r^2.s_y^2$$

$$= s_y^2 - r^2.s_y^2 = s_y^2(1 - r^2)$$

Now, the standard deviation of ε (= the residuals) called the "standard error of estimate" (of Y relative to its linear regression on X) is denoted by $s_{\hat{y}} = s_y\sqrt{1 - r^2}$

This standard error estimate can also be calculated as the standard deviation of residuals, $Y - \hat{Y}$

$$s_{\hat{y}} = \sqrt{\Sigma(Y - \hat{Y})^2 / n} \quad ...(21.24)$$

It can be seen from Eq. (21.23) or (21.24), that for $r = \pm 1.00$ (i.e. when $Y = \hat{Y}$), there is no standard error of estimate. However, when $r = 0.00$, the standard error of estimate, $s_{\hat{y}}$ reaches its maximum possible value, becoming as much variable as the original values of Y to be equal to s_y. Thus the standard error of estimate of Y can be precisely taken as the variability of Y about the regression line for a particular Y value averaged over all values of X. For a good predictability, the total variability of all Y, which is given s_y will be fairly larger than $s_{\hat{y}}$ – in this case, a high score value of Y will have a high score on X and *vice versa*. However, when $s_y < s_{\hat{y}}$, the predictability will be poor. In Example 21.2 above the correlation coefficient is high, i.e. $r = 0.9265$. Hence if we compute the $\hat{Y}$ scores following the regression equation $\hat{Y} = 0.2189\,X - 0.3528$, we can see that s_y is fairly large as compared to standard error estimate of Y on X, i.e. $s_{\hat{y}}$. The calculation is shown below.

X	Y	$\hat{Y}$	$(Y-\bar{Y})^2$	$(Y-\hat{Y})^2$
20	3.5	4.0252	107.6406	0.2758
30	7.5	6.2142	40.6406	1.6533
40	7.0	8.4032	47.2656	1.9690
50	15.0	10.5922	1.2656	19.4287
70	11.0	14.9702	8.2656	15.7625
90	15.0	19.3482	1.2656	18.9068
100	25.0	21.5372	123.7656	11.9910
120	27.0	25.9152	172.2656	1.1768
$\Sigma = 520$	111.0	111.0056	502.3748	71.1639

With $\bar{Y} = \dfrac{111.0}{8} = 13.875$,

we find

$$s_y = \sqrt{\frac{\Sigma(Y-\bar{Y})^2}{n}} = \sqrt{\frac{502.3748}{8}} = 7.9244$$

Now $s_{\hat{y}}$ is either equal to

$\sqrt{\dfrac{\Sigma(Y-\hat{Y})^2}{n}}$ [as in Eq. 21.24)]

or

$s_y\sqrt{1-r}$ [as in Eq. (21.23)]

and we get $s_{\hat{y}} = 2.9825$. Hence we find $s_{\hat{y}} < s_y$ for the given data of Example 21.2.

Thus standard error of a fitted curve (or, of a regression coefficient), $s_{\hat{y}}$ in Eq. (21.24) is found by determining the differences of the calculated values from their observed ones, squaring the differences, taking their sum, dividing the sum by the number of the differences and then extracting the square roots. This standard error estimate, $s_{\hat{y}}$ can be used in regression lines on the scatter diagram (of the *X*, *Y* data) at any limit of the confidence level. This is possible since the use of the linear regression model requires that the variables have normal distributions (in fact, the requirement is that the raw data may not be normally distributed but the conditional distributions of residuals for every values of *X* must be normally distributed). Therefore, if the ±1 $s_{\hat{y}}$ bands are drawn around the regression line, like the standard deviation of normal distribution, they will enclose 68.26 per cent of all the distributions. Similarly, ±2 $s_{\hat{y}}$ bands and ±3 $s_{\hat{y}}$ bands will enclose 95.46 per cent and 99.72 per cent of all the distributions. Another advantage of these error bands (+ or –) is that since the two standardized residual bands run parallel to the regression line (each being drawn with reference to two points estimated from the regression line), the standard error of estimates, $s_{\hat{y}}$ have a wide margin.

It can be seen from the plotting of the standard error of the regression line on the scatter diagram of the set of data (*X*, *Y*) of Example 21.1 in Fig. 21.6 that the $s_{\hat{y}}$ bands have very wide margins regarding their error estimates and hence they are having little use in making estimates.

21.4 RESIDUALS

The difference between the observed value and its predicted (or, estimated) values is known as the "absolute residual". The *absolute residual* of a particular observation of *X* (or, *Y*) does not have much use in research because of the problem of units of measurement when comparison is done between two or more sets of observations. In this context, the "standardised" residual is preferred which expresses the value of the absolute residual in terms of a normal distribution of residuals. *Standardized residuals* can now be defined as

$$S.\ Res.Y' = \frac{\text{Absolute residual}}{\text{Standard error of estimates}} = \frac{\Sigma|Y - \hat{Y}}{s_{\hat{y}}} \qquad \text{... (21.25)}$$

where *Y* is a particular value, $\hat{Y}$ its corresponding predicted value and $s_{\hat{y}}$ is the standard deviation of the predicted $\hat{Y}$ values.

Standardized residual bands (like standard error bands) run parallel to the regression lines. Therefore, they do not give undue emphasis to the residuals in *Y* related to either the large or small values of *X* (as do the absolute residuals). The standardized residuals at one standard error is often incorporated into the regression equation [Eq. (21.2)]. Thus to indicate the relative accuracy of a particular predicted value $\hat{Y}$ we replace the residual, ε by this *S. Res.* y' and the full equation will read as

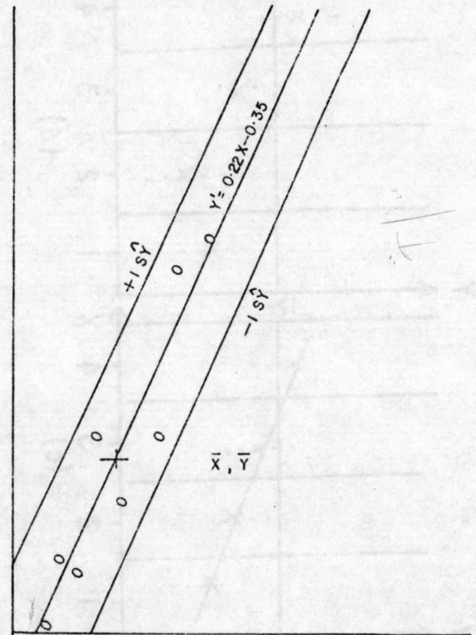

Fig. 21.6 : Standard Error Estimates and the Regression Line

$$\hat{Y} = a + bX \pm S.\ Res.\ y' \qquad ...(21.26)$$

Absolute residual values, $(Y - \hat{Y})$ are often used in a map in geography to show their spatial pattern in terms of their positive and negative values. For mapping the absolute residuals, the residual values can be arranged in an array and class intervals are selected on the basis of some percentile measure. The residual can also be expressed as relative residuals which for a given unit of observation (area-wise) is the absolute residual of the area relative to its magnitude of the observed value, i.e. $(Y - \hat{Y})/Y$.

Residual maps are helpful in identifying the left out positive or negative factors for inclusion in the analysis. Residuals on plotting when give an uniform band of equal variance for all the values of the variable Y is interpreted as a desired one – any other form indicates deficiency in the linear regression model. Thus, residual mapping is used as an aid to identify or delineate regional boundaries of areas. These maps help also in finding out the more effective spatial variables for further explanation of the variations in the dependent variable Y.

Residuals, especially in form of standardised residuals, *S. Res.* y' when plotted on a map, provide an excellent graphic indication of fitness/failure of the regression models to explain all the variations in Y.

Example 21.4 If the regression equation is $\hat{Y}$ = 21.98 + 0.90X and one standardised

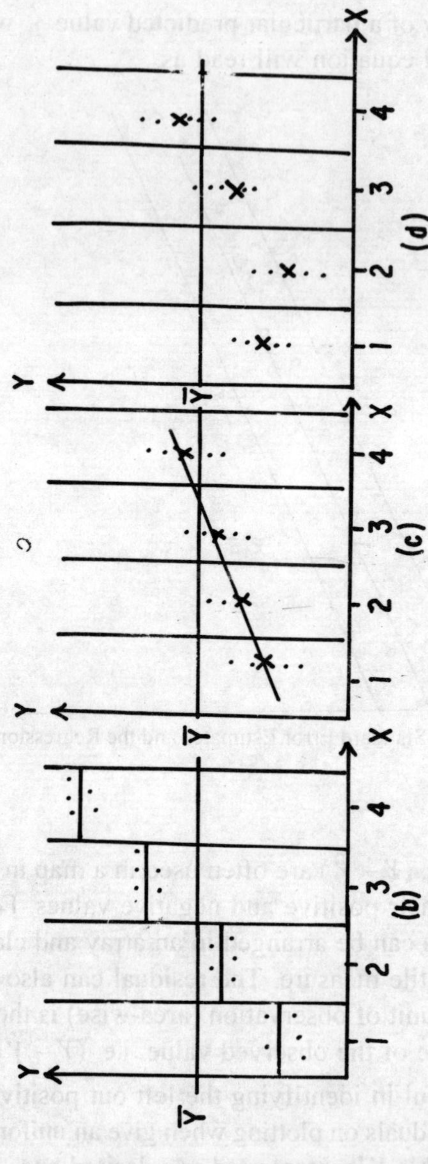

Fig. 21.7 : Rationale of ANOVA for Linear Regression

residual error is 8.20, then for $X_i = 30.0$, calculate the corresponding value of $\hat{Y}$.

Since a computed value $\hat{Y}$ is estimated equal to the sum of a constant, a (related to the mean) plus a linear function of X plus (or minus) a standardised residual error, we can write: $\hat{Y} = 21.98 + 0.90\ (30.0) \pm 8.20$, i.e. $\hat{Y}$ will lie within 40.78 and 57.18 in 68.26% of all cases of $X_i = 30.0$.

21.5 COEFFICIENT OF DETERMINATION

The standard error of estimate, $s_{\hat{y}}$ (for Y-on-X) is also helpful in interpreting the meaning of a given numerical value of r_{xy}. It can be shown from Eq. (21.23) that

$$s_{\hat{y}}^2 = s_y^2 \cdot (1 - r^2)$$

from which on rearrangement we have :

$$r^2 = 1 - s_{\hat{y}}^2 / s_y^2 \qquad \text{...(21.27)}$$

From the above equation, it can be seen that when the variance is around the regression line $s_{\hat{y}}^2 = s_y^2$ then $r \to 0$. On the other hand, when s_y^2 is equal to zero, then r will be one and in this latter case the regression line will account for all the variance of the dependent variables so that all the observed data will lie on regression line.

The squared correlation coefficient r^2 is known as the *coefficient of determination* [mentioned in earlier chapter, Eq. (20.10)] which shows by what proportion the reduction in the variance in predicting for Y has been achieved by knowing the set of scores for X. Thus, in Example 21.2 for the given data $r = 0.9265$ then $r^2 = 0.8584$ and it can be said that 85.84% of the variance of Y is explained by the variance in X. Therefore, in using the correlation coefficient as a measure of the goodness of the relationship, it is best to square it in order to obtain a realistic estimate of the amount of variability which the linear relationship explains. This of course, will always be less than the correlation coefficient (except when $r = 1.0$ exact).

21.6 RATIONALE OF ANOVA FOR REGRESSION

Before we go into the test of linearity we should discuss the rationality for involving the ANOVA for regression. We know that if regression of Y-on-X is taken as a linear function $Y = f(X)$ to estimate (or to predict) Y, we conceive a series of data scores. We begin with a distribution of Y-scores and in this respect, the best guide is a horizontal line representing the grand mean of Y-scores, $\overline{Y}$ (Fig. 21.7a).

Now, if we are given a knowledge of each individual X-score as well as the corresponding Y-score we could compute a series of $\overline{Y}_i$'s one for each category of X (Fig. 21.7b). It will be seen from Fig. 21.7b that if we draw new horizontal line representing the mean of Y_j scores within each category of X, our average error in guessing Y scores will be smaller than what it was when we use a single horizontal line $\overline{Y}$ in Fig. 21.7a.

When the X, Y data are linearly related, then we can see that all the $\overline{Y}_j$'s for the categories fall on the regression equation $Y = a + bX$ and this trend line passes through the grand X and Y means, $\overline{X}$, $\overline{Y}$. Figure 21.7c gives the more accurate graphical representation of this situation.

Now, suppose that the X, Y data are not linearly related, then the category means, $\overline{Y}_j$ will not fall on the regression line so that the $\hat{Y}$ estimates that we could calculate for each X_j category could not be equal to the category means, $\overline{Y}_j$. Figure 21.7*d* illustrates this situation. In this case, we cannot draw the regression line $\hat{Y} = a + bX$ since $\overline{Y}_j$'s are computed directly (Fig. 21.7d).

Coming back to Fig. 21.7c, where we have a linear relationship between X and Y data by computing the $\hat{Y}$-values from $\hat{Y} = a + bX$ we can conceive a "within category means". This is done by computing the variance within each category by finding the sum of squares within each category and then summing these over all the categories.

Now the total variance can be considered as our original amount of error.

Since in Fig. 21.7 we can see that there are a series of Y-scores around the various $\overline{Y}_j$ so on taking into account each j category we can have a sum of square for total variance, SS_T equal to $\Sigma(Y - \overline{Y})^2$. The sum of squares within the categories is smaller than the total variance, SS_T and it is considered as the amount of variation remaining unexplained by regression of X-on-Y. Hence the difference between total variance and within variance can be considered as the amount of variance explained by regression of X-on-Y. The sum of squares of this explained variance, SS_R is named as between-category variance [which in one-way ANOVA is squared difference between the each category mean and the grand mean multiplied with the number of scores in j categories, see Eq. (12.13) in analysis of variance in Chapter 12]. Finally, the amount of variance remaining unexplained is the within-category variance, and hence this can be given the notation of SS_U.

21.6.1. Test of Regression for Linearity

It is often found important to test the fitted regression line for linearity. This is best done by performing the analysis of variance test (which is a technique discussed in Chapter 12) to analyse the mean squares (or estimates of variance) due to several components of the variation. For the linear regression given above it is stated in Eq. (21.2) that $\hat{Y} = Y \pm \varepsilon$ which on rearranging can be put up as:

$$Y = \hat{Y} \pm \varepsilon = \hat{Y} \pm (Y - \hat{Y}) \qquad \text{...(21.28)}$$

i.e. Observed value = Predicted value with a definite pattern ± Error of prediction having a random pattern.

From this we can write : $Y - \bar{Y} = (\hat{Y} - \bar{Y}) \pm (Y - \hat{Y})$.

We square both the sides and, summing for all Y values, we get:

$$\Sigma(Y-\bar{Y})^2 = \Sigma(\hat{Y}-\bar{Y})^2 + \Sigma(Y-\hat{Y})^2 \quad ...(21.29)$$

i.e. Total sum of square of Y, SS_T = Sum of squares explained by the linear regression of Y on X, SS_R + Sum of squares unexplained by the linear regression of Y on X, SS_U

Note in the above squaring. the product term $2\Sigma(\hat{Y}-\bar{Y})\Sigma(Y-\bar{Y})$ is absent on the r.h.s of the Eq. (21.29) because it is zero owing to the least square condition implied in Eq. (21.5).

Hence for linear regression following analysis of variance, it may be observed that there is a total variability SS_T of the dependent variable Y which is divided into a variability accounted for by the regression SS_R and a variability unaccounted for by the regression or residual variability SS_U. This can be conveniently expressed alongwith the degrees of freedom (*d.f.*) by an ANOVA Table 21.1.:

Table 21.1 : ANOVA Table for Simple Linear Regression

Sum of squares of variance	Degrees of freedom *d.f.*	*Mean square *MS*	*F* Ratio
Accounted for by regression $SS_R = \Sigma(\hat{Y}-\bar{Y})^2$	1	$SS_R/1 = MS_R$	$F(1, n-2) =$
Unacounted for by regression (absolute residual) $SS_U = \Sigma(Y-\hat{Y}) = SS_T - SS_R$	$n-2$	$\frac{SS_T - SS_R}{n-2} = MS_U$	$\frac{MS_R}{MS_U}$
Accounted for by the mean (total variance). $SS_T = \Sigma(Y-\bar{Y})^2$	$n-1$		

The degrees of freedom (*d.f*) will be $(n-1)$ for the total variation SS_T since the mean is fixed and $n-1$ of the observations can vary. Since another degree of freedom is lost in estimating the coefficient in regression, there are $(n-1) - 1 = n-2$ degree of freedom left

* Note to derive variances, the sum of squares are divided by n and to derive mean squares. they are divided by *d.f.*

for estimating the absolute residual mean square, MS_U. For estimating the sum of squares explained by regression, SS_R the *d.f.* should be one since the *d.f.* of the components of variation in ANOVA table always add to the total *d.f.* that is, $n-1$.

Now for computations, the *sum of squares of total deviation.*

$$SS_T = \Sigma (Y-\bar{Y})^2 = \Sigma Y^2 - \frac{(\Sigma Y)^2}{n} \quad ...(21.30)$$

Again *the sum of squares for regression,*

$$SS_R = \Sigma ((\hat{Y}-\bar{Y})^2 = \Sigma \hat{Y}^2 - \frac{(\Sigma \hat{Y})^2}{n} \quad ...(21.31)$$

Since, SS_R we obtain as sum of squares of absolute residual, so *sum of squares due to residuals* (or, deviation from the trend) is

$$SS_U = SS_T - SS_R \quad ...(21.32)$$

which is also equal to $\Sigma(Y-\hat{Y})^2$

A test of significance of linearity can be conducted from the analysis of variance by the F statistic which is as follows :

$$F(1, n-2) = \frac{\text{Estimate of the variance accounted for}}{\text{Estimate of the residual or the variance unaccounted for}}$$

$$= \frac{SS_R / 1}{(SS_T - SS_R) / (n-2)} \quad ...(21.33)^*$$

Equation (21.33) can also be written for computational convenience as

$$F(1, n-2) = \frac{(r^2 . s_y^2 / 1)}{(1-r^2)(s_y^2) / (n-2)} \quad ...(21.34)^*$$

---

* Note the numerator in Eq. (21.33) is $SS_R/1$ where $SS_R = \Sigma(\hat{Y}-\bar{Y})^2$: from Eq. (21.19) we have $\hat{Y}-\bar{Y} = r.\frac{s_y}{s_x}(X-\bar{X})$.

If we square this equation on both sides, multiply by $1/n$ and then sum with respect to different values of X and Y from 1 to n, we get :

$$\frac{1}{n}\Sigma(\hat{Y}-\bar{Y})^2 = r^2 . \frac{s_y^2}{s_x^2} . \frac{1}{n}\Sigma(X-\bar{X})^2$$

$$= r^2 . \frac{s_y^2}{s_x^2} . s_x^2 = r^2 . s_y^2 \text{ in numerator of Eq. (21.34)}$$

Again, the denominator in Eq. (21.33) is concerned with the residual variance

$$= \text{var}(\varepsilon) = \frac{1}{n}\Sigma \varepsilon^2 = \frac{1}{n}\Sigma(\hat{Y}-Y)^2$$

(Contd.....)

which reduces to

$$F\,(1,\, n-2) = \frac{r^2(n-2)}{(1-r^2)} \qquad \text{...(21.35)}$$

Note : Eq. (21.35) is the observed t_{n-2} for r^2 in Eq. (20.11).

This computed value of F (1, n–2) by Eq. (21.33) or (21.34) or (21.35) is to be compared to an F or variance ratio table with 1 and (n–2) degrees of freedom at 0.10 or 0.05 significance level to determine whether a linear regression really exists or not, i.e. whether the mean square explained by linear regression is large enough in comparison to the residual mean square to decide that the linearity of regression is due to a real effect rather than to random sampling. Hence if the resulting F value falls in the critical region we can conclude that the linear regression is significant and it can be retained.

Before carrying out this statistical test of significance of regression for its linearity, it is necessary, to see whether the regression is practically significant. In the case of linear regression, practical significance is measured by the "squared correlation" i.e. by the coefficient of determination (discussed earlier). Once the practical significance of regression analysis is tested by the coefficient of determination in finding the proportion of the total variability explained by regression and a positive answer is found, the test of hypothesis for linearity must be made in order to test for reality. For the data given in Example 21.1 the correlation between X and Y data sets is found to be high (r = 0.9265) and hence 85.84% of the variance of Y is explained by the regression of Y-on-X. For a test of linearity we find

$$SS_T = \Sigma Y^2 - \frac{(\Sigma Y)^2}{n} = 2042.50 - \frac{(111.0)^2}{8} = 502.3750$$

$$SS_R = \Sigma \hat{Y}^2 - \frac{(\Sigma \hat{Y})^2}{n} = 1971.5353 - \frac{(111.5056)^2}{8} = 431.2549$$

So, $SS_T - SS_R = SS_U = 71.1201$

Now from Eq. (21.20), we have $\hat{Y} = \overline{Y} + r.\dfrac{s_y}{s_x}(X - \overline{X})$ so we can write

$$\text{var}(\varepsilon) = \frac{1}{n}\Sigma\,[(Y-\overline{Y}) - r.\frac{s_y}{s_x}(X-\overline{X})]^2$$

$$= \frac{1}{n}\Sigma(Y-\overline{Y})^2 - 2r.\frac{s_y}{s_x}.\frac{1}{n}\Sigma(X-\overline{X})(Y-\overline{Y}) + r^2.\frac{s_y^2}{s_x^2}.\frac{1}{n}\Sigma(X-X)^2$$

$$= s_y^2 - 2r.\frac{s_y}{s_x}.(r.s_x.s_y) + r^2.\frac{s_y^2}{s_x^2}.s_x^2$$

$$= s_y^2 - 2r^2.s_y^2 + r^2.s_y^2$$

$$= s_y^2 - r^2.s_y^2 = s_y^2(1-r^2) \text{ in denominator of Eq. (21.34).}$$

Now, from Eq. (21.33) we have $F\,(1,\,6) = \dfrac{431.2549/1}{71.1201/6} = 36.3825$

Again, if we compute this F ratio by either Eq. (21.34) or Eq. (21.35) we get $F\,(1{,}6) = 36.3735$ i.e. practically the same value as we got earlier. Since the F statistic for *d.f.* (1, 6) at 0.5 significance level is 5.99 and the same at 0.01 significance level is 13.7 which are both less than the computed value so we state that null hypothesis has to be rejected to conclude that the linear regression of the data in Example 21.1 does explain a proportion of the total variation and it is significant. Hence the regression equation of $\hat{Y} = 0.2189X - 0.3528$ can be retained.

This analysis of variance method for significance test of linearity of regression tells us a great deal about how regression analysis itself works. Thus, for example, when 2 variables are zero correlated the best fit line through the scatter of points will be horizontal (b = 0.0) passing through the data centroid $(\overline{X}, \overline{Y})$. Under these conditions, the residual or error (unexplained) sum of square SS_U is clearly equal to the total sum of squares, SS_T and the regression sum of squares, SS_R is zero since $\hat{Y}$ is always equal to $\overline{Y}$. However, as the regression coëfficient, b develops slopes the equality between the SS_T and SS_U starts decreasing and the difference is made by SS_R. The problem is one of deciding at what point the proportion of regression variance is statistically significant. The F ratio reflects the balance between the SS_R and the SS_U ones and we use the F table (as done before) to determine if the proportion of SS_R is significant.

Note for the above linearity test of regression, we could also use the correlation coefficient significance test by the Eq. (20.11) and calculate

$$t_{n-2} = \frac{r\sqrt{n-2}}{\sqrt{1-r^2}} = \frac{(0.9265)(\sqrt{6})}{\sqrt{1-0.8584}} = \frac{2.2695}{\sqrt{0.1416}} = 6.0310$$

and we observe that at p = 0.001 this computed t value for *d.f.* = 6 *is* higher than the critical value of 5.96. Hence we can conclude that the correlation coefficient of the data is significantly different from zero. Note also that $(t = 6.0310)^2 \cong (F = 36.3735)$.

The coefficient of determination which itself be taken as a measure of usefulness of the linear regression equation for prediction, can also be determined from the table of analysis of variance by

$$r^2 = \frac{SS_R}{SS_T} = \frac{\Sigma(\hat{Y} - \overline{Y})^2}{\Sigma(Y - \overline{Y})^2} \qquad \text{...(21.36)}$$

Thus the coefficient of determination is a ratio between the sum of squares explained by regression and the total sum of squares. This squared correlation coefficient is always to be preferred as an index of "best fit" for a true measure of relationship.

Note this definition is algebraically equivalent to the definition of the correlation coefficient given in terms of moment measures in Eq. (20.1a).

$$r=\frac{M_{XY}}{\sqrt{M_{XX}}\cdot\sqrt{M_{YY}}}$$

21.7 PREDICTING UNKNOWN VALUE OF DEPENDENT VARIABLE BY REGRESSION

Regression of a set of paired variables is computed for a description of the relationships between the variables and in particular, for predicting the unknown values of Y. For this latter purpose which is *extrapolation*, we need to know the accuracy of our predictions and for this the concept of the standard errors of estimate, discussed earlier, need to be expanded.

Thus to forecast the value of Y for an observation of X not included in the original sample, we use the "standard error of the forecast," $S_{\hat{Y}}$ which is a combination of the following three variance sources :

1. The standard error of the estimate of Y on X $s_{\hat{y}}$ is already given in Eq. (21.23) or (21.24).
2. The standard error of the mean of the dependent variable Y, the mean Y being subjected to sampling error, i.e.

$$s_{\bar{y}}=\frac{s_y}{\sqrt{n}} \qquad ...(21.37)$$

3. The standard error of the regression coefficient of Y on X, s_b is the error we can expect at the intersection of the regression line with the mean of the variables, $\overline{X}$ and $\overline{Y}$. The further one moves from the mean of the variables, ($\overline{X}$, $\overline{Y}$), the greater will be the likely error in Y because of the variability in the slope of the forecast band (Fig. 21.8). Thus the standard error of the regression coefficient b for every value of X (with the origin as $\overline{X}$, $\overline{Y}$) is given by

$$s_b=[(s_{\hat{y}})/(s_x\sqrt{(n-2)}](X-\overline{X})^2 \qquad ...(21.38)$$

Finally, for any value of X_i, the standard error of the forecast $S_{\hat{y}}$ is calculated by

$$S_{\hat{y}}=\sqrt{s_{\hat{y}}^2+s_{\bar{y}}^2+s_b} \qquad ...(21.39)$$

For any value of X, the S_y value thus calculated in the Eq. (21.39) can be taken as one $S_{\hat{y}}$ or two $S_{\hat{y}}$ s. These values can be added/subtracted to the predicted value $\hat{Y}_i$ (= $a + bX$, the regression equation) so that we can know the range within which $\hat{Y}$ will lie for 68.26 or 95.46% of the observations.

Since the standard error of the regression coefficient, s_b is different for every value of X, the calculation of the value of $S_{\hat{y}}$ is a tedious task. Generally, the error bands for the standard error of the forecast are estimated from a few estimates and the lines are drawn.

Figure 21.8 illustrates the results of this sampling errors. In Fig. 21.8, (a) variations in estimates of the intercept term b cause the regression line to move in a vertical yet parallel fashion represented by two parallel dashed lines, (b) sampling variations in the regression coefficient (the slope of the line b) cause a rotational effect about the data centroid $(\bar{X}, \bar{Y})$ through which all lines will pass, but with different slopes. Finally, (c) the combined effect of these above two sources of errors is to create curves, to be precise hyperbolic ones.

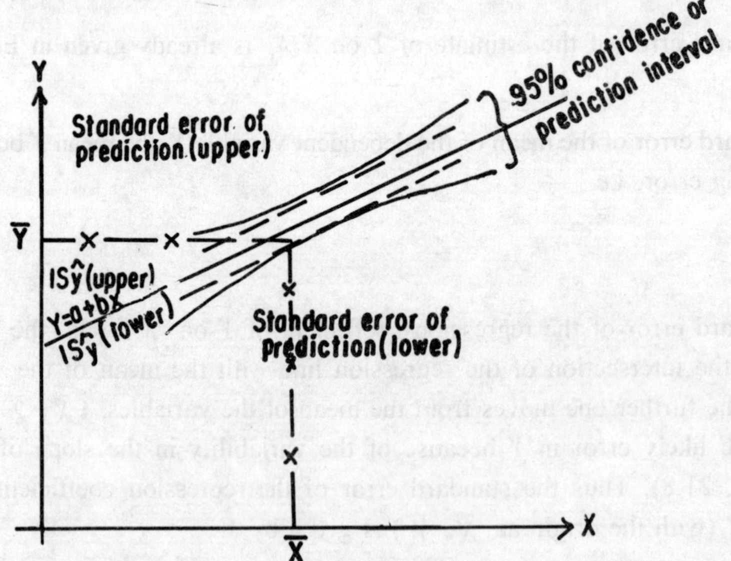

Fig. 21.8 : Confidence Limits about a Regression Line and about the Individual Values

Example 21.5 With the data in Example 21.1 for which we have

$s_{\hat{y}} = 2.9825$, $s_y = 7.9244$, $s_{\bar{y}} = (7.9244/\sqrt{8}) = 2.8017$, $s_x = 33.5410$ and $\bar{X} = 65$ and

the regression equation : $\hat{Y} = 0.2189X - 0.3532$, calculate the predicted values of Y for X

= 20, 30, 40, 50, 70, 90, 100, 120 and 150 so that 68.26% of the observations are included.

We know for 68.26% of the observations to get included within ± 1, standard error of the forecast, the predicted values of Y will be:

$$\hat{y} = \hat{Y} \pm 1.S_{\hat{y}}$$

where $\hat{Y} = 0.2189X - 0.3532$ and $S_{\hat{y}} = \sqrt{s_{\hat{y}}^2 + s_{\bar{y}}^2 + s_b}$ with $s_b = [s_{\hat{y}}/(s_x.\sqrt{n-2}]\,(X-\bar{X})^2$ and $s_{\bar{y}}^2 = (7.9244)^2 = 62.7961$, $s_{\bar{y}}^2 = (2.8017)^2 = 7.8495$. Hence the computation for s_b can be simplified to $[2.9825/(33.5410)\,(\sqrt{6})]\,(X-65)^2 = (0.0363)\,(X-65)^2$. Finally, the entire computation is tabulated in Table 21.2:

Table 21.2 Standard Estimate

X	$X-65$	$S_b = 0.0363\,(X-65)^2$	$s_{\hat{y}}^2$	$s_{\bar{y}}^2$	$S_{\hat{y}}\,(in\pm)$	$\hat{Y}$	$\hat{y} = \hat{Y} \pm 1.S_{\hat{y}}$
20	–45	73.5112	8.8953	7.8495	9.5003	4.0252	13.5255 to –5.4751
30	–35	44.4697			7.8240	6.2142	14.0382 to –1.6098
40	–25	22.6887			6.2796	8.402	14.6816 to 2.1224
50	–15	8.1679			4.9912	10.5822	15.5734 to 5.5910
70	5	0.9075			4.2015	14.9702	19.1717 to 10.7687
90	25	22.6887			6.2796	19.3482	25.6270 to 13.0686
100	35	44.4697			7.8240	21.5372	29.3612 to 13.7132
120	55	109.8131			11.2498	25.9152	37.1650 to 14.6654
150	85	262.2809			16.7041	32.4818	49.1859 to 15.7777

Note the further one moves from the origin $\bar{X}, \bar{Y}$ in either direction, the wider will be the bands $S_{\hat{y}}$ to curve away towards the extremes of the distribution and the bands are closest to the regression line around $\bar{X}, \bar{Y}$ (65.0, 13.875). Therefore, it can be said that the forecast becomes least certain when the values of X are taken for calculating $\hat{y}$ much away from the origin. Thus for $X = 150$ when we extrapolate it for the $\hat{Y}$ value projected (as 32.4818) we require to give another range of ± 16.7041 from the regression point. To have these error bands first we draw the regression line $\hat{Y} = 0.2189X - 0.3532$ and next we plot one $S_{\hat{y}}$ value (±) with respect to each X-score on either side of the regression line and finally the error bands $\pm S_{\hat{y}}$ are drawn through these plotted points.

Thus the curved confidence limits (in Fig. 21.8) can be regarded as an "envelope" within which the true (population) regression line lies with selected standard error (with 95% probability in Fig. 21.8). This arises despite the close correlation found between the two variables and it is a timely reminder that if we do the standard error estimates with small samples, the curved lines will inevitably result since X cannot completely determine the behaviour of Y. Hence the curved prediction limits will result. In some respects when the sample sizes are large and correlation too also is high, the degrees of curvature (if it results) are overlooked and the standard error lines are drawn as straight and parallel to the regression lines (as in Fig. 21.6).

21.8 CORRELATION AND REGRESSION MODEL : A COMPARISON

From the discussion above we see that in regression analysis our purpose is to explain the variation of X with respect to Y by fitting a straight line through the scatter of points to provide the best description of the trend in the scatter and the best set of estimates. Compared to regression, the correlation coefficient is a ratio of the variation accounted for by the two variables X and Y and the product of their respective standard deviations.

In computing the regression equation and the correlation coefficient for the bivariate distributions, moments are useful. We know if $M_{\bar{X}}$ and $M_{\bar{Y}}$ denote the mean for the X and Y values, first order mon its about the means are

$$M_X = \frac{\Sigma(X-\bar{X})}{n} = 0 \quad \text{and} \quad M_Y = \frac{\Sigma(Y-\bar{Y})}{n} = 0$$

Now the second order moments about the mean are the variances

$$M_{XX} = \frac{\Sigma(X-\bar{X})(X-\bar{X})}{n}$$

$$M_{YY} = \frac{\Sigma(Y-\bar{Y})(Y-\bar{Y})}{n}$$

and, the co-variance: $M_{XY} = \dfrac{\Sigma(X-\bar{X})(Y-\bar{Y})}{n}$

To solve for a and b of regression and r of correlation coefficients, in terms of moments, we can write:

$$b = \frac{M_{XY}}{M_{XX}}$$

$$a = (M_{\bar{Y}}) - \left(\frac{M_{XY}}{M_{XX}}\right)(M_{\bar{X}})$$

and $r = \frac{M_{XY}}{\sqrt{(M_{XX})(M_{YY})}}$

Below we compare the correlation and the regression models in which we can see that both of them are about somewhat different bivariate relationship, their theoretical rationale is of the same though it is more clear in context of linear regression.

Correlation Model	Regression Model
1. Y and X are random variables.	1. Y is random variable and X a fixed variable.
2. The correlation coefficient is symmetric ($r_{xy} = r_{yx}$). Hence it is immaterial whether $Y = f(X)$ or $X = f(Y)$ *i.e.* whether X or Y is designated as a dependent or independent variable.	2. Regression equation is assymmetric thus regression of Y-on-X $\neq$ regression of X-on-Y until and unless $r = 1.00$. Hence for regression, a functional relation must be known or assumed i.e. either $Y = f(X)$ or $X = f(Y)$ to identify the dependent and independent variable.
3. A study of interdependence and mutual variation.	3. A study of dependence of one variable upon the other variable.
4. The key measure is the coefficient of correlation which is a pure number and independent of the units of measurement of X and Y. This coefficient must lie between − 1 through zero to + 1.	4. The key measure is the β coefficient which, in linear regression is the increase in Y if X is increased by one unit.
5. Mainly used to explain the relationship between the bivariate data and thus serves as a complement to regression, Coefficient of determination is a common link between one to other.	5. Mainly used for functional relationship between the bivariate data, their estimation and prediction serves as a complement to correlation coefficient.

21.9 CONCLUSION

Finally, it can be said that for any related set of paired data, it is possible to fit a large number of possible trend lines, each of which will provide the best fit to the original data in terms of the relationships assumed by the mathematical functions taken. To find out which of these really provide the best fit, one requires to calculate the residuals from each of the regression lines in the first step and to find out the standard error of the estimate. The function which yields the smallest standard error estimate will provide the closest approximation to the real relationships between the two variables. Thus out of the many types of regression lines considered in this chapter, the regression line by the least square principle is found to provide best fit to the data. Again the regression line is a mathematical one to express the relationship between the two variables studied and it is a statement of average relation too between the series. But for best estimates of regression line, one broad assumption is that the bivariate data should follow normal distribution and the functional relationships between the variables should be linear. Otherwise, the regression estimates at worst become completely invalid and do not provide any answer to the question. This issue is quite broad and we will discuss about the relevance of geographical data in regression analysis again in the next chapter.

LIST OF FORMULAE

I. *Linear Regression*

1.Linear Regression Model, $\hat{Y} = a + bX \pm \varepsilon$

where $a = \overline{Y} - b\overline{X}$

$$b = \frac{\Sigma XY - n\overline{X}.\overline{Y}}{\Sigma X^2 - n(\overline{X})^2}$$ *for absolute values of X and Y*

or

1a. $a = \overline{Y}$

$$b = \frac{\Sigma xY}{\Sigma x^2}$$ where $x = X - \overline{X}$

2a. Linear Regression Equation of X on Y,

$$\hat{X} = \overline{X} + r.\frac{s_x}{s_y}(Y - \overline{Y})$$

2b. Linear Regression Equation of Y on X,

$$\hat{Y} = \overline{Y} + r.\frac{s_y}{s_x}(X - \overline{X})$$

3a. Standard Error Estimate of Y-on-X, $s_{\hat{y}} = s_y^2\sqrt{1-r^2}$

3b. Standard Error Estimate of X- on-Y, $s_{\hat{x}} = s_x^2\sqrt{1-r^2}$

4. Coefficient of Determination, $r^2 = 1 - \dfrac{s_{\hat{y}}^2}{s_y^2}$

5. Standardised Residuals, *S. Res.* $Y' = \dfrac{\Sigma|Y - \hat{Y}|}{s_{\hat{y}}}$

II. *Test of Linearity of Regression,* $F(1, n-2) = \dfrac{r^2(n-2)}{1-r^2}$

Coefficient of Determination, $r^2 = \dfrac{\Sigma(\hat{Y} - \overline{Y})^2}{\Sigma(Y - \overline{Y})^2}$

(from Analysis of variance)

III. *Standard Error of Forecast,* $S_{\hat{y}} = \sqrt{s_{\hat{y}}^2 + s_{\overline{y}}^2 + s_b}$.

where $s_b = [(s_{\hat{y}}^2)/(s_x \cdot \sqrt{n-2})]\,(X - \overline{X})^2$

EXERCISES

21.1 Determine the regression equation for assessment of annual runoff in Exercise 20.4. Also determine the confidence limits of the regression equation at least at 0.05 significance level.

21.2 The following is a climatological data to study the relationship between the altitude and temperature of six stations:

Station	:	1	2	3	4	5	6
Altitude in metres	:	270	2100	210	120	1500	600
Temperature in °C	:	30	8	32	35	12	27

To what general class of curves does the regression of the above data belong? Find out the best fitting regression equation expressing temperature in terms of altitude. What would be the best estimate of temperature for a place having an altitude of 300 m?

21.3 Fit a straight line to the following data :

X	0	1	2	3	4	5	6	7	8
Y	100	86	67	57	47	52	34	26	10

Test also the linearity of the equation fitting the data.

21.4 Find the most likely annual rainfall of one year in a place Y corresponding to that of 70 cm at a place X from the following :

	X	Y
*Normal	65	67
Std. deviation	2.5	3.5

(*that is mean for 40 years of data at both the places)

21.5 If b_{yx} and b_{xy} coefficients are 0.89 and 0.85 respectively and $s_x = 3$, find

(a) Standard errors of the regression coefficients i.e. of Y-on-X and of X-on-Y
(b) Standard error of coefficient of correlation, and
(c) Limits of the true value of "r" in the universe.

22
Regression Analysis in Geo-science Systems III
(Spatial Correlation)

22.1 SPATIAL CORRELATION ANALYSIS

In geography one is motivated by a desire to elucidate and describe attributes in spatial dimension and hence spatial viewpoint of the attributes becomes the primary concern of a geographer rather than formally testing *a priori* hypotheses. Thus geographical data matrix summarise data sets (large or small) of a number of attributes along rows taken on a number of regions (as sampling units or quadrats) distributed over space. Similar to geography, ecology in its observational approach is concerned with observations in communities or species made over space (and/or time) so that it can study (and estimate) spatial interactions (or no interactions whatsoever), in which a certain pattern of interspecific association may be positive, negative or absent. In this chapter we describe methods for detecting the existence of association between the region/species and present indices for measuring the degree of association (or, correlation). These techniques are based solely on the presence or absence of geographical attributes/the ecological species in the regions (or, the sampling units) and since the regions are the objects of the association/correlation analyses rather than the attributes, in the present chapter, the resulting coefficients are termed *spatial* or *ecological correlation*.

One of the most pioneering interpretations of spatial (or, ecological) correlation coefficients is the analysis of the race and literacy data of 1930 USA Census by W.S. Robinson (1950), "Ecological correlation and the Behaviour of Individuals," in, *Am. Sociological Review,* Vol 15. This work has been extended by H.R. Alker (1969) in his article, "A typology of Ecological Fallacies" in, *Quantitative Ecological Analysis in the Social Sciences* edited by M. Dogan and S. Rokkan, M.I.T. Press.

The predominant aim in Robinson's paper was to illustrate the fallacious argument that can be developed from correlations that are based on regions as object of analysis rather than individuals so that it can be that blacks and the illiteracy in 1930 U.S. Census were found in the same location. His study on ecological correlation between race and illiteracy was based on $N = 97.272$ million persons in U.S.A distributed over nine major census divisions in U.S.A. The first relationship Robinson analysed was the individual correlation between race ($X = 0$ if white, $X = 1$ if black) and illiteracy ($Y = 0$, if literate, $Y = 1$, if illiterate); for $N = 97.272$ million persons in the 1930 Census over 10 years old. The individual correlation is calculated irrespective of the census division, j with the Pearsonian correlation formula:

$$r_{XY} = \frac{M_{XY}}{\sqrt{(M_{XX})(M_{YY})}} \qquad ...(22.1)$$

where M_{XY} is the covariance of the blacks and illiterates of the census divisions, j; M_{XX} and M_{YY} are the second-order moments i.e. standard deviations of X and Y respectively.

The individual correlation thus obtained was + 0.2032 which does not provide any evidence about the black and illiterates interspecific association.

To get the ecological correlation, the total number of blacks in the jth region X_j, and the total number of blacks in the jth region, Y_j, are obtained from the following aggregation formula

$$X_j = \sum_{i \to j}^{n_j} X_{ij} \text{ and } Y_j = \sum_{i \to j}^{n_j} Y_{ij}$$

where subscript i identifies the 1st individual in the set of individuals living in the jth region and n_j defines the subscript value i for the last individual in the set belonging to the jth region all by the race, X (= 0 if white and = 1 if black) and by literacy, Y (= 0 if literate and = 1 if illiterate). Thus if jth region is 2 then the above summation takes the following values:

$$X_2 = \sum_{6,702,000}^{28,247,000} X_{i,2}$$

$$= X_{6,702,000}\ 1 + X_{6,702,001}\ 0 + ... + X_{28,247,000}\ 1$$

$$= \text{black} \quad + \text{white} \quad + ... + \text{black}$$

$$= 868,000 \text{ blacks in the } j = 2\text{nd region}$$

similarly, for Y_2, 751,000 illiterates in the $j = 2$nd region.

Now the above individual values of X and Y for both race and literacy are measured on a nominal scale (0, or, 1). When both the variables are nominal, the lengthy calculations implied in correlation formula of Eq. (22.1) can be simplified by arranging the table in a 2 × 2 contingency table. For the list of 21,545,000 paired observations for each individual living in the $j = 2$nd region of U.S.A. (in 1930), we have the following contingency table of row and column frequencies for $j = 2$nd region.

Table 22.1a : 2×2 Contingency Table of Race and Literacy: Enumeration

Enumeration	Black ($X = 1$)	White ($X = 0$)	Total
Illiterate ($Y = 1$)	709,000	42,000	751,000
Literate ($Y = 0$)	159,000	20,635,000	20,794,000
Total	868,000	20,677,000	21,545,000

This information of the individual cell frequencies is conveniently summarised in the form of a 2×2 Table 22.1b below.

Note: Table 22.1b follows closely the Eq. (13.7).

Table 22.1b

	Black	White	Total
Illiterate	a	b	$a + b = m$
Literate	c	d	$c + d = n$
Total	$(a + c) = r$	$(b + d) = s$	

Following closely the formula in Eq. (13.7), we can have the value of the *Within-Region Correlation Coefficient* (for $j = 2$nd region). expressed as

$$r_{XY} = \frac{a.d - b.c}{\sqrt{(a+b)(c+d)(a+c)(b+d)}}$$

or,

$$= \frac{a.d - b.c}{\sqrt{m.n.r.s}} \qquad \text{...(22.2)}$$

which for the region $j = 2$, we have

$$r_{XY} = \frac{(709{,}000)(20{,}635{,}000) - (159{,}000)(42{,}000)}{\sqrt{(751{,}000)(20{,}794{,}000)(868{,}000)(20{,}677{,}000)}}$$

$= 0.8739$ which is an impressive result.

Thus for each pair (X_{ij}, Y_{ij}) we find that in Table 22.1, a = No. of individuals in the j = 2nd region who are both black and illiterate, b = No. of individuals who are illiterate but not black, c = No. of individuals who are black but not illiterate and d = No. of individuals who are neither black nor illiterate.

The individual correlation between race and literacy what Robinson attempted with respect to the 9 census divisions ($j = 9$) by the correlation equation (Eq. 22.1) cannot be interpreted without knowing the value of all of the covariances M_{XY} [equal to $\frac{1}{J} \cdot \sum_{j=1}^{J} (X_j - \overline{X}_j)(Y_j - \overline{Y}_j)$].

The required data in this context are easily available from census publications but the low correlation coefficient of + 0.2032 between race and illiteracy does not appear to contain a geographical component. This low correlation seems to be spurious since the variables, race and illiteracy are correlated with regional urbanisation levels of the census divisions of U.S.A. In this respect, the geographical component in race-illiteracy relationship appears in within-region correlation analysis for few census divisions like the j = 2nd region in the above example.

Spatial correlation analysis as reflected by above ecological-correlation studies (e.g. race-illiteracy, sex-illiteracy) has been one of the mainstream traditions in geography and these 2-variable correlation studies are required to be strengthened by adding more variables. Like habitat status (i.e. rural or urban) etc. to the above nominal data, there is the necessity of involving the multiple group association in the studies. In this respect the community ecologists who are having the observational (in addition to the experimental) approach extend the scope of the analysis to multiple species case and in chapter on analysis of variance we have the variance ratio test. But geography which depends more on the enumerated data by other agencies and do not adopt primary data correlation through field studies rigorously, often there has been no information regarding "within-region correlation" so that the analyses become totally dependent on the arbitrary system of census regions which act as an unspecified rogue.

22.2 SPATIAL AUTOCORRELATION

"Spatial autocorrelation" involves analyzing the way in which a single variable measured over a set of regions is correlated with itself in space. Essentially, *spatial autocorrelation* is concerned with establishing whether the presence of a variable in one region in a regional system makes the presence of that variable in neighbouring regions more or less likely. If the presence is not likely to occur in neighbouring regions we state "negative spatial autocorrelation". Expressed otherwise, in the case of spatially organised data, such as states, districts or blocks, residuals may be spatially autocorrelated. It appears that a considerable number of geographical applications do not have independence because of spatial contiguity effects e.g. involving recording of the same observations many times at many places so that the residuals of regression may tend to group together the negative residuals in one area and the positives in another, we can state that the assumption of residual independence has been infringed. Thus spatial autocorrelation measures the arrangement of a pattern over space and in that way it differs from "point patterns in quadrat analysis" because the latter measures the point patterns in terms of a frequency distribution. Thus if we use a quadrat space of 4 × 4 cells with the number of points distributed in Figs. 22.1a and 22.1b that variance in both the Figs. 22.1a and b are the same ($s_x^2 = 1.0$) but distribution pattern of the points in the former are unrelated to one another (random spatial autocorrelation) whereas in the latter they are in a highly clustered arrangement indicative of "positive spatial autocorrelation".

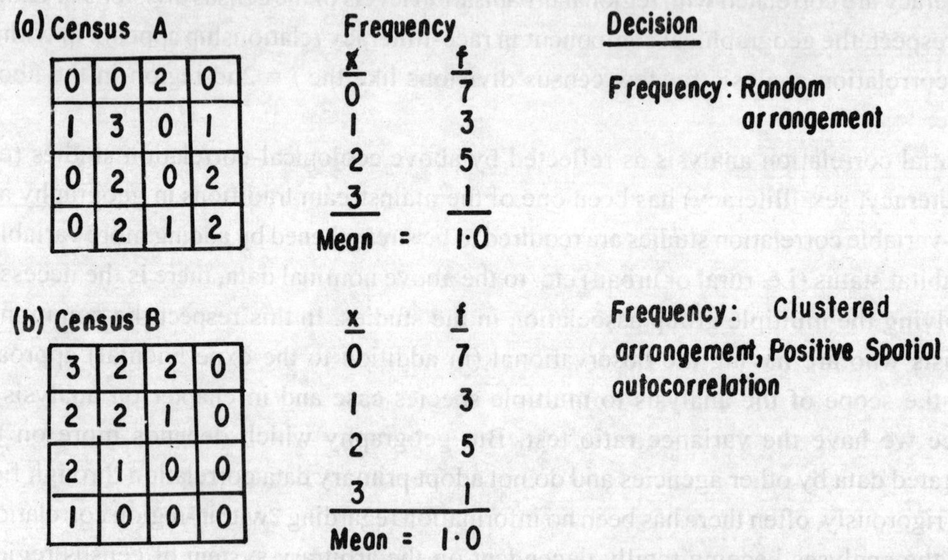

Fig. 22.1 : The Distinction between Frequency and Arrangement

To specify the simple autocorrelation process from the spatial arrangement of a geographical variable, we require a contiguity matrix, $A = \delta_{ij}$ which by its mode of contact can specify the type: random, positive or negative autocorrelation. In Fig 22.2, we use a chess-board pattern of 4 × 4 = 16 cells for showing the distribution of a nominal variable X. In this chessboard pattern we index the cells coloured white as $x_i = 0$ and the cells coloured black as $x_i = 1$. Now if we define the contact of the cells in terms of contiguity of the cells (with values $x_i = 1$ only) as (1) edge-to-edge, (2) vertex-to-vertex and (3) both edge-to-edge and vertex-to-vertex contact, we can record the contacts in the $n \times n$ contiguity matrix (here it is 4 × 4). If cell i is contiguous with cell j, then $\delta_{ij} = 1$ and if cell i is non-contiguous with cell j then $\delta_{ij} = 0$. Thus if we take the edge-to-edge contiguity, then for row 1, the cell 1 is having white with black contacts (= WB, the same as black with white contact, BW) with the cells 2 and 5 therefore $\delta_{1,2} = 1$ and $\delta_{1,5} = 1$. In this way edge to edge contiguity can be recorded for all the other 15 cells each and we find the edge-to-edge has 24 contact points all of which belong to BW contacts. Similarly, if we do the same for the other two contiguities we tabulate δ_{ij} as follows:

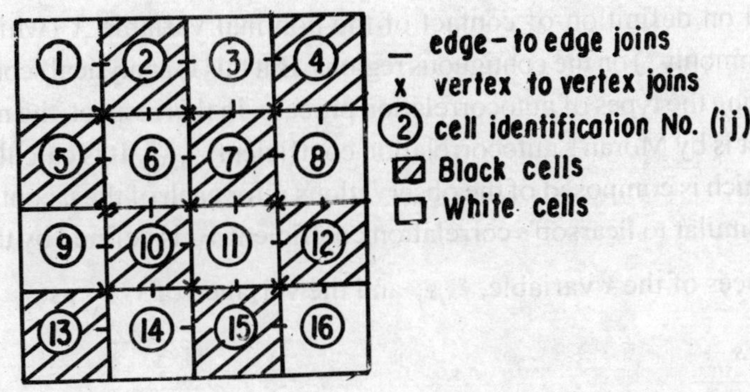

Fig. 22.2 : Spatial Autocorrelation on chessboard pattern (i,j) for the Nominal Variable, x_i

(I) Edge-to-edge contact
No. of joins for δ_{ij} = 24, BW = 24, BB = 0, WW = 0
(II) Vertex-to-vertex contact
No. of joins for δ_{ij} = 18, BW = 0, BB = 9, WW = 9, and,
(III) Edge-to-edge and vertex-to-vertex
No. of joins 42, BW = 24, BB = 9, WW = 9

From the above δ_{ij} contiguity records, we find for (I) the edge-to-edge contiguity, all the 24 joins are having the common edge between a white and a black cell with no BB or WW being reported. Hence this contiguity character of edge-to-edge contact indicates a deterministic limiting case of negative spatial autocorrelation.

For the (II) vertex-to-vertex contact with contiguities reported between a black cell with another black cell as its vertex neighbour on nine occasions and *vice versa,* we can say there is a positive spatial autocorrelation in this vertex-to-vertex contact.

Finally, for the (III) contiguity matrix A with both edge-to-edge and vertex-to-vertex contacts, the total number of joins are 42 out of which BW = 24, BB and WW each of which equal the same number of joins (= 9) and similarly type of the contacts for negative autocorrelation from edge-to-edge joins plus those of the vertex-to-vertex joins are 42. Thus we can say that the contiguity matrix,

$$A = \frac{1}{2}\sum_{i}^{n}\sum_{j}^{n}\delta_{ij} \qquad \text{...(22.3)}$$

are having 42 contacts and they give rise to a random spatial autocorrelation pattern for the above 4 × 4 grid size quadrat area. Note in Eq. (22.3), the summation of the elements δ_{ij} is divided by 2 because each join in the contiguity matrix is recorded twice as $\delta_{1,2} = 1$ and $\delta_{2,1} = 1$.

22.3 MORAN'S AUTOCORRELATION COEFFICIENT

Above we have used a contiguity matrix A whose elements δ_{ij} records the contiguity scores based on definition of contact of the nominal variable X (which is unlikely to be observed commonly*) on the contiguous regions. But this is analytical score and not a statistical one for defining the types of autocorrelation process. In this respect, the most commonly used statistical test is by Moran's autocorrelation coefficient, $r_{x_ix_j}$. It is calculated from a nominal variable x which is composed of the observations x_i for each of the n contiguous regions. This coefficient (similar to Pearson's correlation coefficient, r_{xy} is defined by the ratio of the spatial autocovariances of the x variable, $s_{x_ix_j}$ and the variance of x, s_x^2 as:

$$r_{x_ix_j} = \frac{s_{x_ix_j}}{s_x^2} \qquad \text{...(22.4)}$$

where $s_{x_ix_j}$ is defined by

$$s_{x_ix_j} = \frac{1}{2A}\sum_i^n\sum_j^n (x_i - \bar{x})(x_j - \bar{x})\delta_{ij} \qquad \text{...(22.5)}$$

where x_i, x_j are the values in the (i, j) regions and

$$\bar{x} = \frac{\Sigma x_i}{\text{Total No.of contacts}} \text{ and } s_x^2 = \sum_i^n (x_i - \bar{x})^2 / n \qquad \text{...(22.6)}$$

As usual, δ_{ij} in Eq. (22.5) refers to the elements of the contiguity matrix as before. Therefore the regions which are unconnected will have $\delta_{ij} = 0$ and so for them the individual terms $(x_i - \bar{x})(x_j - \bar{x})\delta_{ij}$ will be zero. For pairs of contiguous regions (where $\delta_{ij} = 1$) depending on the x_i (positive or negative) values and $\bar{x}$ mean, $(x_i - \bar{x})(x_j - \bar{x})\delta_{ij}$ will be relatively large and positive/negative. Note as before in Eq. (22.3) for the contiguity matrix, the sum of the terms $\sum_i^n\sum_j^n (x_i - \bar{x})(x_j - \bar{x})\delta_{ij}$ is divided by twice the no. of joins, $2A$ to give the autocovariance. The autocovariance will be large and positive if the variable x is positively spatially autocorrelated and negative when it is negatively autocorrelated.

The calculation of the autocovariance for a variable x distributed over $n = 4$ regions is shown in Fig. 22.3. The value of the variance, s_x^2 for the map pattern shown in Fig. 22.3a is calculated as usual:

---

* Maps are always based on data enumerated or on measured scales of interval or ratio, for example, the maps showing polling behaviour for a party over constituencies in an area is by percentage of the votes in favour of the party. Now to use these maps for autocorrelation analysis, legends can be prepared for showing the party votes scored in terms of nominal values 1, 2, 3 etc. so that we can index the variable in the map as a nominal variable x.

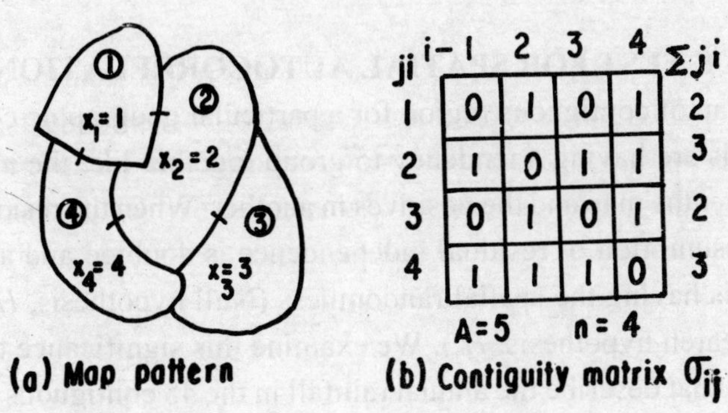

Fig 22.3 : Spatial Arrangement of a Geographical Variable, x and its Contiguity Matrix

$s_x^2 = [(1-2)^2 + (2-2)^2 + (3-2)^2 + (2-2)^2]/4 = 0.5$

The work table for auto-sums-of-squares of the individual terms for each (i, j) of the $n_i = 4$ regions are calculated alongwith below:

i	j	x_i	x_j	δ_{ij}	$(x_i - \bar{x})\ (x_j - \bar{x}).\delta_{ij}$	= Total	
1	1	1	1	0	(1–2) (1–2).0	0	
1	2	1	2	1	(1–2) (2–2).1	0	
1	3	1	3	0	(1–2) (3–2).0	0	
1	4	1	2	1	(2–2) (2–2).0	0	
2	1	2	1	1	(2–2) (1–2).1	0	We have
2	2	2	2	0	(2–2) (2–2).0	0	$\sum x_i = 32$
2	3	2	3	1	(2–2) (3–2).1	0	$\sum x_j = 32$
2	4	2	2	1	(2–2) (2–2).0	0	
3	1	3	1	0	(3–2) (1–2).0	0	So, $\bar{x}_i = 2$
3	2	3	2	1	(3–2) (1–2).0	0	
3	3	3	3	0	(3–2) (3–2).0	0	and $\bar{x}_j = 2$
3	4	3	2	1	(3–1) (2–2).1	0	
4	1	2	1	1	(3–1) (1–2).1	0	
4	2	2	2	1	(2–2) (2–2).1	0	
4	3	2	3	1	(2–2) (3–2).1	0	
4	4	2	2	0	(2–2) (2–2).0	0	

Therefore, the value of the spatial autocorrelation coefficient for the example is :

$$r_{x_i x_j} = \frac{s_{x_i . x_j}}{s_x^2} = \frac{0.0}{0.5} = 0.0$$

So in the absence of spatial autocorrelation we conclude that the regions in the example above is having a random map pattern. Note Moran's autocorrelation index, like correlation coefficient can take on values only between ± 1.

22.4 SIGNIFICANCE TEST FOR SPATIAL AUTOCORRELATION

In the residual map of contiguous region for a particular geographic correlate, often it is found that residuals are having a tendency to group together like the negative residual regions in one area over the map and the positives in another. When the residuals are ordered in this manner the assumption of residual independence is doubted and a test is required whether the regions is having the spatial randomness (Null hypothesis, H_0) or the spatial autocorrelation (Research hypothesis, H_1). We examine this significance test with respect to a regression model that describe the annual rainfall in the 48 contiguous states of U.S.A. by reference to the state's respective mean altitudes (Fig. 22.4). The regression equation which was found significant at the 0.01 level is

$$Y = 115.14 + 0.15X$$

where X is mean altitude in metre and Y is the annual rainfall in cm.

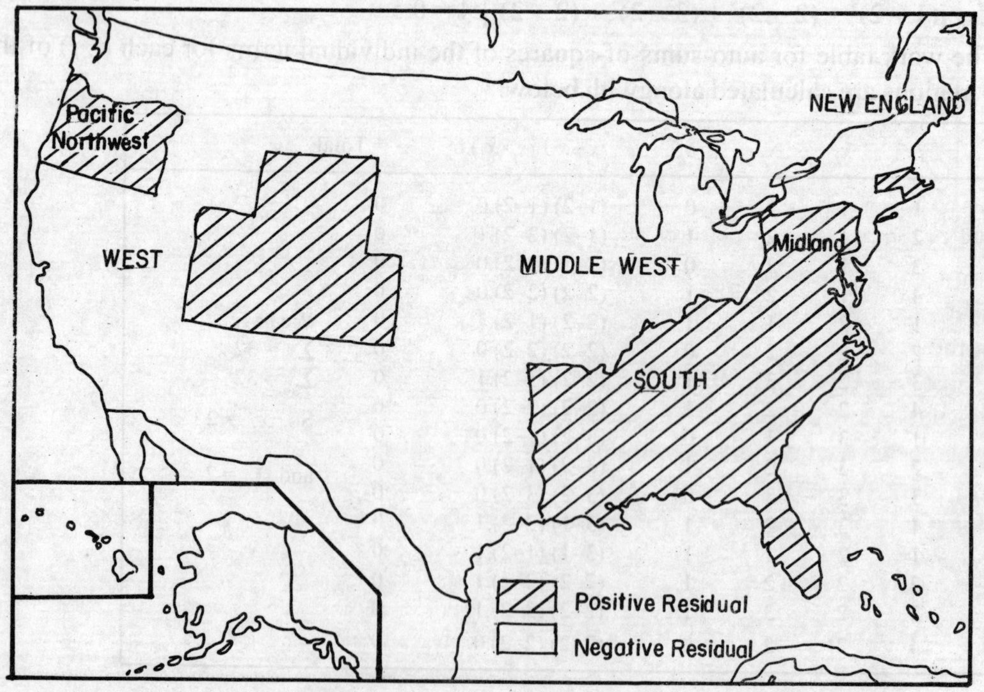

Fig. 22.4 : Residual Regression Map of USA for a Rainfall-Altitude Model.

Now to examine the question of spatial autocorrelation, the residuals are mapped for the positives and for the negatives. Now for the three types of contacts: (a) state with +ve residual itself in contiguity with another state having +ve residual, (b) state with +ve residual having contact with one state with –ve residual and (c) state with –ve residual in contact with another state of –ve residual, the necessary contiguity data elements, $\frac{1}{2}\sum_i^n\sum_j^n \delta_{ij}$ (with n = 48 states) are abstracted manually and the contacts recorded for these three are : 30, 37 and 38, which total 105.

It is an easy matter now to estimate the expected values for a, b and c under an H_0 of no special autocorrelation i.e. spatial randomness by using the binomial theorem and distribution. If the residuals are indeed random then the probability of a positive and negative residuals (p and q respectively) must be equal, and, as only these two outcomes are possible, we have : $p = q = 0.5$

It follows that the probabilities (p) for the 3 twofold connections are:

$p(a) = p^2 = 0.25$; $p(b) = 2pq = 0.5$ and $p(c) = q^2 = 0.25$

Now the expected values for a, b and c are: $E(a) = 0.25 \times 105 = 26.25$; $E(b) = 0.50 \times 105 = 52.50$; $E(c) = 0.25 \times 105 = 26.25$.

Now with this information, we can perform the one-way Chi-square test as given below:

	a	b	c
Observed frequencies (O)	30	37	38
Expected frequencies (E)	26.25	52.50	26.25
$(O–E)^2/E$	0.54	4.58	5.26
$\Sigma(O–E)^2/E$ = calculated 10.38, $\chi^2_{\alpha=0.5}$ = 5.99 with $d.f.$ = 2			

Thus we can conclude that chi-square is significant and the state residuals are arranged in non-random grouped pattern and are not independent of each other. Thus this test of significance points out about the spatial autocorrelation which could be purely guessed from the residual map.

Finally, it should be mentioned that the presence of non-random residuals should not be viewed as a wholly negative conclusion. Note in our discussion on residual maps in the chapter earlier that regional grouping which arise out of the residual mapping is used as an aid to identify or to delineate regional boundaries of areas as well as help in further explanations of the variations in the dependent variable.

22.5 USE OF NON-CONTIGUOUS WEIGHTS AND CLIFF AND ORD'S AUTOCORRELATION INDEX

It is often the case in autocorrelation analysis that insufficient information is either available or known about the spatial process controlling the arrangement of the variable. For example, in a practical application of spatial autocorrelation analysis of party voting in an area it may be seen that the percentage of votes polled by the party is having a wide variation in general but the contiguous constituencies which are small ones have a common above-average support for the party. This gives a clue that the small size of the constituencies (in this case) have given the electors the scope to establish political contacts over short distances and this brings the context of positive spatial autocorrelation for the pattern. In other case there may be the context of the influence of a region *i* is contiguous to a region *j* and *j* shares a large proportion of *i*'s common boundary and the *vice versa* which may be another geographical attribute. Again certain spatial variables like the spread of infectious diseases in studying the incidence of the diseases in the ward/blocks should be distinguished on the basis of their habitat, rural or urban or otherwise. Hence in view of these diversities of spatial units, their sizes, shapes and other characters it is unwise to assign the contiguity element δ_{ij} a binary weight of 0 and 1 only. The reason is that since the values of δ_{ij} represent our numerical specification of the spatial process controlling the arrangement of the spatial pattern, the zero-one representation of contiguity δ_{ij} restricts the amount of detail contained in the specification. It is indeed that even with regular regions δ_{ij} gives the same values for different regional systems. Note Fig. 22.5 with different natural systems of spatial patterns gives the same join structures.

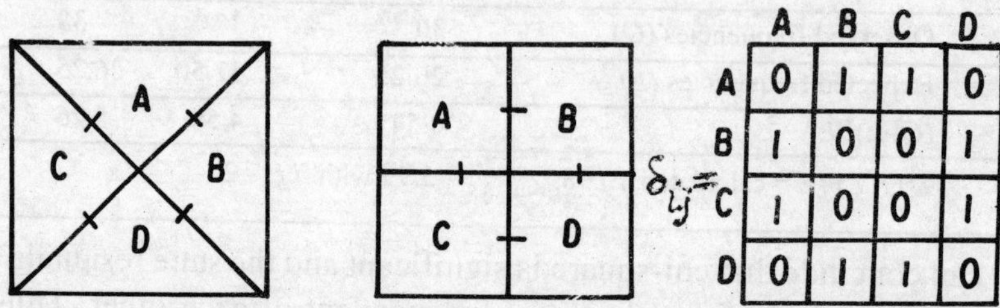

Fig. 22.5 : Different Natural Systems of Spatial Patterns

To circumvent this problem of δ_{ij}, *Cliff and Ord's spatial autocorrelation* devises a generalized weight index, W_{ij} in place of δ_{ij} in autocovariance part [Eq. (22.5)] of Moran's autocorrelation coefficient [in Eq. (22.4)]. Thus, Cliff and Ord's coefficient of autocorrelation,

$$wr_{x_i x_j} = \frac{w.s_{x_i x_j}}{s_x^2} \qquad \text{...(22.7)}$$

where s_x^2 is as before in Eq. (22.6) and $w.s_{x_i x_j}$ is the weighted spatial covariance given by

$$w.s_{x_i x_j} = \frac{1}{W}\sum_{i}^{n}\sum_{j}^{n}(x_i - \bar{x})(x_j - \bar{x}).w_{ij} \qquad \text{...(22.8)}$$

where w_{ij} is the standard weight which depending on the size, shape, nearness of the spatial units can vary widely though in general it follows a simple gravity model. For example, for studying the spatial autocorrelation of the voting pattern of the constituencies, we can have

$$w_{ij} = a_i\, a_j\, d_{ij}^{-1} \quad \text{...(22.9)}$$

where a_i and a_j are the size of the electorates in the ith and jth constituencies and d_{ij} is the distance betwen the centroids of the constituencies. We can have w_{ij} also as :

$$w_{ij} = \beta_{i(j)} \cdot d_{ij}^{-1} \quad \text{...(22.10)}$$

where $\beta_{i(j)}$ is the proportion of the length of boundary of region i common with j to the length of common boundaries of region i with j as well as with any other j region, d_{ij} is as before in

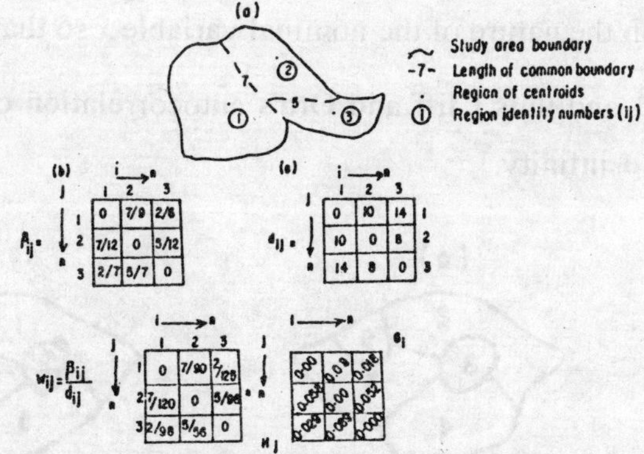

Fig. 22.6: Standard Weights for a 3-Region System

(a) Map of the 3-Regions
(b) Matrix of β_{ij}
(c) Matrix of d_{ij}
(d) Matrix of Standard Weights, w_{ij}
(e) H_j

Eq. (22.9). The idea incorporated in Eq. (22.10) is that the influence of region i on region j will be large when i is a close neighbour to j and shares a large proportion of i's common boundary and *vice versa.* Figure 22.6 provides the calculation of the standard weights for a hypothetical 3-region system.

Again for constructing standard weights for spatial pattern of infectious diseases, the incidence of the disease can be worked out by giving values of 1 when both i and jth units are either rural or urban and 0 when i is rural, j is urban and *vice versa.* Figure 22.7 provides the w_{ij} matrix with reference to $n = 5$ blocks of which 2 and 3 are classified as urban and others as rural.

elements, w_{ij} i.e.

$$W = \sum_{i}^{n}\sum_{j}^{n} w_{ij} \tag{22.11}$$

It is to be seen that W replaces the divisor $2A\ \Sigma\Sigma\ \delta_{ij}$ in Eq. (22.5). Since w_{ij} is no longer the binary weights, δ_{ij}, so there is no double counting since unlike $\delta_{ij} = \delta_{ji}$, w_{ij} is not equal to w_{ji}.

Thus if we compare the Moran's autocorrelation index with the Cliff and Ord's index here in Eqs. (22.7) and (22.8), we find no major difference excepting the fact that the latter replaces the contiguity matrix, A by a standard weight matrix, W. This difference brings a difference in the autocorrelation coefficient value: in Moran's context we know that this coefficient can take values ±1 only with 0 in between but since the weights w_{ij} can give any value depending on the nature of the nominal variable x so that the value of $w.s_{x_i} \cdot x_j$ often exceeds that of s_x^2 and thus Cliff and Ord's autocorrelation coefficient can have values extended between ± infinity.

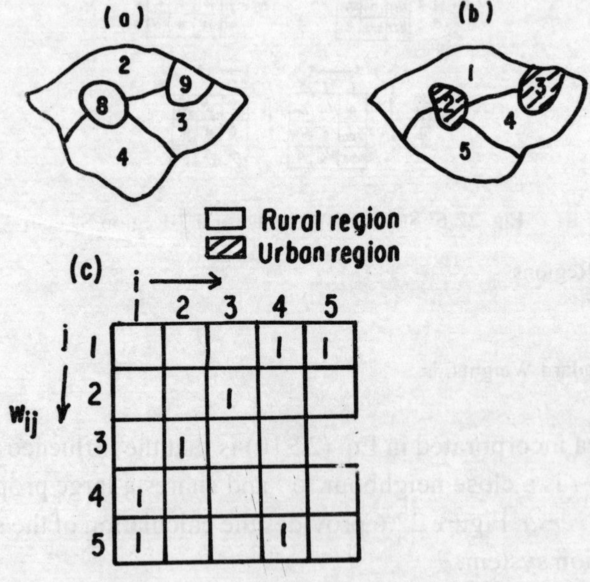

Fig. 22.7: Standard Weights for Spatial Parterns

(a) Incidence of an attribute, x_i

(b) Region Numbers (ij)

(c) Weighting Matrix, w_{ij}

22.6 SERIAL AUTOCORRELATION

If the regression of population in an area is a time series data (that is, the data on time using years as the dependent variable), the same people (who remain alive) would appear not one but many times, thus the observations would not be independent of each other and by enumerating the same people serveral times, the regression cofficient could be biased.

This is frequently termed "serial autocorrelation". Thus if the residuals $Y_t - \hat{Y}$ resulting from a regression model constructed on time series are correlated, serial correlation of residuals exist. It should be stated what is autocorrelation for biased estimates of the regression coefficients of the spatial data, serial autocorrelation is the same for the time series data. If serial autocorrelation exists, the least squares estimation techniques will not provide unbiased estimates of the regression coefficients and the sample standard error of the regression slopes may seriously underestimate the standard deviation of the population regression slope. Hence in case autocorrelation exists for time series data, then the data should not be used in forecasting.

To test whether serial autocorrelation exists in the time series data, there is a widely-used "Durbin-Watson test" which is calculated by a D statistic. This is

$$D = \frac{\sum_{t=2}^{n} (\varepsilon_t - \varepsilon_{t-1})^2}{\sum_{t=1}^{n} \varepsilon_t^2} \quad \text{with } \varepsilon_t = y_t - \hat{y}_t \tag{22.12}$$

where ε_t = residual at time period t and ε_{t-1} = residual at time t-1.

This *Durbin-Watson test* in its D statistic provides the lower and upper limits, d_l and d_u to which we compare the calculated D value in Eq. (22.12). If D is less than d_l we reject H and conclude that we have positive serial correlation but if D is more than d_u we accept H_0 and conclude that no serial autocorrelation exists. However, if D falls in between d_l and d_u no definite conclusion can be reached with respect to serial autocorrelation and hint is given that a large sample size is required.

Appendix Table XV provides the d_l and d_u values of D at two significance levels of 0.05 and 0.01 with reference to a number of independent variables ranging from x_i = 1 to x_i = 5. We test a time series data below by its regression equation to test the significance of any serial autocorrelation problem in the regression equation.

Example 22.1 Following is a time series data with a linear regression model fitted to the data: $\hat{Y} = -663.274 + 546.574t$.

Year	Data	Year	Data	Year	Data	Year	Data
1.	1,459	6.	2,423	11.	4,550	16.	8,092
2.	1,620	7.	2,677	12.	5,166	17.	9,426
3.	1,783	8.	2,815	13.	5,820	18.	10,554
4.	1,794	9.	3,093	14.	5,809	19.	10,998
5.	2,191	10.	3,190	15.	6,622	20.	11,433

Test whether the regression model do have any serial correlation problem. Assume the regression model gives a linear trend to the data.

We work out the $\hat{Y}_t$ values to the data, find out the residual ε_t $(= Y_t - \hat{Y}_t)$, have the difference between $\varepsilon_t - \varepsilon_{t-1}$ and square the differences in the following table:

Year	Data : Y_t	$\hat{Y}_t$	$\varepsilon_t = Y_t - \hat{Y}_t$	$\varepsilon_t - \varepsilon_{t-1}$	ε_t^2	$(\varepsilon_t - \varepsilon_{t-1})^2$
1.	1,459	–116.70	1.575.70	.....	2.482.830.50	......
2.	1.620	429.87	1.190.13	385.57	1.416.409.40	148.664.22
3.	1.783	976.45	806.55	383.58	650.522.90	147.133.62
4.	1.794	1.523.02	270.98	535.57	73.430.16	286.835.22
5.	2,191	2.069.60	121.40	149.58	14.737.96	22.374.18
6.	2.423	2,616.17	–193.17	314.57	37.314.65	98.954.29
7.	2,677	3,162.74	–485.74	292.57	235.943.35	85.597.21
8.	2.815	3.709.32	–894.32	408.58	799.808.26	166.937.62
9.	3.093	4,255.89	–1,162.89	268.57	1.352.313.20	72.129.85
10.	3.190	4.802.46	–1.612.46	449.57	2.600.027.30	202.113.18
11.	4.550	5,349.04	–799.04	–813.40	638.464.92	661.652.10
12.	5.166	5.895.61	–729.61	–69.43	532.330.75	4.820.52
13.	5,820	6.442.19	–622.42	–107.42	387.120.40	11.539.06
14.	5,809	6,988.76	–1,179.76	557.57	1.391.833.70	310.884.30
15.	6.622	7,535.33	–913.33	–266.43	834.171.69	70.984.95
16.	8.092	8,081.91	10.09	–923.42	101.81	852.704.50
17.	9.426	8.628.48	797.52	–787.43	636.038.15	620.046.00
18.	10,554	9.175.05	1,378.95	–581.43	1.901.503.10	338.060.84
19.	10.998	9.721.63	1,276.37	102.58	1,629.120.40	10.522.66
20.	11.433	10.268.20	1.164.80	111.57	1.356.759.00	12.447.87
				Σ	18.970.852.00	4,124.402.19

$$\text{So, } D = \frac{4,124,402.19}{18,970,852.00} = 0.217$$

Now, from Appendix Table XV for $n = 20$, one independent variable X and at $\alpha = 0.05$, d_l is 1.20 and d_u is 1.41. But since D calculated is 0.217 which is less than $d_l = 1.20$, so we must reject the hypothesis and conclude that significant serial correlation exists in the model.

22.7 CONCLUSION

The statistical validity of interpretation of both the spatial and the serial autocorrelation relates to logical validity of the linearity assumption. With time series data, the problem of serial autocorrelation can be fairly well tackled because of the fact that not only the well-known Durbin-Watson test clearly determines about the existence of their problem in the regression model of the data from the statistical point of view but also the fact that time moves in one direction only but spatial autocorrelation is not unidirectional in terms of interdependence among the data. In fact in many geographical works, the interdependence may be in all directions on a two-dimensional plane which brings problems of much complexity. Below we discuss more about the geographical contexts of serial autocorrelation.

An example of serial autocorrelation is the volume of discharge or pollutants measured at irregularly spaced sampling points downstream a river. Figure 22.8a illustrates a linear relationship of volume of a pollutant downstream a river. Now if we have to test the linearity i.e. the increase in volume is constant we replace X and Y by Δx (the change in X) and Δy (the change in pollutant level) by taking the differences between the reading of pairs of adjacent stations by all the possible pairs. In Fig, 22.8b, on plotting these Δx and Δy pairs if we get a different scatter diagram that the pairs give different patterns as compared to the scatter diagram of the original X and Y data we can presume that the assumption of the logical linearity has been infringed. Figure 22.8c illustrates a trend line fitted to a scatter of data, the high correlation coefficient of this trend line ($r = 0.93$) indicates a very close relationship and we find that the negative residuals (the data plotted below the trend line) are clustered in the values of X between 10 and 20 which balance the positive residuals in the values (the data lying above the trend line) of X between 20 and 30. Now if we take the differences between adjacent values of X (i.e. $X_i - X_{i-1}$) and that for $Y (= Y_i - Y_{i-1})$, the scatter diagram of these pairs (Δx, Δy) suggest the contrary with a correlation of $-$ 0.05 (Fig.22.8d). In this context it has to be said that the positive correlation between X and Y as in Fig.22.8b indicate that the adjacent values, on the X-axis are not independent in their values of Y, the reason being that what we interpret as negative residuals balancing the positive residuals in the scatter diagram of Fig.22.8a, the scatter of points themselves is s-shaped and not linear as we thought of.

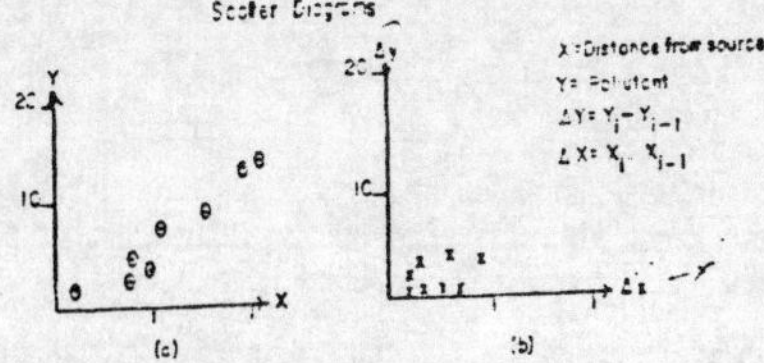

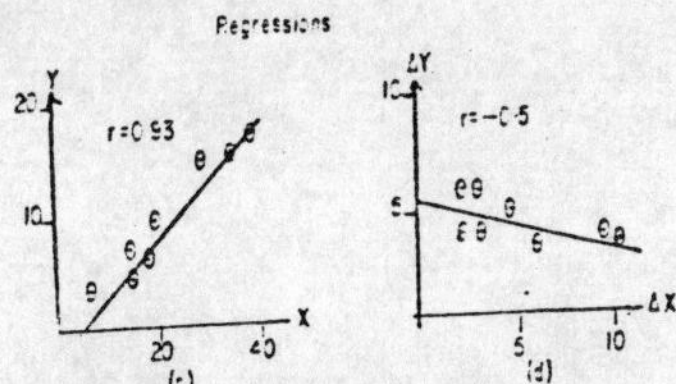

Fig. 22.8 : Serial Correlation

LIST OF FORMULAE

I. *Spatial Autocorrelation*

Pearsonian Correlation $r_{XY} = \dfrac{M_{XY}}{\sqrt{(M_{XX})(M_{YY})}}$

For 2×2 Contingency Table, $r_{XY} = \dfrac{ad - bc}{\sqrt{m.n.r.s}}$

Moran's Autocorrelation Coefficient

$$r_{x_i x_j} = \frac{s_{x_i x_j}}{s_x^2}$$

where $s_{x_i x_j} = \dfrac{1}{2A} \sum \sum (x_i - \bar{x})(x_j - \bar{x}).\delta_{ij}$ with A, the Contiguity

Matrix $= \dfrac{1}{2} \sum_i^n \sum_j^n \delta_{ij}$ and $s_x^2 = \sum (x_i - \bar{x})^2 / n$ with

$$\bar{x} = \frac{\sum x_i}{\text{Total No. of contacts}}$$

Cliff and Ord's Autocorrelation Coefficient

$$wr_{x_i x_j} = \frac{w.s_{x_i x_j}}{s_x^2}$$

where $w.s_{x_i x_j} = \dfrac{1}{W} \sum \sum (x_i - \bar{x})(x_j - \bar{x}).w_{ij}$ with w_{ij} either equal to $a_i a_j.\ d_{ij}^{-1}$ or $\beta_{i(j)}.\ d_{ij}^{-1}$ and

$W = \sum \sum w_{ij}$

II. *Serial Autocorrelation,*

$$D \text{ statistic} = \frac{\sum_{t=2}^{n} (\varepsilon_t - \varepsilon_{t-1})^2}{\sum_{t=1}^{n} \varepsilon_t^2}$$

where $\varepsilon_t = y_t - \hat{y}_t$

EXERCISES

22.1 Following are the production figures (in tons) for 7 consecutive years:
80, 90, 92, 83, 94, 99 and 92.
Fit a regression line to the data by (a) taking the middle year as the origin and by (b) not assuming any midddle years as the origin. Test whether the serial autocorrelation problems does exist in the regression model.

22.2 The time series data in Example 22.1 give a non-linear trend. For linearising the data, it is suggested that the time period in the data should be squared.
Provide the trend equation to this revised data and test whether the serial autocorrelation problem exist in this revised regression model also.

23

Regression Analysis in Geo-science Systems IV
(Linear Multiple Regression and Correlation)

23.1 LINEAR MULTIPLE REGRESSION

Simple linear correlation and regression analysis as illustrated in the previous two chapters deal with the case of the relationship between two variables of which one is dependent variable Y and the other an independent explanatory variable X. However, in the analysis of relationships among two real variables over a set of observations (data), it is rare to discover an association so obvious and so comprehensive that the coefficient of correlation represents a perfect relationship, i.e. = ± 1.0, and all the variations in the independent variable X is accounted for in the dependent variable Y. A simple linear regression line between two variables indicates a general nature of their degree of association and provides a basis for estimating the dependent variable Y by using the value of the independent variable X as the predictor. Hence, a single analysis of this bivariate model does not result in sufficiently accurate estimates because in the real world any dependent phenomenon is a function of several independent variables. Therefore, in order to have a better estimate of a phenomenon it is necessary to take more than one explanatory variable into consideration to combine into a new regression equation. In this chapter, a simple linear regression analysis is extended to the linear multiple regression by considering two or more independent variables taken together for estimating the dependent variable. Thus for k number of independent variables like $X_1, X_2, ..., X_k$, the estimated multiple regression equation is having a general form

$$\hat{Y} = a + b_1 X_1 + b_2 X_2 + ... + b_k X_k \pm \varepsilon$$

$$= a + \sum_{i=1}^{k} b_i X_i \pm \varepsilon \qquad ... (23.1)$$

where X_i represent the values for the k number of independent variables, $\hat{Y}$ is the value estimated for the given dependent variable, a is the Y- intercept when the X_is are all zero as in the linear regression equation and ε is the error term as in the earlier chapter. The b_is (i.e. b_i, ..., b_k) in the above regression model are the estimated partial regression coefficient related to the independent variables X_is because each gives the rate of change (or, slope) in the dependent variable, for a unit change in that particular variable, provided all other independent variables are held constant. Hence we can rewrite the Eq. (23.1) by using the notations as :

$$\hat{Y} = a + b_{y1.23...k}X_1 + b_{y2.13...k}X_2 + ... + b_{y3.12..k}X_k + \varepsilon \qquad ...(23.2)$$

The coefficient $b_{y1.23...k}$ for example is read, "the regression coefficient of the independent variable X_i on Y" (the subscript y stands for the dependent variable Y) as variables 2, 3, ..., k remain constant. In general, these coefficients will differ from the "total regression coefficients" which are the simple regressions of each individual X_i variable on the Y_i variable. We ordinarily expect multiple regression to account for more of the total variation in Y than with any of the total regression coefficient. This is because the *multiple regression* considers all possible interactions within combinations of the X_i variables as well as the variables themselves. Again note in Eq. (23.2) for multiple regression we extend the linear regression equation $\hat{Y} = a + bX$ by adding a term for each of the additional variables.

Thus in Eq. (23.2) when $k = 1$, the equation reverts back to its simple regression form. Hence multiple regression is the extension of simple regression to take into account of the effect of more than one independent X variable on the dependent variable Y. We illustrate below the development of a linear multiple regression model by using two independent variables, $k = 2$.

23.1.1 Linear Multiple Regression Model, $k = 2$

Mathematically, in the case of one independent variable we were concerned with the fiting of one linear regression line by involving the method of least squares. Thus for the bivariate data (X, Y), we have the linear equation $\hat{Y} = a + bX$ (on assuming that $\Sigma\varepsilon = 0$) for which the least square (or, normal) equations are:

$$\Sigma Y = na + b\Sigma X \qquad \text{as in Eq. (21.11)}$$
$$\Sigma XY = a\Sigma X + b\Sigma X^2 \qquad \text{as in Eq. (21.12)}$$

Now for present purpose of developing the *linear multiple regression model,* we go by casting the least square equations (for the model) in matrix forms. Matrices and linear algebra provide us not only with a general notation and a powerful system for computation, but also lead conceptual clarity to linear models. For example, the above two least square (or, normal) equations can be written in matrix notation as

$$\begin{bmatrix} n & \Sigma X \\ \Sigma X & \Sigma X^2 \end{bmatrix} \begin{bmatrix} a \\ b \end{bmatrix} = \begin{bmatrix} \Sigma Y \\ \Sigma XY \end{bmatrix} \qquad ...(23.3)$$

or

$[A]\,[x] = [b]$ ... (23.3a)

that is, for the above two least square equations, we can have the matrix solution for two constants *a, b* by placing the coefficients of the constants in matrix form and on separating the constants in the vector form we get the matrix notation

$$\underset{2\times 2}{[A]}\quad \underset{2\times 1}{[x]} = \underset{2\times 1}{[b]} \qquad \text{... (23.3b)}$$

The matrix solution we will explain later on with the help of the Example 23.1 for a linear trend surface.

Now for a linear trend surface where there are two mutually perpendicular geographical coordinates, X_1 and X_2 with respect to the dependent variable, height Y, we have the linear trend (or linear multiple regression) equation:

$$Y = a + bX_1 + cX_2 \qquad \text{... (23.4)}$$

where Y is an attribute (say, height) assumed to be a linear function of some constant value, a related to the mean of the observations; plus an east-west component, b; and a north-south component, c. Because Eq. (23.4) contains 3 unknowns (= a, b and c) and to solve them, we need a set of three normal equations by the least square technique:

$$\Sigma Y = na + b\Sigma X_1 + c\Sigma X_2 \qquad \text{... (23.5)}$$

$$\Sigma X_1 Y = a\Sigma X_1 + b\Sigma X_1^2 + c\Sigma X_1 X_2 \qquad \text{... (23.6)}$$

$$\Sigma X_2 Y = a\,\Sigma X_2 + b\Sigma X_1 X_2 + c\Sigma X_2^2 \qquad \text{... (23.7)}$$

Like the two normal equations in linear regression (see footnote for Eqs (21.11) and (21.12)), the first normal equation, Eq. (23.5) here is found by summing each term of the regression equation by X_1 and using the fact that $\Sigma a = na$. Multiplying the regression equation by X_1 and then summing the resultant products, one gets the second normal equation [(Eq. 23.6)]. Similarly, the third equation, Eq. (23.7) is found by multiplying each term by X_2 and then summing up.

The above three equations can be written in matrix notation as

$$\begin{bmatrix} n & \Sigma X & \Sigma Y \\ \Sigma X & \Sigma X^2 & \Sigma XY \\ \Sigma Y & \Sigma XY & \Sigma Y^2 \end{bmatrix} \begin{bmatrix} a \\ b \\ c \end{bmatrix} = \begin{bmatrix} \Sigma Y \\ \Sigma X_1 Y \\ \Sigma X_2 Y \end{bmatrix} \qquad \text{... (23.8)}$$

or, which can be simplified as

$$\underset{3\times 3}{[A]}\quad \underset{3\times 1}{[x]} = \underset{3\times 1}{[b]} \qquad \text{... (23.8a)}$$

and the unknown vector is found by $[A]^{-1}[b] = [x]$ ... (23.8b)

To solve these three normal equations we can set the data for each variable in terms of deviations from the mean (rather than the raw scores themselves) i.e. we can measure all observations of the variables Y, X_1 and X_2 from their means and define new set of variables as $y = Y - \bar{Y}$, $x_1 = X_1 - \bar{X}_1$ and $x_2 = X_2 - \bar{X}_2$. This reduces the absolute magnitude of the variables and centres them about a common mean of zero.

It may be recalled from Chapter 3 that the sum of deviations from the mean is zero. Therefore, obviously

$$\begin{aligned}\Sigma(X_1 - \bar{X}_1) &= \Sigma x_1 = 0\\ \Sigma(X_2 - \bar{X}_2) &= \Sigma x_2 = 0\\ \Sigma(Y - \bar{Y}) &= \Sigma y = 0\end{aligned} \quad \text{... (23.9)}$$

Thus the simplification occurs in the normal equations in Eqs. (23.5) to (23.7) in terms of new variables and they will reduce to determining two estimates, b_1 and b_2 only

$$na = 0 \quad \text{... (23.10)}$$

$$\Sigma x_1 y = b_1 \Sigma x_1^2 + b_2 \Sigma x_1 x_2 \quad \text{... (23.11)}$$

$$\Sigma x_2 y = b_1 \Sigma x_1 x_2 + b_2 \Sigma x_2^2 \quad \text{... (23.12)}$$

Solving Eqs. (23.11) and (23.12) algebraically, we have

$$b_1 = \frac{\Sigma x_2^2 \, \Sigma x_1 y - \Sigma x_1 x_2 \, \Sigma x_2 y}{\Sigma x_1^2 \, \Sigma x_2^2 - (\Sigma x_1 x_2)^2} \quad \text{... (23.13)}$$

and $$b_2 = \frac{\Sigma x_1^2 \, \Sigma x_2 y - \Sigma x_1 x_2 \, \Sigma x_1 y}{\Sigma x_1^2 \, \Sigma x_2^2 - (\Sigma x_1 x_2)^2} \quad \text{... (23.13a)}$$

Now since we know b_1 and b_2 we can calculate a from Eq. (23.5) as

$$a = \bar{Y} - b_1 \bar{X}_1 - b_2 \bar{X}_2 \quad \text{... (23.14)}$$

Note this formula for a in multiple regression is close to that of in linear regression $a = \bar{Y} - b\bar{X}$.

To show the working of one linear multiple regression by both the least square method and the matrix solutions we will now expand the Example 21.1 we used in Chapter 21.

Example 23.1 The following is a data set of eight values on income in cattle-farming economy for which farm income (in '000 Rs.), Y, No. of cattle raised in the farms (X_1) and farm size in acres (X_2).

X_1	20	30	40	50	70	90	100	120
X_2	5	3	9	18	5	25	9	6
Y	3.5	7.5	7.0	15.0	11.0	15.0	25.0	27.0

Estimate the multiple linear regression for these three variables by the (a) least square and (b) matrix solution.

We carry out the intermediate calculations necessary (like the sums, sums of powers and sums of cross-products required) before we involve the two methods. These are as follows:

Y	X_1	X_2	X_2Y	X_1X_2	X_1Y	Y^2	X_1^2	X_2^2
3.5	20	5	17.5	100	70	12.25	400	25
7.5	30	3	22.5	90	225	56.25	900	9
7.0	40	9	63.0	360	280	49.00	1600	81
15.0	50	18	270.0	900	750	225.00	2500	324
11.0	70	5	55.0	350	770	121.00	4900	25
15.0	90	25	375.0	2250	1350	225.00	8100	625
25.0	100	9	225.0	900	2500	625.00	10000	81
27.0	120	6	162.0	720	3240	729.00	14400	36
Σ111.0	Σ520	Σ80	Σ1190.0	Σ5670	Σ9185	Σ2042.50	Σ42800	Σ1206

From above, we have $\overline{Y} = 13.875$, $\overline{X}_1 = 65$, $\overline{X}_2 = 10$

(a) Least Square Solution by Deviations from the Mean

Now, substituting the appropriate sum values into Eqs. (23.5) to (23.7), we have the following normal equations:

$$
\begin{aligned}
111 &= 8a + 520b_1 + 80b_2 \\
9185 &= 520a + 42800b_1 + 5670b_2 \\
1190 &= 80a + 5670b_1 + 1206b_2
\end{aligned}
$$

Now for solving b_1 and b_2 we require Eqs. (23.13) and (23.14), we do the computations for X_1 and X_2 and Y transformed into x_1, x_2 and y respectively below :

$y = Y - \overline{Y}$	$x_1 = X_i - \overline{X}_1$	$x_2 = X_2 - \overline{X}_2$	x_1^2	x_2^2	x_1y	x_2y	x_1x_2
− 10.375	−45	− 5	2025	25	466.875	51.875	225
− 6.375	− 35	− 7	1225	49	223.125	44.625	245
− 6.875	−25	− 1	625	1	171.875	6.875	25
1.125	−15	8	225	64	− 16.875	9.000	−120
− 2.875	5	− 5	25	25	−14.375	14.375	− 25
1.125	25	15	625	225	28.125	16.875	375
11.125	35	− 1	1225	1	389.375	−11.125	− 35
13.125	55	− 4	3025	16	721.875	− 52.500	− 220
Total :			9000	406	1970.000	80.000	470

So, we have:

$$b_1 = \frac{(406)(1970)-(470)(80)}{(9000)(406)-(470)^2} = \frac{762220}{3433100} = 0.2220$$

Similarly

$$b_2 = \frac{(9000)(80)-(1970)(470)}{(9000)(406)-(470)^2} = \frac{-205900}{3433100} = -0.0599$$

Again from Eq. (23.14), we have

$a = \bar{Y} - b_1\bar{X}_1 - b_2\bar{X}_2 = 13.875 - (0.2220)(65) - (-0.0599)(10) = 0.0440.$

Thus the estimated linear multiple expression:

$\hat{Y} = 0.0440 + 0.2220X_1 - 0.0599X_2$

This expression can be interpreted as that beginning with a value of $a = 0.0440$, the estimated value of Y increases by 0.02220 for each unit change of X_1 and decreases by 0.0599 for each additional unit in X_2. This can now be used to predict any value of Y with respect to values of X_1 and X_2. For example, when number of cattle raised in the farm is $X_1 = 20$ in a farm size of $X_2 = 5$ acres, we have the farm income estimated as $\hat{y} = 0.0440 + 0.2220(20) - (0.0599)(5) = 2.9865$ units that is Rs.2986.5.

(b) Matrix Solution

For the matrix solution we have to determine the unknown vector $[x]$:

$$[x] = [A]^{-1}[b] \qquad \text{[as in Eq. (23.8b)*]}$$

For which we have the co-factor matrix,

$$C = \begin{bmatrix} \begin{vmatrix} 42800 & 5670 \\ 5670 & 1206 \end{vmatrix} & -\begin{vmatrix} 520 & 5670 \\ 80 & 1206 \end{vmatrix} & \begin{vmatrix} 520 & 42800 \\ 80 & 5670 \end{vmatrix} \\ -\begin{vmatrix} 520 & 80 \\ 5670 & 1206 \end{vmatrix} & \begin{vmatrix} 8 & 80 \\ 80 & 1206 \end{vmatrix} & -\begin{vmatrix} 8 & 520 \\ 80 & 5670 \end{vmatrix} \\ \begin{vmatrix} 520 & 80 \\ 42800 & 5670 \end{vmatrix} & -\begin{vmatrix} 8 & 80 \\ 520 & 5670 \end{vmatrix} & \begin{vmatrix} 8 & 520 \\ 520 & 42800 \end{vmatrix} \end{bmatrix}$$

$$= \begin{bmatrix} 19467900 & -73520 & -475600 \\ -173520 & 3248 & -3760 \\ -475600 & -3760 & 72000 \end{bmatrix}$$

Now, the determinant of the matrix, $= [8(42800 \times 1206 - 5670 \times 5670) - 520(520 \times 1206 - 80 \times 5670) + 80(520 \times 5670 - 80 \times 42800)] = 27464800$. Hence the matrix $|A|$ is non-

* If all the observations of the three variables here are converted to deviations from the mean, since the coefficient of a becomes zero so the coefficient matrix $[A]$ is reduced by one row and one column.

singular and we can have the adjoint matrix, Adj A. Note for the given matrix of the least square matrix, the adjoint matrix is the same as the co-factor matrix, C (since the matrix is square and symmetrical) and we can have the inverse matrix,

$$A^{-1} = \frac{1}{27464800}\begin{bmatrix} 19467900 & -73520 & -415600 \\ -173520 & 3248 & -3760 \\ -475600 & -3768 & 72000 \end{bmatrix}$$

$$\begin{bmatrix} 0.7088309 & -0.0063179051 & -0.0173167 \\ -0.0063179051 & 0.00011826046 & -0.00013690251 \\ -0.0173167 & -0.00013690251 & 0.0026215374 \end{bmatrix}$$

Now, $[A^{-1}].[b] = [x]$, so

$$\begin{bmatrix} 0.7088309 & -0.0063179051 & -0.0173167 \\ -0.0063179051 & 0.00011826046 & -0.00013690251 \\ -0.0173167 & -0.00013690251 & 0.0026215374 \end{bmatrix}\begin{bmatrix} 111 \\ 9185 \\ 1190 \end{bmatrix}$$

Hence on multiplying the vector b with the matrix A^{-1} we get

$$[x] = \begin{bmatrix} 0.0434454 \\ 0.2220208 \\ -0.0599737 \end{bmatrix}$$

and, our regression equation is $\hat{Y} \approx 0.0434 + 0.2220X_1 - 0.0599X_2$.

Note. The above solution tallies well with that of least square solution done except the difference in the a value. The reason for this difference (of 0.0006) is that entries in the $|A|$ matrix differ widely and some sums of squares are so large that significant digits get lost because of their truncation. Extension to greater than three independents in the linear multiple regression would simply add extra rows of normal equations to the least square solution and similarly extra rows and extra columns to each of the matrix and the vectors.

23.2 MULTIPLE REGRESSION : A GRAPHICAL INTERPRETATION

In the linear multiple regression, of two independent variables, X_1 and X_2, the b_1 coefficient with respect to X_1 is the increase in Y if X_1 is increased by one unit while X_2 is held constant, similarly b_2 coefficient is also having a similar interpretation. Since the effect of simultaneous change in X_1 and X_2 is the sum of their separate (or, partial) changes, this model is additive.*

---

*This difference/increase in Y can be shown for linear multiple regression equation as :

initial $Y = a + b_1 X_1 + b_2 X_2$

which after increase by one unit in X_1 holding X_2 constant, new $Y = a + b_1 (X_1 + 1) + b_2 X_2$

So, the difference: Increase in $Y = b_1$

This statement is true for a general linear multiple regression model : $Y = a + b_1 X_1 + b_2 X_2 + ... + b_k X_k$

Thus b_1 and b_2 are called *partial regression coefficients* and these are written as $b_{y1.2}$ and $b_{y2.1}$ in the regression equation so that the Eq. (23.4) for two independent variables is

$$\hat{Y} = a + b_{y1.2} X_1 + b_{y2.1} X_2 \qquad \text{... (23.15)}$$

These three variables cannot be plotted on ordinary graph but they can be shown in a 3-dimensional space with 3 coordinates in which Y occupies the vertical axis and X_1 and X_2 are the two horizontal axes (Fig. 23.1a). Since the regression of the points here can no longer be shown as a trend line on a scatter graph because the points are spread over a volume so what is needed is to summarize the positions of the regression surface $Y = a + b_{y1.2} X_1 + b_{y2.1} X_2$ so that they represent not a trend line but a flat plane in the 3-dimensional variable space, as shown in Fig. 23.1b.

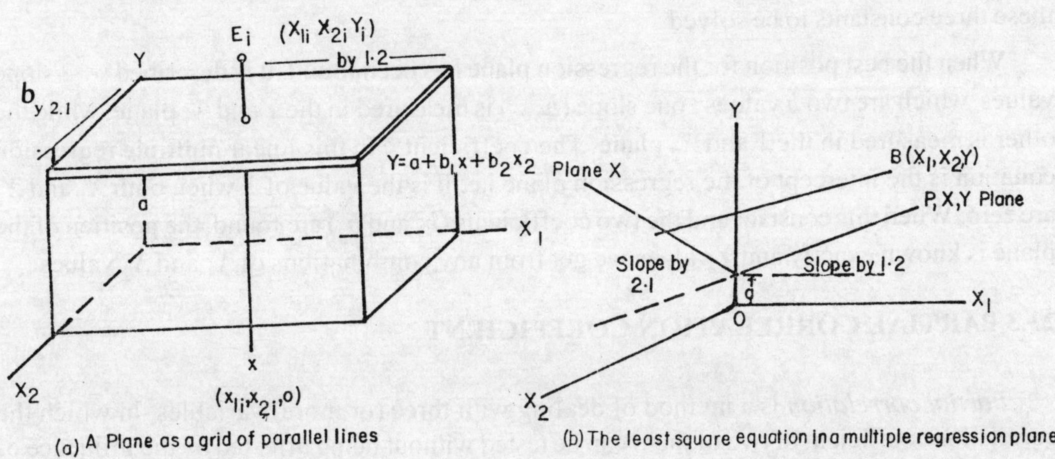

Fig. 23.1 : Least Square Equations in 3-dimension

The intuitive rationale for the least square fittings for linear multiple regression is that this regression plane that comes as close to the points as possible in the sense that some of the points are expected to be a little above the plane and others will fall below it so that though there are vertical distances of the points to the plane as *regression residuals,* but the sum of the squared distances from points to the planes is at a minimum. Thus:

$$\varepsilon = Y - \hat{Y} = Y - a - b_1 x_1 - b_2 x_2$$

On squaring and adding, $S(a, b_1, b_2) = \Sigma\, \varepsilon^2 = \Sigma\,(Y - a - b_1 x_1 - b_2 x_2)^2$

To minimize $S(a, b_1, b_2)$ we have to get the partial derivative of Y with respect to the coefficients a, b_1 and b_2 as

$$\frac{\sigma_s(a,b_1,b_2)}{\sigma a} = \Sigma\,(-1)\,(2)\,(Y - a - b_1 x_1 - b_2 x_2)$$

$$\frac{\sigma_s(a,b_1,b_2)}{\sigma b_1} = \Sigma\,(-x_1)\,(2)\,(Y - a - b_1 x_1 - b_2 x_2)$$

$$\frac{\sigma_s(a,b_1,b_2)}{\sigma b_2} = \Sigma\,(-x_2)\,(2)\,(Y - a - b_1 x_1 - b_2 x_2)$$

Setting these partial derivatives to zero, one gets the simultaneous linear equations for these three constants to be solved.

When the best position for the regression plane has been found, it is described by 2 slope values, which are two b values : one slope ($b_{01.2}$) is measured in the Y and X_1 plane, while the other is measured in the Y and X_2 plane. The coefficient a in this linear multiple regression equation is the intercept of the regression plane i.e. it is the value of Y when both X_1 and X_2 are zero. When this constant and the two coefficients (b_1 and b_2) are found, the position of the plane is known exactly and Y values we get from any combinations of X_1 and X_2 values.

23.3 PARTIAL CORRELATION COEFFICIENT

Partial correlation is a method of dealing with three (or more) variables, in which the correlation coefficient of two of them can be tested without being affected by the influence of the third (or other) variables which is (are) controlled. Thus basically, it is the correlation between two variables, X_1 and X_2 holding the third variable Y constant. It helps in assessing the influence of each of the factors where three or more variables are involved. In this way it is very helpful for determining the extent to which the apparent relationship between the two variables can be explained by the effect of outside variables. Thus $r_{y1.2}$ indicates the correlation between Y and X_1 having removed the effects of both the functional relationships $Y = f(X_2)$ and similarly $X_1 = f(X_2)$. These removals are done by regressing Y on X_2 and X_1 and then regressing the residuals from these two analyses on each other. However, computation of *partial correlation coefficients* does not involve obtaining these relevant residuals and regressing them. Instead, the calculation of coefficient of partial correlation can proceed from that of the net regression coefficients. Consequently, it is more convenient to utilize a method that combines the simple correlation coefficients to determine the partial correlation coefficient. Thus when three variables are involved in the study, the *product moment simple* (or, *zero-order*) *correlation coefficients* are determined. So for partial correlation coefficient for 2 variables X_1, X_2 for their relation to the variable Y, we have:

$$r_{y1.2} = \frac{r_{y1} - (r_{y2})(r_{12})}{\sqrt{1-r_{y_2}^2}\sqrt{1-r_{12}^2}} \quad \text{... (23.16)}$$

where X_1, X_2 are independent variables, Y is the dependent and $r_{y1.2}$ is the partial correlation coefficient between the variables Y and X_1 while the influence of the third variable X_2 is controlled. In this way, any particular correlation can be worked out from the above general formula [Eq. (23.16)] by substituting the variables in turn. Note in Eq. (23.16), if we take a close look at the terms of the formula (for partial correlation) in this equation, we find the numerator $[r_{y1} - (r_{y2})(r_{12})]$ is subtracting the combined measure of the effects of X_2 on both X_1 and Y in terms of a cross product $(r_{y2})(r_{12})$ from the simple (or, zero-order) correlation coefficient for the X_1Y relationship, r_{y1} which suggests that in the numerator we consider the co-variance of X_1 and Y for the portion of their respective variances that is "left over" after X_2 has operated on X_1 and Y. Now if we look at the denominator of this formula (Eq. 23.16), we see that it is a square-root of proportion of variation in X_1 and in Y not explained by X_2 separately. What emerges from the interpretation is that the partial correlation between Y and X_1 (Y controlling for X_2) can be defined as the "correlation between the residuals of the regressions of X_1 on X_2 and Y on X_2". At first this interpretation may sound strange and a little complicated too but the graphic presentation in Fig. 23.2 provides a clarification for what we say for "partial r" that X_2 is held constant. Figs. 23.2a and b represent the regression of X_1 on X_2 and Y on X_2 in which the lines drawn from the points to the regression line represent residuals i.e. variations not explained by X_2. We now plot new points for X_1 and Y taking the difference between the $\bar{Y}$ and Y i.e. the distance of points from the regression line as "new scores" for X_1 and Y. These new scores in Figs. 23.2c and 23.2d are, of course, residuals. Finally, if we plot these residuals Y on residuals X_1, we get Fig. 23.2e.

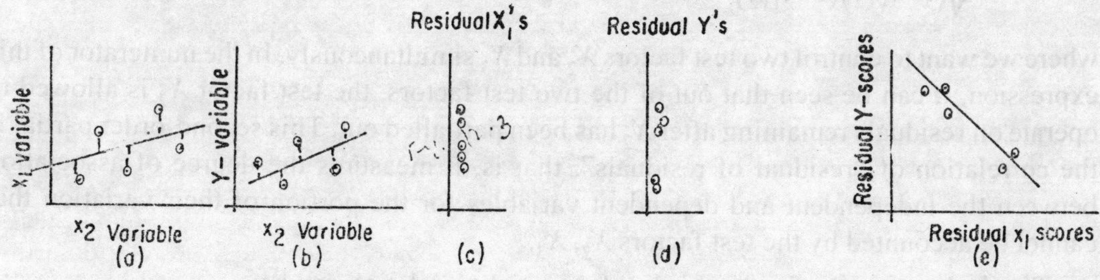

Fig. 23.2 : Rationale for Partial Correlation

Partial correlation can be obtained with more than one controlled variable. Thus, a *first-order partial* will have one control (as above), a *second-order partial* will have two controls and so on. In keeping with this terminology a simple correlation (with no variable held controlled) between the two variables is often referred to as a *zero-order correlation* and their correlation is termed as *total correlation.* Now let us proceed to some more applied considerations for "partial r". The sign and the absolute value of partial correlation depends on the zero-order correlation. We can have following three interpretations:

(1) If the patrial r is much lower in value than the zero-order coefficient ($r_{y\,1.2} < r_{xy}$), the bivariate relationships between Y and X_1 may be spurious or X_2 may intervene between Y and X_1.

(2) In first-order partial r, if we start with two variables which are positively related and if the control variable Y is negatively related to one of them but positively related to the other, the resulting partial correlation between the two variables will be larger than the zero-order correlation. The same result will be attained when the zero-order correlation of the control variable, Y with one of the other variables happens to be zero.

(3) When the control variable Y is unrelated to both of the other variables, the partial variable would, of course, equal the total correlation i.e. $r_{y\,1.2} \cong r_{y1}$ or nearly equal. Thus after we have discussed the relationship between partial correlation and casual interpretations, we shall be able to get an intuitive justification for the behaviour of partial r. Unfortunately, the obtained partial r does not always lend itself to clear-cut interpretation listed above. We often get a partial r that is smaller than the zero-order, but is not exactly zero. In such cases, spuriousness becomes "relative" rather than an "absolute" term.

The equation for the *first-order partial correlation* can be extended to equation for the *second-order partial correlation* which is:

$$r_{y1.23} = \frac{r_{y1.2} - (r_{13.1})(r_{y3.1})}{\sqrt{(1 - r_{y3.2}^2)(1 - r_{13.2}^2)}} \qquad \text{... (23.17)}$$

where we want to control two test factors X_2 and X_3 simultaneously. In the numerator of this expression, it can be seen that out of the two test factors, the test factor X_2 is allowed to operate on residuals remaining after X_2 has been partialled out. This second-order partial is the correlation of "residual of residuals", that is, it measures the degree of association between the independent and dependent variables for the portion of their variation that cannot be accounted by the test factors X_2, X_3.

Similarly, equation for the third-order partial correlation can be

$$r_{y1.234} = \frac{r_{y1.23} - (r_{14.23})(r_{y4.23})}{\sqrt{(1 - r_{y4.23}^2)(1 - r_{14.23}^2)}} \qquad \text{... (23.18)}$$

In the same way (as in above) we can have partial correlation coefficient $r_{y1.234...n}$ indicates that this is the correlation between the dependent variable Y and one independent (X_1) variable, holding constant the effects of other specified independent variable (X_2, X_3, and all others up to X_n).

(1) Partial correlations are very helpful for determining the extent to which apparent relationship between the two (or, more) variables where three (or, more) variables are involved. But partial correlations are having limitations too which are mainly three: (1) the gross or zero-order correlations must have linear regressions, that is, it is not applicable to curvilinear regression where the marginal increase in Y (the dependent variable) and that of test factors also is no longer simply the coefficient of b (as in linear multiple regression mentioned earlier).

(2) From Eq. (23.16), that is, first-order partial correlation, it is quite probable that $r_{y1.2}$ will differ markedly from r_{y1} and similarly $r_{y2.1}$ will differ from r_{y2}. It may even be the case that the partial correlations between two variables with a third held constant will differ in sign from simple correlation between the same two variables. Thus the effects of the independent variables must be additively and not jointly related.

(3) It can be seen that with higher-order partial correlations, the reliability of partial coefficients decreases as well as the work involved becomes tedious rapidly. For example, to obtain a third-order partial correlation, one must have obtained three second-order partials, each of which in turn has to be obtained by computing the first-order partials from zero-order correlations.

Using the Example 23.1 in this chapter we have $r_{y1} = 0.9265$, $r_{y2} = 0.1770$ and $r_{12} = 0.1405$ which gives us a partial correlation coefficient of $r_{y1.2} = 0.9256$ between the dependent variable Y (= farm income) and the active independent variable X_1 (= No. of cattle) as if the other independent variable X_2 (= farm acres) is not there.

Finally, using Example 23.1 we compare the results of the simple correlation analysis with those of the partial correlations, we have:

	r_{y1}	= 0.9265	$r_{y1.2}$	= 0.9256
so that	r^2_{y1}	= 0.8584	$r^2_{y1.2}$	= 0.8566
and	r_{y2}	= 0.1770	$r_{y2.1}$	= – 0.1393
so that	r^2_{y2}	= 0.0313	$r^2_{y2.1}$	= 0.0194

It can be seen that the two independent variables are not so strongly interrelated with each other so that once this inter-relationship is controlled, the independent variable X_2 (farm size) accounts for less than two per cent of the variance in the dependent variable Y not associated with the other independent variable X_1. However, the latter variable accounts for more than eighty five per cent of the variance remaining in Y once the variable X_2 removed.

23.4 MULTIPLE CORRELATION

Partial correlation coefficients as stated above indicate the strength of the relationship between two variables once the effects of others have been held constant which is associated with the residual variance in the independent variable. But it does not say that what proportion of the total variance in the dependent variable can be accounted for by all of the independent variables together? Clearly it is not the sum of the squared zero-order correlations since this would involve double counting. For example, if $r^2_{y1.2} = 0.550$ and $r^2_{y2.1} = 0.308$ the sum of these two does not provide the information on the total variance in Y associated with X_1 and X_2. Now if in the example, since X_1 and X_2 are correlated, some of the variance in Y is associated with the joint variance of X_1 and X_2 and this is not explained in partial correlation.

On the other hand, the multiple correlation coefficient indexes the strength of the relationship in terms of linear correlation between a dependent variable and a set of independents. Always denoted by subscripts, $R^2_{y.12}$ indicates the multiple correlation between the dependent variable, Y (before the point) and the two independent variables, X_1 and X_2 (after the point). $R^2_{y.12}$ is the proportion the variance of Y accounted for by these two independent variables, both jointly and separately. Thus the concept of the coefficient of determination, r^2 which is the square of the correlation coefficient can be extended here from the simple linear correlation. So $R^2_{y.12}$ (which is the square of $R_{y.12}$ just like r^2 is the square of r) is the coefficient of multiple determination and is the ratio of the variance of Y accounted for by the two independent variables and the total variance. Thus

$$R^2_{y.12} = \frac{\text{Explained variance of } Y \text{ by all regressors}}{\text{Total variance of } Y}$$

$$= 1 - \frac{\Sigma(Y-\hat{Y})^2}{\Sigma(Y-\bar{Y})^2} = 1 - \frac{s^2_{y.12}}{s^2_y} \qquad \text{...(23.19)}$$

We obtain $R^2_{y.12} = 0$ when $b_1 = b_2 = 0$ and $R^2_{y.12} = 1$ when all the observed values of Y directly fall on the estimated regression surface. The positive square root of $R^2_{y.12}$ is the *multiple correlation coefficient.* Note unlike the simple correlation coefficient, r which shows the direction of the relationship of the two variables by its sign (+ or –), $R^2_{y.12}$ i.e. the multiple correlation coefficient only takes the positive square root value of $R^2_{y.12}$. Hence we always have $0 \le R^2_{y.12} \le 1$. Now going back to our Example 23.1, we can have $R^2_{y.12} =$ 0.8611, which indicates that the regression plane, $\hat{Y} = 0.0440 + 0.2220\ X_1 - 0.0599\ X_2$ explains 86.11 per cent of the variation in Y as compared to 85.84 per cent of the variation explained by the simple correlation/regression between X and Y.

$R^2_{y.12}$ can also be built up by using the zero-order and partial correlation coefficient as given below :

$$R^2_{y.12} = r^2_{y1} + r^2_{y2.1}\,(1 - r^2_{y1}) \qquad ...(23.20)$$

If the order in which the independent variable enters the Eq. (23.20) is changed equivalently the same result would be obtained. Thus

$$R^2_{y.12} = r^2_{y2} + r^2_{y1.2}\,(1 - r^2_{y2}) \qquad ...(23.21)$$

For three independent variables, the coefficient of multiple determination will add one more term to become $R^2_{y.123}$. This will cover the variance in Y not associated with the other two independent variables, X_1 and X_2. However, this involves an expression which becomes more and more complex (like higher-order partial correlation), requiring it to be run on the computer. Again since the consideration of many independent variables do not contribute much to the multiple regression analysis, more than two independent variables are not considered further.

In Example 23.1 earlier, we know some of the variance in Y is associated with the joint variances of X_1 and X_2 but squared values of the partial correlation (i.e. $r^2_{y1.2}$ for example) omit this joint variance. Therefore, multiple correlation between the dependent variable and the two independent variables, when squared, $R^2_{y.12}$ is the proportion of the variance of Y accounted for by the two independent variables, X_1 and X_2 both jointly and separately. The multiple correlation $R^2_{y.12}$ in the Example 23.1 can now be calculated by either Eq. (23.20) or (23.21). Thus

$$R^2_{y.12} = r^2_{y1} + r^2_{y2.1}\,(1 - r^2_{y1})$$

(Proportion of variance explained by X_1 and X_2) = (Proportion explained by X_1) + (Additional proportion explained by X_2) (Proportion not explained by X_1)

$$= 0.8584 + (0.0194)\,(1 - 0.8584)$$

$$= 0.8584 + 0.0027 = 0.8611^*$$

Again, we know $R^2_{y.12} = r^2_{y2} + r^2_{y1.2}\,(1 - r^2_{y2})$

$$= 0.0313 + 0.8566\,(1 - 0.0313)$$

$$= 0.0313 + 0.8298 = 0.8611^*$$

We can thus conclude that when the addition of an independent variable in this multiple correlation of a given series results in a reduction of the unexplained variance, this reduction

---

*This value of $R^2_{y.12}$ is quite close to its value of 0.8595 calculated earlier by Eq. (23.19).

of the total unexplained variance in the simple correlation is 0.9687. The multiple coefficient of determination, $R^2_{y.12}$ does explain variance to the extent of 0.8611 and the variance unaccounted by it is 0.1389. Thus the reduction of the unexplained variance is 0.8298/ 0.9687 = 0.8566, which is itself the coefficient of the partial determination here.

23.5 STANDARD ERROR OF ESTIMATE OF MULTIPLE REGRESSION

The standard error of estimate of multiple regression provides some measure of the degree of reliability of the multiple regression expression. The error term for estimate of the value of Y about the regression line is defined in the same way as it is done in the simple regression earlier.

This standard error of estimate of multiple regression is denoted as $s_{y.12}$ and it is one of the following:

$$s_{y.12} = \sqrt{\frac{\Sigma(Y-\hat{Y})^2}{n}} \quad \text{...(23.22)}$$

or $$s_{y.12} = s_y \cdot \sqrt{1 - R^2_{y.12}} \quad \text{...(23.23)}$$

In Eq. (23.22), the standard error of the estimate of multiple regression is found by taking the square root of the mean squared deviations of observed Y values from the estimated regression plane.

The relation in Eq. (23.23) shows that larger the value of the multiple correlation the smaller is the standard error of estimate. If $R^2_{y.12} = 1$, then $s_{y.12} = 0$ so that the observed Y values then coincide with the corresponding predicted values, $\hat{Y}$. In this case, the multiple regression equation may be said to be a perfect prediction formula. On the other hand, if $R^2_{y.12} = 0$, then $s_{y.12} = 0$ and in this case the multiple regression equation fails to throw any light on the value of Y when X_1 and X_2 are known and is, therefore, of no use whatsoever. Because of this Eq. (23.23) can be used by the coefficient of multiple determination $R^2_{y.12}$.

Again, from linear regression since we know from Eq. (23.23) that $s_{y1} \cdot \sqrt{1 - r^2_{y1}}$ and $s_{y2} \cdot \sqrt{1 - r^2_{y2}}$ are the residual variances, if Y is estimated from X_1 and X_2 individually, while $s^2_y (1 - R^2_{y.12})$ are the residual variance if Y is estimated from X_1 and X_2 taken together; it is obvious that the inclusion of an additional independent variable can only reduce the residual variance. Now, the inclusion of X_2, when X_1 has already been taken, for predicting Y is worthwhile only when the resultant reduction in the residual variance is substantial. This will be the case when the numerical value of $r_{y2.1}$ is sufficiently large. This illustrates the importance of partial correlation in deciding whether to include or not an additional independent variable in a regression equation.

Example 23.2 Using the data in Example 23.1 and the multiple regression equation of it being given as $\hat{Y} = 0.0440 + 0.2220\, X_1 - 0.0599X_2$ calculate the standard error estimate of Y.

Following the above linear multiple regression equation, we calculate below the $\hat{Y}$ values and the deviations square.

Y	$\hat{Y}$	$(\hat{Y} - Y)^2$
3.5	4.2	0.49
7.5	6.5	1.00
7.0	8.4	1.96
15.0	10.0	25.00
11.0	15.3	18.49
15.0	18.6	12.96
25.0	21.7	10.89
27.0	26.3	0.49
	$\Sigma(\hat{Y} - Y)^2$	$= 71.28$

Therefore, $s_{y.12}$ is 2.99 which can be compared to the earlier standard error, i.e. $s_{\hat{Y}}$ = 2.98 in Chapter 21 for simple regression analysis. Since the standard error of estimate of multiple regression expresses the amount of variation in Y that is left unexplained by regression, $s_{y1.2}$ is generally more than $s_{\hat{Y}}$. However, in this case $s_{y.12}$ is about the same as $s_{\hat{Y}}$. This indicates that the inclusion of the second independent variable X_2 has not improved the predictions of Y. We now have the complete regression equation

$$\hat{Y} = 0.0440 + 0.2220X_1 - 0.0599\,X_2 \pm 2.99$$

into which we can substitute values of X_1 and X_2 to estimate $\hat{Y}$ at a significance level. Thus for $X_1 = 20$ and $X_2 = 5$ the estimate of $\hat{Y}$ is $\hat{Y} = 0.0440 + 0.2220\,(20) - 0.0599\,(5) \pm 2.99 = 4.16 \pm 2.99$.

Therefore 68.26 per cent of the values of X_1 and X_2 should have a value of Y between 1.17 and 7.15.

23.6 STATISTICAL SIGNIFICANCE OF MULTIPLE CORRELATION

As with zero-correlation coefficient, multiple correlation may be used to infer the probable relationship in a population from a properly selected sample. For this purpose, Snecdor's F test can be used as described in Chapter 21 on linear regression earlier, except that the degrees of freedom lost in producing the explained variance (in the numerator) will be the number of independent variables, m and not 1.

Now since we know the degrees of freedom ($d.f.$) will be always $n-1$ for the total variance and this $n-1$ is the total $d.f.$ so for estimating the absolute residual (i.e. the variance unexplained) it will be $n-m-1$, where m is the number of independent variables. Since m is 2 in the present context so for the F ratio, $d.f.$ is 2 in estimating the explained variance, SS_R in the numerator and $d.f.$ is $n-3$ for the same with respect to the unexplained one, SS_U, in the denominator (Eq. 23.24).

A statistically significant value of multiple correlation coefficient, $R^2_{y.12}$ would indicate that a correlation with the observed sign almost certainly occurs in the population.

Thus the complete formula for testing multiple correlation is :

$$F = \frac{\text{Estimate of the variance explained}}{\text{Estimate of the variance not explained}}$$

$$= \frac{R^2_{y.12}/2}{(1-R^2_{y.12})/(n-3)} = \frac{R^2_{y.12}/(n-3)}{2(1-R^2_{y.12})} \qquad \text{...(23.24)}$$

Note, increasing the number of independent variables in the regression equation will always increase the SS_R (except in the situation where a new X_1 variable is completely correlated with a previous variable). However, the increase may not be significant. The loss of $d.f.$ for deviations may offset the reduction in SS_U and actually increase the mean squares due to SS_U. If this happens, the F ratio for the significance of the regression will decrease, and the addition of another variable may act as a detractor for the regression.

23.7 STANDARDISED PARTIAL REGRESSION COEFFICIENTS

As it has been shown above, the partial regression coefficient, b_1 and b_2 indicate the absolute change in the dependent variable Y associated with unit change in X_1 or X_2. However, when the independent variables are measured in different units, the comparison of the coefficients, b_1 and b_2 become difficult. For instance, in Example 23.1, X_1 had been measured in number of cattle and X_2 the farm size in acres, it is not possible to say whether b_1 = 0.2220 indicates a more rapid rate of change (i.e. a relatively steeper slope) as compared to $b_2 = -0.0599$. Since the relative inportance of the independent variables are desirable for study, the b coefficients, are required to be standardised. Therefore, we turn to the "beta coefficients" by converting them to units of standard deviation :

$$\beta_1 = b_1 \frac{s_{x1}}{s_y} \qquad \text{...(23.25a)}$$

and $$\beta_2 = b_2 \cdot \frac{s_{x2}}{s_y} \qquad \text{...(23.25b)}$$

Thus, the standardised partial regression coefficients, β_k are found by

$$\beta_k = b_k \cdot \frac{s_k}{s_y} \qquad \text{...(23.25c)}$$

In all these Eqs. (23.25a to 23.25 c), s_y, s_{x1}, s_{x2} and s_k are the standard deviations of the variables Y, X_1, X_2 and X_k respectively.

Because the standard partial regression coefficients are all expressed in units of standard deviation, they may be compared directly with each other and the most effective variables determined. In case the standard deviation is more for the variable which has less β coefficient value, in relative terms more change in Y can be associated with the less β coefficient value and thus it can reverse the interpretation of the result. For Example 23.1, we know s_{x1} = 33.54, s_y = 7.91, b_1 = 0.2220, s_{x2} = 7.12 and b_2 = – 0.0599 so we find Y will increase at a higher (β_1 = 0.9413) rate with respect to X_1 and decrease at about the rate (β_2 = – 0.0539) with respect to X_2 than what it apparently indicates.

23.8 MULTICOLLINEARITY

In regression applications, a major objective is to use the parameters of the regression equation for predictive purposes. For this purpose plot of the independent variables should have a good spread so that they are uncorrelated. When the observations of the independent variables X_1 and X_2 are highly correlated among themselves, the multicollinearity is said to exist. Two structural problems arise in this case: (i) the partial regression coefficients can vary substantially and thus their effect as independent variables, on the dependent variable cannot be an absolute one, and (ii) their standard errors can be very large, indicating that they vary widely in repeated samples. As a result, if the independent variables are correlated among them, they will give very imprecise information about their separate effects on the dependent variable. Thus multicollinearity is undesirable if the objective is to measure the effects of the independent variables on the dependent variable. In the limiting case when the variables are perfectly correlated, the adjoint matrix will have no inverse, i.e. the determinant $|A|$ of the matrix would be zero. Hence the selection of variables for a regression equation is very important. For example, if we take observations on yield of wheat in relation to rainfall and temperature at a place for a number of years, the correlation between rainfall and yield will be negative giving a decrease in yield with increase in rainfall but if we go for a multiple regression we get a much positive direct relation. This happens because temperature has a negative correlation with rainfall and again low temperature in turn results in low yield. This is the negative direct relation which is corrected when we go for the multiple regression. Thus the great advantage of multiple regression that it avoids bias.

For detection of multicollinearity in a set of independent variables, (a) an arbitrary value of zero-order correlation, $r = 0.8$ or 0.9 is assigned beyond which the collinearity is said to be severe, or (b) to relate the simple correlation, r_{yx} to the overall multiple correlation, R^2_{y12} which is severe if $r_{yx} \geq R^2_{y.12}$. Note since time series variables tend to grow together multicollinearity is more common with time series data than are cross-section data.

Example 23.3. The following is a set of scores on three variables X_1, X_2 and Y of which X_1 and X_2 are supposed to be independent variables:

Y	X_1	X_2
3.5	20	10
7.5	30	16
7.0	40	20
15.0	50	25
11.0	70	30
15.0	90	40
25.0	100	50
27.0	120	65

Establish that collinearity exists in the multiple linear regression and it is confirmed by the regression of residuals.

Bivariate regression equations and correlation coefficients are worked out between each of the independent variables and Y. The intermediate calculation necessary for these estimates are:

$\Sigma Y = 111.0 \qquad \overline{Y} = 13.875 \qquad \Sigma Y^2 = 2102.5$

$\Sigma X_1 = 520.0 \qquad \overline{X}_1 = 65.0 \qquad \Sigma X_1^2 = 42800$

$\Sigma X_2 = 256.0 \qquad \overline{X}_2 = 32.0 \qquad \Sigma X_2^2 = 10606$

$\Sigma X_1 Y = 9185.0 \qquad \Sigma X_2 Y = 4605 \qquad \Sigma X_1 X_2 = 21230$

The relationships between each of the independent variables and the dependent variable are both strong and positive with equations

$$\hat{Y} = -0.3535 + 0.2189\, X_1,\ r_{y1} = 0.8757$$

$$\hat{Y} = -0.0834 + 0.4362\, X_2,\ r_{y2} = 0.9037$$

Similarly, the regression equation and the correlation coefficient between the independent variables are worked out as

$$\hat{X}_1 = 4.1552 + 1.9014\, X_2,\ r_{12} = 0.9847$$

$$\hat{X}_2 = -1.1500 + 0.5100\, X_1,\ r_{21} = 0.9847$$

This high correlation between the two independent variables is by itself indicative of collinearity.

The complete multiple regression equation and the coefficient of multiple correlation are

$$\hat{Y} = 0.0859 - 0.1080\, X_1 + 0.6503\, X_2,\ R^2_{y.12} = 0.9071$$

The multiple linear correlation is only marginally higher than the bivariate correlation, r_{y1}. The variable X_1 is negatively related in the multiple linear regression equation and the partial correlation coefficients are $r_{y1.2} = -0.1889$ and $r_{y2.1} = 0.4917$.

Finally, to work on the regression of residuals, the residuals of $(Y - \hat{Y}_1)$ and of $(Y - \hat{Y}_2)$ are calculated as

$(Y - \hat{Y}_1)$: – 0.52, 1.29, – 1.40, 4.41, – 3.97, – 4.35, 3.46 and 1.08

$(Y - \hat{Y}_2)$: – 0.78, 0.60, – 1.64, 4.18, – 2.00, – 2.36, 3.27 and – 1.27

The intermediate calculations for the correlation coefficient of the residuals and their regression trend lines are given in terms of their absolute values as:

$$\Sigma (Y - \hat{Y}_1) = 20.48,\ \Sigma (Y - \hat{Y}_2) = 16.10,\ \Sigma (Y - \hat{Y}_1)^2 = 71.1640$$

$$\Sigma (Y - \hat{Y}_2)^2 = 43.0058,\ \Sigma (Y - \hat{Y}_1)(Y - \hat{Y}_2) = 52.80$$

Thus $(Y - \hat{Y}_1) = 0.36 + 1.0925\,(Y - \hat{Y}_2)$ and $r = 0.9219$

This high correlation between the residuals of X_1 and X_2 confirms the true measure of collinearity.

23.9 CONCLUSION

In this chapter we have discussed multiple regression which is the obvious choice for considering the joint influence of more than one independent variable on a dependent variable, so that we can use this relationship for predicting the dependent variable for values of the other variables given. For example, in estimating the yield of wheat at a place we may require to know its relationship with temperature, rainfall over a number of years. Thus, we require to examine the techniques for the analysis of "multivariate data" in which each observational unit is characterized by several variables. For the sake of simplicity in this chapter we have considered only two independent veriables, X_1 and X_2 alongwith the dependent variable Y and have assumed linear relationship. Again majority of the relationships are not linear for which we require the knowledge on curve fitting so that their regressions approximate the trend of points across the graph. Hence we take up the multivariate data again and discuss the curve fittings in the next chapter.

LIST OF FORMULAE

I. *Equations for Linear Multiple Regression*

Linear Multiple Regression Model $\hat{Y} = a + \sum_{i=1}^{k} b_i X_i$

that is, on expanding, $\hat{y} = a + b_{y1.23\cdots k} X_1 + b_{y2.13\ldots k} X_2 + \ldots + b_{yk.12\ldots\ldots k} X_k$

Linear Multiple Equations in Matrix Notation

$$[A]\,[x] = [b]$$

For $k = 2$, Linear Multiple Regression Equation

$Y = a + bx_1 + bx_2$ for which

$$b = \frac{\sum x_2^2 \sum x_1 y - \sum x_1 x_2 \sum x_2 y}{\sum x_1^2 \sum x_2^2 - (\sum x_1 x_2)^2}$$

$$c = \frac{\sum x_1^2 \sum x_2 y - \sum x_1 x_2 \sum x_1 y}{\sum x_1^2 \sum x_2^2 - (\sum x_1 x_2)^2}$$

and $a = \overline{Y} - b_1\ \overline{X}_1 - b_2 X_2$

For $k = 2$, Linear Multiple Regression Equation $\hat{Y} = a + b_{y1.2}\ X_1 + b_{y2.1}\ X_2$ where $b_{y1.2}$ and $b_{y2.1}$ are partial regression coefficients

II. *Equations for Partial Correlation and Multiple Correlation*

First-order partial correlation coefficient

$$r_{y1.2} = \frac{(r_{y1}) - (r_{y2})(r_{12})}{\sqrt{1 - r_{y2}^2} \cdot \sqrt{1 - r_{12}^2}}$$

Second-order partial correlation coefficient

$$r_{y1.23} = \frac{(r_{y1.2}) - (r_{y13.1})(r_{y3.1})}{\sqrt{(1 - r_{13.2}^2)(1 - r_{y3.2}^2)}}$$

Third-order partial correlation coefficient

$$r_{y1.234} = \frac{r_{y1.23} - (r_{14.23})(r_{y4.23})}{\sqrt{(1 - r_{14.23}^2)(1 - r_{y4.23}^2)}}$$

Multiple Correlation, $R_{y.12}^2 = 1 - \dfrac{s_{y.12}^2}{s_y^2}$

or

$$R_{y.12}^2 = r_{y1}^2 + r_{y2.1}^2 (1 - r_{y1}^2) = r_{y2}^2 + r_{y1.2}^2 (1 - r_{y2}^2)$$

Standard Error of Estimate

$$s_{y.12} = \sqrt{\frac{\Sigma(Y - \hat{Y})^2}{n}}$$

or

$$s_{y.12} = s_y \cdot \sqrt{1 - R_{y.12}^2}$$

Test of Linearity of Multiple Regression

$$F = \frac{R_{y.12}^2 / (n-3)}{2(1 - R_{y.12}^2)}$$

Standardised Partial Regression Coefficients

$$\beta_{i=k} = b_k \cdot \frac{s_k}{s_y} \quad .$$

EXERCISES

23.1 Following is the yield of wheat with relation to rainfall and temperature at a place over eight years:

Yield (in metric tonnes/hectare)	5.39	4.50	6.29	6.29	7.19	4.50	5.39	3.59
Rainfall (in cm)	20.32	25.40	27.94	25.40	22.86	22.86	30.48	27.94
Temperature (in °C)	13	8	12	12	13	8	7	7

Establish that though individual relationships of temperature and rainfall with yield of wheat are contrary to each other the latter get a much positive direct relation with yield in multiple regression.

23.2 Following are the 20 sets of scores for X, Y and Z :

X: 53, 32, 38, 51, 64, 32, 56, 34, 35, 22, 29, 59, 61, 31, 40, 37, 45, 47, 57 and 57.

Y : 136, 132, 131, 112, 137, 134, 165, 124, 108, 144, 127, 138, 145, 143, 132, 152, 123, 125, 130, and 163.

Z: 119, 101, 89, 118, 88, 86, 98, 96, 94, 95, 29, 59, 61, 31, 40, 37, 45, 47, 57 and 57.

If we are interested in finding out the importance of Y and Z in predicting X, show how the partial correlation coefficient will be important in concluding that the inclusion of Z will not be worth of itself for improving the predictability of X.

24

Regression Analysis in Geo-science Systems v

(Curve Fitting and Multiple Regression)

In Chapter 23 we have considered the theory and computation of multiple regression in great detail. We know, so long the form of functional relationship between two variables is essentially that of a straight line with one single direction and two constants, as shown by the regression equation $\hat{Y} = a + bX$ all its statements remain validated for its multiple model: $\hat{Y} = a + \Sigma bX \pm \varepsilon$ (see Equations 23.1 and 23.2 of Chapter 23). But in many instances functions are non-linear and the slope of these curves changes so their relationships are not straight-forward. Hence a good knowledge of non-linear functions and their graphical expressions are necessary so that the resulting regression equations approximate the trend of points across the scatter diagram and provide comparison between similar relationship from different data populations. In the present context of studying the curve fitting and regression we discuss them by the : (i) popular linearizable order, and the (ii) higher order polynomial curves. Finally, we end this chapter with a discussion on trend surfaces as large-scale systematic changes that extend smoothly and predictably from one map-edge to the other.

24.1 CURVE FITTINGS

Curve fitting is concerned with determining the relations between the variables by appropriate regression equations so that when the sets of observations are plotted they are straightened. For straightening the curves a large number of specialist graph papers are also available – for example, semi-logarithmic graph or double log graph with one or both axes of the graph in logarithm form. These graphs facilitate the curve fitting of the data.

Three popular curve fittings with their general forms can be considered:

(1) Simple power curve : $Y = aX^b$ ...(24.1)
(2) Simple exponential curve : $Y = ae^{bX}$ ...(24.2)
and (3) Simple logarithmic curve : $Y = a + b\log X$ ...(24.3)

It can be seen that all these three curves (or, functions) deal with two variables of which the dependent variable Y is controlled by the behaviour of the independent variable, X. Speaking mathematically, when the functions of $Y = f(X)$ are plotted on sectional paper with X on the horizontal axis and Y on the vertical axis, the resulting curve which is defined by the corresponding function shows how Y varies as X varies. This helps to explain why X is called the independent variable and Y is called a dependent variable.

In the above three Eqs. (24.1 to 24.3), *logarithmic* and *exponential equations* produce curves of a form which differ markedly from those produced by power equations. This is due to the reason that in general forms of the logarithmic and power equations, the slope of their curves vary as a varies as it represents a base 10 or a base e (e = a constant = 2.71828). The logarithm of one to any base is always zero, thus all curves of the form $Y = \log_a X$ intersect at the common point (1, 0) irrespective of their base (Fig. 24.1).

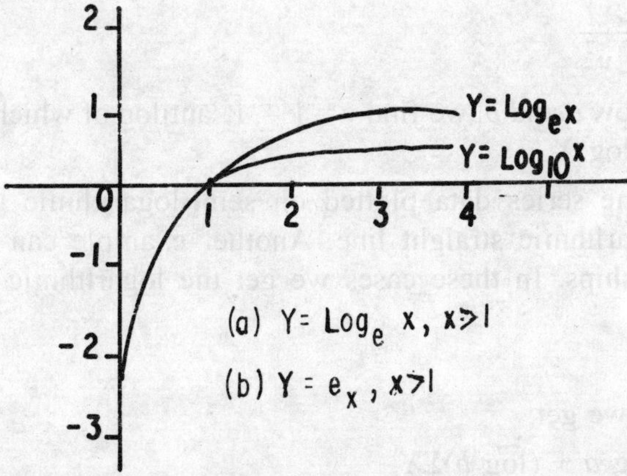

Fig. 24.1 : Curves of the Form $Y = \text{Log}_e X$

Power equations take the general form $Y = aX^b$. At their most simple form i.e. when $b = 0$, Y becomes a as the coefficient of X which itself reduces to 1 (since $X^0 = 1$) or when $b = 1$ (i.e. $Y = X$). However, for values of b other than zero or one, not only curves are produced due to their changes in slope but the curves reverse their direction also at least once which can yield more than one value of Y for any single value of X or *vice versa.* For example, the most elementary curvilinear equation is

$$Y = X^2 \qquad \text{...(24.4)}$$

which yields two solutions of Y for every value of X, since Y may take the value of $+\sqrt{X}$ or $-\sqrt{X}$.

These three curves – the *power, exponent* and *logarithmic curves* – share one important characteristic of being "linearisable" through the transformation of either one or both of the variables. More importantly, the required transformation having been made traditional (linear), so that least square regression can be used to find the values for the two coefficients

a and b. This process produces a *least square trend line* but by *detransforming* the variables, the curve corresponding to the straight line can be studied.

In the logarithmic curve $Y = a + b \log_{10}(X)$, Y is plotted against the logarithms of X (i.e. $\log X$) instead of raw score X. The logarithmic curve corresponds to a relationship in which a unit change in one variable X is associated with a large change in the Y variable at small values, but small changes at large values. Thus if we plot any bivariate data in environmental sciences like river discharge and sediment load, the latter will show many points clustering near the origin and relatively few observations with very high values will occupy the rest of the graph.

On the semi–logarithmic graph paper, it will be a straight line. This happens because $\log X$ is assumed as V and we get a linear trend equation $Y = a + bV$ in which transforming the $\log X$ values into deviations from its mean by a new variable v we can have from the Chapter 21 (on linear regression) following Eqs (21.13) and (21.14) as

$$a = \bar{Y} \text{ and } b = \frac{\Sigma vY}{\Sigma v^2} \qquad \text{...(24.5)}$$

Thus once we know a and b, we find $v = V - \bar{V}$ antilog of which gives us the resulting equation $Y = a + b \log X$.

Sometimes a time series data plotted on semi-logarithmic graph paper produces approximately a logarithmic straight line. Another example can be the cultivator and landholding relationships. In these cases we get the logarithmic curve in the form of equation of

$$Y = ab^X \qquad \text{...(24.6)}$$

By least square, we get:

$$\Sigma \log Y = N \log a + (\log b)\Sigma X \qquad \text{...(24.7)}$$

and $\Sigma X \log Y = (\log a)\Sigma X + (\log b)\Sigma X^2$

Example 24.1 Following is the data regarding number of cultivators (X) per 50 hectares of cultivated area and the average area of land holdings in hectares (Y).

X:	8.1	11.5	7.6	17.3	11.2	26.8	27.5	29.1	35.2
Y:	12.4	8.7	13.2	5.8	9.0	3.7	3.6	3.4	2.8

Fit a regression curve to the data.

We calculate $\Sigma X = 174.3$, $\Sigma X^2 = 4240.69$, $\Sigma \log Y = 6.9743265$, $\Sigma X \log Y = 113.80449$. Now substituting these values in Eq. (24.7), we get

$$6.9743265 = 9 \log a + 174.3 \log b$$

$$113.80449 = 174.3 \log a + 4240.69 \log b$$

Now treating these two as simultaneous equations, we solve $\log b = -0.0245815$ and $\log a = 1.2509869$ that is, $b = 0.9450$ and $a = 17.8232$ so that the curve which fits the data is $(17.8232)(0.9450)^X$.

For the power and exponential curves, the linearizing procedure differ only slightly from the above scheme. The relationship in various hydrological situations like run off and drainage area belong to power functions. For the *power curve,* the general form is $Y = aX^b$ which when taken by logarithm becomes:

$$\log Y = \log a + b \log X \qquad ...(24.8)$$

Thus if the data of power function is plotted on double-logarithmic graph it will approximate a straight line. Hence if we take the least square method of Eq. (24.8), we get the normal equations as:

$$\Sigma \log Y = n \log a + b \Sigma \log X \qquad ...(24.9)$$

$$\Sigma (\log X)(\log Y) = (\log a)(\Sigma \log X) + b\Sigma (\log X)^2$$

Now the linearity of this power curve, $Y = aX^b$ becomes clearer, if the terms are designated with $\log Y = U$, $\log a = c$ and $\log X = V$, so that Eq. (24.8) is $U = c + bV$ and the normal equations in Eq. (24.9) become :

$$\Sigma U = nc + b\Sigma V \text{ and } \Sigma UV = c\Sigma V + b\Sigma V^2$$

Now we have:

$$b = \frac{\Sigma UV - n\bar{U}\bar{V}}{\Sigma V^2 - n\bar{V}^2} \text{ and } \log a = \bar{U} - b\bar{V} \qquad ...(24.10)$$

It can be noted that the power curve in its equation has a variable at the base and a constant as its exponent, the constant can have any value, positive or negative, greater or less than one.

We explain the fitting of a power curve to a *distance-decay function* in the data below.

Example 24.2 Following is the set of a number of visitors, Y on distance in km (X) who travelled to a National Park. Plot the data and fit a regression to the data.

X : 5, 10, 15, 20, 25, 30, 35, 40, 45, 50, 55, 60, 65, 70, 75, 80, 85, 90, 95, 100

Y : 424, 77, 32, 36, 17, 13, 7, 6, 4, 4, 5, 3, 3, 2, 1, 1, 1, 1, 1, 1

We plot the data : Number of visitors (Y) against distance from the park on a graph (Fig. 24.2a). By transforming these two variables into logarithms we get a double logarithm to which these transformed data are plotted which provide the linear relationship (Fig. 24.2b).

Now we have, from the log-transformed data : $\Sigma V = 32.3655$, $\bar{V} = 1.6183$, $\Sigma U = 14.7013$, $\bar{U} = 0.7351$, $\Sigma V^2 = 54.7401$ and $\Sigma UV = 18.94$, so that $b = -2.06$ and hence $\log a = 4.07$.

We write now the regression equation as: $\log \hat{Y} = 4.07 - 2.06 \log X$, that is, $Y = 11750X^{-2.06}$.

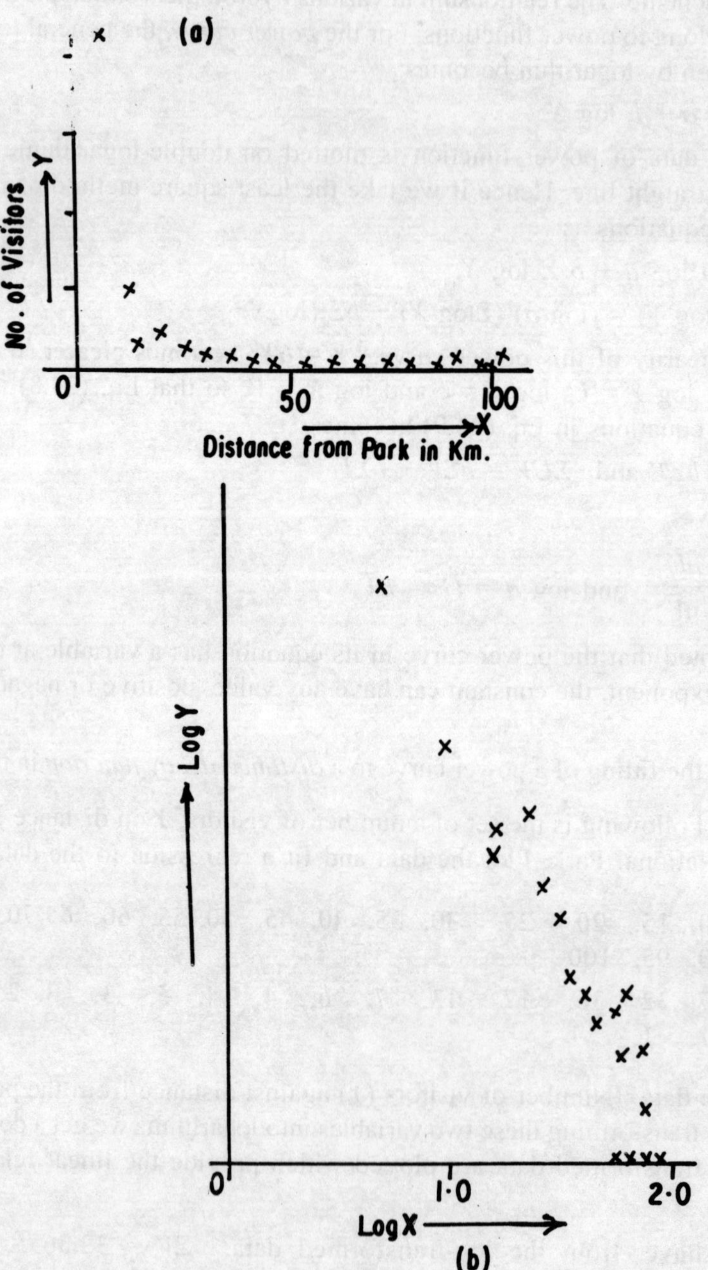

Fig. 24.2 : Curve Fitting

Exponential curve with the general form of $Y = ae^{bX}$ is essentially a curve in contrast to power curve $Y = X^2$ (for example) that has a constant base and a variable exponent. It is a growth (or, conversely decay) curve as they are defined the one in which the amount of growth (or decay) at any point of time (or, during any period of time) is a function of the level attained at that point of time. Any curve of the form of $Y = ca^{bX}$ where the variable X is the exponent, is also known as exponential curve. But when a is replaced by e, we get an exponential curve, for example, $Y = e^X$ where it is taken as having a constant, a number = 2.71828 which is the sum of a convergent series

$$\left(1+\frac{1}{n}\right)^n = 1+\frac{1}{1!}+\frac{1}{2!}+\frac{1}{3!}+...+\frac{1}{n!} = e \qquad ...(24.11)$$

This "e" can be best explained by the growth of the invested money at compound interest so that the growth at any instant is proportional to the total sum of money at that instant. For example, if the sum initially invested is I_0 and if it is subjected to an annual growth rate k then after t year, I_0 will grow exponentially and it is given by

$$I = I_0 e^{kt} \qquad ...(24.12)^*$$

The graph of Eq. (24.12) will be straight line with a slope of k. Note in the general form of exponential curve $Y = ae^{bX}$ where a equals the value of the dependent variable initially, X is the time, the independent variable, b is the present rate of growth, e the exponent (= 2.71828) and Y equals the value of the dependent variable at time X, and again it can be written as:

$$bX = \log_e \frac{Y}{a} \qquad ...(24.13)$$

which in other words is the inverse of a logarithmic function to the base e. Hence, an exponential relationship plotted on a graph with the dependent variable on a logarithmic scale and the independent variable on a linear scale, the plotted data will appear as a straight line whose slope is the constant b in the general case.

Again, if we take an exponential equation of $Y = 1.0e^{1.047X}$ the base e logarithm of 1.047 is 0.0459 which is the same as multiplying the base_{10} logarithm of 1.047 by 2.3026. This tells us that for every increase of one unit in X, Y increases by 0.0459 of itself.

Thus the value of e relates itself to compound or proportional growth processes but it is not just limited to growth (or decay) in time series for finance and population only. Distance-decay function is another important application of exponential curves.

If this standard exponential equation $Y = ae^{bX}$ is written in logarithms we get

$$\log_e Y = \log_e a + bX \log_e^e \qquad ...(24.14)^{**}$$

---

* Assuming the rate of interest is 100% per year then after one year of investment Re. 1 will have grown to Rs. 2 but if we allow the interest to be added continuously rather than at discrete intervals of time [and as the number of compounding periods grow, the duration of a period approaches zero] the sum of Re. 1 at 100% compound interest will have grown to Rs. 2.71828. Hence the exponential curve is also called compound interest curve.

but since $\log_e^e = 1$, then $\log_e Y = \log_e a + bX$ ...(24.15)**

Thus by using the natural logarithm to the base e in exponential curve (instead of the common logarithm to the base 10), we can see from Eq. (24.15) that the coefficient b as the slope of the line is interpreted more easily – it is the constant rate of increase, k in Eq. 24.12 (or change) in Y per unit of X. This is not true of slope of graph if common logarithms are used. In the example below we are examining the population growth over time by the exponential curve.

Example 24.3 Following are the data on population in Chile between 1835 and 1960:

Year	Population
1835	1010336
1843	1083801
1854	1439120
1865	1819223
1875	2075971
1885	2507005
1895	2695625
1907	3231022
1920	3730235
1930	4287445
1940	5023539
1952	5932995
1960	7374115

Plot the data. Fit a curve to the data and estimate population in 1970.

Note : We consider the year 1835 as the year 1. year 1843 is the year 9 ... similarly. .... 1960 is the 126th year after 1835. Now we use Eq. (24.15) for applying the exponential growth to the population data given.

For the exponential growth formulation we get two normal equations for $\log_e Y = \log_e a + bX$ which are by least square principle:

$$\Sigma \log_e Y = n \log_e a + b\Sigma X$$

$$\Sigma X \log_e Y = \log_e a \Sigma X + b\Sigma X^2 \qquad ...(24.16)$$

** Equation (24.15) can also be written as

$\ln Y = \ln a + bX. \ln e = \ln a + bX$

Now we do the necessary computations for the given population data below:

Year	Serial No. X	Population. Y	$\log_e$ (Population)
1835	1	1010336	13.826
1843	9	1083801	13.896
1854	20	1439120	14.179
1865	31	1819223	14.414
1875	41	2075971	14.546
1885	51	2507005	14.735
1895	61	2695625	14.807
1907	73	3231022	14.988
1920	86	3730235	15.132
1930	96	4287445	15.271
1940	106	5023539	15.430
1952	118	5932995	15.596
1960	126	7374115	15.814
Total	Σ819		Σ192.634

We also compute : $\Sigma X^2 = 72423$, $\Sigma X \log_e Y = 12449.136$ and we know $n = 13$.

Now substituting all these values in the above two normal equations we get

$$192.634 = 13 \log_e a + 819b$$

$$12449.136 = 819 \log_e a + 72423b$$

On solving them we get $b = 0.0150$ and $\log_e a = 13.870568$ from which we have $a = 1056601.5$. So the equation can be written as: $\hat{Y} = 1056601.5e^{0.0150X}$ where $e = 2.71828$.

It can be seen from plotting the raw score population data of 13 years for Chile (from 1835 to 1960) that the plotted points denote a curve (not a straight line) and this curve bends also forward twice before it leads to a concave curve (Fig. 24.3a). In this situation, a useful strategy is to go by the *semi-logarithmic transformation* (that is, of converting the dependent variable, Y into logarithms and leaving the independent/predictor variable, X into its usual normal numbers). This is done by using the exponential equation (Eq. 24.15) in Fig. 24.3b and we find that the transformed data (X, $\log_e Y$) points oscillate initially in the downward but finally upward. If the trend line is drawn by plotting three values of X and their corresponding computed Y's, this trend passes through time and the points are found to oscillate close to it downward and upward though in the later years the plotted points (X, $\log_e Y$) coincide with the trend line $\hat{Y} = 1056601.15\ e^{0.0150X}$. Accordingly, we can extrapolate the population with respect to the year 1970 and this predicted population is 8,125,898.

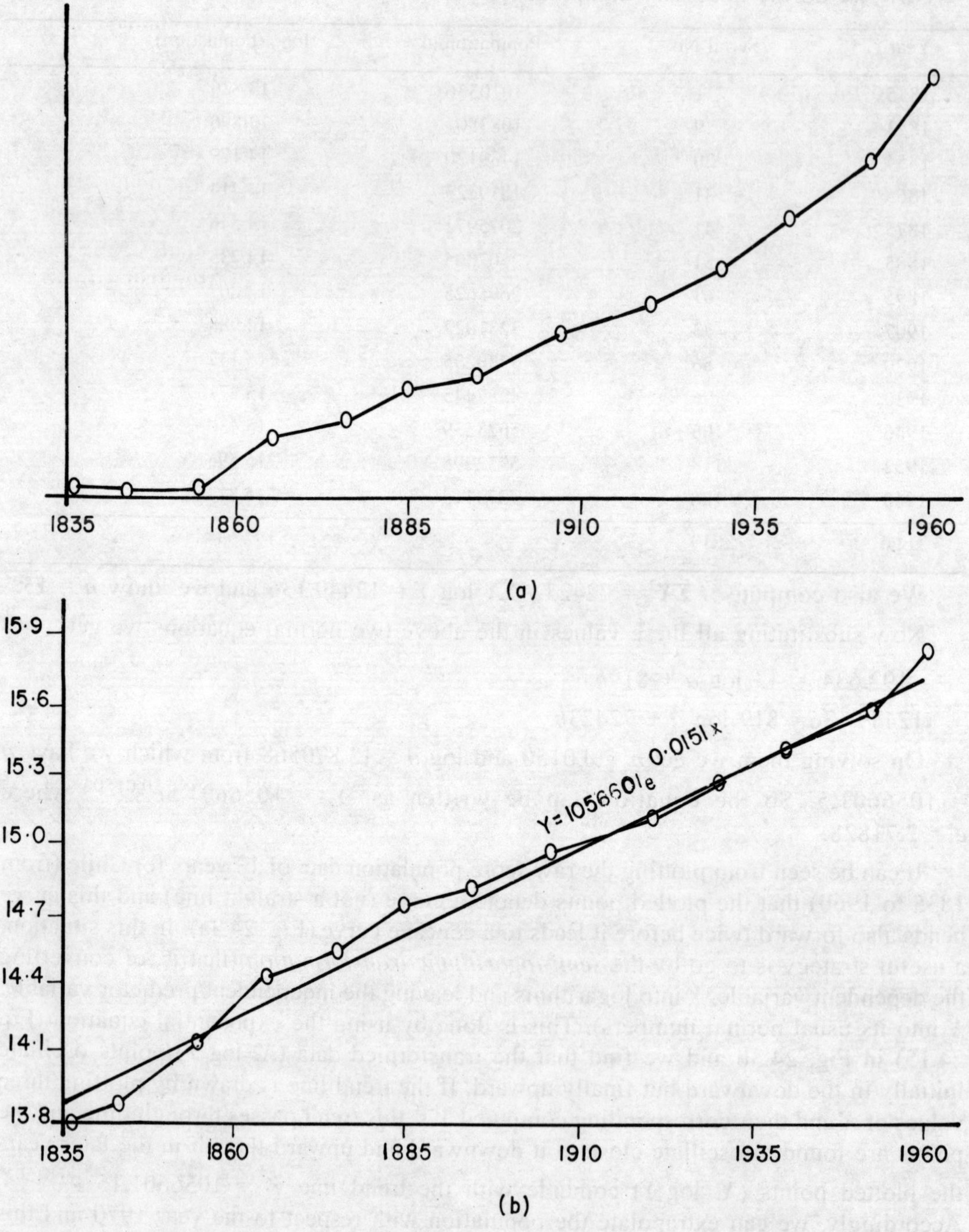

Fig. 24.3 : Exponential Curve and Its Linearised Form

Earlier, we have defined the exponential curve as a curve of growth process characterised by a constant per cent increase in value over time and its general form $Y = ae^{bX}$ can be reduced to its simplest $Y = e^X$ with $a = b = 1$. With $b \neq$ zero but less than 1 (i.e. $0 < b < 1$), we get the decay function $Y = e^{-X}$ which observes all the characteristics of the exponential growth curve except that Y becomes the decreasing function of X so that any increase in X is accompanied by a decrease in the value of Y so that for $X_1 < X_2$, $Y_1 > Y_2$. Figure 24.4 illustrates this role of exponential curve both as a growth as well as a decay curve also. As an exponential decay, apart from the distance-decay relationship in urban population, the rate of decay of slope in stream development can be suggested as another example.

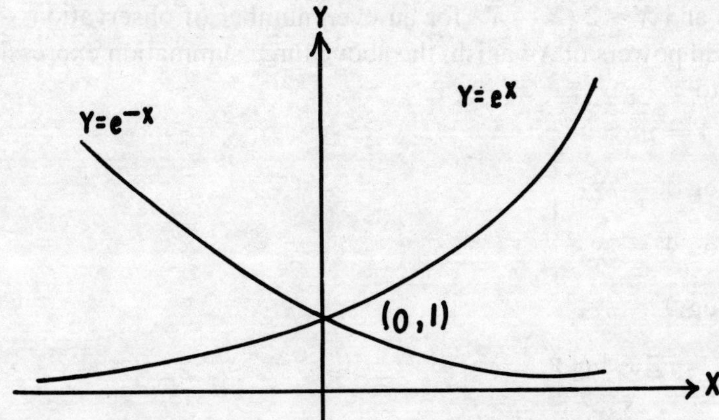

Fig. 24.4 : Exponential Curves $Y = e^x$ and e^{-x}

Finally, we should say that any exponential curve which is having a base other than e (like $Y = ca^{bX}$) can be converted to the base–*e*– exponential function. For example, $Y = 3^X$ can be changed to base –*e*– function, provided we find the expression for n in $e^n = 3$ and this $n = 1.1$. Hence we can write the original function :

$$Y = 3^X = (e^{1.1})^X = e^{1.1X} \quad ...(24.17)$$

Exponential curves because of wide applications are used in different models. One form is known as "second degree exponential curve", which is having an equation

$$Y = AB^{X_1} C^{X_2} \quad ...(24.18)$$

which has the logarithmic form:

$$\log Y = \log A + X_1 \log B + X_2 \log C \quad ...(24.18a)$$

where X_1 and X_2 are the independent variables for which Y is the dependent variable; A, B and C are the constants to be determined. Now in Eq. (24.18a), if we substitute $\log A = a$, $\log B = b$ and $\log C = c$, it reduces to

$$\log Y = a + bX_1 + cX_2 \qquad ...(24.19)$$

These three constants *a*, *b* and *c* can be estimated by three least square equations as

$$\left.\begin{aligned} \Sigma\log Y &= na + b\Sigma X_1 + c\Sigma X_2 \\ \Sigma X_1 \log Y &= a\Sigma X_1 + b\Sigma X_1^2 + c\Sigma X_1X_2 \\ \Sigma X_2 \log Y &= a\Sigma X_2 + b\Sigma X_1X_2 + c\Sigma X_2^2 \end{aligned}\right\} \qquad ...(24.20)$$

Now to reduce the computation of these three equations we can have the X_1 and X_2 values transformed in terms of deviations that is $x = X - \bar{X}$ for an odd number of observations and $X = 2(X - \bar{X})$ for an even number of observations so that we can see that all the odd powers of X vanish, the above three summation expressions in Eq. (24.20) get reduced to

$$\left.\begin{aligned} \Sigma\log Y &= na \\ \Sigma x_1 \log Y &= b\Sigma x_1^2 \\ \text{and} \quad \Sigma x_2 \log Y &= c\Sigma x_2^2 \end{aligned}\right\} \qquad ...(24.21)$$

From $\Sigma x_1 \log Y = b\Sigma x_1^2$

$$\text{we get } b = \frac{\Sigma x_1 \log Y}{\Sigma x_1^2} \qquad ...(24.22)$$

This known value of *b* on getting substituted in any two expressions of Eq. (24.20), we can calculate *a* and *c*. Now the antilog on both sides of Eq. (24.19) that is $\log Y = a + bX_1 + cX_2$ we get the solution for the second degree exponential equation $Y = AB^{X_1}C^{X_2}$ where *B* and *C* are positives, more than 1. It is to be noted that if we take the ratios of the successive values of *Y* computed (= $\hat{Y}$) by this equation and then take the successive ratios of these ratios (= second ratios) we will find these second ratios to have a constant exponent c^2 and hence this curve is known as *second degree exponential curve*.

There are other models of exponential curves : the model which is close to the above second degree exponential model is the *modified exponential curve*, which is having an equation

$$Y = k + AB^X \text{ where } 0 < B < 1 \qquad ...(24.23)$$

which by logarithm on both sides and on rearranging becomes:

$$\log (Y - k) = \log A + X \log B \qquad ...(24.24)$$

where *k* is an "asymptote", a constant which the curve reaches (as $X \to \infty$) an upper limit when *A* is negative, $A < 0$ and a lower limit when *A* is positive, $A > 0$. From Eq.

(24.24), it can be seen that the ratios of the first differences of computed Y values from k (i.e. $\hat{Y} - k$) will have a constant base, B. Again though Eq. (24.24) can be put in linear form but the method of least square can no longer be used for fitting this modified exponential curve. We are not providing any clue in this respect. Instead, we give an example for *modified exponential curve.*

Example 24.4 Following are the data regarding 11 watersheds with respect to the time of rise of flood in minutes, T which are supposed to be correlated with their stream lengths in km, M and their average slope N in gradient, 1/1000.

Watershed No.	T (in minutes)	M (in km)	N (1/1000)
1.	75	2.5	15.0
2.	140	6.0	8.0
3.	30	0.4	20.0
4.	60	7.6	13.0
5.	110	3.0	10.0
6.	60	3.5	12.0
7.	90	1.3	18.0
8.	30	1.0	150.0
9.	40	1.1	65.0
10.	30	0.3	25.0
11.	50	1.0	30.0

The time of rise of flood is defined as the time for a stream to rise from low water to flow above the flood level for the stream due to a storm occurring in the watershed of the stream. Find the functional relationship between T, M and N of the above data.

We take the second degree exponential equation which in the example have: $T = P.\ M^b.\ N^c$.

This by logarithm is $\log T = \log P + b \log M + c \log N$, now if $\log P = a$, $\log M = L$ and $\log N = S$ we write :

$\log T = a + bL + cS$ for which the least square equations are

$$\Sigma \log T = na + b\Sigma L + c\Sigma S$$

$$\Sigma L \log T = a\Sigma L + b\Sigma L^2 + c\Sigma LS$$

$$\Sigma S \log T = a\Sigma S + b\Sigma LS + c\Sigma S^2$$

We do the summations as required in these least square equations : $\Sigma \log T = 19.3055$, $\Sigma L = 2.31$, $\Sigma S = 14.70$, $\Sigma L^2 = 2.52$, $\Sigma S^2 = 21.03$, $\Sigma LS = 2.19$, $\Sigma L \log T = 4.85$, $\Sigma S \text{ Log } T = 24.11$ and $n = 11$.

We substitute the above summations in the least square equations to get

$$11a + 2.31b + 14.70c = 19.3055$$

$$2.31a + 2.52b + 2.19c = 4.85$$

$14.7a + 2.19b + 21.03c = 24.11$

which is written in matrix form

$$\begin{bmatrix} 11 & 2.31 & 14.7 \\ 2.31 & 2.52 & 2.19 \\ 14.7 & 2.19 & 21.03 \end{bmatrix} \begin{bmatrix} a \\ b \\ c \end{bmatrix} = \begin{bmatrix} 19.3055 \\ 4.85 \\ 24.11 \end{bmatrix}$$

Solving this above matrix, we have

$b = 0.2698$, $c = -0.2718$, $a = 2.0629$.

But since $P = \log^{-1}(2.0629) = 115.5846$ and hence we write:

$T = 115.5846\, M^{0.2698}\, N^{-0.2718}$

24.2 HIGHER ORDER CURVES

If we take the equation $Y = a + bX$ for linear fitting to the data and add to it another term cX^2, where c is a parameter, we get a quadratic (or second-degree) curve with a regression:

$$Y = a + bX + cX^2 \qquad \text{...(24.25)}$$

which is a non-linear equation because provided c is not zero, cX^2 will increase with increase in X in a geometrical manner. If we add another term dX^3 to this quadratic equation we will have a cubic curve with two inflexions in it as opposed to one in the quadratic. Depending on number of terms, any non-linear curve will have a number of inflexions, the minimum of which will be one in quadratic curve. A non-linear curve is also known as a "polynomial curve," a *polynomial* (or, *multiple) regression of degree k* in the variable X can be represented by a model equation

$$Y = a + bX + cX^2 + \ldots\; kX^k \qquad \text{...(24.25a)}$$

In applications of numerical methods in geomorphology the use of quadratic and cubic polynomials have been suggested for providing a means to fit a line of best fit using least square method to the shape of a river profile, a beach profile etc.

In these applications, height of the profile above some suitable datum line, the distance from an origin measured along the datum are taken. For example, the U-shape of glaciated valleys can be described by a quadratic curve $Y = a + bX + cX^2$ in which a and b are zero (with the value of X, the distance taken as zero at the lowermost point in valley cross profile, so that at ($X = 0$, $Y = 0$) we get the point of inflexion and the equation is reduced to $Y = cX^2$ where Y = height of the valley and X is the distance from the point of inflexion and we deduce that height of the glaciated valley is proportional to the distance from the inflexion point (Fig. 24.5), with c, the constant which is less than 1, varying from valley to valley. This quadratic curve is actually having a parabolic trend which can be seen in Fig. 24.5. We give below the least square equations to this curve $Y = a + bX + cX^2$.

We get

$$\left.\begin{aligned}\Sigma Y &= na + b\Sigma X + c\Sigma X^2\\ \Sigma XY &= a\Sigma X + b\Sigma X^2 + c\Sigma X^3\\ \Sigma X^2 Y &= a\Sigma X^2 + b\Sigma X^3 + c\Sigma X^4\end{aligned}\right\} \quad ...(24.26)$$

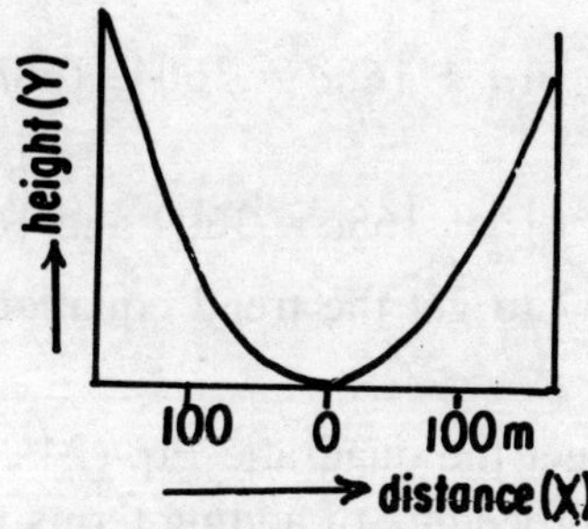

Fig. 24.5 : Glaciated Valley $Y = cX^2$

If we take the middle point of the period covered by the data as the origin for the odd number of observations so that $\Sigma x = \Sigma\,(X - \overline{X}) = 0$ and $\Sigma x = 2\Sigma\,(X - \overline{X}) = 0$, for the even number of observations, the above normal equations become

$$\left.\begin{aligned}\Sigma Y &= na + c\Sigma x^2\\ \Sigma xY &= b\Sigma x^2\\ \Sigma x^2 Y &= a\Sigma x^2 + c\Sigma x^4\end{aligned}\right\} \quad ...(24.27)$$

We illustrate this linear fitting of the quadratic equation by Example 24.5 below.

Example 24.5 Fit a linear trend to the following bivariate data:

X (= Year)	1	2	3	4	5	6	7	8
Y (= Nos.)	249	299	363	415	461	537	582	698

For these data of even number of values of X, we set x as the new value to X by $2(X - 4.5)$ where 4.5 is the mid-period, $\overline{X}$. So we rewrite the data as follows with respect to x.

x	−7	−5	−3	−1	1	3	5	7	$\Sigma x = 0$
Y	249	299	363	415	461	537	582	698	$\Sigma Y = 3604$
xY	−1743	−1495	−1089	−415	461	1611	2910	4886	$\Sigma xY = 5126$
x^2	49	25	9	1	1	9	25	49	$\Sigma x^2 = 168$
x^2Y	12201	7475	3267	415	461	4833	14550	34202	$\Sigma x^2Y = 77404$
x^4	2401	625	81	1	1	81	625	2401	$\Sigma x^4 = 6216$

So, with $n = 8$, we get : $11a + 168c = 3604$, $168b = 5126$, or $b = 30.5119$ and, $168a + 6216c = 77404$

From the linear equations $11a + 168c = 3604$ and $168a + 6216c = 77404$, we solve $a = 150.0491$ and $c = 11.6277$ to get the trend equation: $\hat{Y} = 150.0491 + 30.5119X + 11.6277X^2$.

It is to be noted that we get the quadratic Eq. (24.25) by adding a term cX^2 to the linear equation. Whether this procedure of adding terms for higher power of curve fitting is significant or not in the polynomial regression is required to be tested by ANOVA Table 24.1.

Table 24.1 : Testing a Quadratic Curve Fitting by ANOVA

Source of variation	Sums of squares	*d.f.*	Mean square	*F*-test
Linear regression,	SSR_1	1	MSR_1	
Quadratic regression	SSR_2	2	MSR_2	$\frac{MSR_2}{MSR_1}$
Additional regression by quadratic,	SSR_{2-1}	1	MSR_{2-1}	
Quadratic deviation,	SSU_2	$n - 3$	MSD_2	$\frac{MSR_{2-1}}{MSD_2}$
Total variance	SST	$n - 1$		

Note in Table 24.1 a new sum of squares of regression is created by subtracting the sum of squares of regression for a linear fit (SSR_1) from the quadratic fit (SSR_2). This new sum of squares is a measure of the increase in fit resulting from the additional regression term by quadratic fitting. Now the test of MSR_2/MSR_1 is the test examining the significance of the linear and quadratic terms combined.

The linear fit may be highly significant but the quadratic combination may be low. Hence this test may be significant but the test of the additional regression variance,

$MSR_{2\text{-}1}/MSD_2$ may not be significant. If the test is not significant the additional power in the curvilinear regression by quadratic equation is not contributing to the regression. In different situation, we may find that either, or both, or neither terms may be significant.

This above procedure for testing the significance of added terms can be extended to successively higher powers in the polynomial regression provided the statistical assumptions are fulfilled.

For a cubic curve we have mentioned about two points of inflexion in the curve. If we add another term dX^3 to the quadratic equation Eq. (24.25) it will be a cubic equation

$$Y = a + bX + cX^2 + dX^3 \quad \text{...(24.28)}$$

The least square equations for fitting this curve will be

$$\left.\begin{aligned} \Sigma Y &= na + b\Sigma X + c\Sigma X^2 + d\Sigma X^3 \\ \Sigma XY &= a\Sigma X + b\Sigma X^2 + c\Sigma X^3 + d\Sigma X^4 \\ \Sigma X^2 Y &= a\Sigma X^2 + b\Sigma X^3 + c\Sigma X^4 + d\Sigma X^5 \\ \text{and} \quad \Sigma X^3 Y &= a\Sigma X^3 + b\Sigma X^4 + c\Sigma X^5 + d\Sigma X^6 \end{aligned}\right\} \quad \text{...(24.29)}$$

As before, if the origin is taken in the middle, the odd summations ΣX, ΣX^3 and ΣX^5 becomes zero, so that the above equations get reduced to:

$$\left.\begin{aligned} \Sigma Y &= na + c\Sigma x^2 \\ \Sigma xY &= b\Sigma x^2 + d\Sigma x^4 \\ \Sigma x^2 Y &= a\Sigma x^2 + c\Sigma x^4 \\ \text{and} \quad \Sigma x^3 Y &= b\Sigma x^4 + d\Sigma x^6 \end{aligned}\right\} \quad \text{...(24.30)}$$

In substituting the summation values, we can solve $\Sigma Y = na + c\Sigma x^2$ and $\Sigma x^2 Y = a\Sigma x^2 + c\Sigma x^4$ to find a and c and then the other two equations $\Sigma xY = b\Sigma x^2 + d\Sigma x^4$ and $\Sigma x^3 Y = b\Sigma x^4 + d\Sigma x^6$ to find b and d.

Higher order (or polynomial) curve fitting requires polynomial regression. Any observed variable can be considered to be a function of any other variable measured on the same sample, for example, on a number of third order river basins several variables can be measured like

1. Basin magnitude based on the number of sources or joining streams, taken as Y (dependent) variable
2. Elevation of basin outlet (in m), X_1
3. Relief of the basin (in m), X_2
4. Basin areas (in sq. km), X_3
5. Total stream length in the basin (in km), X_4
6. Drainage density, defined as the total length of streams in the basin area, X_5
7. Basin shape, measured as the basin circularity ratio i.e. the ratios of inscribed to circumscribed circles, X_6

The problem is to determine the influence of the six variables from the second to the seventh above with the first one, basin magnitude as the dependent variable Y to which all others have to be related as $X_1, X_2 ..., X_6$ independent variables. Multiple regression in this case will be the technique and Eq. (23.1) or (23.2) of Chapter 23 will be the equation to determine the effectiveness of the independent variables as predictor of the dependent variable. Unlike polynomial regression, in multiple regression we cannot determine the relative effectiveness of the independent variables from the regression coefficient because their magnitudes are dependent upon the magnitudes of the variables themselves. Neither multiple regression of m number of variables ($m = 6$ in the example cited before) can be graphically worked out because of multivariate character of the data which is possible in polynomial regression. It is indeed that there is a wide range of choice to do curve fitting to a data. In polynomial regression, it is advised that given data should be plotted on the arithmetic paper. If there is one bend so that straight line fitting does not look obvious from the free hand plotting of the data, the ratios of successive values are taken and if they appear linear on a semi-logarithmic paper, an exponential curve is indicated. In case of bend persisting, the data should be plotted on log-log (or, double logarithmic) paper, if a linear trend appears power or logarithmic curves will be indicated. Finally, if there are one (or more) bends we get the idea of polynomial curves and they require to be fitted by second-degree (or high-degree) trend equations.

Finally, we discuss the trend surface analysis as a special case of multiple regression analysis.

24.3 TREND SURFACE ANALYSIS

Trend surface analysis is an analysis of treating the mapping surface mathematically so that the trends in mapping attributes like the U-shaped valley, the domal character of population/atmospheric pollution over a city and so on can be detected and separated.

For surface mapping, a mathematical concept "scalar field" is introduced which is any graph showing the value of an attribute, z_i (i.e. a scalar) as a function of its position, i.e.

$$z_i = f(x_i, y_i) \quad ...(24.31)$$

where z_i is a scalar magnitude which attributewise can be height, soil pH etc. relating itself to geographic position (x_i, y_i) at each and every data point. Seen this way the trend surface analysis is a special case of multiple regression procedures in thematic mapping for which the independent variables are the location coordinates (X_i, Y_i). Mathematically, it tries to separate the map data (or the data) for attribute, like height, soil pH etc. into two components : (1) that of a regional character, and (2) local functions based on the belief that any spatial observation is the outcome of two interacting processes (or a set of processes) that shape a regional setting and that which caused small areas to deviate from the regional pattern. But what we consider to be "regional" or "local" is largely subjective. It depends in part (a) upon the size of the region being examined, (b) upon the availability of the sample size, their frequency of data and (c) upon the emphasis given on defining and isolating the "local deviations" for spatial components.

Thus the obvious question in any trend surface analysis is how can the data be objectively separated into two components since the definition of components is entirely subjective? This problem arises because in Eq. (24.31), the function remains unspecified and there are enormous range of functions and to decide upon a particular function for the trend part of the equation and the analysis of the data is an operational problem. If we give an operational definition of trend "as a linear function of geographical coordinates of a set of observations so constructed that the squared deviations from the trend are minimized," the trend surface we can imagine is an inclined plane which can be specified as:

$$z = f(x, y) + \varepsilon \qquad \text{...(24.32)}$$

$$= a + bx + cy + \varepsilon \qquad \text{...(24.33)}$$

which is analogous to the linear or first degree trend curve fitting of $y = a + bx$, the only difference is in $f(x, y)$ we are concerned with two mutually perpendicular geographic coordinates so that in Eq. (24.33), we regard z, a spatial attribute as a linear function of the constants a, b and c, ε represents the error term or residual as the difference between z observed and $\hat{z}$ predicted. The constants have a simple physical interpretation : a represents the attribute (e.g. the height of the inclined plane surface at the map origin where $x = y = 0$, the coefficient b is the surface slope in the x - (or, east-west) direction and the coefficient c gives its slope in the y- (or, north-south) direction. The different values for these three components give different gradients for the linear trend surface. All that a least-square regression does, is to ensure that the constants a, b and c have values which collectively make the sum of squares of residuals as small as it possibly can be. These best-fit constants can be found by solving a set of three simultaneous equations, called the *normal equations:*

$$\left.\begin{aligned} \Sigma z &= na + \Sigma xb + \Sigma yc \\ \Sigma zx &= \Sigma xa + \Sigma x^2 b + \Sigma xyc \\ \Sigma zy &= \Sigma ya + \Sigma xyb + \Sigma y^2 c \end{aligned}\right\} \qquad \text{...(24.34)}$$

where n = total number of data points, $i = 1, ..., n$

For solving a, b and c we can have the necessary summations for substitutions in the above normal equations and then to solve them algebraically or we can put the entire set into matrix notation and get a matrix solution as explained in the earlier chapter on linear multiple regression. We will provide an example afterwards on fitting a linear trend surface to a set of spatial data and to provide its fitness test.

In cases, the above linear trend modelling does not fit and the actual trend is not an inclined plane but a complex one. In this context, since there is no prior knowledge about what the fundamental form of the trend should be so one cannot speak with authority about what form a particular geographical attribute can take spatially. Hence one can adduce several explanations that (1) there is really no trend of any sort in the data, (2) that there is a trend in the underlying surface but sampling frame and sample size are inadequate to detect it and finally, (3) a possibility is that the data is fitted to wrong sort of function.

In the above case, no matter how we change the value of the constants a, b, and c in our linear trend equation, the result will always be a simple inclined plane. When these do not provide a significant fit, or where geographical theory leads us to expect a different shape, then other more complex surfaces may have to be fitted. For example, if we wish to fit a dome or a U-shaped trough like trend surface across the study area, the least-square-linear trend may have to be expanded to a quadratic or second-degree-trend line by adding a squared term to the linear equation of Eq. (24.33) giving a second-degree-trend surface.

$$z = \underbrace{a + bx + cy}_{\text{Ist degree}} \quad \underbrace{+ dx^2 + exy + fy^2}_{\text{2nd degree}} + \quad \underset{\text{Terms Residual}}{\varepsilon_t} \qquad ...(24.35)$$

There are now six constants, a to f and note the second degree terms include the square of the geographical coordinates (x, y) and a cross-product of them. Now with six constants in second-degree there will be six simultaneous equations. Thus the addition of further terms like cubic surface, there will be nine constants whereas a four term quadratic (= a fourth order) regression will contain 14 constant, for its corresponding trend surface expression and so on. For these complex polynomial trend surfaces, computer involvements are necessary. Ultimately, when there are as many terms in the equation as there are observations, perfect description may be achieved, but the results are rendered uninterpretable as the fundamental spatial patterns are lost in a welter of detail.

Example 24.6 Following is the set of datum heights, z of 10 locations (x, y) in a 4 sq. km area.

Serial No.	x-coordinate (in km)	y-coordinate (in km)	Height, z (in 100 metres)
1.	0	0	12
2.	0	4	6
3.	1	3	8
4.	1	1	14
5.	2	2	12
6.	2	3	11
7.	3	3	12
8.	3	0	18
9.	4	4	14
10.	4	0	22

Determine the goodness-of-fit of a linear trend to the above data. Prepare the trend surface and the residual map of topographic height of the given area. Find out the gradient and direction of isolines of the trend map.

We do the necessary computations of the summations below for the *x*, *y*, and *z* data along the rows as follows :

x-coordinate:	0	0	1	1	2	2	3	3	4	4	Σ20
y-coordinate:	0	4	3	1	2	3	3	0	4	0	Σ20
z-coordinate:	12	6	8	14	12	11	12	18	14	22	Σ129
x^2	0	0	1	1	4	4	9	9	16	16	Σ60
y^2	0	16	9	1	4	9	9	0	16	0	Σ64
xy	0	0	3	1	4	6	9	0	16	0	Σ39
xz	0	0	8	14	24	22	36	54	56	88	Σ302
yz	0	24	24	14	24	33	36	0	56	0	Σ211
z^2	144	36	64	196	144	121	144	324	196	484	Σ1853

Now the solved normal equations for linear trend surface fitting to the data are

$$129 = 10a + 20b + 20c$$

$$302 = 20a + 60b + 39c$$

$$211 = 20a + 39b + 64c$$

We solve the above three equations to get a = 12.4282, b = 2.1065 and $c = -1.8706$. Hence the complete best fitting, linear trend surface equation is

$$z = 12.4282 + 2.1065x - 1.8706y + \varepsilon$$

Based on the above equation, the heights $\hat{z}$ are estimated with respect to the corresponding actual heights in the given data. The residuals ε are also worked out below.

Observed z	Computed $\hat{z}$	Residual, ε $z-\hat{z}$	(Trend values)^2 $=\hat{z}^2$
12	12.4282	– 0.4282	154.4601
6	4.9459	1.0541	24.4619
8	8.9229	– 0.9229	79.6181
14	12.6641	1.3359	160.3794
12	12.9000	– 0.9000	166.41
11	11.0294	– 0.0294	121.6477
12	13.1359	– 1.1359	172.5519
18	18.7476	– 0.7476	351.4725
14	13.3718	0.6282	178.8050
22	20.8541	1.1459	434.8936

We find the sum of the computed trend values, $\hat{z}$ = 128.9999 which practically equals the sum of the observed z values (= 129). Sum of the residuals also equals zero and we summed $\hat{z}^2$ values = 1844.7002. So these checks provide the clue for the arithmetical accuracy of the above linear trend equation computed.

To derive the index of *goodness-of-fit of the trend surface* to the original data we carry on an F test. But before we go for the F test, we calculate the coefficient of determination in linear correlation by following Eq. (23.19) in the earlier chapter as

$$R^2_{y.12} = \frac{\text{Explained variance by all regressors}}{\text{Total variance}}$$

$$= \frac{\Sigma\hat{z}^2 - (\Sigma\hat{z})^2/n}{\Sigma z^2 - (\Sigma z)^2/n} \quad \text{...(24.36)}$$

$$= \frac{[1844.7002 - (129)^2/10]}{[1853 - (129)^2/10]}$$

$$= 0.9561 \text{ or } 95.61 \text{ per cent}$$

Now, whether or not this fit is significantly different from the linear fit or not is by the usual F test (in Eq. 23.24)

$$F = \frac{R^2_{y.12}/d\,f\,1}{(1-R^2_{y.12})/d\,f\,2}$$

where *d.f.* 1 = degrees of freedom associated with the fitted surface equal to the number of constants used, less one for the base term that is 3 – 1 = 2; *d.f.* 2 is the degrees of freedom associated with the variance unexplained, which is found as the total degrees of

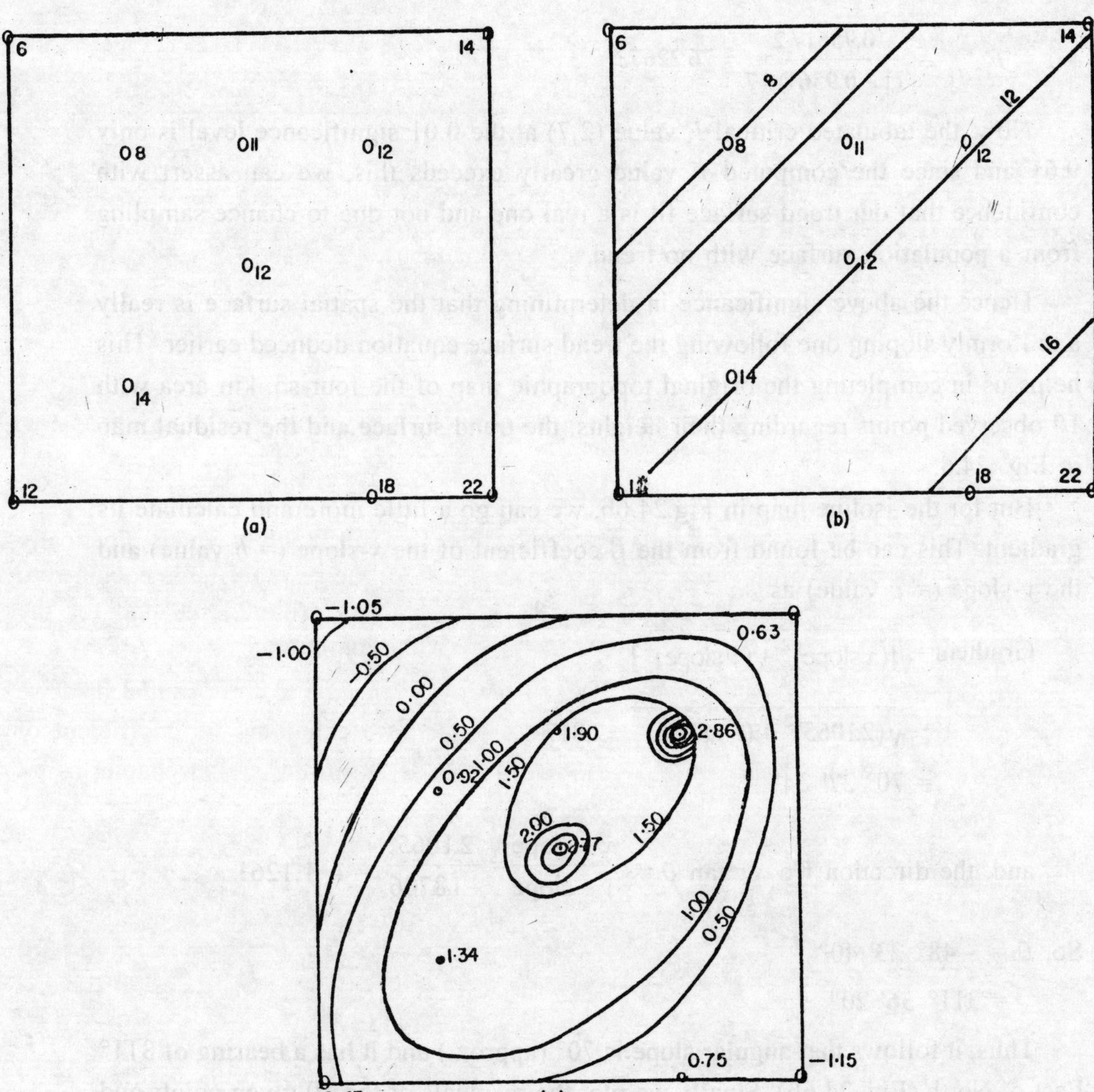

Fig. 24.6 : Linear Trend Surface for Topographic Data (in metre) (a) Topographical Map, (b) Linear Trend Surface Map, (c) Residual Map

freedom ($n - 1$), less those already assigned with *d.f.* 1, that is equal to 10 − 1 − 2 = 7, so that*

*Followings are the *d.f.* for Trend Surface Significance Testing

Trend surface	For explained variance	For unexplained variance
Linear	2	$n - 2 - 1$
Quadratic	5	$n - 5 - 1$
Cubic	9	$n - 9 - 1$

where n is the number of spatial data, z

$$F_{(2,7)} = \frac{0.9561/2}{(1-0.9561)/7} = 76.2267$$

Now, the tabulated critical F value (2,7) at the 0.01 significance level is only 9.61 and since the computed F value greatly exceeds this, we can assert with confidence that our trend surface fit is a real one and not due to chance sampling from a population surface with no trend.

Hence the above significance in determining that the spatial surface is really a uniformly sloping one following the trend surface equation deduced earlier. This helps us in completing the original topographic map of the four sq. km area with 10 observed points regarding their heights, the trend surface and the residual map in Fig. 24.6.

But for the isoline map in Fig.24.6b, we can go a little more and calculate its gradient. This can be found from the β coefficient of the x-slope (= b value) and the y-slope (= c value) as

$$\text{Gradient} = \sqrt{[(x\,\text{slope})^2 + (y\,\text{slope})^2]}$$

$$= \sqrt{(2.1065)^2 + (-1.8706)^2} = 2.817$$

$$= 70° \; 27' \; 24''$$

and, the direction fro : $\tan\theta = \dfrac{x-\text{slope}}{y-\text{slope}} = \dfrac{2.1065}{-1.8706} = -1.1261$

So, $\theta = -48° \; 23' \; 40''$

$= 311° \; 36' \; 20''$

Thus, it follows that angular slope is 70° (approx.) and it has a bearing of 311° East of North (Fig. 24.6b). Finally we plot the residuals of the 10 given points and plot the isoline for the residuals at ± 0.5 metre level (Fig. 24.6c).

Lastly in concluding on the chapter, it can be said that polynomial curves have been used for trend surface analysis merely as a matter of convenience. But use of the polynomials in no way intimates that processes themselves in geo-system sciences are polynomial functions or are linear. Their nature, unknown and perhaps unknowable, can only be approximated by polynomial expansion.

LIST OF FORMULAE

I. ***Equations for Curve Fittings***

Power Curve, $Y = aX^b$

Exponential Curve, $Y = ae^{bX}$ or $bX = \log_e \frac{Y}{a}$

Logarithmic Curve, $Y = a + b. \log X$

Modified Exponential Curve, $Y = k + AB^X$

Higher Order Curve : general equation

$Y = a + bX + cX^2 + ... + kX^k$

II. ***Trend Surface Equations***

For mapping surfaces (x, y as locational coordinates) with z as an attribute, general equation for first-order regression is $\hat{z} = f(x, y) = a + bx + cy$

For second-order regression plane add second-order terms $dx^2 + exy + fy^2$ to the above equation.

EXERCISES

24.1 Following are the data on peak flow and volume of water of a river for 10 successive years.

Serial No. (years)	Volume (in cubic metre per sq. km.), V	Peak Flow, Q (in cumecs)
1.	4114	612
2.	13196	1248
3.	13164	1340
4.	5734	915
5.	2543	400
6.	7841	1070
7.	2401	520
8.	2746	415
9.	2254	370
10.	14894	2240

Establish a functional equation between volume, V and peak flow, Q by the least square method, Justify the significance of the relationship.

24.2 Following are the data on rainfall intensity, *R* and duration, *T* in hours recorded for a number of storms. The data are given for descending order of rainfall intensity.

Serial No.	Rainfall intensity, *R*	Duration in hour, *T*
1.	52.8	0.0853
2.	36.5	0.1667
3.	25.6	0.25
4.	22.8	0.5
5.	13.4	1.0
6.	7.4	2.0
7.	6.8	3.0
8.	4.9	6.0
9.	3.6	12.0

Plot data on arithmetic paper and interpret the graph. Establish the relationship between *R* and *T* by least square.

24.3 Following are the data on suspended sediment load and discharge of a river over a period of time:

Suspended sediment load, *Y* (in tonnes/day)	Discharge, *Q* (in cumecs)
288	1.3
339	2.3
741	2.2
1000	5.0
1660	4.1
2692	6.2
2951	3.9
7244	5.6
7762	11.0
18200	20.9
25120	21.4
60268	28.2
63100	54.9
288400	58.9
323600	85.1

Illustrate how the above paired data when plotted on a log-log graph gives a better distribution than a skewed one when plotted on arithmetic paper. Insert the trend line equation you calculate on the graph.

24.4 Following are the data on a variable *Y* against a number of successive years, *X*.

X.	1st	2nd	3rd	4th	5th	6th	7th
Y.	230	241	263	282	295	304	316

Plot the data and in case there is a curvilinear trend, fit a quadratic trend to the data.

24.5 Following are the spot height data (z_i) with respect to regular grid-cell locations:

x_i	y_i	z_i	x_i	y_i	z_i
1	1	3.8	1	2	3.9
2	1	4.0	2	2	4.0
3	1	4.0	3	2	4.1
4	1	4.1	4	2	4.1
5	1	4.2	5	2	4.3
6	1	4.2	6	2	4.2
7	1	4.3	7	2	4.3
1	3	4.1	1	4	4.0
2	3	4.1	2	4	4.2
3	3	4.3	3	4	4.4
4	3	4.2	4	4	4.4
5	3	4.6	5	4	4.8
6	3	4.5	6	4	4.5
7	3	4.4	7	4	4.6
1	5	4.3	1	6	4.6
2	5	4.4	2	6	4.5
3	5	4.6	3	6	4.8
4	5	4.5	4	6	5.0
5.	5	4.8	5	6	5.2
6	5	4.7	6	6	5.0
7	5	5.0	7	6	4.8

Calculate the ground elevations so that the concerned area is to be made perfectly horizontal and uniform slopes are provided in both directions.

WORKED ANSWERS

Chapter 2

2.1 The altimetric frequency curve is bimodal with peak for frequency of 15 at 876-925 metre and 13 at 726-775 metre.

2.2 Cumulative frequency, "less than" expressed in percentages are : 8.95, 16.42, 35.82, 44.78, 65.67, 88.06, 95.52 and 100.00

2.3 Q_1 = 749.07 metre; Q_3 = 899.33 metre; Q_D = 75.13 metre. The 4th decile = 801 and 9th decile = 946 metre; The 3rd and 50th percentiles are 642 and 840 metre respectively.

2.4 For the data arranged in a frequency table with 40.00 cm as the lower boundary of the class interval of 12 cm, the mean, median and mode are 69.2, 69.56 and 69.14 cm. respectively.

For the listed data mean is 67.54 and median 67.08 cm.

2.5 Interpolated mean = 69 cm

2.6 Median 165.4 cm

2.7 Ethnic segregation index is 34.75 per cent

2.8 Correct mean 5.1

2.9 The 1st series

2.10 Missing values are 34 and 45; coded mean 48.22

2.11 Arithmetic mean 10.74, geometric mean 10.61 and harmonic mean 10.52

2.12 Mean intensity of storm rainfall per hour is 3.53 cm.

2.13 Mode 5052 cumec, coefficient of quartile deviation 38.22 percent

2.14

Data	12		14		15		18		25		22
Progressive average in 2s	12		13		13.7		14.75		16.80		17.67
Moving average in 2s		13		14.5		16.5		21.5		23.5	

2.15 Value of Mode is 5 and not 8 which is an irregular number

2.16 Average speed 192 km. p.h.

Chapter-3

3.1 Four crop combinations : groundnut, bajra, jowar and pulses.

3.2 Surface height is of more varied character in Palamau district.

3.3 Coefficients of quartile deviation for Udaipur and Palamau districts are 0.0999 and 0.1873 but since the coefficients of variation are 0.1591 and 0.2105 for these two districts so the quartile deviation index is strictly not comparable with coefficient of variation.

3.4 Region 1 is moderately skewed with a negative character (=-0.1768) but Region 2 is more skewed with a positive character (=0.5463). The Kurtosis value for both the distribution is less than 3, so both the distributions are strongly peaked.

3.5 Standard deviation of the second group 2.59

3.6 Missing values : $n_2 = 60$, $\bar{x}_2 = 20$, $\sigma_3 = 7.26$

3.7 Minimum number of rain-gauge sites required is 13

3.8 $\mu_1 = 2.1$ $\mu_2 = 6.09$, $\mu_3 = 20.13$ and $\mu_4 = 71.23$

3.9 (a) Mean 1264.1mm, median 1174.5mm, mode 1150mm

(b) Mean 1250mm, median 1200mm, standard deviation 480mm, skewness measure, $S_k = 0.31$

Chapter 4

4.1 Probability values, p with respect to rainfall values are

Rainfall (in cm)	18	20	22	24	26	28	30	32
Probability, (p)	0.04	0.11	0.22	0.38	0.56	0.74	0.86	0.94

4.2 (i) $p = 0.4147$

(ii) $p = 0.0832$

(iii) $p = 0.4988$

(iv) $p = 0.6664$

(v) $p = 0.6311$

4.3 Lowest value 866.5

4.4 Mean 158.142 and Standard deviation 4.95

4.5 (a) 2 buses (b) 2720 hours

4.6

Given Data	Cumulative Proportions	Normalized raw scores
50	1.0000	100.00
49	0.9090	54.05
43	0.8181	48.87
48	0.7272	45.22
45	0.6363	42.14
43	0.5454	39.29
40	0.4545	36.47
35	0.3636	33.85
30	0.2727	30.58
20	0.1818	26.89
10	0.0909	21.75

4.7 First rank.................................Student B (Scores 1.89)
Second rank..............................Student A (Scores 1.79)
Third rank.................................Student C (Scores 1.75)

Chapter 5

5.1 Probability for 0, 1, 2, 3, 4 and 5 years with rainfall below normal are 0.5905, 0.3281, 0.0729, 0.0081, 0.00045 and 0.00001 respectively.

5.2 Probability of game be played on a ground soaked with rain water is (a) none, 0.9040; (b) exactly one, 0.0913; (c) not more than one, 0.9953; (d) for more than one, 0.0047 and (e) at least one game, 0.0960

5.3 (a) 0.25, (b) 0.50 and (c) 0.25

5.4 Mean 6 and variance 4.2

5.5 Probability of having annual rainfall more than normal in more than 25 years is 0.9032.

5.6 (a) 0.10, (b) 0.49 and (c) 0.96

5.7 Probability of change in ownership for atleast one house is 0.0070 by binomial and 0.0096 by Poisson distribution method.

5.8 Probable limits of foggy days in the place is either 46 or 34.

5.9 Probability of sample containing 2 rainy days, without replacement is 0.5 and the same with replacement is 0.3241.

5.10 For (a) $p\,(1, 20) = 0.072$
(b) $p\,(2, 15) = 0.0292$
(c) $p\,(1) = 0332$

5.11 Probability of 0.081

Chapter 6

6.1 Probability of flooding in a year : none, 0.1999; once, 0.3218; twice, 0.2591; thrice, 0.1390, four times, 0.0560 and five times, 0.0180

6.2 Poisson distribution : 42, 53, 34, 15 and 4

6.3 (a) for binomial, $p\,(2) = 0.194$, (b) for Poisson, $p(2) = 0.184$

6.4 Frequencies for 0, 1, 2 and 3 floods per year are 156, 39, 5 and 0 respectively.

6.5 Probability for 4 floods in the next 10 years is 0.0000966 and no flood in the next 10 years is 0.792.

Chapter 7

7.1 (a) 14.55 cm and 18 cm
(b) p(m) = 0.417

Chapter 8

8.1 Mean centre (5.08, 3.96)
Median centre (6.3, 3.5)
Modal centre (7.0, 3.0)
Standard distance deviation 4.13

8.2 Standard distance deviation 3.98

8.3 Mean distance 0.34 km

8.4 Random distribution pattern with points independently located; sample points are randomly distributed; Yes, the point pattern may well have been generated by Poisson spatial process.

8.5 $E(r)$ = 0.3536 Km, $\sigma_r = 0.0289$, z observed is – 1.64 which does not vary more than chance of $z_\alpha = 0.5$ equal to 1.96. Again with C = 1.197 and $y = 0.545$ z is found to be 0.82 which is close to expected. Hence the point pattern supports Poisson spatial process.

Chapter 9

9.2 Random numbers : 33, 03, 21, 07 and 09

9.4 Sample means are not grouped around population mean of 44.8 but the individual sample means get compressed with increasing sample size.

9.5 (i) 10 percent as samples for each of the 12 strata making a total of 830.
(ii) Samples weighted for category a : 166, 125, 83 and 41
similarly for category b : 100, 75, 50 and 24 and, for category c : 66, 50, 33 and 17

Chapter 10

10.1 Population mean 4.5, sampling mean 4.29.

10.2 Boring depth for 95 percent confidence limit : 804 to 960 cm., for 99 percent confidence limit : 756 to 1044 cm.

10.3 z calculated is significant

10.4 Since z_c is 3.90 so samples are drawn from different populations

10.5 Field size and aspect are not related but since z_c is 1.92 for which $p = 0.057$

decision is not very conclusive which requires more field data

10.6 z calculated at $\alpha = 0.05 = 3.02$: Difference significant

10.7 Farm size ranges 4.10 to 4.89 ha. and 4.19 to 4.80 ha, this decreases : 4.24 to 4.46 ha. and 4.30 to 4.69 ha. respectively.

CHAPTER 11

11.1 95 percent limit of annual rainfall : 114.67 to 125.33 cm

11.2 t at $d.f. = 19$ is 1.79, so H_0 accepted

11.3 t at $d.f. = 25$ is 2.1, so H_0 rejected

11.4 t at $d.f. = 5$ calculated as 2.80 which is higher than $t_a = 0.05$, 2.57, H_0 rejected

11.5 t calculated at $d.f. = 9$ is 10.04 which is beyond $p = 0.001$, so we must have $p < 0.0005$.

11.6 Significant difference in productivity

CHAPTER 12

12.1 $F_{(12, 11)} = 3.25$, yield of crop more variable

12.2 At $\alpha = 0.05$ $F_{(3, 3)} = 9.28$, homogeneity retained

12.3 $F_{(3, 16)} = 6.31$, routes are significantly different

12.4 $F_{(3.16)} = 1.85$, H_0 is retained

12.5 $F_{(2, 24)} = 4.48$, H_0 is rejected

12.6 $F_{(4, 8)} = 151.72$, so appraiser designing into 3 calegories is appropriate

12.7 $F_{(3, 52)} = 39.71$ and $F_{(4, 52)} = 2.34$: Productivity variations between the groups of industries significant but not amongst the regions.

12.8 $F_{1\ (3, 15)} = 3.30$ & $F_{2\ (5, 15)} = 2.96$: Treatments and blocks do not indicate any difference in yields

CHAPTER 13

13.1 χ^2 at $\alpha = 0.05$ is 9.49, H_1 not accepted

13.2 χ^2 calculated is 23.33 which is more > $\chi^2_{\alpha = 0.05}$ at $d.f. = 4$, hence H_0 rejected

13.3 Phi coefficient = 0.24, this contingency coefficient and the χ^2 test beyond 0.05 level of significance do not suggest association.

13.4 χ^2 at $d.f. = 4$ calculated is $11.69 < \chi^2_{\alpha=0.05} = 9.49$; H_0 retained

13.5 Yate's correction $\chi^2 = 20.27$ (that for continuity is 17.74), Fisher's exact probability test, p gives a very high value, $\rho = 0.79$. Both these tests lead to a conclusion of significant difference between soil types and parent materials.

13.6 χ^2 value at *d.f.* 4 is 5.48 which gives a probability of more than 0.20.

13.7 (i) χ^2 calculated = 12.87 at *d.f.* = 9, H_0 is retained.

(ii) χ^2 calculated at *d.f.* = 3 is 9.06 which is close to $\chi^2_{\alpha = 0.02}$ is 9.8

13.8 χ^2 calculated is 18.021 with *d.f.* = 2, H_0 is rejected

13.9 At $k = 4$, $\chi^2 = 10.16 > \chi_c^2 = 15.51$. The data can be fitted to a Poisson distribution

13.10 With $\bar{x} = 7.85$ thousand kgs. and $\sigma = 81.89$. The data cannot be fitted to a normal distribution

13.11 At *d.f.* = 6 x^2 is 12.55 which is lower than the critical χ^2 value of 12.59. Hence the observed distribution can be described adequately by normal distribution.

13.12 χ^2 calculated for 3 *d.f.* = 15.033 which is highly significant. Hence irrigation and manuring are not independent

CHAPTER 14

14.1 χ^2 for *D* statistic at 2 *d.f.* is 0.2064 which is less than the corresponding χ^2 critical value. Settlements are randomly distributed

14.2 *D* Statistic = 0.273 > $D_{\alpha = 0.05}$ which is 0.391. Hence no significant difference between the industries located close to raw materials and industries not located at raw material sources

14.3 Computed $D = 0.2179 > D_{\alpha=0.05}$ and the distribution follows normal distribution

CHAPTER 15

15.1 $U = 97$, z transformation of U score = 3.2 and it is supported that the pebbles at a river confluence is supplied not from the river banks but from higher up the river channel.

15.2 $U = 10$ and H_0 is rejected for $\rho = 0.05$ and more

CHAPTER 16

16.1 $T = 10$, $z = -2.48$ and H_0 is rejected

16.2 $T = 8$ which is more than that of 7 at $\alpha = 0.01$, hence H_0 is rejected

16.3 $T = 3$, $z = 3.41$ and H_0 is rejected

CHAPTER 17

17.1 $H = 4.62$, $\chi^2 = 5.99$ and H_0 is accepted at $\alpha = 0.05$ for d.f. 2

17.2 $H = 4.99$, $\chi^2 = 11.34$ and H_0 is retained at $\alpha = 0.01$ for d.f. 3

17.3 $H = 8.30$, χ^2 corrected for tie = 9.88 so H_1 is accepted at $\alpha = 0.01$ for d.f. 2

CHAPTER 18

18.1 $\rho_{XY} = -0.21$, $\rho_{XZ} = 0.64$ and $\rho_{YZ} = -0.30$. Hence the environmentalists X and Z have the nearest approach to common lineage in environmental pollution perception.

18.2 $\rho = 0.85$, $t = 6.04$ and correlations is significant

18.3 $\rho = -0.55$, population and GNP growth are negatively related but this relationship is not significantly higher than 0.05 level

CHAPTER 19

19.1 z transformation for Kandall's τ (tau) is 3.15, the probability of occurrence of which is 0.002 or around 2 per cent. Hence rainfall and runoff data of the basins are positively related and this relationship is significant at 0.05 level.

CHAPTER 20

20.1 r_{xy} of 0.25 for $n = 50$ is not significant at any level but p is 0.82 which indicate that the paired data do not have any linear relationship.

20.2 (a) $r_{XY} = 0.80$, (b) $r_{XY} = 0.02$ and for (c) $r_{XY} = -0.99$

20.3 $r_{XY} = 0.0153$, by phi coefficient it is 0.1429

20.4 by raw score method, $r_{XY} = 0.4053$, by z score method, $r_{XY} = 0.4054$

20.6 $r_{XY} = -0.60$

20.7 corrected $r_{XY} = 0.67$

20.8 Coefficient of correlation between basin area and total stream length is 0.92

20.9 Correlation coefficient 0.79, computed t value 6.89 indicates significance of correlation and hence H_0 is rejected.

CHAPTER 21

21.1 $\hat{Y} = 1.7017 + 0.010778\,X$ where X = rainfall and Y = runoff with confidence limits of points placed 5.19 above and below the regression line

21.2 $Y = 34.9150 - 0.0136\,X$ where with $X = 300$ m. estimated temperature will be 30.8°C.

21.3 $\hat{y} = 93.95 - 10.18X$; Test of linearity $F_{\alpha=0.1}(1,7) = 12.3$; linearity of the data significant

21.4 For $X = 70$ cm, Y is 72.6 cm

21.5 (a) 0.025, (b) 0.0243, (c) limits of r : 0.7971 to 0.9429

CHAPTER 22

22.1 (a) With the origin at the middle year, the trend equation is $\hat{y} = 90 + 2t$.

(b) The trend equation without assuming any middle year $\hat{y} = 84 + 2t$. A significant autocorrelation problem exists in the regression model.

22.2 The new trend equation is $\hat{y} = 1302.3 + 26.296t^2$.

D statistic calculated to the revised data is 1.041 and we conclude that there is significant positive autocorrelation in the revised regression model.

CHAPTER 23

23.2 $r_{XY.Z} = 0.248$, $r_{XZ.Y} = -0.023$ against
$r_{XY} = 0.248$ and $r_{XZ} = -0.025$

CHAPTER 24

24.1 Volume, $V = 1.7492\ Q^{1.207}$, Q is peak flow, $r = 0.9653$

24.2 Asymptotic curve; Rainfall intensity, $R = 13.1T^{-0.56}$

24.3 Trend line equation, $Y = 162.2Q^{1.67}$

24.4 $X = 279.286 + 14.857Y - 0.857Y^2$

24.5 $z = 3.555 + 0.08x + 0.15y$

APPENDICES
Statistical Tables

APPENDIX TABLE I(a)

NORMAL DISTRIBUTION TABLE

Area under the normal curve: *z* curve – Fractional parts of the total area (equal to 1). Values in the body of the table are proportions of the normal distribution between the mean and given *z* score points. The proportions are given to four decimal points under the normal curve, corresponding to distances between the mean and the ordinates which are *z* standard deviation units from the mean.

To read desired area of any value of *z* say *z* = 1.57, read down the first column unitill the *z* value 1.5 is reached then across the entry in the column headed .07 to find the desired area 0.4418. *Note* the table entry is the shaded area in the curve below.

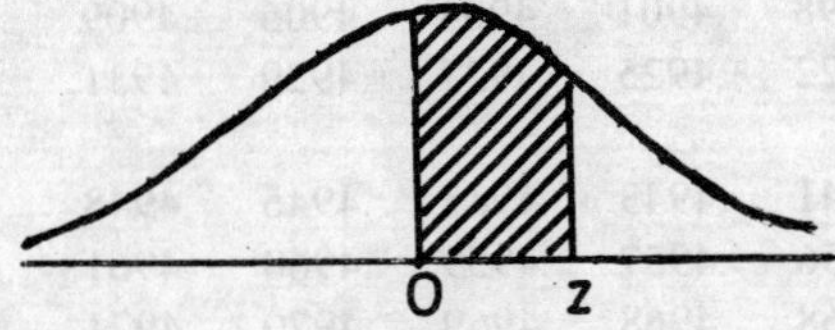

z_i	Second		Decimal			Place			Value of z_i	
	.00	.01	.02	.03	.04	.05	.06	.07	.08	.09
0.0	.0000	.0040	.0080	.0120	.0159	.0199	0.239	.0279	.0319	.0359
0.1	.0398	.0438	.0478	.0517	.0557	.0596	.0636	.0675	.0714	.0753
0.2	.0793	.0832	.0871	.0910	.0948	.0987	.1026	.1064	.1108	.1141
0.3	.1179	.1217	.1255	.1293	.1331	.1368	.1406	.1443	.1480	.1517
0.4	.1554	.1591	.1623	.1664	.1700	.1736	.1772	.1808	.1844	.1879
0.5	.1915	.1950	.1985	.2019	.2054	.2088	.2123	.2157	.2190	.2224
0.6	.2257	.2291	.2324	.2357	.2389	.2422	.2454	.2486	.2518	.2549
0.7	.2580	.2612	.2642	.2673	.2704	.2734	.2764	.2794	.2823	.2852
0.8	.2881	.2910	.2939	.2967	.2995	.3023	.3051	3073	.3106	.3135
0.9	.3159	.3186	.3212	.3238	.3264	.3289	.3315	.3340	.3365	.3389
1.0	.3413	.3438	.3461	.3485	.3508	.3531	.3554	.3577	.3599	.3621

Appendix Table I(a) (contd.)

z_i	Second		Decimal			Place			Value of z_i	
	.00	.01	.02	.03	.04	.05	.06	.07	.08	.09
1.1	.3643	.3665	.3686	.3718	.3729	.3749	.3770	.3790	.3810	.3830
1.2	.3849	.3869	.3888	.3907	.3925	.3944	.3962	.3980	.3997	.4015
1.3	.4032	.4049	.4066	.4083	.4099	.4115	.4131	.4147	.4162	.4177
1.4	.4192	.4207	.4222	.4236	.4251	.4265	.4279	.4292	.4306	.4319
1.5	.4332	.4345	.4357	.4370	.4382	.4394	.4406	4418	.4430	.4441
1.6	.4452	.4463	.4474	.4485	.4495	.4505	.4515	.4525	.4535	.4545
1.7	.4554	.4564	.4573	.4582	.4591	.4599	.4608	.4616	.4625	.4633
1.8	.4641	.4649	.4656	.4664	.4671	.4678	.4686	.4693	.4699	.4706
1.9	.4713	.4719	.4726	.4732	.4738	.4744	.4750	.4758	.4762	.4767
2.0	.4773	.4778	.4783	.4788	.4793	.4798	.4803	.4808	.4812	.4817
2.1	.4821	.4826	.4830	.4834	.4838	.4842	4846	.4850	.4854	.4857
2.2	.4861	.4865	.4868	.4871	.4875	.4878	.4881	.4884	.4887	.4890
2.3	.4893	.4896	.4898	.4901	.4904	.4906	.4909	.4911	.4913	.4916
2.4	.4918	.4920	.4922	.4925	.4927	.4929	.4931	.4932	.4934	.4936
2.5	.4938	.4940	.4941	.4943	.4945	.4946	.4948	.4949	.4951	.4952
2.6	.4953	.4955	.4956	.4957	.4959	.4960	.4961	.4962	.4963	.4964
2.7	.4966	.4967	.4968	.4968	.4969	.4970	.4971	.4972	.4973	.4974
2.8	.4974	.4975	.4976	.4977	.4977	.4978	.4979	.4980	.4980	.4981
2.9	.4981	.4982	.4883	.4984	.4984	.4984	.4985	.4985	.4986	.4986
3.0	.49865	.4987	.4987	.4988	.4988	.4988	.4989	.4989	.4989	.4990
3.1	.49900	.4991	.4991	.4991	.4992	.4992	.4992	.4992	.4993	.4993
3.2	.4993129									
3.3	.4995166									
3.4	.4996631									
3.5	.4997674									
3.6	.4998409									
3.7	.49989922									
3.8	.4999277									
3.9	.4999519									
4.0	.4999683									
4.5	.4999966									
5.0	.4999997133									

APPENDIX TABLE I(b)

NORMAL DISTRIBUTION TABLE

Table of Probabilities Associated with Values of z in a Normal Distribution (Correct to 4 Decimal Places)

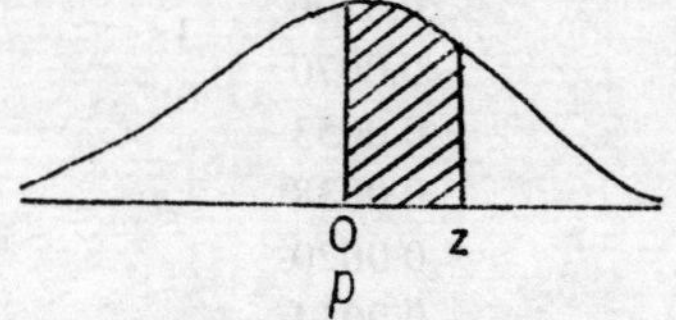

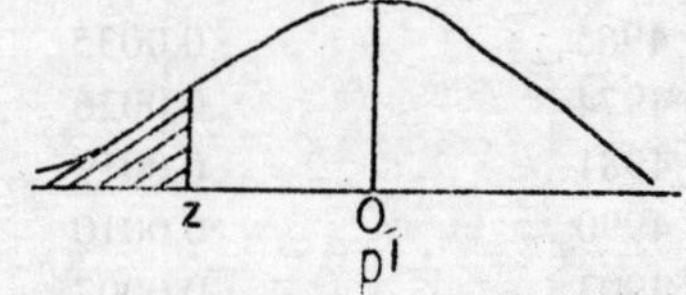

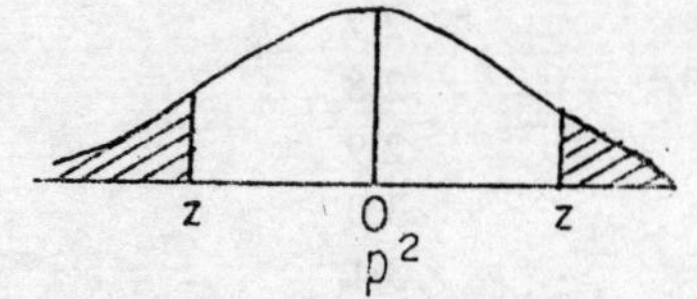

z	Column A p	Column B p^1	Column C p^2
0.0	0.0000	0.5000	1.0000
0.1	0.0398	0.4602	0.9204
0.2	0.0793	0.4207	0.8414
0.3	0.1179	0.3821	0.7642
0.4	0.1554	0.3446	0.6892
0.5	0.1915	0.3085	0.6170
0.6	0.2257	0.2743	0.5486
0.7	0.2580	0.2420	0.4840
0.8	0.2881	0.2119	0.4238
0.9	0.3159	0.1841	0.3682
1.0	0.3413	0.1587	0.3174
1.1	0.3645	0.1357	0.2714
1.2	0.3849	0.1151	0.2302
1.3	0.4032	0.0968	0.1936
1.4	0.4192	0.0808	0.1616
1.5	0.4332	0.0668	0.1336
1.6	0.4452	0.0548	0.1096
1.7	0.4554	0.0446	0.0892
1.8	0.4641	0.0359	0.0718
1.9	0.4713	0.0287	0.0574
1.96	0.4750	0.0250	0.0500
2.0	0.4773	0.0227	0.0454

Appendix Table 1(b) (contd.)

z	Column A p	Column B p^1	Column C p^2
2.1	0.4821	0.0179	0.0358
2.2	0.4861	0.0139	0.0278
2.3	0.4893	0.0107	0.0214
2.4	0.4918	0.0082	0.0164
2.5	0.4938	0.0062	0.0124
2.6	0.4950	0.0050	0.0100
2.7	0.4965	0.0035	0.0070
2.8	0.4974	0.0026	0.0053
2.9	0.4981	0.0019	0.0038
3.0	0.4990	0.0010	0.0020
3.1	0.4993	0.0007	0.0014
3.2	0.4995	0.0005	0.0010
3.3	0.4996	0.0004	0.0008
3.4	0.4997	0.0003	0.0006
3.5	0.4998	0.0002	0.0004
3.6	0.4998409		
3.7	0.4998922		
3.8	0.4999277		
3.9	0.4999519		
4.0	0.4999683		
4.5	0.4999966		
5.0	0.499997133		

Note p: The probability of a value lying between the axis of symmetry (i.e. $z = 0$), and given value of z (an one-tailed probability) above or below the axis of symmetry.)

p^1 The probability of a value being more extreme than z (an one-tailed probability).

p^2 The probability of a value being more extreme than either $+z$ or $-z$ (a two-tailed probability).

APPENDIX TABLE II

COEFFICIENTS OF BINOMIAL DISTRIBUTION

Following table give the Binomial Coefficients $\frac{n!}{r!(n-r)!}$, i.e. the r number of events occuring in n number of values from 2 to 20

n ↓ / r →	2	3	4	5	6	7	8	9	10
2	1								
3	3	1							
4	6	4	1						
5	10	10	5	1					
6	15	20	15	6	1				
7	21	35	35	21	7	1			
8	28	56	70	56	28	8	1		
9	36	84	126	126	84	36	9	1	
10	45	120	210	252	210	120	45	10	1
11	55	165	330	462	462	330	165	55	11
12	66	220	495	792	924	792	495	220	66
13	78	286	715	1287	1716	1716	1287	715	286
14	91	364	1001	2002	3003	3432	3003	2002	1001
15	105	455	1365	3003	5005	6435	6435	5005	3003
16	120	560	1820	4368	8008	11440	12870	11440	8008
17	135	680	2380	6188	12376	19448	24310	24310	19448
18	153	816	3060	8568	18564	31824	43758	48620	43758
19	171	969	3876	11628	27132	50388	75582	92378	92378
20	190	1140	4845	15504	38760	77520	125970	167960	184756

APPENDIX TABLE III

VALUE OF e^{-z} IN POISSON DISTRIBUTION FUNCTION

z	e^{-z}	z	e^{-z}	z	e^{-z}
.0	1.000	1.5	.223	3.0	.050
.1	.905	1.6	.202	3.1	.045
.2	.819	1.7	.183	3.2	.041
.3	.741	1.8	.165	3.3	.037
.4	.670	1.9	.150	3.4	.033
.5	.607	2.0	.135	3.5	.030
.6	.549	2.1	.122	3.6	.027
.7	.497	2.2	.111	3.7	.025
.8	.449	2.3	.100	3.8	.022
.9	.407	2.4	.091	3.9	.020
1.0	.368	2.5	.082	4.0	.018
1.1	.333	2.6	.074	4.5	.011
1.2	.301	2.7	.067	5.0	.007
1.3	.273	2.8	.061	5.0	.002
1.4	.247	2.9	.055	7.0	.001

The above table gives the value of e^{-z} for numbers in one place of decimal. For e^{-z} of the above numbers in their second place of decimal multiply the entry in the column by 0.99005. For example e^{-z} for $z = 0.70$ is given in the above table as 0.497. Now to get the above for e^{-z} for $z = 0.73$, we have to multiply 0.497 by 0.99005 three times = 0.497 × 0.99005 × 0.99005 × 0.99005 equal to 0.482. So e^{-z} for $z = 0.73$, the exponential value will be 0.482.

Also, if $z = 0.70$ then $e^{-z} = \ln^{-(0.70)} = 0.49658$ where "ln" denotes a logarithm to the base e (natural logarithm), "log" denotes logarithm to the base 10 (common logarithm), and "log z" denotes logarithm to any other base, specified by the numerical value of z.

APPENDIX TABLE IV

LOG PEARSON TYPE III DISTRIBUTION

Following is the table of values of the frequency factor *k* following the Log Pearson Type III Distribution

Skew co-efficient β_1	Recurrence Interval, *T*-years						
	2	5	10	25	50	100	200
	Annual probability of occurrence %						
	50	20	10	4	2	1	0.5
3.5	–0.396	0.420	1.180	2.270	3.152	4.051	4.970
2.5	–0.360	0.518	1.250	2.162	3.048	3.845	4.652
2.0	–0.307	0.609	1.302	2.219	2.912	3.605	4.298
1.8	–0.282	0.643	1.318	2.193	2.848	3.499	4.147
1.6	–0.254	0.675	1.329	2.163	2.780	3.388	3.990
1.4	–0.225	0.705	1.337	2.128	2.706	3.271	3.828
1.2	–0.195	0.732	1.338	2.083	2.626	3.172	3.656
1.0	–0.164	0.758	1.338	2.043	2.542	3.022	3.489
0.9	–0.148	0.769	1.339	2.018	2.498	2.957	3.401
0.8	–0.132	0.780	1.336	1.993	2.453	2.891	3.312
0.7	–0.116	0.790	1.333	1.967	2.407	2.824	3.223
0.6	–0.099	0.800	1.328	1.939	2.359	2.755	3.132
0.5	–0.083	0.808	1.323	1.910	2.311	2.686	3.041
0.4	–0.066	0.816	1.317	1.880	2.261	2.615	2.949
0.3	–0.050	0.823	1.309	1.849	2.211	2.544	2.856
0.2	–0.033	0.830	1.301	1.818	2.159	2.472	2.763
0.1	–0.017	0.836	1.292	1.785	2.107	2.400	2.670
0	–0.000	0.841	1.282	1.751	2.054	2.326	2.576
–0.1	–0.017	0.846	1.270	1.716	2.000	2.252	2.482
–0.2	–0.033	0.850	1.258	1.680	1.954	2.178	2.388
–0.3	–0.050	0.853	1.245	1.643	1.890	2.104	2.294
–0.4	–0.066	0.855	1.231	1.606	1.834	2.029	2.201
–0.5	–0.083	0.860	1.216	1.567	1.777	1.955	2.108
–0.6	–0.099	0.857	1.200	1.528	1.720	1.880	2.016

Appendix Table IV (*contd.*)

Skew co-efficient	Recurrence Interval, T-years						
	2	5	10	25	50	100	200
	Annual probability of occurrence %						
	50	20	10	4	2	1	0.5
–0.7	–0.115	0.857	1.183	1.488	1.663	1.806	1.926
–0.8	–0.132	0.856	1.166	1.448	1.606	1.733	1.837
–0.9	–0.148	0.854	1.147	1.407	1.549	1.660	1.749
–1.0	–0.164	0.852	1.128	1.366	1.492	1.588	1.664
–1.2	–0.195	0.843	0.086	1.282	1.379	1.449	1.501
–1.4	–0.225	0.832	1.041	1.198	1.270	1.318	1.351
–1.6	–0.245	0.817	0.994	1.116	1.116	1.197	1.216
–1.8	–0.282	0.799	0.945	1.035	1.069	1.087	1.097
–2.0	–0.307	0.777	0.895	0.959	0.980	0.990	0.995

APPENDIX TABLE V

RANDOM SAMPLING NUMBERS TABLE

Following is the table of two digit random numbers.

20	17	42	28	23	17	59	66	38	61	02	10	86	10	51	55	92	52
74	49	04	49	03	04	10	33	53	70	11	54	48	63	94	60	94	49
94	70	49	31	38	67	23	42	29	65	40	88	78	71	37	18	48	64
22	15	78	15	69	84	32	52	32	54	15	12	54	02	01	37	38	37
93	29	12	18	27	30	30	55	91	87	50	57	58	51	49	36	12	53
45	04	77	97	36	14	99	45	52	95	69	95	03	83	51	87	85	56
44	91	99	49	89	39	94	60	48	49	08	77	64	72	59	26	08	51
16	23	91	02	19	96	47	59	89	65	27	84	30	92	63	37	26	24
04	50	65	04	99	34	65	65	82	42	70	51	55	04	61	87	88	83
32	70	17	72	03	61	66	26	24	71	22	77	88	33	17	78	08	92
03	64	59	07	42	95	81	39	06	41	20	81	92	34	51	90	39	08
62	49	00	90	67	86	83	48	31	83	19	07	67	68	49	03	27	47
61	00	95	86	98	36	14	03	48	88	51	07	33	40	06	86	33	76
89	03	90	49	28	74	21	04	09	96	60	45	22	03	52	80	01	79
01	72	33	85	52	40	60	07	06	71	89	27	14	29	55	24	85	79
27	56	49	79	34	32	32	32	60	53	91	17	33	26	44	70	93	14
49	05	74	48	10	15	35	25	24	28	20	22	35	66	66	34	26	35
49	74	37	25	97	26	33	94	42	23	01	28	59	58	92	69	03	66
20	26	22	43	88	08	19	85	08	12	47	65	65	63	56	07	97	85
48	87	77	96	43	39	76	93	08	79	22	18	54	55	93	75	97	26
08	72	87	46	75	73	00	11	27	07	05	20	30	85	22	21	04	67
95	97	98	62	17	27	31	42	64	71	46	22	32	75	19	32	20	99
37	99	57	31	70	40	46	55	46	12	24	32	36	74	69	20	72	10
05	79	58	37	85	33	75	18	88	71	23	44	54	28	00	48	96	23
55	85	63	42	00	79	91	22	29	01	41	39	51	50	36	65	26	11
67	28	96	25	68	36	24	72	03	85	49	24	05	69	64	86	08	19
85	86	94	78	32	59	51	82	86	43	73	84	45	60	89	57	06	87
40	10	60	09	05	88	78	44	63	13	58	25	37	11	18	47	75	62
94	55	89	48	90	80	77	80	26	89	87	44	44	23	74	66	20	20
19	11	63	77	77	20	33	62	62	19	29	03	94	15	56	37	14	09
11	63	77	77	23	20	33	62	62	19	29	03	94	15	56	37	14	09
64	00	26	04	54	55	38	57	94	62	68	40	26	04	24	25	03	61
50	94	13	23	78	41	60	58	10	60	88	46	30	21	45	98	70	96
66	98	37	96	44	13	45	05	34	59	75	85	48	97	27	19	17	85
66	91	42	83	60	77	90	91	60	90	79	62	57	66	72	28	03	70
33	58	12	18	02	07	19	40	21	29	39	45	90	42	58	84	85	43
52	49	70	16	72	40	73	05	50	90	02	04	98	24	05	30	27	25
74	98	93	99	78	30	79	47	96	62	45	58	40	37	89	76	84	41
50	26	54	30	01	88	69	57	54	45	69	88	23	21	05	69	93	44

Appendix Table V (*contd.*)

49	46	61	89	33	79	96	84	28	3	19	35	28	73	39	59	56	34
19	64	13	44	78	39	73	88	62	03	36	00	25	96	86	76	67	90
64	17	47	67	87	59	81	40	72	61	14	09	28	28	55	86	23	38
18	43	97	37	68	97	56	56	57	95	01	88	11	89	48	07	42	07
65	58	60	87	51	09	96	61	15	53	66	81	66	88	44	75	37	01
79	90	31	00	91	14	85	65	31	75	43	15	45	93	64	78	34	11
07	23	00	15	59	05	16	09	94	42	20	40	63	76	65	67	34	11
90	98	14	24	01	51	95	46	30	32	33	19	00	14	19	28	40	51
53	82	62	02	21	82	34	13	41	03	12	85	65	30	00	97	56	30
98	17	26	15	04	50	76	25	20	33	54	84	39	31	23	33	59	64
08	91	12	44	82	40	30	62	45	50	64	54	65	17	89	25	59	44
37	21	46	77	84	87	67	39	85	54	97	37	33	41	11	74	90	50

APPENDIX TABLE VI

STUDENT'S *t* DISTRIBUTION

Table of critical value of *t* corresponding to specified two-tailed probabilities and degrees of feedom

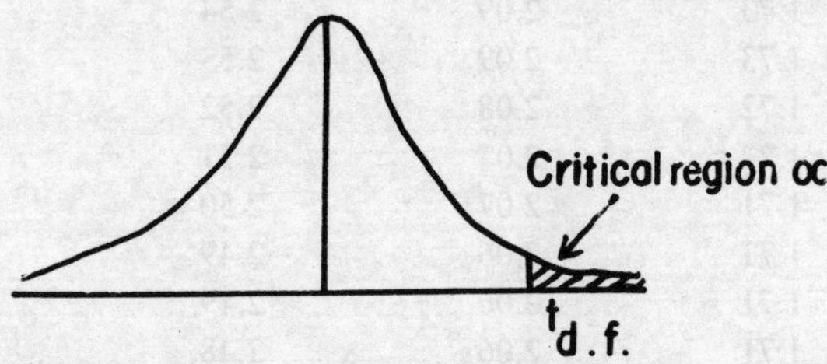

The computed values of *t* should exceed the following values of *t* in order to reject H_o. On the Figure here the probability column indicates the area in the right tail of *t* distribution. For a particular *d.f.* double the level of significance of probability and then have the critical value for a specified upper tail area α from the entries given in the table.

Degree of	Level of Significance				
Freedom (*d.f.*)	$p = 0.1$ $p' = 90\%$	$P = 0.05$ $p' = 95\%$	$p = 0.02$ $p' = 98\%$	$p = 0.01$ $p' = 99\%$	$p = 0.001$ $p' = 99.9\%$
1	6.31	12.71	31.92	63.66	636.62
2	2.92	4.30	6.97	9.93	31.60
3	2.35	3.18	4.54	5.84	12.94
4	2.13	2.78	3.75	4.60	8.61
5	2.02	2.57	3.37	4.03	6.86
6	1.94	2.45	3.14	3.71	5.96
7	1.90	2.37	3.00	3.50	5.41
8	1.86	2.31	2.90	3.36	5.04
9	1.83	2.26	2.82	3.25	4.78
10	1.81	2.26	2.82	3.36	4.59
11	1.80	2.20	2.72	3.11	4.44
12	1.78	2.18	2.68	3.06	4.32
13	1.77	2.16	2.68	3.06	4.22

Appendix Table VI (*Contd.*)

Degree of Freedom (*d.f.*)	Level of Significance				
	$p = 0.1$ $p' = 90\%$	$P = 0.05$ $p' = 95\%$	$p = 0.02$ $p' = 98\%$	$p = 0.01$ $p' = 99\%$	$p = 0.001$ $p' = 99.9\%$
14	1.76	2.15	2.62	3.01	4.14
15	1.75	2.13	2.60	2.95	4.07
16	1.75	2.12	2.58	2.95	4.02
17	1.74	2.11	2.57	2.90	3.97
18	1.73	2.10	2.55	2.88	3.92
19	1.73	2.09	2.54	2.86	3.88
20	1.73	2.09	2.53	2.85	3.85
21	1.72	2.08	2.52	2.83	3.82
22.	1.72	2.07	2.51	2.82	3.79
23	1.71	2.07	2.50	2.81	3.77
24	1.71	2.06	2.49	2.80	3.75
25	1.71	2.06	2.49	2.79	3.73
26	1.71	2.06	2.48	2.78	3.71
27	1.70	2.05	2.47	2.77	3.69
28	1.70	2.05	2.47	2.76	3.67
29.	1.70	2.05	2.46	2.76	3.66
30	1.70	2.04	2.46	2.75	3.65
40	1.68	2.02	2.42	2.70	3.55
60	1.67	2.00	2.39	2.66	3.46
120	1.66	1.98	2.36	2.62	3.37
α (infinity)	1.64	1.96	2.33	2.58	3.29

Note: p' is the two-tailed probability (expressed as a percentage) of a value being less extreme than t.

p is the two-tailed probability of a value being more extreme than t.

APPENDIX TABLE VII

THE *F* DISTRIBUTION TABLE

Following are the values of *F* distribution at 0.05 significance level. To find out the tabulated value of *F* (5,10), first we look for the column headed 5 at the top of the *F* table and then look down this column to read 3.33 at the row 10. If the calculated value of *F* is found to be higher than the tabulated value of *F* at a significance level, the differences within the samples are taken to be significant.

v_2 ↓ / → v_1 (Degrees of Freedom for Denominator)	Degrees of Freedom for Numerator												
	1	2	3	4	5	6	7	8	9	10	30	60	∞
1	161	200	216	225	230	234	237	239	242	244	248	252	254
2	18.5	19.0	19.2	19.3	19.3	19.3	19.4	19.4	19.4	19.4	19.4	19.5	19.5
3	10.1	9.55	9.28	9.12	9.01	8.94	8.89	8.85	8.79	8.74	8.66	8.57	8.53
4	7.71	6.94	6.59	6.39	6.26	6.16	6.09	6.04	5.96	5.91	5.80	5.69	5.63
5	6.61	5.79	5.41	5.19	5.05	4.95	4.88	4.82	4.74	4.68	4.56	4.43	4.37
6	5.99	5.14	4.76	4.53	4.39	4.82	4.21	4.15	4.06	4.00	3.87	3.74	3.67
7	5.99	4.74	4.35	4.12	3.97	3.87	3.79	3.73	3.64	3.57	3.44	3.30	3.23
8	5.32	4.46	4.07	3.84	3.69	3.58	3.50	3.44	3.35	3.28	3.15	3.01	2.93
9	5.12	4.26	3.36	3.63	3.48	3.37	3.29	3.23	3.14	3.07	2.94	2.79	2.71
10	4.96	4.10	3.71	3.48	3.33	3.22	3.14	3.07	2.98	2.91	2.77	2.62	2.54
11	4.84	3.98	3.59	3.36	3.20	3.09	3.01	2.95	2.85	2.79	2.65	2.49	2.40
12	4.75	3.89	3.49	3.26	3.11	3.00	2.91	2.85	2.75	2.69	2.54	2.38	2.30
13	4.67	3.81	3.41	3.18	3.03	2.92	2.83	2.77	2.67	2.60	2.46	2.30	2.21
14	4.60	3.74	3.34	3.11	2.96	2.85	2.76	2.70	2.60	2.53	2.39	2.22	2.13
15	4.54	3.68	3.29	3.06	2.90	2.79	2.71	2.64	2.54	2.48	2.33	2.16	2.07
16	4.49	3.63	3.24	3.01	2.85	2.74	2.66	2.59	2.49	2.42	2.28	2.11	1.01
17	4.45	3.59	3.20	2.96	2.81	2.70	2.67	2.55	2.45	2.38	2.23	2.06	1.96
18	4.41	3.55	3.26	2.93	2.77	2.66	2.58	2.51	2.41	2.34	2.19	2.02	1.92
19	4.38	3.52	3.13	2.90	2.74	2.63	2.54	2.48	2.38	2.31	2.11	1.98	1.88
20	4.35	3.49	3.10	2.87	2.71	2.60	2.51	2.45	2.35	2.28	2.12	1.95	1.84
21	4.32	3.47	3.07	3.84	2.68	2.57	2.49	2.42	2.32	2.25	2.10	1.92	1.81
22	4.30	3.44	3.05	2.82	2.66	2.55	2.46	2.40	2.30	2.23	2.07	1.89	1.78
23	4.28	3.42	3.03	2.80	2.64	2.53	2.44	2.37	2.27	2.20	2.05	1.86	1.76

Appendix Table VII (*Contd.*)

$v_2 \downarrow$ \ $\rightarrow v_1$	Degrees of Freedom for Numerator 1	2	3	4	5	6	7	8	9	10	30	60	∞
Degrees of Freedom for Denominator 24	4.26	3.40	3.01	2.78	2.62	2.51	2.42	2.36	2.25	2.18	2.03	1.84	1.73
25	4.24	3.39	2.99	2.76	2.60	2.49	2.40	2.34	2.24	2.16	2.01	1.82	1.71
30	4.17	3.32	2.92	2.69	2.53	2.42	2.33	2.27	2.16	2.09	1.93	1.74	1.62
40	4.08	3.23	2.84	2.61	2.45	2.34	2.25	2.16	2.12	2.00	1.84	1.64	1.51
60	4.00	3.15	2.76	2.53	2.37	2.25	2.18	2.10	2.04	1.92	1.75	1.53	1.39
120	3.92	3.07	2.68	2.45	2.29	2.18	2.10	2.04	1.96	1.83	1.66	1.43	1.25
∞	3.84	3.00	2.60	2.37	2.21	2.10	2.01	1.94	1.83	1.75	1.57	1.32	1.00

Values of v_1 and v_2 represent the degree of freedom associated with the larger (i.e. the numerator) and smaller (i.e. the denominator) estimates of variance respectively.

Following are the Values of *F* Distribution at 0.01 Significance Level

$\nu_2 \downarrow$ \ $\rightarrow \nu_1$	Degrees of Freedom for Numerator											
	1	2	3	4	5	6	7	8	10	12	20	60
1	4,952	5,000	5,403	5,625	5,764	5,860	5,930	5,982	6,056	6,106	6,209	6,313
2	98.5	99.0	99.2	99.2	99.3	99.3	99.4	99.4	99.4	99.4	99.4	99.5
3	34.1	30.8	29.5	28.7	28.2	27.9	27.7	27.5	27.2	27.1	26.7	26.3
4	21.2	18.0	16.7	16.0	15.5	15.2	15.0	14.8	14.7	14.4	14.0	13.7
5	16.3	13.3	12.1	11.4	11.0	10.7	10.5	10.3	10.1	9.9	9.5	9.2
6	13.7	10.9	9.8	9.1	8.7	8.5	8.3	8.1	7.9	7.7	7.4	7.0
7	12.3	9.6	8.5	7.9	7.5	7.2	7.0	6.8	6.8	6.5	6.2	5.8
8	11.3	8.6	7.6	7.0	6.6	6.4	6.2	6.0	5.8	5.7	5.4	5.0
9	10.6	8.0	7.0	6.4	6.1	5.8	5.6	5.5	5.4	5.1	4.8	4.5
10	10.0	7.6	6.5	6.0	5.6	5.4	5.2	5.0	4.8	4.7	4.4	4.1
11	9.6	7.2	6.2	5.7	5.3	5.1	4.9	4.7	4.5	4.4	4.1	3.8
12	9.3	6.9	5.9	5.4	5.1	4.8	4.6	4.4	4.3	4.2	3.9	3.5
13	9.1	6.7	5.7	5.2	4.9	4.6	4.4	4.3	4.2	4.1	3.7	3.3
14	8.9	6.5	5.6	5.0	4.7	4.5	4.3	4.1	3.9	3.8	3.5	3.2
15	8.7	6.4	5.4	4.9	4.6	4.3	4.1	4.0	3.9	3.7	3.5	3.0
16	8.5	6.2	5.3	4.8	4.4	4.2	4.0	3.9	3.7	3.6	3.2	2.9
17	8.4	6.1	5.2	4.7	4.3	4.1	3.9	3.8	3.7	3.6	3.2	2.8
18	8.3	6.0	5.1	4.6	4.3	4.0	3.8	3.7	3.5	3.4	3.1	2.7
19	8.2	5.9	5.0	4.5	4.2	3.9	3.8	3.6	3.5	3.3	3.0	2.7
20	8.1	5.8	4.9	4.4	4.1	3.8	3.7	3.6	3.4	3.2	2.9	2.6
21	8.0	5.8	4.9	4.4	4.0	3.8	3.6	3.5	3.4	3.2	2.9	2.5
22	7.9	5.7	4.8	4.3	3.9	3.7	3.6	3.4	3.3	3.1	2.8	2.5
23	7.9	5.7	4.8	4.3	3.9	3.7	3.5	3.4	3.2	3.1	2.8	2.5
24	7.8	5.6	4.7	4.2	3.9	3.7	3.5	3.4	3.2	3.0	2.7	2.4
25	7.7	5.6	4.7	4.2	3.8	3.6	3.5	3.3	3.1	2.9	2.7	2.3
30	7.6	5.4	4.5	4.0	3.7	3.5	3.3	3.2	3.0	2.8	2.6	2.2
40	7.3	5.2	4.3	3.8	3.5	3.3	3.1	3.0	2.9	2.7	2.4	2.0
60	7.1	5.0	4.1	3.6	3.3	3.1	2.9	2.8	2.7	2.5	2.2	1.8
120	6.9	4.8	3.9	3.5	3.2	2.9	2.8	2.7	2.6	2.3	2.0	1.7
∞	6.6	4.6	3.8	3.3	3.0	2.8	2.6	2.5	2.4	2.2	1.9	1.5

(Rows: Degrees of Freedom for Denominator)

APPENDIX TABLE VIII

CHI-SQUARE DISTRIBUTION

Table of Value of χ^2 with probability p of being exceeded in random sampling, the first column lists the number of degrees of freedom and the other columns give probabilities p.

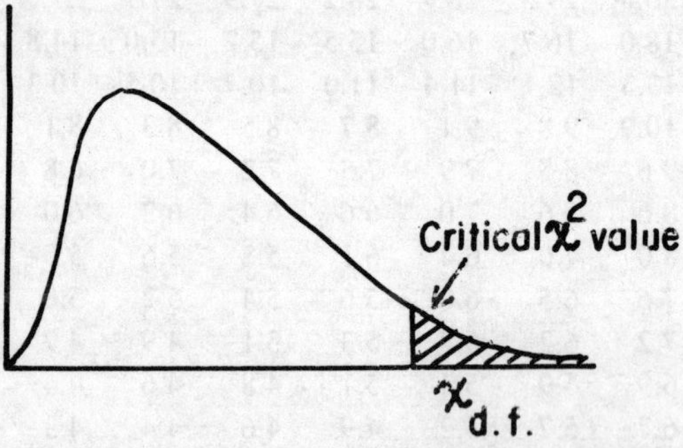

The χ^2 distribution is not symmetrical and all the values are positive. *Note* the smaller the *d.f.*, the more skewed is the distribution.

Degrees of Freedom	0.20	Probabilities (p) 0.10	0.05	0.02	0.01	0.001
1	1.64	2.71	3.84	5.41	6.63	10.83
2	3.32	4.61	5.99	7.82	9.21	13.81
3	4.64	6.25	7.81	9.84	11.67	16.27
4	5.90	7.73	9.49	11.67	13.28	18.47
5	7.29	9.24	11.07	13.39	15.03	20.52
6	9.80	12.02	14.07	16.3	16.81	22.46
7	9.80	11.03	13.36	15.51	18.17	24.32
8	11.03	13.36	15.51	18.17	20.09	26.12
9	12.24	14.68	16.92	19.68	21.67	27.88
10	13.44	15.99	18.31	21.16	23.21	29.59
11	14.63	17.28	19.63	22.62	24.72	31.26
12	15.81	18.55	21.03	24.05	26.22	32.91
13	16.98	19.81	23.36	25.47	27.69	34.53

Appendix Table VII (*Contd.*)

Degrees of Freedom	Probabilities (p) 0.20	0.10	0.05	0.02	0.01	0.001
14	18.15	21.06	23.68	26.87	29.14	36.12
15	19.31	22.31	25.00	28.26	30.58	37.70
16	20.46	23.54	26.30	29.63	32.00	39.25
17	21.62	24.77	27.59	30.99	33.41	40.79
18	22.76	25.99	28.87	32.35	34.81	42.31
19	23.09	27.20	30.14	33.69	36.19	43.82
20	25.04	28.41	31.41	35.02	37.57	45.31
21	26.17	29.62	32.67	36.34	38.93	46.80
22	27.30	30.81	33.92	37.66	40.29	48.27
23	28.43	32.01	35.17	38.97	41.64	49.73
24	29.55	33.20	36.42	40.27	42.98	51.18
25	30.68	34.38	37.65	41.57	44.31	52.62
26	31.80	35.56	38.88	41.14	45.64	54.05
27	32.91	36.74	40.11	44.14	46.96	55.48
28	35.03	37.92	41.11	45.42	48.28	56.89
29	35.14	38.09	42.56	46.69	49.59	58.30
30	36.25	40.26	43.77	47.96	56.89	59.70
40		51.81	55.76		63.69	73.40
50		63.17	67.50		76.15	86.66
60		74.40	79.08		88.38	99.61
70		85.53	90.53		100.4	112.3
80		96.58	101.9		112.3	124.8
90		107.6	113.1		124.1	137.2
100		118.5	124.3		135.8	149.4
120		140.2	146.6		159.0	173.6
150		172.6	179.6		193.2	209.3

Note : for *d.f.* greater than 30, the χ^2 distribution approximates the normal distribution that is,

$z = \sqrt{2\chi^2} - \sqrt{2(d.f.) - 1}$ given in Eq. (13.2)

APPENDIX TABLE IX

KOLMOGOROV-SMIRNOV TEST

The computed values of D must exceed the following critical values of D in order to reject at the following significant levels for two-tail tests only.

Sample Size	Significance Levels				
(n)	0.20	0.15	0.10	0.05	0.01
1	0.900	0.935	0.950	0.975	0.995
2	0.684	0.726	0.776	0.842	0.929
3	0.565	0.597	0.642	0.708	0.828
4	0.494	0.525	0.564	0.624	0.733
5	0.446	0.474	0.510	0.565	0.669
6	0.410	0.436	0.470	0.521	0.618
7	0.381	0.405	0.438	0.486	0.577
8	0.358	0.381	0.411	0.457	0.543
9	0.339	0.360	0.388	0.432	0.514
10	0.322	0.342	0.368	0.410	0.490
11	0.307	0.326	0.352	0.391	0.468
12	0.295	0.313	0.338	0.375	0.450
13	0.284	0.302	0.325	0.361	0.433
14	0.274	0.292	0.314	0.349	0.418
15	0.266	0.283	0.304	0.338	0.404
16	0.258	0.274	0.295	0.328	0.392
17	0.250	0.266	0.286	0.318	0.381
18	0.244	0.259	0.278	0.309	0.371
19	0.237	0.252	0.272	0.301	0.363
20	0.231	0.246	0.264	0.294	0.356
25	0.21	0.22	0.24	0.27	0.32
30	0.19	0.20	0.22	0.24	0.29
35	0.18	0.19	0.21	0.23	0.27
Over 35	$\frac{1.07}{\sqrt{n}}$	$\frac{1.14}{\sqrt{n}}$	$\frac{1.22}{\sqrt{n}}$	$\frac{1.36}{\sqrt{n}}$	$\frac{1.63}{\sqrt{n}}$

(For equal sizes in a two-sample test)

Over 35	$1.07\sqrt{\frac{n_1+n_2}{n_1 n_2}}$	$1.14\sqrt{\frac{n_1+n_2}{n_1 n_2}}$	$1.22\sqrt{\frac{n_1+n_2}{n_1 n_2}}$	$1.36\sqrt{\frac{n_1+n_2}{n_1 n_2}}$	$1.63\sqrt{\frac{n_1+n_2}{n_1 n_2}}$

(For unequal sizes in a two-sample test)

APPENDIX TABLE X

THE MANN-WHITNEY *U* TEST

For Tables A *to* F: The probability read off in the table must be less than the rejection level (α) if *U* is to be regarded as significant.

For Tables G *to* I: The calculated value of *U* must be less than the critical value at the specified rejection level if *U* is to be regarded as significant by rejecting H_0.

Note : All probability values in Tables A to F are two-tailed. For one-tailed valued, simply halve the probabilities.

Table A

$n_2 = 3$

↓ U \ n_1→	1	2	3
0	0.250	0.100	0.050
1	0.500	0.200	0.100

Table B

$n_2 = 4$

↓ U \ n_1→	1	2	3	4
0	0.200	0.067	0.028	0.014
1	0.400	0.133	0.057	0.029
2	0.600	0.267	0.114	0.057
3		0.400	0.200	0.100

Table C

$n_2 = 5$

↓ U \ n_1→	1	2	3	4	5
0	0.167	0.047	0.018	0.008	0.004
1	0.333	0.095	0.036	0.016	0.008
2	0.500	0.190	0.071	0.032	0.016
3	0.667	0.286	0.125	0.056	0.028
4		0.429	0.196	0.095	0.048
5		0.571	0.286	0.143	0.075

Table D
$n_2 = 6$

$n_1 \rightarrow$ / $\downarrow U$	1	2	3	4	5	6
0	0.143	0.036	0.012	0.005	0.002	0.001
1	0.286	0.071	0.024	0.010	0.004	0.002
2	0.428	0.143	0.048	0.019	0.009	0.004
3	0.571	0.214	0.083	0.033	0.015	0.008
4		0.321	0.131	0.057	0.026	0.013
5		0.429	0.190	0.086	0.041	0.021
6		0.571	0.274	0.129	0.063	0.032
7			0.357	0.176	0.089	0.047
8			0.452	0.238	0.123	0.066
9			0.548	0.305	0.165	0.090

Table E
$n_2 = 7$

$n_1 \rightarrow$ / $\downarrow U$	1	2	3	4	5	6	7
0	0.125	0.028	0.008	0.003	0.001	0.001	0.000
1	0.250	0.056	0.017	0.008	0.003	0.001	0.000
2	0.375	0.111	0.033	0.012	0.005	0.002	0.001
3	0.500	0.167	0.058	0.021	0.009	0.004	0.002
4	0.625	0.250	0.092	0.036	0.015	0.007	0.003
5		0.333	0.133	0.055	0.024	0.011	0.006
6		0.444	0.192	0.082	0.037	0.017	0.009
7		0.556	0.258	0.115	0.053	0.026	0.013
8			0.333	0.158	0.074	0.037	0.019
9			0.444	0.206	0.101	0.051	0.027
10			0.500	0.264	0.134	0.069	0.036
11			0.583	0.324	0.172	0.090	0.049
12				0.394	0.216	0.117	0.064
13				0.464	0.265	0.147	0.082

Table F

$n_2 = 8$

↓ U \ n_1 →	1	2	3	4	5	6	7	8
0	0.111	0.022	0.006	0.002	0.001	0.000	0.000	0.000
1	0.222	0.044	0.012	0.004	0.002	0.001	0.000	0.000
2	0.333	0.089	0.024	0.008	0.003	0.001	0.001	0.000
3	0.444	0.133	0.042	0.014	0.005	0.002	0.001	0.000
4	0.556	0.020	0.067	0.024	0.009	0.004	0.002	0.001
5		0.267	0.097	0.036	0.015	0.006	0.003	0.001
6		0.356	0.139	0.055	0.023	0.010	0.005	0.001
7		0.444	0.188	0.077	0.033	0.015	0.007	0.003
8		0.556	0.248	0.107	0.047	0.021	0.010	0.005
9			0.315	0.141	0.064	0.030	0.014	0.007
10			0.387	0.184	0.085	0.041	0.020	0.010
11			0.461	0.230	0.111	0.054	0.027	0.014
12			0.539	0.285	0.442	0.071	0.036	0.010
13				0.341	0.177	0.091	0.047	0.025
14				0.404	0.217	0.114	0.056	0.032
15				0.467	0.262	0.441	0.076	0.041
16				0.533	0.311	0.172	0.095	0.052
17					0.362	0.207	0.166	0.065
18					0.416	0.245	0.140	0.080
19					0.472	0.286	0.168	0.097
20								
21								
22						0.426		

Table G

Critical values of U for a two-tailed test at $p = 0.002$ and a one-tailed test at $p = 0.001$

$n_2 \rightarrow$ / $\downarrow n_1$	9	10	11	12	13	14	15	16	17	18	19	20
1												
2												
3									0	0	0	0
4		0	0	0	1	1	1	2	2	2	3	3
5	1	1	2	2	3	3	4	5	5	6	7	7
6	2	3	4	4	5	6	7	8	9	10	11	12
7	3	5	6	7	8	9	10	11	13	14	15	16
8	5	6	8	9	11	14	15	17	18	19	20	21
9	7	8	10	12	14	15	17	19	21	23	25	26
10	8	10	12	14	17	19	21	23	25	27	29	32
11	10	12	15	17	19	22	24	27	29	32	34	37
12	12	14	17	20	23	25	28	31	34	37	40	42
13	14	17	20	23	26	29	32	35	38	42	45	48
14	15	19	22	25	29	32	36	39	43	46	50	54
15	15	21	24	28	32	36	40	43	47	51	55	59
16	10	23	27	31	35	39	43	48	52	56	60	65
17	21	25	29	34	38	43	47	52	57	61	66	70
18	23	27	32	37	42	46	51	56	51	66	71	76
19	25	29	34	40	45	50	55	60	66	71	77	82
20	26	32	37	42	48	54	59	65	70	73	82	88

Table H

Critical values of U for a two-tailed test at $p = 0.05$ and a one-tailed test at $p = 0.025$

$n_2 \rightarrow$ / $\downarrow n$	9	10	11	12	13	14	15	16	17	18	19	20
1												
2	0	0	0	1	1	1	1	1	2	2	2	2
3	2	3	3	4	4	5	5	6	6	7	7	8
4	4	5	6	7	8	9	10	11	11	12	13	13
5	7	8	9	11	12	13	14	15	17	18	19	20
6	10	11	13	14	16	17	19	21	22	24	25	27
7	12	14	15	18	20	22	24	26	28	30	32	34
9	17	20	23	26	28	31	34	37	39	42	45	48
10	20	23	26	30	33	37	40	42	45	48	52	55
11	23	26	30	33	37	40	44	47	51	55	58	62
12	26	29	33	37	41	45	49	53	57	61	65	69
13	28	33	37	41	45	49	53	57	61	67	74	76
14	31	36	40	45	50	54	59	64	67	74	78	83
15	34	39	44	49	54	59	64	70	75	80	85	90
16	37	42	47	53	59	64	70	75	81	86	92	98
17	39	45	51	57	63	67	74	80	86	93	99	105
18	42	48	55	61	67	74	80	86	93	99	106	112
19	45	52	58	65	72	78	85	92	99	106	113	119
20	48	55	62	69	76	83	90	98	105	112	119	127

Table I

Critical values of U for a two-tailed test at $p = 0.10$ and a one-tailed test at $p = 0.05$

$\downarrow n_1$ \ $n_2 \rightarrow$		10	11	12	13	14	15	16	17	18	19	20
1											0	0
2	1	1	1	2	2	2	3	3	3	4	4	4
3	3	4	5	5	6	7	7	8	9	9	10	11
4	6	7	8	9	10	11	12	14	15	16	17	18
5	9	11	12	13	15	16	18	19	20	22	23	25
6	12	14	16	17	19	21	23	25	26	28	30	32
7	15	17	19	21	24	26	28	30	33	35	37	39
8	18	20	23	26	28	31	33	36	39	41	44	47
9	21	24	27	30	33	36	39	42	45	48	51	54
10	24	27	31	34	37	41	44	48	51	55	58	62
11	27	31	34	38	42	46	50	54	57	61	65	69
12	31	34	38	42	46	50	55	57	61	65	69	80
14	36	41	46	51	56	61	66	71	77	82	87	92
15	39	44	50	55	61	66	72	77	83	88	94	100
16	42	48	54	60	65	71	77	83	89	95	101	107
18	48	55	61	68	75	82	88	95	102	109	116	123
19	51	58	65	72	80	87	94	101	109	116	123	130
20	54	62	69	77	84	92	100	107	115	123	130	138

APPENDIX TABLE XI

WILCOXON-RANKED PAIR TEST

The table of critical values of *T* in the Wilcoxon Ranked-Pair Significance test. *T* must be less than critical value to be significant at specified level.

N=No. of pairs of samples with non-zero differences	Level of Significance for one-tailed test		
	0.025	0.01	0.005
	Level of Sifnificance for two-tailed test		
	0.05	0.02	0.01
6	0	–	–
7	2	0	–
8	4	2	0
10	8	5	3
11	11	7	5
12	14	10	7
13	17	13	13
14	21	16	13
15	25	20	16
16	30	24	20
17	35	28	23
18	40	33	28
19	46	38	32
20	52	43	38
21	59	49	43
22	66	56	49
23	73	62	55
24	81	69	61
25	89	77	68
26	98	85	76
28	117	102	100
30	137	130	118
35	195	186	171
40	264	238	221
50	424	398	373
100	1955	1850	1779

Note : the above test cannot be used with samples of less than 6.

APPENDIX TABLE XII

THE KRUSKAL-WALLIS ANALYSIS OF VARIANCE

Probabilities (p) associated with calculated values of H for three samples from $n_1 = 2, n_2 = 1, n_3 = 1$ to each equal to 5. The larger the value of H the smaller the probability that the null hypothesis is correct.

Sample Size			H	p	Sample Size			H	p
n_1	n_2	n_3			n_1	n_2	n_3		
2	1	1	2.7000	0.500	4	1	1	3.5714	0.200
2	2	1	3.6000	0.200	4	2	1	4.8214	0.057
2	2	2	4.5714	0.067				4.0179	0.114
			3.7143	0.200					
					4	2	2	6.0000	0.114
3	1	1	3.2000	0.300				5.3333	0.033
								5.1250	0.052
3	2	1	4.2857	0.100				4.4583	0.100
								4.1667	0.105
3	2	2	5.3572	0.029					
			4.7143	0.048	4	3	1	5.8333	0.021
			4.5000	0.067				5.0000	0.057
								4.0556	0.093
								3.8889	0.129
3	3	1	5.1429	0.043					
			4.5714	0.100					
			4.0000	0.129	4	3	2	6.4444	0.008
								6.3000	0.011
3	3	2	6.2500	0.011				5.4444	0.046
			5.3611	0.032				5.4000	0.051
			4.5556	0.100				4.5111	0.098
			4.2500	0.121				4.4444	0.102

Appendix Table XII (*Contd.*)

Sample Size			H	p	Sample Size			H	p
n_1	n_2	n_3			n_1	n_2	n_3		
3	3	3	7.2000	0.004	4	3	3	6.7455	0.010
			6.4889	0.011				6.7091	0.013
			5.6889	0.029				5.7909	0.046
			5.6000	0.050				5.7273	0.050
			5.0667	0.086				4.7091	0.092
			4.6222	0.100				4.7000	0.101
4	4	1	6.6667	0.010	5	3	1	6.4000	0.012
			6.1667	0.022				4.9600	0.048
			4.9667	0.048				4.8711	0.052
			4.8667	0.054				4.1178	0.095
			4.1667	0.082				3.8400	0.123
			4.0667	0.102					
4	4	2	7.0364	0.006	5	3	2	6.9091	0.009
			6.8727	0.011				6.8218	0.010
			5.4545	0.046				5.2509	0.049
			5.2364	0.052				4.6509	0.091
			4.5545	0.103					
4	4	3	7.1439	0.010	5	3	3	7.0788	0.009
			7.1364	0.011				6.9818	0.011
			5.5985	0.049				5.6485	0.049
			5.5758	0.051				5.5152	0.051
			4.5455	0.099				4.5333	0.097
			4.4773	0.102				4.4121	0.109
4	4	4	7.6538	0.008	5	4	1	6.9545	0.008
			7.5385	0.011				6.8400	0.011
			5.6923	0.049				4.9855	0.044
			5.6538	0.054				4.8600	0.056

Appendix Table XII (*Contd.*)

Sample Size n_1	n_2	n_3	H	p	Sample Size n_1	n_2	n_3	H	p
			4.6539	0.097				3.9873	0.098
			4.5001	0.104				3.9600	0.102
5	1	1	3.8571	0.143	5	4	2	7.2045	0.009
5	2	1	5.2500	0.036				7.1182	0.010
			5.0000	0.048				5.2727	0.049
			4.4500	0.071				5.2682	0.050
			4.2000	0.095				4.5409	0.098
			4.0500	0.119				4.5182	0.101
5	2	2	6.5333	0.008	5	4	3	7.4449	0.010
			6.1333	0.013				7.3049	0.011
			5.1600	0.034				5.6564	0.049
			5.0400	0.056				5.6308	0.050
			4.3733	0.090				4.5487	0.099
			4.2933	0.122				4.5231	0.103
5	4	4	7.7440	0.011	5	5	3	7.5780	0.010
			5.6571	0.050				5.6264	0.051
			4.6187	0.100				4.5451	0.100
			4.5527	0.102				4.5363	0.102
5	5	1	7.3091	0.009	5	5	4	7.8229	0.010
			6.8364	0.011				7.7914	0.013
			5.1273	0.046				5.6657	0.049
			4.9091	0.053				5.6429	0.050
			4.1091	0.086				4.5229	0.099
			4.0364	0.105				4.5200	0.101
5	5	2	7.3385	0.010	5	5	5	8.0000	0.009
			7.2692	0.013				7.9800	0.010
			5.3385	0.047				5.7800	0.049
			5.2462	0.051				5.6600	0.051
			4.6231	0.097				4.5000	0.102

APPENDIX TABLE XIII

SPEARMAN'S RANK CORRELATION COEFFICIENT

The following table gives the critical values of ρ for $n = 4$ to $n = 10$ at the 0.05 and 0.01 levels of significance. The larger the value of ρ the more significant is the result.

For number of pairs greater than $n = 30$ the critical values of ρ alters only slightly.

	Significance Level (two-tailed test)			
n	0.10	0.05	0.02	0.01
4	1.000			
5	0.900	1.000	1.000	
6	0.829	0.886	0.943	1.000
7	0.714	0.786	0.893	0.929
8	0.643	0.738	0.833	0.881
9	0.600	0.683	0.783	0.833
10	0.564	0.591	0.746	0.794
12	0.506	0.544	0.712	0.777
14	0.456	0.506	0.645	0.715
16	0.425	0.475	0.601	0.665
18	0.399	0.450	0.564	0.625
20	0.377	0.428	0.534	0.591
22	0.359	0.409	0.508	0.562
24	0.343	0.392	0.485	0.537
28	0.317	0.364	0.448	0.496
30	0.306	0.352	0.432	0.478
35	0.283	0.335	0.394	0.433
40	0.264	0.313	0.368	0.405
45	0.248	0.294	0.347	0.382
50	0.235	0.279	0.329	0.363
100	0.165	0.197	0.234	0.259

APPENDIX TABLE XIV

KENDALL'S RANK CORRELATION COEFFIEIENT

Probabilities associated with calculated values of *S* for $n = 4$ to $n = 10$. The larger the value of *S* for any given value of *n* the smaller the probability that the association of values tested is due to chance or random variations.

	Values of *n*					Values of *n*		
S	4	5	8	9	*S*	6	7	10
0	0.625	0.592	0.548	0.540	1	0.500	0.500	0.500
2	0.375	0.408	0.452	0.460	3	0.360	0.386	0.431
4	0.167	0.242	0.360	0.381	5	0.235	0.281	0.364
6	0.042	0.117	0.274	0.306	7	0.136	0.191	0.300
8		0.042	0.199	0.238	9	0.068	0.119	0.242
10		0.0083	0.138	0.179	11	0.028	0.068	0.190
12			0.089	0.130	13	0.0083	0.035	0.146
13			0.054	0.090	15	0.0014	0.015	0.108
16			0.031	0.060	17		0.0054	0.078
18			0.016	0.038	19		0.0014	0.054
20			0.0071	0.022	21		0.00020	0.036
22			0.0071	0.022	23			0.023
24			0.00087	0.0063	25			0.014
26			0.00019	0.0029	27			0.0083
28			0.000025	0.0012	29			0.0046
					31			0.0023
					33			0.0011

APPENDIX TABLE XV

DURBIN-WATSON *D* TEST

In the following table of Durbin Watson d_l and d_u values are given by different sample sizes and for regression models with different numbers of independent variables for significance levels of 0.05 and 0.01 (The latter are given in brackets)

Note. p – 1 is the number of regression coefficiently estimated excluding the constant.

n	$p-1=1$		$p-1=2$		$p-1=3$		$p-1=4$		$p-1=5$	
	d_l	d_u	d_l	d_u	d_l	d_u	d_l	d_u	d_l	d_u
6	0.61	1.40								
	(0.39)	(1.14)								
7	0.70	1.36	0.47	1.90						
	(0.44)	(1.04)	(0.29)	(1.68)						
8	0.76	1.33	0.56	1.78	0.37	2.29				
	(0.50)	(1.00)	(0.35)	(1.49)	(0.23)	(2.10)				
9	0.82	1.32	1.63	1.70	0.46	2.13	0.30	2.59		
	(.55)	(.99)	(0.41)	(1.39)	(0.28)	(1.88)	(0.18)	(2.43)		
10	0.88	1.32	0.70	1.64	0.53	2.01	0.38	2.41	0.24	2.82
	(0.60)	(1.00)	(0.47)	(1.33)	(0.34)	(1.73)	(0.23)	(2.19)	(0.15)	(2.69)
11	0.93	1.32	0.66	1.60	0.59	1.93	0.44	2.28	0.32	2.65
	(0.65)	(1.01)	(0.52)	(1.30)	(0.40)	(1.64)	(0.29)	(2.03)	(0.19)	(2.45)
12	0.97	1.33	0.81	1.58	0.66	1.86	0.51	2.18	0.38	2.51
	(0.69)	(1.02)	(0.57)	(1.27)	(0.45)	(1.58)	(0.34)	(1.91)	(0.24)	(2.28)
13	1.01	1.34	0.86	1.56	0.72	1.82	0.57	2.09	0.45	2.39
	(0.74)	(1.04)	(0.62)	(1.26)	(0.50)	(1.53)	(0.39)	(1.83)	(0.29)	(2.15)
14	1.05	1.35	0.91	1.55	0.77	1.78	0.63	2.03	0.51	2.30
	(0.78)	(1.05)	(0.66)	(1.25)	(0.55)	(1.49)	(0.44)	(1.76)	(0.34)	(2.05)
15	1.08	1.36	0.95	1.54	0.81	1.75	0.69	1.98	0.56	2.22
	(0.81)	(1.07)	(0.70)	(1.25)	(0.59)	(1.41)	(0.49)	(1.70)	(0.39)	(1.96)
16	1.10	1.37	0.98	1.59	0.86	1.73	0.74	1.93	0.62	2.15
	(0.84)	(1.09)	(0.74)	(1.25)	(0.63)	(1.44)	(0.53)	(1.66)	(0.44)	(1.90)

n	p−1=1 d_l	p−1=1 d_u	p−1=2 d_l	p−1=2 d_u	p−1=3 d_l	p−1=3 d_u	p−1=4 d_l	p−1=4 d_u	p−1=5 d_l	p−1=5 d_u
17	1.13	1.38	1.02	1.54	0.90	1.71	0.78	1.90	0.67	2.10
	(0.87)	(1.10)	(0.77)	(1.25)	(0.67)	(1.43)	(0.57)	(1.63)	(0.48)	(1.85)
18	1.16	1.39	1.5	1.53	0.39	1.69	0.82	1.85	0.71	2.06
	(0.90)	(1.12)	(0.80)	(1.26)	(0.71)	(1.42)	(0.61)	(1.58)	(0.52)	(1.80)
19	1.18	1.40	1.08	1.53	0.97	1.68	0.86	1.83	0.75	2.02
	(0.93)	(1.13)	(0.83)	(1.26)	(074)	(1.41)	(0.65)	(1.57)	(0.56)	(1.77)
20	1.20	1.41	1.10	1.54	1.00	1.68	0.90	1.81	0.79	1.99
	(0.95)	(1.15)	(0.89)	(1.27)	(0.77)	(1.41)	(0.69)	(1.57)	(0.60)	(1.74)
25	1.29	1.45	1.21	1.55	1.12	1.66	1.04	1.77	0.95	1.89
	(1.05)	(1.21)	(0.98)	(1.30)	(0.90)	(1.41)	(0.83)	(1.52)	(0.75)	(1.75)
30	1.35	1.49	1.28	1.57	1.21	1.65	1.14	1.74	1.07	1.83
	(1.13)	(1.26)	(1.07)	(1.34)	(1.01)	(1.42)	).94)	(1.51)	(0.88)	(1.61)
35	1.40	1.52	1.34	1.58	1.28	1.65	1.22	1.73	1.16	1.80
	(1.19)	(1.31)	(1.14)	(1.37)	(1.08)	(1.44)	(1.03)	(1.51)	(0.97)	(1.59)
40	1.44	1.54	1.39	1.60	1.34	1.66	1.29	1.72	1.23	1.78
	(1.25)	(1.34)	(1.20)	(1.40)	(1.15)	(1.46)	(1.10)	(1.52)	(1.05)	(1.58)
45	1.48	1.57	1.43	1.62	1.38	1.67	1.34	1.72	1.29	1.78
	((1.29)	(1.38)	(1.24)	(1.42)	(1.20)	(1.48)	(1.16)	(1.53)	(1.11)	(1.58)
50	1.50	1.59	1.46	1.63	1.42	1.67	1.38	1.72	1.34	1.77
	(1.32)	(1.40)	(1.28)	(1.45)	(1.24)	(1.49)	(1.20)	(1.54)	(1.16)	(1.59)
55	1.53	1.60	1.49	1.64	1.45	1.68	1.41	1.72	1.38	1.77
	(1.36)	(1.43)	(1.32)	(1.47)	(1.28)	(1.51)	(1.25)	(1.55)	(1.21)	(1.59)
60	1.55	1.62	1.51	1.65	1.48	1.69	1.44	1.73	1.41	1.77
	(1.38)	(1.45)	(1.35)	(1.48)	(1.32)	(1.52)	(1.28)	(1.56)	(1.25)	(1.60)
65	1.57	1.63	1.54	1.66	1.50	1.70	1.47	1.73	1.49	1.77
	(1.41)	(1.47)	(1.38)	(1.50)	(1.35)	(1.53)	(1.31)	(1.57)	(1.28)	(1.61)
70	1.58	1.64	1.55	1.67	1.52	1.70	1.49	1.74	1.46	1.77
	(1.43)	(1.49)	(1.40)	(1.52)	(1.37)	(1.55)	(1.34)	(1.59)	(1.31)	(1.61)

n	$p-1=1$		$p-1=2$		$p-1=3$		$p-1=4$		$p-1=5$	
	d_l	d_u	d_l	d_u	d_l	d_u	d_l	d_u	d_l	d_u
75	1.60	1.65	1.57	1.68	1.54	1.71	1.51	1.74	1.49	1.77
	(1.45)	(1.50)	(1.42)	(1.53)	(1.39)	(1.56)	(1.37)	(1.60)	(1.34)	(1.62)
85	1.62	1.67	1.60	1.70	1.57	1.72	1.55	1.75	1.52	1.77
	(1.48)	(1.53)	(1.46)	(1.55)	(1.43)	(1.58)	(1.41)	(1.61)	(1.39)	(1.63)
90	1.63	1.68	1.61	1.70	1.59	1.73	1.57	1.75	1.54	1.78
	(1.50)	(1.54)	(1.47)	(1.56)	(1.45)	(1.59)	(1.43)	(1.61)	(1.41)	(1.64)
95	1.64	1.69	1.62	1.71	1.60	1.73	1.58	1.75	1.56	1.78
	(1.51)	(1.55)	(1.49)	(1.57)	(1.47)	(1.60)	(1.45)	(1.62)	(1.42)	(1.64)
100	1.65	1.69	1.63	1.72	1.61	1.74	1.59	1.76	1.57	1.78
	(1.52)	(1.56)	(1.50)	(1.58)	(1.48)	(1.60)	(1.46)	(1.63)	(1.44)	(1.65)

Note: *p*-1 is the number of regression coefficients estimated, excluding the constant. Thus in a linear trend line fitting to the time series data of aggregate variable's γ_t (exhibiting a steadily increasing or decreasing pattern), the equation involved is: $\gamma_t = \alpha + \beta_1 t \pm \varepsilon$, in which only β_1 is estimated, hence $p-1 = 1$. For a quadration pattern the trend equation is $\gamma_t = \alpha + \beta_1 t + \beta_1 t^2 \pm \varepsilon$, $p-1$ is 2. Similarly, following is given the other trend line equations for the relation of dependent variable over time:

Cubic pattern	:	$\gamma_t = \alpha + \beta_1 t + \beta_2 t^2 + \beta_3 t^3 \pm \varepsilon$
Linear-log	:	$\gamma_t = \alpha + \beta_1 \ln(t) \pm \varepsilon$
Log-linear	:	$\ln(\gamma_t) = \alpha + \beta_1 t \pm \varepsilon$
Double-log	:	$\ln(\gamma_t) = \alpha + \beta_1 \ln(t) \pm \varepsilon$
Reciprocal	:	$\gamma_t = \alpha + \beta_1 (1/t) \pm \varepsilon$
Logistic	:	$\ln\left[\dfrac{\Upsilon_t}{1-\Upsilon_t}\right] = \alpha + \beta t \pm \varepsilon$ for forecasted γ_t to be $0 < \gamma_t < 1$.

LIST OF MATHEMATICAL AND STATISTICAL SYMBOLS

Types of Number : This book deals with two types of numbers: INTEGER and REAL NUMBER. An INTEGER is simply a whole number, with no fraction or decimal point. Integers can be either positive, negative or zero, e.g., 47, –77, 0. A REAL NUMBER is any number which has a fractional or decinal part with or without a negative sign, e.g. 10.905, 0.302, –450.92, $7\frac{5}{7}$.

SUBSCRIPT (or, SUFFIX) : When there are repeated measurements/observations of the same quantity/phenomena, suffixces are used, e.g. $x_1, x_2, x_3, ...x_i, ...x_n$ where there are n measurements of type x and x_i is the general form of x.

SUMMATION : Often we require the sum or aggregate of a number of quantities in the same series, e.g. $x_1 + x_2 + x_3 ..... + x_n$ where n denotes the quantities of the type x added together.

Σ : the upper case Greek letter (sigma), meaning "have the sum of"

$\sum_i^n$: the summation over all items from $i = 1$ to $i = n$, the same as Σ.

The full notation for the above sum would be, $x_1 + x_2 + x_3 + ... + x_n = \sum_{i=k}^{i=n} x_i$ (or, simply Σx). This is known as the **Ist Rule** for Summation over a variable. x_i it is taken as any value in a set of x-scores. $\sum_{i=1}^{n} x_i$ is an instruction in the observations on the variable x, beginning with the ith observation equal to 1 and ending with the nth observation.

2nd Rule for Summation of a constant: If p is a constant and we are concerned with a set of scores that remain constant, then $\Sigma p = np$; similarly, $\Sigma px = p\Sigma x$, also $\Sigma\left(\frac{x_i}{c}\right) = \frac{\Sigma x_i}{c} = \frac{1}{c}(\Sigma x_i)$.

3rd Rule of Summation over the product or quotient of two or more variables. For example, Σxy = Product of all the x and y pairs and then sum all xy products. Similarly

$\cdot\ \Sigma\left(\frac{x_i}{y_i}\right) \neq \frac{\Sigma x_i}{\Sigma y_i}$. Again $\Sigma xy \neq \Sigma x\, \Sigma y$.

4th Rule of Summation of a variable with a constant multiplier or divisor, e.g. if k is a constant, then $\Sigma(x + k) = \Sigma x + nk$; similarly when a constant p is subtracted from each variable value x_i is $\Sigma(x - p) = \Sigma x - np$

5th Rule of Summation of powers and roots of a variable, e.g. squaring of the different values of x and then adding is Σx^2. If the values are added and then squared it is $(\Sigma x)^2$.

Following summation rules are useful

(1) $\Sigma(x + y) = \Sigma x + \Sigma y$ [since $\Sigma(x + y) = (x_1 + y_1) + (x_2 + y_2) + ... + (x_n + y_n) = (x_1 + x_2 + ... + x_n) + (y_1 + y_2 + ...+y_n) = \Sigma x + \Sigma y$

(2) $\Sigma\,(x - y) = \Sigma x - \Sigma y$

(3) $\Sigma xy \neq \Sigma x\, \Sigma y$

(4) $\Sigma cx_i = c\Sigma x_i$ where c is a constant, similarly $\Sigma c = nc$

MULTIPLICATION : It is denoted by a cross × or a stop . or by quantities placed together. Thus a. Log$b$ equals to $a \times$ Log$b$, quantities like pq mean $p \times q$

INEQUALITIES : $x > y$ means x is greater than y; $x \geq y$ means x is greater than or equal to y.

Again $x < y$ means x is less than y. $x \leqslant y$ (written also as $x \leq y$) means x is less than or equal to x. In $4x \geq 4x - 8$ it is an inequality true all the time for all the real values of x.

In the inequality sign, the open end of the sign is always placed towards the larger item, so $4 > x > 8$ means that x has a value greater than 4 but less than 8 and $4 \leq x \leq 8$ means that x is greater than or equal to but less than or equal to 8.

PROPORTIONS AND PERCENTAGES: Both are for standardizing the data: PERCENTAGE written as % with respect to the base and PROPORTIONS are with respect to the base 1.00, e.g. percentages, $p\,(\%) = \frac{f}{n} \times 100$ and proprotions, $(p) = \frac{f}{n}$ where f is the No. of cases in any category and n is the total number

RATIO AND RATE: These are the two additional ways to standardise the distribution of a variable. Example for ratio $= \frac{f_1}{f_2}$ and rate is birth rate $= \frac{\text{No. of births}}{\text{Total population}} \times 1000$ per year

LOGARITHMS (or, in short 'Log') : Logarithm of a number consists of an integral part (i.e. the part before decinal) called the characteristic and a decimal part called the 'mantissa'. The logaritem tables give the mantissa of a logarithm. The characteristic is to be supplied in each case and it is always one less than the number of digits in the actual number, e.g.

6254 is Log 3.7959 (characteristic is 3)

62.54 is Log 1.7959 (characteristic is 1)

0.6254 is Log $\bar{1}$. 7959 (characteristic is minus 1)

0.06254 is Log $\bar{2}$.7959 (characteristic is minus 2)

$\bar{1}$ and $\bar{2}$ one pronounced as 'bar one' and 'bar two' respectively. Thus, logarithms are transformation of number which are used to work out the multiplication, divisions and exponents. Following is an example of the use of logarithms to work out an exponent. In general terms therefore :

If $x^m = a$ then $\log_x a = m$

$x^n = b$ then $\log_x b = n$

For example, if $y = 2.32$ find the value of $y^{4.25}$. Let the value be assumed as x, then $x = 2.32^{4.25}$. Now taking logarithm on both sides, we have Log x = 4.25 Log 2.32 = 4.25 × 0.36595= Log 1.55332

By antilog, the exponential product will be 35.75394, thus, logarithm of a number x is the exponent or the power to which a number constant, y here in the equation $y^{4.25} = x$ must be raised in order to produce the natural number or antilog x. Any positive number can be used as base; hence base 10 is the common log. Common log thus a number is the power to which 10 must be raised, so that $\log_{10} 100 = 2.0$ similarly $0.1 = 10^{-1}$ so that $\log_{10} 0.1 = -1.0$

Natural (or, Naperian) logarithm, are based on the number 2.71828 which is designated as e. For example from Poisson distribution the value of e^{-z} where $z = 1.40$ can be found by natural logarithm as follows : Log e^{-z} (or, $\ln^{-z}$) = log $2.71828^{-1.40}$.

It is, however, easy to convert from one 'system' of logarithm to the other as given below : For any number x,

Log $(x) = \ln(x) \times 0.4343$

or

$\ln(x) = \log(x) \times 2.303$

Logarithms are used in : (1) multiplication, (2) division, (3) raising a number to a certain power and in (4) extracting square root of a given number.

FACTORIAL : Factorial is the product of the numbers down to 1. It is the number of the ways that all n objects can be arranged in order, e.g. $n! = n(n-1)(n-2)\ldots 3, 2, 1$; $4! = 4 \times 3 \times 2 \times 1 = 24$.

PERMUTATION : If we have to pick up two different letters out of the word, 'man' then we can have the following permutations – ma, an, mn, na, nm, and am. These ordered arrangements (of a set of letters in this example) are known as permutation. Thus in this case there are 3 words in which if we like to have two words, then the number of ways of permuting them is given by a general formula of permutation which for permutting r objects out of n number of different objects is

$$n_{p_r} = p(n,r) = \frac{n!}{(n-r)!}$$

For example, for $n = 3$, $r = 2$ it is $3_{p_2} = p(3,2) = \frac{3!}{(3-2)!} = 6$

COMBINATION : In many situations, we are not interested in the order of the outcomes (as permutations) but only in the number of ways that r objects can be related out of n objects irrespective of the order. This rule is called the rule of COMBINATIONS. For example, the number of combinations of n things taken n at a time is obviously 1, because there is only one way of picking all n objects if the order of arrangement is ignored. while in permutations the number of arrangements of objects will be $n!$.

Thus by a general formula, the number of ways n items can be chosen r at a time is

$$n_{c_r} = \frac{n_{p_r}}{r!} = \frac{n!}{(n-r)!r!}$$

For $n = 3$ and $r = 2$ we have $3_{c_2} = \frac{3!}{(3-2)!2} = 3$

Miscellaneous Mathematical Symbols

$=$: equals

$\neq$: not equal

$\rightarrow$: arrow indicating quantity/value becomes or tends to

∞ : infinity, e.g. $a \rightarrow \infty$, i.e. value of "a" approaches infinity

α : proportional to (or, varies as)

$\pm$: plus or minus, as used to give range of values on either side of a given number

$\sim$: similar to

$\parallel$: parallel to

$\approx$: approximate to

$\doteqdot, \approx, \cong$: a sign of approximation

$\perp$: perpendicular to

$|\times|$: This sign $|\ |$ known as modulus takes the absolute values of the quantity inside the sign and the negative sign is ignored e.g. $2 = 2, |-2|$ but $\neq (-2)$

π : a mathematical contant = 3.14159

e : base of natural logarithms = 2.718281

$\sqrt{\ }$: square roots, e.g. $\sqrt{16} = 4$

$\sqrt[3]{\ }$: cubic roots, e.g. $\sqrt[3]{64} = 4$

$\sqrt[n]{\ }$: nth root, e.g., nth root of $\sqrt[4]{16} = 2$

BASIC DATA : CONVERSION FACTORS

Metric System : The metre is the unit of length, the smallest unit in the linear measure for metre generally is millimetre, being 10^{-3} metre. Another very small length unit is the Angstrom (A), which equals 0.0001 micron, the micron being 10^{-6} metre. Following is the linear measure of metric system

Linear Measure Multiple

Picometre = 10^{-12} metre
Angstrom, A = 10^{-10} metre
Nanometre, n = 10^{-9} metre
Micrometre, μ = 10^{-6} metre
(or, Micron)
Millimetre, mm = 10^{-3} metre
Centimetre, cm = 10^{-2} metre
Decimetre, dm = 10^{-1} metre = 10 cm
Metre, m = 10^{0} = 10 dm
Dekametre = 10^{1}m = 10m
Hectometre = 10^{2}m = 100 m
Kilometre = 10^{3}m = 1,1000 m

The litre, l is the unit of volume or capacity – it equals 1,000 cu cm. The gram, g is the unit of weight. Other units are the carat which equals 200 milligrams (mg), and the metric ton which equals 1,000 Kilograms (Kg).

Conversion Factors : Metric, with English Equivalents and English with Metric

Length Measure

1 mm = 0.039 inch
1 cm = 0.394 inch
1 dm = 3.94 inches
(or 3.281 ft or 39.37 in)
1 km = 0.6214 mile

1 inch = 2.54001 cm
1 ft = 30.4801 cm
1 yd = 91.4403 cm
1 mile = 1.609 km

Square Measure : (Area)

1 sq cm = 0.155 sq inch
1 sq m = 1.96 sq yd
(or, 10.7639 sq ft)
1 hectare = 2.471 acres
1 sq. km = 0.386 sq mile

1 sq ft = 0.0929 sq m
1 sq m = 10.7639 sq ft
1 acre = 4, 046.86 sq m
= 0.4047 ha
1 hectare, ha = 0.00 3861 Sq mile
1 Sq mile = 258.9998 hectares

Cubic Measure : (Volume)

1 cu cm = 0.061 cu. inch

1 cu cm = 1.308 cu yds

1 US gallon = 3.785 litres

1 cu in = 0.01638716 litre

1 cu ft = 10,000 cu cm

1 litre = 0.2642 = US gallon

= 0.0275 bushels

Weight Measure

1 kg = 10 mg

1 kg = 10 cg

1 g = 10 dg

1 kg = 1,000 g

1 metric ton = 1,000 kg

1 pound, lb = 0.45349 kg

1 kg = 2.205 kg

1 ton = 0.9072 metric ton

1 metric ton = 2, 204.62 lb.

Fundamental Constants

$\pi = 3.14159$

$e = 2.71828$

1 degree (°) = 0.01745 radian

1 radian = 57.29578°

Temperature Units

$$°F = \frac{9}{5}C + 32$$

$$°C = \frac{5}{9}(F - 32)$$

$$°K = °C + 273°$$

THE GREEK ALPHABETS

Name	Capital	Lower Case	Uses
Alpha	Α	α	Used as Constant
Beta	Β	β	-do-
Gamma	Γ	γ	Used in Gamma distribution
Delta	Δ	σ	Used for small changes or increments in Calculus
Epsilon	Ε	ε	
Zeta	Η	η	Used as angle
Theta	Θ	θ	-do-
Iota	Ι	ι	
Kappa	Κ	*x*	
Lambda	Δ	λ	Used in wave length
Mu	Μ	μ	Used in micron (=10^{-6}m)
Nu	Ν	ν	Used for *d.f.* in *t* and *F* tests
Xi	Ξ	ζ	Used in Venn diagram for universal set
Omicron	Ο	ο	
Pi	Π	π	Used in Circular measures, a constant (=3.14159)
Rho	Ρ	ρ	Used in Spearman's correlation coefficient
Sigma	Σ	ξ	Used in summation
Tau	Τ	τ	Used in Kendal's test
Upsilon	γ	υ	
Phi	Φ	φ	Represents angle
Chi	χ	ω	Used in Chi-square test
Psi	Ψ	ψ	
Omega	Ω	ω	

SELECT BIBLIOGRAPHY

Agterberg, F.P. (1974) Geomathematics, Elsevier, Amsterdam

Aitchinson, J. & J.A.C. Brown (1957) The Log Normal Distribution with special reference to its use in Economics, Cambridge Univ. Press, Cambridge

Anderson, M.G. (ed) (1988) Modelling Geomorphological Systems, John Wiley, New York

Atkinson, K. (1985) Elementary Numerical Analysis, John Wiley, New York

Bennett, R.J. (1981) (ed) European Progress in Spatial Analysis, Pion Ltd, London

—— (1979) Spatial Time Series : Analysis, Forecasting and Control, Pion Ltd, London

—— (1974) 'Process Identification for Time Series Modelling in Urban and Regional Planning' *Regional Studies,* Vol 8, 157-174

—— and R.J. Chorley (1978) Environmental Systems : Philosophy, Analysis and Control, Methuen, London

Berry, B.J. L & D. Marble (eds) (1968) Spatial Analysis, Prentice Hall Inc., Englewood

—— (1961) 'City, Size Distributions and Economic Development' *Economic Development and Cultural Change,* Vol 9, 573-588

—— and W.L. Garrison (1958) 'The Functional Bases of the Central Place Hierarchy' *Economic Geography,* Vol. 34, 148-158

Bhattacharya, Sudeshna (1994) Representation of Relief in Geomorphology, Unpublished Ph.D. Thesis, Delhi University

Bobee, B. (1975) 'Log-Pearson Type III Distributior and Its Application in Hydrology' *Water Resources Research,* Vol II, No. 5, 681-689

Burton, I. (1963) 'The Quantitative Revolution and Theoretical Geography' *Canadian Geographer,* Vol. 7, 151-162

Cassetti, E. (1966) 'Analysis of Spatial Association by Trigomometric Polynomials' *Canadian Geography,* Vol 10, 199-204

Chapman, G.T. (1977) Human and Environmental Systems, Academic Press, London

Cheeney, R.F. (1983) Statistical Methods in Geology, Allen & Unwin, London

Chorley, R.J. (1973) Directions in Geography, Methuen, London

—— (1969) Introduction to Physical Hydrology, Methuen, London

—— & P. Haggett (1965) 'Trend Surface Mapping in Geographical Research' *Trans. Inst. of British Geographers,* Vol 37, 47-67

Chow, Van, Te *et al* (1988) Applied Hydrology, McGraw Hill, New York

—— (ed) (1964) Handbook of Applied Hydrology (A Compendium of Water Resource Technology), McGraw Hill, New York.

Clark, C. (1951) 'Urban Population Densities' *Journ. Royal Statistical Society,* Vol 114, 490-496.

Clark, M. & Wilson, A.G. (1983) The Dynamics and Evolution of Urban Spatial Structure, Croom Helm, London

Clark, W.A.V. and P.L. Posking (ed) (1986) Statistical Methods for Geographers, John Wiley, New York

Cliff, A.D. (1970) 'Computing the Spatial Correspondence between Geographical Patterns' *Trans. Inst. British Geographers,* Vol 50, 143-154.

—— and J.K. Ord (1980) Spatial Process, Pion Ltd, London

—— and —— (1973) Spatial Autocorrelation, Pion Ltd, London

—— and P. Haggett, J. K. Ord *et al* (1975) Elements of Spatial Structure, Cambridge Univ. Press, Cambridge

Coffey, William, J. (1981) Geography: Towards a General Spatial Systems Approach, Methuen, London

Croxton, F.E., D.J. Cowden and B.W. Bolch (1969) Practical Business Statistics, 4th edition, Prentice-Hall Inc., Englewood

—— and S. Klein (1967) Applied General Statistics, 3rd edition, Prentice-Hall inc., Englewood

Dacey, M.F. (1965) 'Geometry of Central Place Theory' *Geographical Annaler,* Vol 47, 111-119

—— (1964) 'Modified Poisson Probability Law for Point Pattern more Regular than Random' *Annals Assoc. American Geographers,* Vol 54, 559-564

—— (1962a) 'Analysis of Central Place and Point Pattern by a Nearest Neighbour Method' *Lund Studies in Geography,* Vol. 24, 55-75

—— (1962b) 'The Identification of Randomness in Point Pattern' *Journ. Regional Science,* Vol. 4, 83-96

—— (1960) 'A Note on the Derivation of Nearest Neighbour Distances' *Journ. Regional Science,* Vol 2, 81-84

Davis, J.C. (1973) Statistics and Data Analysis in Geology, John Wiley, New York

Doornkamp, J.C. (1972) 'Trend Surface Analysis of Planation Surfaces with an East African Case Study' *Spatial Analysis in Geomorphology* (ed) R.J. Chorley

—— and C.A.M. King (1971) (ed) Numerical Analysis in Geomorphology–An Introduction, Arnold, London

Draper, N.R. and H. Smith (1966) Applied Regression Analysis, John Wiley, New York

Dubey, Alok (1986) Environmental Geomorphology, Inter-India Publications, New Delhi

Dunne Thomas and Luna B. Leopold (1978) Water in Environmental Planning, W.H. Freeman & Co., San Francisco

Durbin, J. and G.S. Watson (1951) 'Testing for Serial Correlation in Least Square Regression', *Biometrika*, Vol 38, 159-178

Ebdon David (1985) Studies in Geography, 2nd edition (revised with 17 Computer programmes) Basil Blackwell. London

Fergusson, J. (1988) Mathematics in Geology, Allen and Unwin, London

Found William, C. (1971) A Theoretical Approach to Rural Land Use Patterns, Edward Arnold, London

Gary, L. Gaile and Cort, J. Willmott (1984) Spatial Statistics and Models, D. Reidel Publishing Co., Boston

Gheorghe, A. (1978) Processing and Synthesis of Hydrogeological Data, Abacus Press, Kent, England

Grobner, D.F. and P.W. Shannon (1989) Business Statistics: A Decision Making Approach, Merill Publishing Co., Columbus, Ohio

Gumbel, E.J. (1954) 'Statistical Theory of Extreme Values for Some Practical Application' National Bureau of Standards, *Applied Mathematical Series* 33.

—— (1941) 'The Return Period of Flood Flows' *The Annals of Mathematical Statistics*, Vol 12, No. 2, 163-190

Gupta, S.C. (1988) Statistical Techniques, Sultan Chand, New Delhi

—— and V.K. Kapoor (1973) Elements of Mathematical Statistics, Sultan Chand, New Delhi

—— and —— (1984) Fundamentals of Applied Statistics (Vols I & II). Sultan Chand, New Delhi

Hammond, R. and P. McCullagh (1974) Quantitative Techniques in Geography : An Introduction. Clarendon Press, Oxford

Healey, Joseph F. (1984) Statistics : A Tool for Social Research, Wadsworth Publishing Co., Belmont, California

Huggett, R.J. (1985) Earth Surface Systems, Springer-Verlagh, Berlin

—— (1980) Systems Analysis in Geography, London

King, C.A.M. (1969) 'Trend Surface Analysis of Erosion Surfaces' *Trans. Inst. of British Geographers*, Vol. 47, 47-59.

King, L.J. (1969) Statistical Analysis in Geography, New York

Kish, Leslie (1965) Sampling Survey, John Wiley, New York

Koch, S. George, Jr and Richard, F. Link (1980), Statistical Analysis of Geological Data (2 vols in one), Dover Publications Inc., New York

Kohoot, Frenk, J. (1974) Statistics for Social Scientists : A Coordinated Learning System, John Wiley, New York

Kranenski, W.J. (1988) Principles of Multivariate Analysis : A User's Perspective, Clarendon Press, Oxford

Krumbein, W.C. and F.A. Graybill (1985) An Introduction to Statistical Models in Geology, McGraw Hill, New York

Ludwig, J.A. and J.F. Reynolds (1988) Statistical Ecology : A Primer on Methods and Computing, John Wiley

Mardia, K.V. (1972) Statistics of Directional Data, Academic Press, London.

Mathews, John, A. (1981) Quantitative and Statistical Approaches to Geography : A Practical Manual, Pergamon Press, Oxford

McCammon, Richard, B. (1975) Concepts in Statistics, Springer-Verlag, Berlin

Mutreja, K.N. (1986) Applied Hydrology, Tata McGraw Hill, New Delhi

Nye, J.F. (1957) 'The Distribution of Stress and Velocity in Glaciers and Ice Sheets' *Proc. Royal Society*, A 239, 112-133, London

Pal, Saroj, K. (1982) Statistical Techniques : A Basic Approach to Geography, Tata McGraw Hill, New Delhi

Panse, V.G. and Sukhatme, P.V. (1985) Statistical Methods for Agricultural Workers, IARI, New Delhi

Raghunath, H.M. (1985) Hydrology, Wiley Eastern, New Delhi

Rodda, J.C. (1970) 'A Trend Surface Analysis Trial to Planation Surfaces' *Trans. Inst. of British Geographers*, Vol 50, 107-114

Schilling, W. (1947) 'A Frequency Distribution Represented as the Sum of Two Poisson Distributions' *Jour. Ame. Statistical Association*, 42, 407-424

Shaw, G. and Wheeler, D. (1985) Statistical Techniques in Geographical Analysis, John Wiley, New York.

Smith, D.E. *et al* (1969) 'Isobases for Raised Shoreline as determimed by Trend Surface Analysis' *Trans. Inst. of British Geographers*, Vol 46, 45-56

Stephenson, G. (1973) Mathematical Methods for Science Students, 2nd edition, Longman, London.

Strahler, A.N. (1980) Systems Theory in Physical Geography, in, Physical Geography, Chap I, 1-27

Subramanyam, V.P. (1983) (ed) Applied Climatology, Heritage, New Delhi

Tan, K.C. and R.J. Bennett (1984) Optimal Control on Spatial Systems, Allen & Unwin, London

Till, Roger (1974) Statistical Methods for the Earth Scientist : An introduction, John Wiley, New York

Thomas, R.W. and R.J. Huggett (1983) Modelling in Geography : A Mathematical Approach, Harper and Row, London

Trimble, S.W. (1981) Changes in Sediment Storage, *Science*, 214, 454-474

Unwin, D.J. and J.A. Dawson (1988) Computer Programming for Geographers, Longman, London.

—— (1975) An Introduction to Trend Surface Analysis, Catmag 5, Geobooks. Norwich.

Waymire, E. and V.K. Gupta (1981) 'Review of Stochastie Rainfall Models. *Water Resources Research*, Vol 17, No. 5, 1261-

1294

Wilson, A.G. and R.J. Bennett (1985) Mathematical Methods in Human Geography and Planning, John Wiley, New York

—— (1981) Geography and the Environment, Systems Analytical Methods, John Wiley, New York

Williams, Peter, J. (1982) The Surface of the Earth: An Introduction to Geological Science, Longman

Wrigley, N. (ed) (1979) Statistical Applications in the Spatial Sciences, Pion Ltd, London

—— and R.J. Bennett (eds) (1981) Quantitative Geography : A British View, Routledge & Kegan, Paul, London

Yevjevich, V. (1971) Probability and Statistics in Hydrology, Water Resources Publications

INDEX